# Trends in Mathematics

Trends in Mathematics is a book series devoted to focused collections of articles arising from conferences, workshops or series of lectures.

Topics in a volume may concentrate on a particular area of mathematics, or may encompass a broad range of related subject matter. The purpose of this series is both progressive and archival, a context in which to make current developments available rapidly to the community as well as to embed them in a recognizable and accessible way.

Volumes of TIMS must be of high scientific quality. Articles without proofs, or which do not contain significantly new results, are not appropriate. High quality survey papers, however, are welcome. Contributions must be submitted to peer review in a process that emulates the best journal procedures, and must be edited for correct use of language. As a rule, the language will be English, but selective exceptions may be made. Articles should conform to the highest standards of bibliographic reference and attribution.

The organizers or editors of each volume are expected to deliver manuscripts in a form that is essentially "ready for reproduction." It is preferable that papers be submitted in one of the various forms of LaTeX in order to achieve a uniform and readable appearance. Ideally, volumes should not exceed 350-400 pages in length.

Proposals to the Publisher are welcomed at either:
Birkhäuser Boston, 675 Massachusetts Avenue, Cambridge, MA 02139, U.S.A.
kostant@birkhauser.com
or
Birkhäuser Publishing, Ltd., 40-44 Viaduktstrasse, CH-4051 Basel, Switzerland
hempfling@birkhauser.ch

# Real and Stochastic Analysis

## *New Perspectives*

M. M. Rao

Editor

Birkhäuser
Boston • Basel • Berlin

M. M. Rao
University of California
Department of Mathematics
Riverside, CA 92521
U.S.A.

Mathematics Subject Classification: Primary: 60B15, 60H07, 60H30, 60H99, 58J65, 58D20, 46H25, 35H10, 26E15, 81P10, 81P15, 81P68; Secondary: 46E30, 43A10

**Library of Congress Cataloging-in-Publication Data**
Real and stochastic analysis : new perspectives / M. M. Rao, editor.
   p. cm. – (Trends in Mathematics)
   Includes bibliographical references and index.
   ISBN-13: 978-1-4612-7397-4   e-ISBN-13:978-1-4612-2054-1
   DOI:10.1007/978-1-4612-2054-1
   1. Stochastic analysis. I. Rao, M. M. (Malempati Madhusudana), 1929- II. Series.

QA274.2.R423 2004
519.2'2–dc22                         2004053658

ISBN-13: 978-1-4612-7397-4
Printed on acid-free paper.

©2004 Birkhäuser Boston      **Birkhäuser**   
Softcover reprint of the hardcover 1st edition 2004

(TXQ/HP)

9 8 7 6 5 4 3 2 1      SPIN 10914342

Birkhäuser is a part of *Springer Science+Business Media*
*www.birkhauser.com*

# Contents

**Noncommutative Probability and Applications**
*Stanley Gudder* .................................................. 199

# Preface

As in the case of the two previous volumes published in 1986 and 1997, the purpose of this monograph is to focus the interplay between real (functional) analysis and stochastic analysis show their mutual benefits and advance the subjects. The presentation of each article, given as a chapter, is in a research–expository style covering the respective topics in depth. In fact, most of the details are included so that each work is essentially self contained and thus will be of use both for advanced graduate students and other researchers interested in the areas considered. Moreover, numerous new problems for future research are suggested in each chapter.

The presented articles contain a substantial number of new results as well as unified and simplified accounts of previously known ones. A large part of the material covered is on stochastic differential equations on various structures, together with some applications. Although Brownian motion plays a key role, (semi-) martingale theory is important for a considerable extent. Moreover, noncommutative analysis and probability have a prominent role in some chapters, with new ideas and results. A more detailed outline of each of the articles appears in the introduction and outline to assist readers in selecting and starting their work. All chapters have been reviewed.

It is expected that the works here will stimulate further research in several directions in the areas covered. I would like to express my appreciation to the authors for their assistance (and some revisions) in this enterprize, and the UCR academic senate research committee for a modest grant for preparing this volume. Also, I am grateful to Jan Carter for her key assistance in formatting the volume, and Toby Bartels for resolving some difficulties with the LATEX files.

*M. M. Rao*
Riverside, CA
March, 2004

# Introduction and Outline

M. M. Rao

The work in the following chapters is given in a research–expository style outlining the current state and containing new perspectives in real and stochastic analysis, essentially with complete details. These articles are prepared by active researchers in the respective areas dealing with problems of current and immediate interest and they are given in an unhurried and detailed manner that will be helpful for senior graduate students engaged in their theses as well as for research mathematicians exploring new problems in the areas covered here. The general point of view is similar to the earlier volumes under the same title, appearing in 1986 and 1997, both edited by me. The subjects covered here complement the earlier volumes and are on stochastic analysis on manifolds (two chapters), noncommutative probability (of interest in quantum mechanics and elsewhere), the non-absolute Feynman integration with applications, stochastic flows of diffeomorphisms with some new aspects of Itô (and Stratonovich) calculus, and structural characterizations of locally compact groups based on symmetric random walks on them—arising from algebraic and spectral properties of the associated convolution operators. An outline of each of the chapters will now be given for a bird's-eye view of the overall volume.

In the classical studies of the abstract Cauchy problem, one starts with a solution $u(\cdot, \cdot)$ of the partial differential equation (PDF)

$$\frac{\partial u}{\partial t} = (\frac{1}{2}\Delta + A_0)u = Pu, t > 0; \quad \lim_{t \downarrow 0} u(x, t) = f(x), \ x \in \mathbb{R}^d, \tag{1}$$

where $A_i = \sum_{j=1}^{d} a_{ij}\frac{\partial}{\partial x_j}, i = 0, 1, \ldots, r, \Delta = \sum_{i=1}^{r} A_i^2$, and $A = (a_{ij})$ is a $d \times d$ positive definite invertible matrix (or its minimum eigenvalue is bounded away from zero). Then (1) is an elliptic PDE. For this there is a smooth (infinitely differentiable) solution which may be expressed as an integral

$$u(x, t) = \int_{\mathbb{R}^d} K(t, x, y) f(y) \, dy, \quad (t, x) \in (0, \infty) \times \mathbb{R}^d, \tag{2}$$

where $K(t, \cdot, \cdot) : \mathbb{R}^d \times \mathbb{R}^d \to \mathbb{R}$ is a smooth kernel, $t > 0$. This conclusion can hold for operators $A_i$ subject to weaker restrictions (and the ellipticity can also be weakened), as seen from Kolmogorov's (1934) paper. Equation (1) was generalized by Hörmander to the case where $u$ is a $C^\infty$-function in a domain (open set) of $\mathbb{R}^d$ where $Pu$ is a $C^\infty$-function in it. Such an operator is called *hypoelliptic*, which seems to have been introduced by L. Schwartz in 1950 in his theory of distributions. It was noted that an equation of the type

$$\frac{\partial^2 u}{\partial x^2} + x\frac{\partial u}{\partial y} - \frac{\partial u}{\partial t} = f, \tag{3}$$

in $\mathbb{R}^2$, does not satisfy the hypoellipticity condition in the full space $\mathbb{R}^2$. But the early work of Kolmogorov already shows that (3) has a (smooth) fundamental solution, by observing that it corresponds to a Fokker-Plank equation of diffusion theory, which admitted solutions to Markov processes (see, e.g., Weber (1951), a thesis under W. Feller). Analyzing the situation closely, Hörmander found that for an extension of (3) to equations of the type

$$-\frac{\partial u}{\partial x_0} + \sum_{ij,k=1}^n a_{jk}\frac{\partial^2 u}{\partial x_j \partial x_k} + \sum_{j,k=1}^n b_{jk}x_j\frac{\partial u}{\partial x_k} + cu = f, \tag{4}$$

with $A = (a_{ij})$ as a symmetric positive definite matrix, fail to be hypoelliptic only if the null space of the matrix $A$ contains a nontrivial subspace that is invariant for $B = (b_{ij})$, $b_{ij}$ being constants. General conditions for hypoellipticity were then obtained by requiring the functions $a_{ij}, b_{ij}$ to have reasonable and verifiable properties. Let $A_i = a_{ij}\frac{\partial}{\partial x_i}, i = 0, 1, \ldots, r$ be first order homogeneous differential operators in a domain $\Omega \subset \mathbb{R}^d$ with $C^\infty$-coefficients $a_{ij}$. The principal part $(a_{ij})$ of the operator $P$ is desired to be positive (semi-)definite where

$$P = \sum_{j=1}^r A_j^2 + A_0 + c, \tag{5}$$

with $a_{ij}, c$ being $C^\infty$-functions. Indeed the positive (semi)-definiteness of $(a_{ij})$ or of its negative is necessary for the hypoellipticity. [See Oleĭnek and Radkevič (1973), Sec. II.5 on the subject.] The best condition given by Hörmander is that the system of operators $A_0, \ldots, A_r$ (also termed vectorfields) should have rank $d$ at each point $x \in \Omega$. To make this statement precise, he considers the vectorspace generated by the $A_i$ and their commutators, i.e., $A_{j_1}, [A_{j_1}, A_{j_2}], [A_{j_1}, [A_{j_2}, A_{j_3}]], \ldots, 0 \le j_i \le r$ where $[A, B] = AB - BA$, containing $d$-linearly independent vectorfields at each point $x$ of $\Omega$ (the collection changing with the points). Then $Pu = f$ has a smooth solution. The condition is suggested by Lie algebra theory and some of its results. Later alternative proofs and generalizations using 'pseudo differential operator' techniques were found, and even a necessary and sufficient condition for hypoellipticity of $P$ was obtained when the coefficients $a_{ij}, b_{ij}, c$ are smooth real analytic, as well as when $P$ is not necessarily the sum of squares as in (5). For a detailed discussion with proofs of these statements, one may refer to a nice presentation of the subject in Oleĭnek and Radkevič ((1973), Chapter II). In this treatment probabilistic arguments play no role, although the original equations (4) and (5) have their origins in probability. This connection, which is important for our work, was successfully brought out by Malliavin (1978) and helped advance stochastic analysis substantially. This will now be sketched.

It is known that the operator $P$ of (1) appears as an infinitesimal generator of the semi-group determined by the solution of a stochastic differential equation (SDE) of Itô's type that is a Markov process that, through the Chapman–Kolmogorov equation, associates the desired above stated set of positive linear operators. Then equations (3) and (4) are direct consequences of this link. But the extension to (5) and an alternative to the Lie algebra (rank) condition is the major advance due to Malliavin (1978). It is apparently suggested by the Lévy inversion of the analytical probability theory, which states that an $\mathbb{R}^d$ valued random variable $X$ with distribution $F$ has a $C^{n-d-1}$- smooth density (relative to the Lebesgue measure) if for each bounded $C^\infty$-function $f : \mathbb{R}^d \to \mathbb{R}$ one has

$$E(|(D^j f)(X)|) \leq c_n \|f\|_\infty, \tag{6}$$

for some absolute constant $c_n > 0, n \geq d+1$ and $|j| = j_1 + \cdots + j_n \leq n - d - 1$ (the $j_i, 0 \leq j_i \leq d$ are integers). This is obtained from the fact that the Fourier transform $\hat{F}$ of $F$ should satisfy (for the desired smoothness of the density) a condition such as

$$\int_{\mathbb{R}^d} |x|^{d+1} |\hat{F}|(x)\, dx < \infty.$$

However, in our stochastic applications $\mathbb{R}^d$ must be replaced by $C([0, 1], \mathbb{R}^d)$, an infinite dimensional space, and the Lebesgue measure must be replaced by the (continuous or diffuse) Wiener measure on it. Thus one considers the process $\{X_t, t \geq 0\}$ as a solution of

$$X_t = x + \int_0^t A_0(X_s)\, ds + \sum_{i=0}^r \int_0^t A_i(X_s) \circ dB_s, \quad X_0 = x, \tag{7}$$

where $\{B_s, s > 0\}$ is the $d$-dimensional Brownian motion (BM) and the SDE (in integrated form) is in the sense of Stratonovich.

To replace the Lie algebra condition above, consider a (nonsingular) $d \times d$-matrix valued process $Z_t, t > 0$, associated to (7), given by

$$Z_t = I + \int_0^t Z_s(DA_0)(X_s)ds - \sum_{i=1}^r \int_0^t Z_s(DA_i)(X_s) \circ dB_s, \quad t \geq 0, \tag{8}$$

and define a new matrix valued process $C_t, t \geq 0$, called the *Malliavin covariance process*, based on (8) as:

$$C_t = Z_t^{-1} \left[ \sum_{i=1}^r \int_0^t [Z_s A_i(X_s)] \otimes [Z_s A_i(X_s)]^* ds \right] (Z_t^{-1})^*,$$

where $Z_t^*$ is the transpose of the matrix $Z_t$. After a number of steps and analysis, Malliavin shows that the solution $X_t$ of (7) has $C^\infty$-smooth (finite dimensional) densities if $[\det(C_t)]^{-1} \in \bigcap_{p \geq 1} L^p$, the $L^p$ on the given probability space, and verified that this condition is automatic if Hörmander's Lie algebra rank condition holds for the vectorfields $A_i$, $i = 0, 1, \ldots, r$, and is somewhat weaker than the latter. This is still a mysterious condition although it does not involve Lie products with the $A_i$. His original 1978 paper, which is about 70 pages, is a tightly written account that lays out a "road map" for future work. Some simplifications and generalizations of Malliavin's work followed. The first two chapters are on this theme, and will be briefly described, and then the results in the other chapters will be discussed in outline.

The first chapter by Denis Bell presents a somewhat more elementary approach of Malliavin's theory with a few simplifications and additions to the original theme, including some by the author. An interesting consequence of this work which presents an approximation procedure for the hypoellipticity condition is that for the SDE

$$dX_t = Q(t, X_t)dt + g(X_{t-r_0})dB_t, \tag{9}$$

where $\{B_t, t \geq 0\}$ is the BM, $Q(t, X_t)$ is an adapted nonanticipating functional, $g$ is matrix valued, which may be singular on a hyper surface, and $r_0$ being time delay. But now the solution process $X_t$ need not be Markovian in contrast to the original Itô-formulation when $g$ is strictly nonsingular, as in (7), so that the solution there is necessarily Markovian. A readable account, with complete details, along with a result on a hypoelliptic equation with smooth but nonanalytic vectorfields for which Hörmander's (Lie algebra) condition is not satisfied, is presented. It is also shown that the solution of such a degenerate SDE as (9) can have $C^\infty$-densities, extending Malliavin's result in some respects. A number of related open problems when, for instance, $Q$ and $g$ in (9) depend on the whole past history $\{X_s, 0 \leq s \leq t\}$, which generalize the above methods, and/or considering quasi-elliptic PDE problems using a connection established, with 'super processes' and PDE by E. B. Dynkin and others are proposed.

Another aspect already considered by Malliavin (1978) is to study the SDEs on subsets of $\mathbb{R}^d$, and more generally submanifolds for which the ideas of Riemannian geometry can be applied. To pursue this in detail on any such manifold is the main focus of Chapter 2 by Bruce Driver. Before presenting stochastic analysis on such manifolds, it is beneficial to have a certain amount of Riemannian geometry directly applicable for the present needs, especially to define the BM process on these spaces. This is done here with a view on having the (nonlinear or) curved Wiener spaces on which deeper aspects of SDEs can be studied. To have an essentially self-contained exposition, Driver spends over 45 pages of the initial part for this so that readers, both new and old, will feel 'at home' with this detail and be pleased with the presentation. It includes relevant parts of Riemannian geometry, some results on flows of vectorfields and 'Cartan's rolling map' along with the stochastic calculus employing these facts. This analysis may be compared (or read simultaneously) with a lucid but a rapid review of the general subject by L. Schwartz (1991) in discussing a book review of M. Emery's volume on the topic. Here Driver's account is specialized to Riemannian manifolds carrying the Wiener measure so that more detailed and finer aspects can be described. In fact logarithmic Sobolev inequalities for the Wiener measure on path spaces, as well as the author's own work on integration by parts formulas in this generality and (quasi-)invariance of heat kernels are included. The work is given with simplified proofs and contains some new developments. Detailed arguments and several figures should be of interest both to the starting and seasoned workers in these areas. A reader friendly presentation is maintained and a major part of the material has recently been exposed by the author in Europe and elsewhere. All these points culminated in making this the longest chapter in the volume. New and unresolved problems are pointed out at various places and there is an extended bibliography on the subject to help the reader. Another recent article, essentially complementing the present one, dealing with recurrence and transience of a Brownian motion along with its connections to PDE problems on Riemannian manifolds, by Grigor'yan (1999), describes the topics in considerable detail and will be of interest to the readers of Driver's exposition of curved Wiener manifolds. It also contains relevant parts of Riemannian geometry together with an extended set of references.

The third chapter considers both a survey and some key detailed descriptions together with applications of noncommutative probability by Stanley Gudder. It starts with a realistic problem for measuring observations of an experiment that cannot give the same values when done in different orders, and hence leads to a noncommutative theory of probability. There are, however, different types of noncommutative theories. For instance, there is a 'free probability theory' based on classical probability applied to infinite random matrices that leads to a new concept of independence and limit theorems that are not readily recognizable unless this new concept called *free independence* is analyzed and exploited. Here the basic probability space is an algebra of self adjoint operators on a Hilbert space so that the spectral theory furnishes a measure in place of the standard probability measure. With this idea a number of results were extended starting with the central limit theorem which, as Wigner first noted in the mid 1950s but which was perfected by Voiculescu in the 1980s, plays a central part in the new analysis that replaces the Gaussian law in the commutative theory. A great deal of infinitely divisible distributions was extended and some analogs from statistics (e.g., Fisher's information and entropy) were considered, cf. Voiculescu (1998). Another type concentrates on defining (after establishing the existence of) conditional expectations, started by H. Umegaki in the mid 1950s, leading to extensions of martingale theory as well as entropy. In either case there is no sampling or path analysis. These are two different aspects of operator theory. For details of the latter type, one may refer to Gudder and Hudson (1978). A still different aspect that has close relations with mathematical physics, especially with quantum mechanics, and other applications is described by Gudder in this chapter. It concentrates on the following problems.

There is a detailed discussion of observables and related statistical maps, presenting a unified treatment of different theories using *sequential effect algebras* after contrasting with *sharp* and *unsharp* probability theories in Hilbert space. The concept of *fuzziness* of earlier times finds an appropriate place in this context. A comparison of classical (or unsharp) probability and its quantum counterpart is also highlighted for a convenient reference. Although noncommutative probability is primary here, a close relation with classical probability is maintained by obtaining analogs of conditioning, a Bayes formula, and Boolean algebra methods. It has a different flavor from free probability noted above that also extends the classical work in another direction. Because of its special *free independence* concept of the latter has a bijective correspondence with traditional infinitely divisible (or convolution operator) analysis. (For an extension of the latter aspect, see the recent work by Barndorff-Nielsen and Thorjørnsen (2002).) The lucid account given in this chapter should be of interest for independent study of the subject as well as for further extensions and applications in quantum physics and elsewhere.

The material of the fourth chapter concerns recent advances on Feynman integrals in both of its mathematical (i.e., stochastic) and physical aspects, presented by Brian Jefferies. In many ways it complements the recent detailed monograph by Johnson and Lapidus (2000) as well as his own volume on the Feynman-Kaç formula. Since the Feynman integral is non-absolute and only finitely additive, its general treatment needs a different set of ideas. For instance, the Henstock–Kurzweil definition of generalized Riemann type integration is pertinent and was discussed by Muldowney (1987). However, many workers prefer approximation with Lebesgue's (absolute) integration (just as Denjoy extended the latter with transfinite induction in 1912 to get a non-absolute case), and in the present theory this involves a use of operator valued measures. This is explained carefully and in detail by Jefferies in the present chapter. Moreover, he presents material beyond the Feynman integral for nonrelativistic quantum mechanics and includes a variety of other problems related to this work. His discussion of the Dirac equation and also renormalization of the Coulomb interaction will be of interest for making explicit calculations as well as appreciating the general ideas involved. The material will be helpful for both the new and seasoned researchers in the subject.

The next chapter examines some aspects of SDEs and flows of diffeomorphisms written by Hiroshi Kunita, concentrating on certain new developments since his monograph on the subject was published in 1990. Here the solutions of SDEs are allowed to have jumps and the diffeomorphic (or just homeomorphic) property for them is not always satisfied. This circumstance leads to stochastic integration relative to semimartingales with jumps, and the relevant theory is first presented that utilizes the Stratonovich integral. Then the author concentrates on Lévy processes that contain the BM and Poisson processes. A complete discussion of the subject as well as the Lévy-Itô decomposition and the representation of $d$-dimensional semimartingales relative to the BM and Poisson processes are given. There he studies differentiability properties of the solutions of the SDEs when spatial parameters are allowed. The uniqueness, continuity, and differentiability of the resulting solutions are detailed. Then homeomorphic and diffeomorphic properties of the flow relative to spatial parameters are studied. This work needs several $L^p$-estimates of the moments of the solution process, and all details are included, which makes it particularly useful for advanced graduate students and others coming into this area.

The final chapter on convolution operators determining amenability of the underlying locally compact group is written by me. It illustrates how random walk problems in classical probability theory lead to some nontrivial functional analysis results related to the spectral radius and other properties of the above operators leading to several structural characterizations of groups. This problem was originally formulated and solved by H. Kesten for countable groups that are range (or state) spaces of symmetric random walks. He showed that the transience and recurrence properties of such walks are closely related to the existence of invariant means on the groups considered. The fact that the latter is a functional analysis problem equivalent to having the spectral radius of the convolution operator determined by the walk in $\ell^2$ was recognized by M. M. Day who thereafter generalized the result for $\ell^p$, $1 < p < \infty$, spaces on countable semi-groups and raised the question of its analog for Orlicz spaces on locally compact groups. A student of his extended the above results for uniformly convex Orlicz spaces $\ell^\varphi$ on countable semi-groups. The present chapter contains a complete solution of Day's problem for Orlicz spaces on a locally compact group $G$. It also discusses a related result due to B. E. Johnson on amenable group algebras $L^1(G)$. The work leads to some open problems as well as new probabilistic applications related to random walks to investigate and analyze some Beurling and Segal (sub)algebras of $L^1(G)$.

The material included in this chapter was originally suggested by me as a thesis problem for a student, P. V. Kumar, in the early 1980s and he started work in the context of $L^p(G)$, $1 < p < \infty$, but unfortunately left the project without pursuing it further. After waiting for over a decade, I have taken up the project and completed it in the context of M. M. Day's problem, and presented it here.

Now, looking at all the chapters, it becomes clear that the works detailed, in this volume give new perspectives and open up several new problems in the areas considered. Since the presentations contain essentially all details, it is hoped that the material stimulates research in new directions for both graduate students and other researchers in these areas.

# References

[1] Barndorff-Nielsen, O. E. and S. Thorbjørnsen, Lévy laws in free probability, and Lévy processes in free probability, *Proc. Nat. Acad. Sci.* **99** (2002), 16568–16575; 16576–16580.

[2] Grigor'yan, A., Analytic and geometric background of recurrence and non-explosion of the Brownian motion on Riemannian manifolds, *Bull. Amer. Math. Soc.* **36** (1999), 135–249.

[3] Gudder, S. P. and R. L. Hudson, A noncommutative probability theory, *Trans. Amer. Math. Soc.* **245** (1978), 1–41.

[4] Hörmander, L., Hypoelliptic second order differential equations, *Acta Math.* **119** (1967), 147–171.

[5] Johnson, G. W. and M. L. Lapidus, *The Feynman Integral and Feynman's Operational Calculus*, Oxford University Press, Oxford, UK, 2000.

[6] Kolmogoroff, A., Zuföllige Bewegungen, (Zur Theorie du Brownschen Bewegung), *Ann. Math.* **35** (1934), 116–117.

[7] Kunita, H., *Stochastic Flows and Stochastic Differential Equations*, Cambridge University Press, Cambridge, UK, 1990.

[8] Malliavin, P., Stochastic calculus of variation and hypoelliptic operators, in: *Proc. International Symp. on Stochastic Differential Equations*, Wiley, New York, 195–263, 1978.

[9] Muldowney, P., *A General Theory of Integration in Function Spaces including Wiener and Feynman Integration*, Longmans Scientific and Technical Notes, Essex, UK, 1987.

[10] Oleĭnek, O. A. and E. V. Radkevič, *Second Order Equations with Nonnegative Characteristic Form*, Amer. Math. Soc., Providence, R.I., and Plenum Press, New York, 1973.

[11] Schwartz, L., Review of 'Stochastic Calculus in Manifolds' by M. Emery, *Bull. Amer. Math. Soc., New Series*, **24** (1991), 451–466.

[12] Voiculescu, D., Lectures on free probability theory, in: *Lect. Notes in Math.*, Springer, **1738**, 280–349, 1998.

[13] Weber, M., The fundamental solution of a degenerate partial differential equation of parabolic type, *Trans. Amer. Math. Soc.* **71** (1951), 24–37.

# Stochastic Differential Equations and Hypoelliptic Operators

Denis R. Bell

Department of Mathematics, University of North Florida, Jacksonville, FL 32224
dbell@unf.edu

## 1 Introduction

The first half of the twentieth century saw some remarkable developments in analytic probability theory. Wiener constructed a rigorous mathematical model of Brownian motion. Kolmogorov discovered that the transition probabilities of a diffusion process define a fundamental solution to an associated heat equation. Itô developed a stochastic calculus which made it possible to represent a diffusion with a given (infinitesimal) generator as the solution of a Stochastic Differential Equation. These developments created a link between the fields of Partial Differential Equations and stochastic analysis whereby results in the former area could be used to prove results in the latter.

More specifically, let $X_0, \ldots, X_n$ denote a collection of smooth vector fields on $\mathbf{R}^d$, regarded also as first-order differential operators, and define the second-order differential operator

$$L \equiv 1/2 \sum_{i=1}^{n} X_i^2 + X_0. \tag{1.1}$$

Consider also the Stratonovich SDE

$$d\xi_t = \sum_{i=1}^{n} X_i(\xi_t) \circ dw_i + X_0(\xi_t)dt \tag{1.2}$$

where $w = (w_1, \ldots, w_n)$ is an $n$-dimensional standard Wiener process. Then the solution $\xi$ to (1.2) is a time-homogeneous Markov process with generator $L$, whose transition probabilities $p(t, x, dy)$ satisfy the following PDE (known as the *Kolmogorov forward equation*) in the weak sense

$$\frac{\partial p}{\partial t} = L_y^* p.$$

A differential operator $G$ is said to be *hypoelliptic* if, whenever $Gu$ is smooth for some distribution $u$ defined on some open subset of the domain of $G$, then $u$ is smooth. In 1967, Hörmander proved that an operator of the form $L$ in (1.1) is hypoelliptic if the

Lie algebra generated by $X_0, \ldots, X_n$ has dimension $d$ at each point (this hypothesis is known as *Hörmander's condition*). It follows immediately that the transition probabilities $p$ of $\xi$ in (1.2) have smooth densities at all positive times provided the parabolic operator $\partial/\partial t - L_y^*$ satisfies Hörmander's condition at $\xi_0$. Of course, this is a rather circuitous route for proving a result of a decidedly probabilistic nature, a fact that had hardly escaped the notice of PDE theorists and stochastic analysts alike.

In the mid-seventies Malliavin ([Ma1], [Ma2]) outlined a striking new method for *directly* proving the smoothness of the transition probabilities $p$. The idea is as follows: the measure $p(t, x, dy)$ is the image of the Wiener measure under the *Itô map* $g : w \mapsto \xi_t$. Now the Wiener measure, which can be considered as an infinite-dimensional analogue of standard Gaussian measure on Euclidean space, has a well-understood analytic structure. If the map $g$ were *smooth*, then regularity properties of $p$ could be studied by a process of integration by parts (cf. Section 2). In fact, this is not the case; $g$ is only defined up to a set of full Wiener measure and is most pathological from the standpoint of classical calculus. Malliavin solved this problem by constructing an extended calculus applicable to solution maps of SDEs. His method (which has since been termed *Malliavin calculus*) was based on the infinite-dimensional Ornstein–Uhlenbeck semigroup and was rather elaborate. It has since been simplified and extended by many authors and has become a powerful tool in stochastic analysis. Following further pioneering work by Kusuoka and Stroock, it has led to a complete probabilistic proof of Hörmander's theorem, as Malliavin originally intended, and has inspired a host of other results. Examples include filtering theorems by Michel [Mi], a deeper understanding of the Skorohod integral and the development of an anticipating stochastic calculus by Nualart and Pardoux [NP], an extension of Clark's formula by Ocone [O], and Bismut's probabilistic analysis of the small-time asymptotics of the heat kernel of the Dirac operator on a Riemannian manifold [Bi2]. In this article, I describe my own contributions to the subject. The reader is encouraged to also study the aforementioned works.

The contents of the article are as follows. In Section 2, we present an elementary derivation of Malliavin's integration by parts formula by which one establishes smoothness of the transition probabilities of the process $\xi$. We also include a result of Bismut that illustrates how this formula is related to Hörmander's condition. Much of the notation, together with the basic tools that will be used later, are introduced in this section of the article.

The material presented in Sections 3 and 4 is joint work with S. Mohammed. In Section 3, we prove a generalization of Hörmánder's theorem. This result asserts that the operator $L$ in (1.1) is hypoelliptic under hypotheses that allow Hörmander's condition to *fail* (at a controlled rate) on hypersurfaces in the domain of $L$. The theorem is sharp in the category of Hörmander operators with smooth (as opposed to analytic) coefficients.

In Section 4, we establish sufficient conditions for the existence of smooth densities for a class of stochastic functional equations of the form

$$dx_t = g(x_{t-r})dw + B(t, x_t)dt \tag{1.3}$$

where $g$ denotes a matrix-valued function, $r$ is a positive time-delay, and $B$ is a non-anticipating functional defined on the space of paths. Our hypotheses allow degeneracy of $g$ on a hypersurface in the ambient space of the equation. Probabilistically, there is an important difference between the solution $\xi$ to equation (1.2) and the solution $x$ to (1.3). Namely, while $\xi$ is a Markov process, $x$ is *non*-Markov. As a consequence, the aforementioned result cannot be proved by appealing to existing results in PDE theory.

Finally, in Section 5 we describe some open problems, which we hope will stimulate further study into this exciting subject.

## 2 Integration by parts and the regularity of induced measures

As in Section 1, let $\xi$ denote the solution to the SDE (1.2). In this section we will give a proof of a key result in Malliavin's paper [Ma1], which gives a criterion under which the random variables $\xi_t$ have absolutely continuous distributions at all positive times $t$. The material in this section is taken from the author's doctoral dissertation [Be1]. Following Malliavin, we will make use of the following result from harmonic analysis (see [Ma1] for a proof).

**Lemma 2.1.** *Let $\nu$ denote a finite Borel measure on $\mathbf{R}^d$. Suppose that for $k \in \mathbb{N}$ and each set of non-negative integers $d_1, \dots, d_k$, there exists a constant $C$ such that for all test ($C^\infty$ compact support) functions $\phi$ on $\mathbf{R}^d$*

$$\left| \int_{\mathbf{R}^d} \frac{\partial^{d_1 + \dots d_k}}{\partial x_1^{d_1} \dots \partial x_d^{d_k}} \phi \, d\nu \right| \leq C \|\phi\|_\infty. \tag{2.1}$$

*If this condition holds for $k = 1$, then the measure $\nu$ is absolutely continuous with respect to the Lebesgue measure on $\mathbf{R}^d$. If the condition holds for every $k \geq 1$, then $\nu$ has a smooth density.*

The following result is included in order to motivate the approach that follows. It treats a finite-dimensional analogue of the infinite-dimensional problem that will be studied later.

**Theorem 2.2.** *Suppose $T : \mathbf{R}^m \mapsto \mathbf{R}^d$ is a $C^2$ map, where $m \geq d$. Let $d\gamma(x) = (2\pi)^{-m/2} \exp(-|x|^2/2)$ denote the standard Gaussian measure on $\mathbf{R}^m$. Let $\nu \equiv T(\gamma)$ denote the induced measure on $\mathbf{R}^d$, and $\nu(B) \equiv \gamma(T^{-1}(B))$ for Borel subsets $B \subseteq \mathbf{R}^d$. Define*

$$N \equiv \{x \in \mathbf{R}^d | DT(x) \in L(\mathbf{R}^m, \mathbf{R}^d) \text{ is not surjective}\}.$$

*where $D$ is a differential operator. Then the condition*

$$\gamma(N) = 0 \tag{2.2}$$

*is necessary and sufficient for absolute continuity of the measure $\nu$.*

*Proof.* The necessity of condition (2.2) follows immediately from Sard's theorem which asserts that the set $T(N)$ has zero Lebesgue measure.

To prove sufficiency, we argue as follows. Define a $d \times d$ matrix

$$\sigma(x) = DT(x)DT(x)^*. \tag{2.3}$$

Let $\{\psi_k\}$ denote a sequence of smooth bump functions from $[0, \infty) \mapsto [0, 1]$ such that

(i)   $\psi_k(t) = 1$ if $0 \leq t \leq k$
(ii)  $\psi_k(t) = 0$ if $t \geq k+1$
(iii) For each $p \geq 1$, $\sup_k |D^p \psi_k(t)| < \infty$.

For each $k \in \mathbf{N}$, define $R_k : \mathbf{R}^d \otimes \mathbf{R}^d \mapsto [0, 1]$ by

$$R_k(\alpha) = \begin{cases} \psi_k(\|\alpha^{-1}\|), & \alpha \in GL(d) \\ 0, & \alpha \notin GL(d). \end{cases}$$

Note that $R_k$ is $C^\infty$, since $GL(d)$ is an open subset of $\mathbf{R}^d \otimes \mathbf{R}^d$. We also define sequences of measures $\{\gamma_k\}$ on $C_0$ and $\{\nu_k\}$ on $\mathbf{R}^d$ by

$$\frac{d\gamma_k}{d\gamma}(x) = R_k(\sigma(x)), \quad \nu_k = T(\gamma_k).$$

Assume now that (2.2) holds. Since $N$ is precisely the set on which $\sigma$ is degenerate, this is equivalent to the condition $\sigma(x) \in GL(d)$ a.s., which implies $\nu_k \to \nu$ in variation norm. Hence it suffices to prove absolute continuity of $\nu_k$ for each $k$. We do this as follows: let $e_1, \ldots, e_d$ denote the standard basis of $\mathbf{R}^d$ and define, for any $1 \leq i \leq d$, a function $h_i : \mathbf{R}^m \mapsto \mathbf{R}^m$ by

$$h_i(x) = \begin{cases} DT(x)^* \sigma^{-1}(x)e_i, & \sigma \in GL(d) \\ 0, & \sigma \notin GL(d). \end{cases}$$

Let $\phi$ be a test function on $\mathbf{R}^d$. By definition of $h_i$, we have

$$\int_{\mathbf{R}^d} \frac{\partial \phi}{\partial x_i} d\nu_k = \int_{\mathbf{R}^m} D\phi(T(x)e_i \psi_k(\|\sigma^{-1}(x)\|)d\gamma(x)$$

$$= \int_{\mathbf{R}^m} D(\phi \circ T)(x)h_i(x)\psi_k(\|\sigma^{-1}(x)\|)d\gamma(x). \tag{2.4}$$

Define the (Gaussian) *divergence* operator $Div$ acting on smooth functions $G : \mathbf{R}^n \mapsto \mathbf{R}^n$ by

$$Div(G)(x) = < G(x), x > -Trace\, DG(x).$$

Integrating by parts in (2.4) yields

$$\int_{\mathbf{R}^d} \frac{\partial \phi}{\partial x_i} d\nu_k = \int_{\mathbf{R}^m} \phi \circ T(x)X_i(x)d\gamma(x)$$

where

$$X_i = Div[\psi_k(\|\sigma^{-1}\|)h_i]. \tag{2.5}$$

Since $X_i$ is a continuous function with compact support, it follows that $X_i \in L^1(\gamma)$. Thus (2.5) yields

$$\left| \int_{\mathbf{R}^d} \frac{\partial \phi}{\partial x_i} d\nu_k \right| \leq \|\phi\|_\infty \|X_i\|_{L^1(\gamma)}$$

and the absolute continuity of $\nu_k$ follows from Lemma 2.1.

With the above finite-dimensional case as motivation, we now move to the heart of the matter. Let $\gamma$ denote the law of the Wiener process. This is a measure defined on the space $C_0$ of continuous paths $\{w : [0, 1] \mapsto \mathbf{R}^n \mid \mathbf{w}(0) = 0\}$. Recall that we wish to study the law $\nu$ of a random variable $\xi_t$, for $t > 0$, where $\xi$ is the solution to a SDE. Let $A$ and $B$ denote bounded smooth maps from $\mathbf{R}^d$ into $\mathbf{R}^d \otimes \mathbf{R}^n$ and $\mathbf{R}^d$ respectively with bounded derivatives of all orders, and let $x \in \mathbf{R}^d$. As before, let $w = (w_1, \ldots, w_n)$ denote a standard Wiener process and consider the Itô SDE

$$\xi_t = x + \int_0^t A(\xi_s)dw_s + \int_0^t B(\xi_s)ds, t \geq 0. \tag{2.6}$$

Denoting by $g$ the map $w \mapsto \xi$ and by $g_t$ the composition of $g$ with evaluation at time $t > 0$, we have $\nu = g_t(\gamma)$. We seek conditions under which the measure $\nu$ is absolutely continuous with respect to the Lebesgue measure on $\mathbf{R}^d$. The setting looks similar to that of Theorem 2.2; there are, however, two important differences:

(i) the measure $\gamma$ is defined on an *infinite-dimensional* vector space
(ii) the map $g$ (and hence $g_t$) is *non-differentiable* in the classical sense.

Point (i) can be handled without too much difficulty. Although there is nothing like a Lebesgue measure on $C_0$, the usual backdrop for an integration by parts calculation, the Wiener measure has a well-developed analytic structure. In particular, there are formulae for integration by parts on Wiener space (for example, the Gross' divergence theorem for abstract Wiener spaces [G]). The second point is more serious. Malliavin developed his stochastic calculus of variations in [Ma1] in order to address this.

We shall adopt here a more elementary approach, based on the following observation. As has been known since the original work of Wiener, there is a Hilbert subspace $H \subset C_0$ of paths, now known as the *Cameron–Martin* space, canonically associated with the Wiener measure. Namely, H is the set of Lebesgue a.e. differentiable paths $h$ such that

$$\int_0^1 |h'(t)|^2 dt < \infty$$

equipped with the inner product

$$< h, k >_H \equiv \int_0^1 < h'(t), k'(t) > dt, h, k \in H.$$

The point is that, *when restricted to $H$*, the map $g$ becomes entirely *regular*. In fact, it is intuitively clear that the restriction of $g$ to $H$, which we denote by $\tilde{g}$, is the map $h \in H \mapsto k$ defined by the ordinary integral equation

$$k_t = x + \int_0^t A(k_s)h'_s ds + \int_0^t B(k_s)ds, t \geq 0 \tag{2.7}$$

and it is easily shown that the map $\tilde{g}$ possesses the same degree of smoothness as the coefficient functions $A$ and $B$. For example, an equation for the derivative $\eta \equiv D_r \tilde{g}(h)$, $r \in H$, is obtained by formally differentiating in (2.7) with respect to $h$. Thus $\eta$ satisfies

$$\eta_t = \int_0^t \left\{ DA(k_s)(\eta_s, h_s') + A(k_s)r_s' + DB(k_s)\eta_s \right\} ds.$$

Integral equations for higher order derivatives of $\tilde{g}$ can be similarly obtained.

Actually, the term "restriction" needs interpretation because $g$ is only defined up to a set of $\gamma$-measure 1, and $\gamma(H) = 0$. It is more correct to say that $g$ *stochastically extends* $\tilde{g}$ in the following sense. For each $m \in \mathbf{N}$, define $P_m : C_0 \mapsto H$ to be the operator that piecewise linearizes on the uniform partition $[0, 1/m, 2/m \ldots, 1]$. It can be shown that, as $m \to \infty$, $P_m$ converges strongly to the identity map on $C_0$, and $\tilde{g}(P_m w) \to g(w)$ a.s.

The foregoing remarks provide an alternative elementary approach to Malliavin's work. As before, let $\phi$ denote a test function on $\mathbf{R}^d$ and let $\tilde{g}_t$ denote the composition of $g$ with evaluation at time $t$. Suppose that for each standard basis vector $e_i$, one can construct a sequence of paths $h_i^m$ such that

$$D\tilde{g}_t(P_m w)h_i^m = e_i.$$

Applying the dominated convergence theorem and integrating by parts with respect to the measure $\gamma$, we will then have

$$\int_{\mathbf{R}^d} \frac{\partial \phi}{\partial x_i} \, d\nu = \lim_m \int_{C_0} D\phi(\tilde{g}_t(P_m w))e_i d\gamma$$

$$= \lim_m \int_{C_0} D(\phi \circ g_t^m)(P_m w)h_i^m d\gamma$$

$$= \lim_m \int_{C_0} \phi(\tilde{g}_t(P_m w)) Div[h_i^m] d\gamma.$$

If we can show that the sequence $\{Div[h_i^m]\}_m$ is bounded in $L^1(\gamma)$, then we obtain

$$\left| \int_{\mathbf{R}^d} \frac{\partial \phi}{\partial x_i} \, d\nu \right| \leq \|\phi\|_\infty \sup_m \|Div[h_i^m]\|_{L^1(\gamma)}$$

and the absolute continuity of $\nu$ follows from Lemma 2.1 as before.

We will actually carry out a modified version of the above procedure, using a sequence of piecewise linear approximations to $g$. Being finite-rank operators, these have the advantage of *finite-dimensionalizing* the problem. Thus, at each level of approximation, we need only perform an elementary integration by parts in Euclidean space, as in the proof of Theorem 2.2. Our approximation scheme is as follows.

For each $m \in \mathbf{N}$ and $w \in C_0$, let $\Delta_j w (= \Delta_j^m w)$ denote $w((j+1)t/m) - w(tj/m)$. Define $v_0, v_{t/m}, \ldots, v_t \in \mathbf{R}^d$ inductively by $v_0 = x$ and

$$v_{kt/m} = x + \sum_{j=0}^{k-1} A(v_{jt/m})\Delta_j w + \frac{t}{m} \sum_{j=0}^{k-1} B(v_{jt/m}), k = 1, \ldots, m. \qquad (2.8)$$

Let $v_t^m : [0, 1] \mapsto \mathbf{R}^d$ denote the path piecewise linear between the points $(kt/m, v_{kt/m})$, $k = 0, \ldots, m$ and constant on $[t, 1]$.

It is easy to see that for each $m$, the map $v^m : C_0 \mapsto H$ is $C^\infty$. We proved in [Be2, 35–37]:

**Theorem 2.3.** *For every $p \in \mathbf{N}$,*

$$\lim_{m \to \infty} \sup_{s \in [0,t]} E\left[|\xi_s - v_s^m|^p\right] = 0$$

*where $\xi$ is the solution to the SDE (2.6).*

We now define an analogue of the matrix $\sigma$ appearing in the proof of Theorem 2.2. Let $\sigma^m(w) \equiv Dg_t^m(w)Dg_t^m(w)^* \in \mathbf{R}^d \otimes \mathbf{R}^d$. We proved in [Be2, 37–39]

**Theorem 2.4.** *As $m \to \infty$, the matrix sequence $\sigma^m$ converges in probability to a limit $\sigma$ in $\mathbf{R}^d \otimes \mathbf{R}^d$. Let $I$ denote the $d \times d$ identity matrix and consider the $d \times d$ matrix-valued equations*

$$Y_s = I + \int_0^s DA(\xi_u)(Y_u, dw_u) + \int_0^s DB(\xi_u)Y_u du$$

*and*

$$Z_s = I - \int_0^s Z_u DA(\xi_u)(., dw_u) - \int_0^s Z_u DB(\xi_u)du.$$

*Then*

$$\sigma = Y_t \left[ \int_0^t Z_s A(\xi_s) A(\xi_s)^* Z_s^* ds \right] Y_t^* \tag{2.9}$$

*where $*$ denotes matrix transpose.*

**Remarks.** The matrix processes $Y_t$ and $Z_t$ have a natural interpretation in terms of the *stochastic flow* of the SDE (2.6), i.e. the random map on $\mathbf{R}^d$, $\phi_t : x \mapsto \xi_t$. It can be shown that $\phi_t$ is a.s. a $C^\infty$ map and an easy computation shows that $Y_t = D\phi_t$. Furthermore, $Y_t \in GL(d)$ for all $t \geq 0$ and $Z_t = Y_t^{-1}$.

The matrix $\sigma$ defined in (2.9) is called the *Malliavin covariance matrix* associated to the random variable $\xi_t$. The next theorem shows that establishing nondegeneracy of this matrix is the key to proving regularity of the distribution of $\xi_t$ (as might be suspected from Theorem 2.2 and its proof).

**Theorem 2.5 (Malliavin).** *Suppose $\sigma \in GL(d)$ a.s. Then the random variable $\xi_t$ is absolutely continuous with respect to the Lebesgue measure on $\mathbf{R}^d$.*

Our proof of Theorem 2.5 will make use of the following technical result, which is tailor-made for the estimations we will need to do later (see [Be2, 34–35] for its proof).

**Lemma 2.6.** *Suppose that $X, U_0, \ldots, U_{m-1}$ are $\mathbf{R}^d$-valued random variables, $V_0, \ldots, V_{m-1}$ and $Y_0, \ldots, Y_{m-1}$ are random linear maps from $\mathbf{R}^n$ to $\mathbf{R}^d$, and $Z_0, \ldots, Z_{m-1}$ are random bilinear maps from $\mathbf{R}^d \times \mathbf{R}^n$ to $\mathbf{R}^d$, satisfying the following conditions:*

(i) *For all $0 \leq i \leq m - 1$, $U_i$, $V_i$, $Y_i$, and $Z_i$ are measurable with respect to $F_{it/m}$, where $\{F_t\}$ is the filtration generated by $\{w_t\}$.*

(ii) *$\max\left\{\|X\|_p, \|U_i\|_p, \|V_i\|_p, \|Y_i\|_p, \|Z_i\|_p, 0 \leq i \leq m - 1\right\} \leq M$, where $\|.\|_p$ denotes the $L^p$ norms of the various quantities in their respective spaces.*

*Let $\eta_{kt/m}, 0 \le k \le m$, be random variables satisfying the equations*

$$\eta_{kt/m} = X + \frac{1}{m}\sum_{j=0}^{k-1} U_j + \sum_{j=0}^{k-1} V_j(\Delta_j w) + \frac{1}{m}\sum_{j=0}^{k-1} Y_j(\eta_{jt/m}) + \sum_{j=0}^{k-1} Z_j(\eta_{jt/m}, \Delta_j w).$$

*Then there exists a constant $N$, depending only on $M$ and $p$, such that*

$$\|\eta_{kt/m}\|_p \le N, \forall\ 0 \le k \le m.$$

*Proof of Theorem 2.5.* Let $V_m$ denote the finite-dimensional subspace of $C_0$ consisting of paths that are piecewise linear between the times $0, t/m, \ldots, t$, and constant on $[t, 1]$. Following the method used to prove Theorem 2.2, we define

$$\frac{d\gamma_k}{d\gamma}(w) = R_k(\sigma(w)), v_k = g_t(\gamma_k)$$

where $R_k$ are as defined previously. As before, the assumption $\sigma \in GL(d)$, a.s. implies $v_k \to v$ in variation, so it suffices to prove that each $v_k$ is absolutely continuous.

Let $e$ denote any unit vector in $\mathbf{R}^d$ and define $h^m : C_0 \mapsto H$ by

$$h^m(w) = \begin{cases} Dg_t^m(w)^*\sigma^m(w)^{-1}e, & \sigma^m \in GL(d) \\ 0, & \sigma^m \notin GL(d). \end{cases}$$

Arguing as before and integrating by parts with respect to the measures $P_m(\gamma)$ (note these are Gaussian measures on the finite-dimensional vector spaces $V_m$) yields

$$\int_{\mathbf{R}^d} D_y\phi\, dv_k = \lim_{m\to\infty} \int_{C_0} \phi \circ g_t^m\, Div\, [h^m R_k \circ \sigma^m]\, d\gamma$$

where $Div$ is now defined with respect to the Cameron–Martin space $H$

$$Div\, G(w) = < G(w), w >_H -Trace_H DG(w).$$

To complete the proof, we must thus show that

$$\sup_m E|Div\, [h^m R_k \circ \sigma^m]| < \infty \tag{2.10}$$

First consider the inner product term in the divergence. This is non-zero only if $\sigma^m \in GL(d)$ and $\|\sigma^m\|^{-1} \le k + 1$. In this case

$$E| < h^m R_k \circ \sigma^m, w >_H | = E| < (\sigma^m)^{-1}e, Dg_t^m(w)w > |$$

$$\le (k+1)E|\eta_t^m| \tag{2.11}$$

where $\eta^m = Dg^m(w)w$ satisfies the equation for $k = 1, \ldots, m$,

$$\eta_{kt/m}^m = \sum_{i=0}^{k-1} \{A(v_{it/m})\Delta_i w + DA(v_{it/m})(\eta_{it/m}^m, \Delta_i w) + t/m DB(v_{it/m})\eta_{it/m}^m\}.$$

Since $A$, $DA$, and $DB$ are bounded, Lemma 2.6 implies that $E|\eta_t^m|$ is bounded and it follows from (2.11) that $E| < h^m R_k \circ \sigma^m, w >_H |$ is bounded.

It remains to show that the same holds for the second term in the divergence, i.e.,

$$\sup_m E|Trace_H D[h^m R_k \circ \sigma^m]| < \infty. \tag{2.12}$$

Let $f_1, \ldots, f_n$ be an orthonormal basis of $\mathbf{R}^n$. For $1 \le r \le n$ and $0 \le l \le m - 1$, define $f^{rl} \in H$ by

$$f_s^{rl} = \begin{cases} 0, & 0 \le s < lt/m \\ \sqrt{m/t}(s - lt/m)f_r, & lt/m \le s < (l+1)t/m \\ \sqrt{t/m} f_r, & (l+1)t/m < s \le 1 \end{cases}$$

The set $S_m \equiv \{f^{rl}, 1 \le r \le n, 0 \le l \le m - 1\}$ is orthonormal in $H$, thus can be extended to an orthonormal basis $B_m$ of $H$. Note that for any $f \in B_m \cap S_m^c$, $Dg^m(w)f = 0$. Evaluating the Trace in (2.12) on the basis $B_m$ gives

$$Trace_H D[h^m R_k \circ \sigma^m] = \sum_{r=1,l=0}^{n,m-1} < D[h^m R_k \circ \sigma^m](w) f^{rl}, f^{rl} >$$

$$= \sum_{r=1,l=0}^{n,m-1} \Big\{ R_k \circ \sigma^m(w)[ < (\sigma^m)^{-1}e, D^2 g_t^m(w)(f^{rl}, f^{rl}) >$$

$$- < (\sigma^m)^{-1} D\sigma^m(w) f^{rl}(\sigma^m)^{-1}e, Dg_t^m(w) f^{rl} > ]$$

$$+ DR_k(\sigma^m) D\sigma^m(w) f^{rl} < (\sigma^m)^{-1}e, Dg_t^m(w) f^{rl} > \Big\}.$$

In view of the definition of $R_k$, it suffices to show that

$$\sup_m E\Big| \sum_{l=0}^{m-1} D^2 g_t^m(w)(f^{rl}, f^{rl})\Big] < \infty \tag{2.13}$$

and

$$\sup_m E\Big[ \sum_{l=0}^{m-1} |D\sigma^m(w) f^{rl}| \times |Dg_t^m(w) f^{rl}|\Big] < \infty. \tag{2.14}$$

Let $\eta^{rl}$ denote the path $\sqrt{m} Dg^m(w) f^{rl}$. Differentiation in (2.8) yields $\eta_{jt/m}^{rl} = 0$ if $j \le l$ and

$$\eta_{jt/m}^{rl} = \sqrt{t} A(v_{lt/m}) f_r + \sum_{p=l}^{j-1} DA(v_{pt/m})(\eta_{pt/m}^{rl}, \Delta_p w)$$

$$+ \frac{t}{m} \sum_{p=l}^{j-1} DB(v_{pt/m}) \eta_{pt/m}^{rl}, j \ge l + 1.$$

It follows from Lemma 2.6 that

$$\sup_{j,k,m} \|\eta^{rk}_{jt/m}\|_4 < \infty \tag{2.15}$$

Now let $\rho^{rl}$ denote the path $mD^2g^m(w)(f^{rl}, f^{rl})$. This satisfies the equation

$$\rho^{rl}_{jt/m} = tDA(v_{lt/m})(\eta^{rl}_{lt/m}, f_r)$$

$$+ \sum_{p=l}^{j-1} \{ tD^2A(v_{pt/m})(\eta^{rl}_{pt/m}, \Delta_p w) + DA(v_{pt/m})(\rho^{rl}_{pt/m}, \Delta_p w) \}$$

$$+ \frac{1}{m} \sum_{p=l}^{j-1} \{ tD^2B(v_{pt/m})(\eta^{rl}_{pt/m}, \eta^{rl}_{pt/m}) + DB(v_{pt/m})\rho^{rl}_{pt/m} \}, \ j \geq l+1$$

$(\rho^{rl}_{jt/m} = 0$ for $j \leq l)$. Lemma 2.6 together with (2.15) now give

$$\sup_{r,j,k,m} \|\rho^{rk}_{jt/m}\|_4 < \infty$$

and this implies (2.13). Condition (2.14) can be established by a similar argument. With this, the proof of the theorem is complete.     □

We now return to the SDE (1.2) defined in terms of vector fields. It follows from (2.9) that the Malliavin covariance matrix $\sigma$ for $\xi_t$ now has the form

$$\sigma = Y_t \sum_{i=1}^{n} \int_0^t [Z_s X_i(\xi_s)] \otimes [Z_s X_i(\xi_s)]^* ds Y_t^* \tag{2.16}$$

where $Z$ satisfies the $d \times d$ matrix-valued SDE

$$Z_s = I - \sum_{i=1}^{n} \int_0^s Z_u DX_i(\xi_u) \circ dw_i - \int_0^s Z_u DX_0(\xi_u)du \tag{2.17}$$

As before, $Y_t$ is the inverse matrix of $Z_t$. Note that the stochastic integral in (2.17) is of Stratonovich type.

The following result and its proof (which I first learned from Bismut's paper [Bi1]) makes transparent the relationship between Hörmander's Lie bracket condition and the non-degeneracy of $\sigma$.

**Theorem 2.7.** *If the Lie algebra generated by the vector fields $X_1, \ldots, X_n$ span $\mathbf{R}^d$ at $\xi_0$, then $\sigma \in GL(d)$, a.s. (hence, by Theorem 2.5 $\xi_t$ is absolutely continuous with respect to Lebesgue measure on $\mathbf{R}^d$).*

The proof will require the following:

**Lemma 2.8.** *Suppose $Y \in \mathbf{R}^d$ and $B$ is a smooth vector field on $\mathbf{R}^d$ such that*

$$< Z_s B(\xi_s), Y >= 0, \forall s \in [0, t]. \tag{2.18}$$

*Then for $i = 1, \ldots, n$,*

$$< Z_s[X_i, B](\xi_s), Y >= 0, \forall s \in [0, t].$$

*Proof.* Applying Itô's formula and using (2.17), we obtain, for $s \in [0, t]$,

$$0 = d < (Z_s B(\xi_s)), Y >$$

$$= \Big( Z_s[B, X_0]ds + \sum_{i=1}^{n} Z_s[B, X_i](\xi_s) \circ dw_i, Y \Big). \tag{2.19}$$

Converting the Stratonovich integrals to the Itô form, we see that the infinitesimal term in (2.19) has the form

$$G(s)ds + \sum_{i=1}^{n} Z_s[B, X_i](\xi_s)dw_i$$

for some continuous adapted process $G$. Thus

$$< G(s)ds + \sum_{i=1}^{n} Z_s[B, X_i](\xi_s)dw_i, Y >= 0, \forall s \in [0, t].$$

The conclusion now follows from the computational rules of the Itô calculus: $dw_i dt = 0$, $dw_i dw_j = \delta_{ij} dt$. $\qquad \square$

*Proof of Theorem 2.7.* Define $V$ to be the span of the set of vectors

$$\{X_i(\xi_0), [X_i, X_j](\xi_0), [X_i, [X_j, X_k]](\xi_0), \ldots, 1 \le i, j, k, \ldots, \le n\}$$

and suppose that $Y \in \mathrm{Span}(\sigma)^{\perp}$. Then for all $i = 1, \ldots, n$ and $s \in [0, t]$

$$< Z_s X_i(\xi_s), Y >= 0. \tag{2.20}$$

Iterating Lemma 2.8 on (2.20) gives

$$< Z_s X_i(\xi_s), Y >, < Z_s[X_i, X_j](\xi_s), Y >, < Z_s[[X_i, X_j], X_k](\xi_s), Y >, \cdots = 0$$

for all $s \in [0, t]$. Setting $s = 0$, we have $Y \in V^{\perp}$. Thus span $V \subseteq \mathrm{Span}(\sigma)$ and the result follows. $\qquad \square$

In [Ma1], Malliavin proves further:

**Theorem 2.9.** *If*

$$(\det \sigma) \in \cap_{p \ge 1} L^p(\gamma) \tag{2.21}$$

*then the density of $\xi_t$ is $C^{\infty}$.*

This is also proved in [Be1], by iterating the argument used to prove Theorem 2.5. Kusuoka and Stroock prove in [KS2] that (2.21) holds under Hörmander's general condition and they use this to give a complete probabilistic proof of Hörmander's theorem. We do not include this work here since we will prove a more general result in the next section.

## 3  A Hörmander theorem for infinitely degenerate operators

The material in this section comes from [BM3]. Let $X_0, \ldots, X_n$ denote smooth vector fields defined on an open subset $D$ of $\mathbf{R}^d$. As before, we consider the second-order differential operator

$$L \equiv \frac{1}{2} \sum_{i=1}^{n} X_i^2 + X_0. \tag{3.1}$$

Let $Lie(X_0, \ldots, X_n)$ be the Lie algebra generated by the $X_0, \ldots, X_n$. According to the theorem of Hörmander ([H], Theorem 1.1), $L$ is hypoelliptic on $D$ if the vector space $Lie(X_0, \ldots, X_n)(x)$ has dimension $d$ at every $x \in D$. It can be shown that this is a *necessary* condition for hypoellipticity for operators of the form (3.1) with analytic coefficients.

This is not the case if the vector fields $X_0, \ldots, X_n$ defining $L$ are allowed to be (smooth) *non*-analytic. This fact was strikingly illustrated by Kusuoka and Stroock, who studied differential operators on $\mathbf{R}^3$ of the form

$$L_a \equiv \frac{\partial^2}{\partial x_1^2} + a^2(x_1)\frac{\partial^2}{\partial x_2^2} + \frac{\partial^2}{\partial x_3^2}. \tag{3.2}$$

They assume $a$ to be a $C^\infty$ real-valued even function, non-decreasing on $[0, \infty)$, which vanishes (only) at zero. It is shown in ([KS2], Theorem 8.41) that $L_a$ is hypoelliptic on $\mathbf{R}^3$ if and only if $a$ *satisfies the condition* $\lim_{s \to 0+} s \log a(s) = 0$. In particular, in the case $a(s) = \exp(-|s|^p)$, $L_a$ is hypoelliptic provided $p \in (-1, 0)$. However, it is clear that any such operator fails to satisfy Hörmander's condition on the hyperplane $x_1 = 0$.

In this section we present a sharp criterion for hypoellipticity that implies Hörmander's theorem and encompasses the class of superdegenerate hypoelliptic operators of Kusuoka and Stroock.

We introduce the following notation. For a positive integer $m$, let $E^{(m)}$ denote any matrix whose columns consist of $X_0, \cdots, X_n$ together with all (iterated) Lie brackets of the form

$$[X_{i_1}, X_{i_2}]_{i_1, i_2=0}^{n}; \cdots; [X_{i_1}, [X_{i_2}, [X_{i_3}, \cdots, [X_{i_{m1}}, X_{i_m}]] \cdots ]]_{i_1, i_2, \cdots, i_m=0}^{n}.$$

For $x \in D$ and $m \geq 1$, define

$$\mu^{(m)} \equiv \text{smallest eigenvalue of } [E^{(m)} E^{(m)*}].$$

Observe that $\mu^{(m)}(x) > 0$ for some $m \geq 1$ if and only if Hörmander's (general) condition holds for the operator $L$ at $x \in D$. In this case we will say that $x$ is a *Hörmander point* for $L$. We denote the set of all such points by $H$ (note that $H$ is an open subset of $D$). The set $D \cap H^c$ of non-Hörmander points of $L$ will be denoted simply by $H^c$ in the sequel.

The main result of this section is the following:

**Theorem 3.1.** *Suppose the non-Hörmander set $H^c$ of $L$ is contained in a $C^2$ hypersurface $S$. Assume that at every point $x$ in $H^c$*

(i)  *at least one of the vector fields* $X_1, \cdots, X_n$ *is transversal to* $S$.

(ii)  *There exists an integer* $m \geq 1$, *an open neighborhood* $U$ *of* $x$, *and an exponent* $p \in (-1, 0)$ *such that*

$$\mu^{(m)}(y) \geq \exp\{-[\rho(y, S)]^p\}, \quad \forall y \in U \tag{3.3}$$

*where* $\rho(y, S)$ *denotes the Euclidean distance between* $y$ *and* $S$.

*Then* $L$ *is hypoelliptic on* $D$.

We note that hypotheses (ii) in Theorem 3.1 controls the rate at which the Hörmander condition fails in a neighborhood of non-Hörmander points. It is clear that some such condition is necessary, since the Kusuoka–Stroock result cited above shows that the operator

$$L_p = \frac{\partial^2}{\partial x_1^2} + \exp\{-|x_1|^p\} \frac{\partial^2}{\partial x_2^2} | \frac{\partial^2}{\partial x_3^2}$$

is non-hypoelliptic if $p \leq -1$. Furthermore, the non-hypoellipticity of the operator $L_{-1}$ shows that the allowed range $(-1, 0)$ for $p$ in (3.3) is *optimal*. Hypothesis (ii) has the following probabilistic interpretation in terms of the diffusion process $\xi_t$ corresponding to $L$, defined in (1.2). It implies that if $\xi$ starts at a non-Hörmander point $x$ on $S$, then it will escape from $x$ at a fast enough rate to acquire a non-singular distribution.

One can see that a hypothesis such as (i) is also necessary for the hypoellipticity of $L$ (at least in the case where $X_0 = 0$) by looking at the probabilistic picture. For if $X_i$ is tangential to $S$ for each $i = 1, \ldots, n$ then, started from a point $x \in S$, $\xi$ *will stay on* $S$. Hence $\xi_t$ will not have a density in $\mathbf{R}^d$ at positive times $t$, which implies that $L$ cannot be hypoelliptic (as we remarked earlier hypoellipticity of $L$ implies the existence of a density for $\xi_t$, $t > 0$).

The ideas underlying the proof of Theorem 3.1 are as follows. In Section 2 we introduced the Malliavin covariance matrix $\sigma$ corresponding to the random variable $\xi_t$ and showed that non-degeneracy of $\sigma$ implies the existence of a density for $\xi_t$. Kusuoka and Stroock have proved that if the inverse moments of $\sigma$ do not explode too quickly as $t \downarrow 0$, then the parabolic operator $L + \partial/\partial t$ is hypoelliptic, they also showed how to deduce hypoellipticity of $L$ from this. More specifically, let $\sigma$ denote the matrix defined in (2.16) and (2.17) (recall that the process $\xi$ in these formulas is defined by equation (1.2)) and let $\Delta(t, \xi_0)$ denote the determinant of $\sigma$. The following result is proved in [KS2].

**Lemma 3.2.** *Let* $D$ *be an open set in* $\mathbf{R}^d$. *Suppose that for every* $q \geq 1$ *and every* $x$ *in* $D$, *there exists a neighborhood* $V \subseteq D$ *of* $x$ *such that*

$$\lim_{t \to 0+} t \log \left\{ \sup_{y \in V} \|\Delta(t, y)^{-1}\|_q \right\} = 0. \tag{3.4}$$

*Then the differential operator* $L + \dfrac{\partial}{\partial t}$ *is hypoelliptic on* $\mathbf{R} \times D$.

We will prove that (3.4) is satisfied under a (parabolic version of) the hypotheses of Theorem 3.1. There are two stages to this argument:

(a) A local parameterization $\phi$ of the hypersurface $S$ is introduced. Throughout this section, let $\xi$ denote the diffusion process defined in (1.2). The quantity $\phi(\xi_t)$ measures the distance between $\xi_t$ and $S$. We obtain probabilistic lower bounds for the $L^q$-norms of the paths $\phi(\xi)$. These lower bounds are asymptotically sharp as $q \to 1$.

(b) We study the way the lower bounds in (a) are degraded under hypothesis (ii) of Theorem 3.1. This allows us to obtain sharp lower bounds on $\Delta$ from which we are able to verify condition (3.4).

The proof of Theorem 3.1 uses basic stochastic analytic tools, e.g., Itô's formula, Girsanov's theorem, and the time-change theorem for stochastic integrals. The theorem is a technical improvement of an important classical result. More significantly, this work establishes a precise connection between two seemingly unrelated phenomena, namely the *maximal class of hypoelliptic operators* of the form (1.1) and the *space-time scaling property* of the Wiener process. This provides new insight into hypoellipticity that could not be acquired through a classical analysis of the problem.

Before proving Theorem 3.1, we will state and prove a parabolic version of the theorem. To this end, denote by $F^{(m)}$ the matrix obtained by deleting the column $X_0$ from the matrix $E^{(m)}$ defined prior to Theorem 3.1 and let $\lambda^{(m)}$ to be the smallest eigenvalue of the matrix $[F^{(m)} F^{(m)*}]$. Define $K \equiv \{x \in D| \lambda^{(m)}(x) > 0\}$ where $D \subset \mathbf{R^d}$ is open.

**Theorem 3.3.** *Suppose the set $K^c$ is contained in a $C^2$ hypersurface $N$ of $D$. Assume that at every point in $K^c$, at least one of the vector fields $X_1, \cdots, X_n$ is transversal to $N$. Assume further that for every $x \in K^c$, there exists an integer $m \geq 1$, an open neighborhood $U$ of $x$, and $p \in (-1, 0)$ such that $\lambda^{(m)}(y) \geq \exp\{-[\rho(y, N)]^p\}$ for all $y \in U$. Then the operator $L + \dfrac{\partial}{\partial t}$ is hypoelliptic on $\mathbf{R} \times \mathbf{D}$.*

The proof of Theorem 3.3 will require a definition and several preliminary results which we now state.

**Definition.** A non-negative random variable $T$ is *exponentially positive* if there exist positive constants $c_1$ and $c_2$ (which we will refer to as the *characteristics of X*) such that

$$P(T < \epsilon) < e^{-c_1/\epsilon}$$

for all $\epsilon \in (0, c_2)$.

We will make frequent use of the following well-known result ([IW], Lemma 10.5).

**Lemma 3.4.** *Let $y : [0, 1] \times \emptyset \to \mathbf{R}^d$ be an Itô process of the form*

$$dy(t) = \sum_{i=1}^{n} a_i(t)\, dW_i(t) + b(t)\, dt, \quad 0 \leq t \leq 1, \tag{3.5}$$

*where $a_1, \ldots, a_n, b : [0, 1] \times \emptyset \to \mathbf{R}^d$ are measurable adapted processes, all bounded a.s. by a deterministic constant $c_3$. Let $r > 0$ and define*

$$\tau \equiv \inf\{s > 0 : |y_s - y_0| = r\} \wedge 1. \tag{3.6}$$

Then $\tau$ is an exponentially positive stopping time, and the characteristics of $\tau$ depend only on $r$, $c_3$, $n$ and $d$.

The next two lemmas are central to our argument (in order to facilitate the exposition, we delay their proofs to the end of the section). The first yields sharp probabilistic lower bounds when applied to diffusion processes with at least one non-zero initial time diffusion coefficient.

**Lemma 3.5.** *Let* $y : [0, 1] \times \varnothing \rightarrow \mathbf{R}^d$ *be the Itô process in (3.5). Suppose that* $\tau \leq T$ *is an exponentially positive stopping time such that at least one diffusion coefficient* $a_i$ *satisfies the condition: a.s.,* $|a_i(s)| \geq \delta$, *for all* $0 \leq s \leq \tau$, *for some deterministic* $\delta > 0$. *Then for every* $m \geq 2$, *there exist positive constants* $c_4$, $c_5$ *and* $T_0$ *such that for all* $t \in (0, T_0)$ *and* $\epsilon \in (0, c_4 t^{m+1})$, *the following holds*

$$P\left(\int_0^{t \wedge \tau} |y(u)|^m \, du < \epsilon\right) < \exp\left\{-c_5 \epsilon^{-\frac{1}{m+1}}\right\}. \tag{3.7}$$

*The constants* $c_4$ *and* $c_5$ *can be chosen to depend only on* $m$, $c_3$, $\delta$, *and the characteristics of* $\tau$. *The constant* $T_0$ *depends only on the characteristics of* $\tau$.

The following result describes how the estimate (3.7) transforms, under composition of the integrand with a function that vanishes at zero, at an appropriate exponential rate.

**Lemma 3.6.** *Let* $\tau$ *be an exponentially positive stopping time. Suppose* $y$ *is an Itô process as in Lemma 3.4. Suppose further that* $y$ *and* $\tau$ *satisfy an estimate of the form* (3.7) *for some* $m > -\dfrac{p}{p+1}$, *where* $p \in (-1, 0)$. *Then there exist positive constants* $T_1$, $c_6$, $c_7$ *and* $q > 1$ *such that for all* $t \in (0, T_1)$ *and all* $\epsilon < \exp\{-c_6 t^{-\frac{1}{q}}\}$, *the following holds*

$$P\left(\int_0^{t \wedge \tau} \exp(-|y_u|^p) \, du < \epsilon\right) < \exp\{-c_7 |\log \epsilon|^q\}. \tag{3.8}$$

*Furthermore, the constants* $T_1$, $c_6$, $c_7$ *and* $q$ *are completely determined by* $c_3$ *in Lemma 3.4,* $c_4$, $c_5$, *and* $m$ *in (3.7),* $p$, *and the characteristics of* $\tau$.

Finally, we will need the following two technical Lemmas (since the proofs are straightforward we omit them)

**Lemma 3.7.** *For every* $q \geq 1$ *and every bounded set* $V \subset \mathbf{R}^d$ *there exists a positive constant* $c_8$ *such that for all* $t \in (0, 1)$ *and* $x \in V$

$$\|\Delta(t, x)^{-1}\|_{2q}^{2q} \leq c_8 \left\{1 + \sum_{j=1}^{\infty} P\left(Q(t, x) < j^{-\frac{1}{2dq}}\right)\right\}, \tag{3.9}$$

where

$$Q(t, x) \equiv \inf\left\{\sum_{i=1}^{n} \int_0^t < Z^x(u) X_i(\xi^x(u)), h >^2 \, du : h \in \mathbf{R}^d, |h| = 1\right\}. \tag{3.10}$$

**Lemma 3.8.** *Suppose that the hypotheses of Theorem 3.3 are satisfied. Then for every* $x \in D \subset \mathbf{R}^d$ *there exists an integer* $m \geq 1$ *such that exactly one of the following two conditions holds:*

*(a)* $\lambda^{(m)}(x) > 0.$
*(b) There exists an open neighborhood* $U \subseteq D$ *of* $x$, *a* $C^2$ *function* $\phi : U \to \mathbf{R}$, *and an exponent* $p \in (-1, 0)$ *such that*
   *(i)* $\phi(x) = 0$ *and* $\nabla\phi(x) \cdot X_i(x) \neq 0$, *for at least one* $i = 1, \ldots, n.$
   *(ii)* $\lambda^{(m)}(y) \geq \exp(-|\phi(y)|^p)$, *for all* $y \in U.$

We now assume the hypotheses and notations of Theorem 3.3. Without loss of generality, the vector fields $X_0, \cdots, X_n$ may be supposed to be defined on *the whole of* $\mathbf{R}^d$ and to have compact support (this follows from a simple argument using a partition of unity and the fact that hypoellipticity is a local property). We will assume this from now on.

Let $x_0 \in D$ and choose $m$ so that either $(a)$ or $(b)$ of Lemma 3.8 hold for $x_0$. Let $t \in (0, 1)$ and suppose $x$ lies in a fixed bounded neighborhood $W$ of $x_0$. Define

$$\tau_1 \equiv \inf\left\{s > 0 : |\xi_s^x - x| \vee \|Z_s^x - I\| = \frac{1}{2}\right\} \wedge 1. \tag{3.11}$$

By Lemma 3.4, $\tau_1$ is an exponentially positive stopping time with characteristics independent of $x \in W$.

Let $S^d \equiv \{h \in \mathbf{R^d} : |\mathbf{h}| = 1\}$ denote the unit sphere in $\mathbf{R}^d$. Suppose $h \in S^d$ and $\alpha = 1/18$, then

$$P\left(\sum_{i=1}^n \int_0^t < Z_u^x X_i(\xi_u^x), h >^2 du < \epsilon\right) \leq P(A \cap E) + P(A \cap E^c),$$

where

$$A \equiv \left(\sum_{i=1}^n \int_0^{t \wedge \tau_1} < Z_u^x X_i(\xi_u^x), h >^2 du < \epsilon\right)$$

and

$$E \equiv \left(\sum_{i=1}^n \int_0^{t \wedge \tau_1} \left[\sum_{j=1}^n < Z_u^x[X_i, X_j](\xi_u^x), h >^2 + \right.\right.$$
$$\left.\left. < Z_u^x\left\{[X_i, X_0] + \frac{1}{2}\sum_{j,k=1}^n [X_i, [X_j, X_k]]\right\}(\xi_u^x), h >^2\right] du < \epsilon^\alpha\right).$$

By a lemma of Kusuoka–Stroock and Norris (cf., e.g., [B], Lemma 6.5; cf[KS2], Theorem A. 24), there exist positive constants $c_9$ and $c_{10}$ such that

$$P(A \cap E^c) \leq c_9 \exp(-c_{10}\epsilon^{-\alpha}). \tag{3.12}$$

The constants $c_9$ and $c_{10}$ are independent of $h \in S^d$. Note that $E \subseteq F \cap G$, where

$$F \equiv \sum_{i,j=1}^{n} \int_{0}^{t\wedge\tau_1} < Z_u^x[X_i, X_j](\xi_u^x), h >^2 \, du < \epsilon^{\alpha}$$

$$G \equiv \sum_{i=1}^{n} \int_{0}^{t\wedge\tau_1} < Z_u^x\left\{[X_i, X_0] + \frac{1}{2}\sum_{j,k=1}^{n} [X_i, [X_j, X_k]]\right\} (\xi_u^x), h >^2 \, du < \epsilon^{\alpha}.$$

Thus

$$P\left(\sum_{i=1}^{n} \int_{0}^{t} < Z_u^x X_i(\xi_u^x), h >^2 \, du < \epsilon\right) \leq c_9 \exp(-c_{10}\epsilon^{-\alpha}) + P(A \cap F \cap G). \quad (3.12)$$

Applying a similar argument to $P(A \cap F \cap G)$ gives

$$P(A \cap F \cap G) \leq c_{11} \exp(-c_{12}\epsilon^{-\alpha^2}) + P(A \cap F \cap G \cap H) \quad (3.13)$$

where

$$H := \left(\sum_{i,j,k=1}^{n} \int_{0}^{t\wedge\tau_1} < Z_u^x[X_i, [X_j, X_k]](\xi_u^x), h >^2 \, du < \epsilon^{\alpha^2}\right).$$

It is easy to check that

$$G \cap H \subseteq \left(\sum_{i=1}^{n} \int_{0}^{t\wedge\tau_1} < Z_u^x[X_i, X_0](\xi_u^x), h >^2 \, du < \epsilon^{r_1}\right)$$

for some $r_1 \in (0, 1)$ and sufficiently small $\epsilon > 0$.

Thus

$$A \cap F \cap G \cap H \subseteq \left(\int_{0}^{t\wedge\tau_1} \left\{ < \sum_{i=1}^{n} (Z_u^x X_i(\xi_u^x), h >^2 \right.\right.$$
$$\left.\left. + \sum_{i,j=0}^{n} < Z_u^x[X_i, X_j](\xi_u^x), h >^2 \right\} du < \epsilon^{r_2}\right)$$

for some $r_2 \in (0, 1)$ and sufficiently small $\epsilon > 0$. Combining this with (3.12) and (3.13), one obtains

$$P\left(\sum_{i=1}^{n} \int_{0}^{t} < Z_u^x X_i(\xi_u^x), h >^2 \, du < \epsilon\right)$$
$$\leq c_9 \exp(-c_{10}\epsilon^{-\alpha}) + c_{11} \exp(-c_{12}\epsilon^{-\alpha^2})$$
$$+ P\left(\int_{0}^{t\wedge\tau_1} \left\{ \sum_{i=1}^{n} < Z_u^x X_i(\xi_u^x), h >^2 \right.\right.$$
$$\left.\left. + \sum_{i,j=0}^{n} < Z_u^x[X_i, X_j](\xi_u^x), h >^2 \right\} du < \epsilon^{r_2}\right). \quad (3.14)$$

Iterating the argument used to derive (3.14) yields the following.

For each $m \geq 1$, there exist positive constants $c_{13}$ and $c_{14}$ and exponents $r_3$ and $r_4 \in (0, 1)$, all independent of $h \in S^d$, such that for all $t \in (0, T)$, $x \in W$, and $\epsilon \in (0, c_{14})$, one has

$$P\left(\sum_{i=1}^{n} \int_0^t < Z_u^x X_i(\xi_u^x), h >^2 du < \epsilon\right)$$

$$\leq \exp(-c_{14}\epsilon^{-r_3}) + P\left(\sum_{j=1}^{N} \int_0^{t \wedge \tau_1} < Z_u^x K_j(\xi_u^x), h >^2 du < \epsilon^{r_4}\right). \quad (3.15)$$

Here the vector fields $K_1, \ldots, K_N$ are the columns of the matrix function $X^{(m)}$. Applying a straightforward compactness argument (cf. [B], Lemma 6.8) to (3.15), one obtains as $h \in S^d$ varies

$$P(Q(t, x) < \epsilon) \leq \exp(-c_{15}\epsilon^{-r_3})$$

$$+ c_{16}\epsilon^{-d} \sup_h \left\{ P\left(\sum_{j=1}^{N} \int_0^{t \wedge \tau_1} < Z_u^x K_j(\xi_u^x), h >^2 du < c_{17}\epsilon^{r_4}\right) \right\} \quad (3.16)$$

for $\epsilon \in (0, c_{18})$ and positive constants $c_{15}, c_{16}, c_{17}, c_{18}$.

Since (3.8) implies $\|Z^x(u) - I\| \leq \frac{1}{2}$, for all $0 \leq u \leq \tau_1$, it is easy to deduce from (3.16) that

$$P(Q(t, x) < \epsilon) \leq \exp(-c_{15}\epsilon^{-r_3}) + c_{16} \epsilon^{-d} P\left(\int_0^{t \wedge \tau_1} \lambda^{(m)}(\xi_u^x) du < c_{18}'\epsilon^{r_4}\right).$$
$$(3.17)$$

We now consider each of the two cases (a) and (b) delineated in the conclusion of Lemma 3.8. Suppose first that (a) holds at $x_0$ for some $m \geq 1$. Then by continuity of $\lambda^{(m)}$ there exist $\rho > 0$ and $\delta > 0$ such that

$$\lambda^{(m)}(y) \geq \delta \quad (3.18)$$

for all $y \in B_\rho(x_0)$, where $B_\rho(x_0)$ denotes the open ball in $\mathbf{R}^d$ with center $x_0$ and radius $\rho$. Let $V \equiv B_{\rho/2}(x_0)$, assume $x \in V$, and let $\tau_2$ denote the first exit time of $\xi^x$ from $V$. Then (3.17) and (3.18) imply

$$P(Q(t, x) < \epsilon) \leq \exp(-c_{15}\epsilon^{-r_3}) + c_{16}\epsilon^{-d} P\left(\tau_1 \wedge \tau_2 \wedge t < \frac{c_{18}\epsilon^{r_4}}{\delta}\right) \quad (3.19)$$

$$\leq c_{19} \exp(-c_{20}\epsilon^{-c_{21}r_5}) \quad (3.20)$$

provided $t > \frac{c_{18}\epsilon^{r_4}}{\delta}$, where $r_5 \equiv r_3 \wedge r_4$ and $c_{19}$, $c_{20}$ and $c_{21}$ are positive constants, independent of $(t, x) \in (0, T) \times V$. Substituting (3.20) into (3.9) yields, for every $q \geq 1$, the following inequality

$$\|\Delta(t, x)^{-1}\|_{2q}^{2q} \leq c_8\left\{ \left(\frac{\delta t}{c_{18}}\right)^{-\frac{2dq}{r_4}} + A(t) \right\},$$

where

$$A(t) \equiv 1 + \sum_{j=k}^{\infty} c_{19} \exp(-c_{20} j^{r_6}),$$

$$\leq 1 + \sum_{j=1}^{\infty} c_{19} \exp(-c_{20} j^{r_6}) < \infty,$$

where $r_6 \equiv c_{21} r_5 / 2dq$ and $k$ is the integer part of $(\delta t / c_{18})^{-2dq/r_4}$. We conclude that $\|\Delta(t,x)^{-1}\|_{2q}$ grows no faster than a power of $t$ as $t \downarrow 0$, uniformly with respect to $x \in V$. Hence (3.4) is satisfied.

We now turn to the case where (b) of Lemma 3.8 holds at the point $x_0$. By the transversality condition (b)(i) we may choose $\rho > 0$ small enough to ensure that $B_\rho(x_0) \subset U$ and such that

$$|\nabla \phi(x).X_i(x)| \geq \frac{1}{2} |\nabla \phi(x_0).X_i(x_0)| > 0$$

for some $1 \leq i \leq n$ and every $x \in B_\rho(x_0)$. Let $V \equiv B_{\rho/2}(x_0)$. Assume $x \in V$ and let $\tau_3$ denote the first exit time of $\xi^x$ from $B_{\rho/2}(x)$. In view of Lemma 3.4 (b)(ii), (3.17) implies

$$P(Q(t,x) < \epsilon) \leq \exp(-c_{15} \epsilon^{-r_3}) + c_{16} \epsilon^{-d} P\left( \int_0^{t \wedge \tau_1 \wedge \tau_3} \exp(-|\eta_u^x|^p) \, du < c_{18} \epsilon^{r_4} \right)$$

$$(3.21)$$

where $\eta^x(t)$ denotes the process $\phi(\xi^x(t))$, $t \leq \tau_3$. Applying Itô's lemma to compute $\eta^x(t)$ gives

$$d\eta^x(t) = \sum_{i=1}^{n} \nabla \phi(\xi^x(t)).X_i(\xi^x(t)) dW_i(t) + (L - c)\phi(\xi^x(t)) \, dt. \qquad (3.22)$$

Lemma 3.8 (b)(i), Lemma 3.4, and (3.22) imply that the process $y \equiv \eta^x$ and the stopping time $\tau \equiv \tau_1 \wedge \tau_2$ satisfy the hypotheses of Lemma 3.5. Hence (3.7) is satisfied for *every* $m > 1$ with $\tau = \tau_1 \wedge \tau_3$ and $y = \eta^x$. Thus, by Lemma 3.6 there exist positive constants $c_6, c_7, T_1$ and $q' > 1$, all independent of $x \in V$, such that for all $t \in (0, T_1)$ and $\epsilon < \exp(-c_6 t^{-\frac{1}{q'}})$

$$P\left( \int_0^{t \wedge \tau_1 \wedge \tau_3} \exp(-|\eta_u^x|^p) \, du < \epsilon \right) < \exp\{-c_7 |\log \epsilon|^{q'}\}. \qquad (3.23)$$

Substituting this into (3.17) gives

$$P(Q(t,x) < \epsilon) \leq \exp(-c_{15} \epsilon^{-r_3}) + c_{16} \epsilon^{-d} \exp(-c_7 |\log \epsilon^{r_4}|^{q'}) \qquad (3.24)$$

for $t \in (0, T_1)$ and $\epsilon < \exp(-c_6 t^{-\frac{1}{q'}})$. Combining (3.21) with (3.5), we arrive at

$$\|\Delta(t,x)^{-1}\|_{2q}^{2q} \leq c_8 \{\exp(2dq c_6 t^{-\frac{1}{q'}}) + c_{22}\}, \quad 0 < t < T_1 \qquad (3.25)$$

where

$$c_{22} \equiv 1 + \sum_{j=1}^{\infty} \left\{ \exp\left(-c_{15} j^{\frac{r_3}{2dq}}\right) + c_{16} j^{1/2q} \exp(-c_7 |\log j^{\frac{r_4}{2dq}}|^{q'}) \right\} < \infty.$$

Note that the constants $c_6$, $c_8$ and $c_{22}$ can all be chosen to be independent of $x \in V$. The right hand side of (3.25) *explodes exponentially fast* as $t \downarrow 0$. However, since $q' > 1$ we conclude that (3.4) holds also for this case, and the proof of Theorem 3.3 is complete.

$\square$

*Proof of Theorem 3.1.* Assume $L$ satisfies the hypotheses of Theorem 3.1. We borrow yet another technique from Kusuoka and Stroock [KS2]. The idea is to imbed the operator $L$ in another operator $\tilde{L}$ defined on a $(d+1)$-dimensional domain and satisfying the hypotheses of Theorem 3.3. This will prove the hypoellipticity of the parabolic operator $\tilde{L} + \frac{\partial}{\partial t}$, which in turn will imply hypoellipticity of $L$.

Choose a smooth non-negative real-valued function $\rho$ on $(0, 1)$ such that both $\rho(s)$ and its derivative $\rho'(s)$ are bounded away from zero for all $s \in (0, 1)$. Define the operator $\tilde{L}$ on the domain $D \times (0, 1)$ by

$$\tilde{L} \equiv \rho(s) L + \frac{1}{2} \frac{\partial^2}{\partial s^2}.$$

Then $\tilde{L}$ has the form

$$\tilde{L} \equiv \frac{1}{2} \sum_{i=1}^{n+1} \tilde{X}_i^2 + \tilde{X}_0,$$

where $\tilde{X}_0(x, s) \equiv \rho(s) X_0(x)$, $\tilde{X}_i(x, s) \equiv \rho(s)^{1/2} X_i(x)$, $1 \le i \le n$, and $\tilde{X}_{n+1}(x, s) \equiv \frac{\partial}{\partial s}$, $(x, s) \in D \times (0, 1)$. Define $\tilde{F}^{(m)}$, $\tilde{\lambda}^{(m)}$, $\tilde{H}$, and $\tilde{K}$ similarly as before, using the vector fields $\tilde{X}_i$ in place of the $X_i$'s. Since $\rho$ and $\rho'$ are bounded away from zero on $(0, 1)$, it is easy to see that there are positive constants $\delta_m$ such that

$$\tilde{\lambda}^{(m)}(x, s) \ge \delta_m \mu^{(m)}(x), \quad \forall (x, s) \in D \times (0, 1). \tag{3.26}$$

This implies $\tilde{K}^c \subseteq H^c \times (0, 1)$. Since $H^c$ is contained in a $C^2$ hypersurface $S$ in $D$, then $\tilde{H}^c$ is contained in the $C^2$ hypersurface $S \times (0, 1)$ in $D \times (0, 1)$. By assumption, at least one of the vector fields $X_1, \cdots, X_n$ is transversal to $S$ at every point of $H^c$ hence one of the vector fields $\tilde{X}_1, \cdots, \tilde{X}_{n+1}$ is transversal to $S \times (0, 1)$ at every point of $\tilde{H}^c$. Hypothesis (ii) of Theorem 3.1 together with (3.26) imply that $\tilde{\lambda}^{(m)}$ satisfies the corresponding hypothesis in Theorem 3.3 (with respect to the set $\tilde{K}^c$). We conclude from Theorem 3.3 that the operator $\tilde{L} + \frac{\partial}{\partial t}$ is hypoelliptic on $\mathbf{R} \times D \times (0, 1)$. Consequently $\tilde{L}$ is hypoelliptic on $D \times (0, 1)$, and this implies $L$ is hypoelliptic on $D$. $\square$

We conclude this section by proving Lemmas 3.5 and 3.6, which provided the key steps in the preceding argument. The proof of Lemma 3.5 requires two preliminary results that we first state and prove.

**Proposition 3.9.** *Suppose $m \ge 2$ and $a > 0$. Let $B : [0, \infty) \times \emptyset \to \mathbf{R}$ be a one-dimensional Brownian motion. Then there exists a positive constant $c_{23}$ such that*

$$P\left(\int_0^a |B(u)|^m du < \epsilon\right) \le \sqrt{2}\exp\left(-c_{23}a^{1+\frac{2}{m}}\,\epsilon^{-\frac{2}{m}}\right)$$

for every $c > 0$. The constant $c_{23}$ may be chosen to be $2^{-7}$.

Proof. The result is known to hold for $m = 2$, with $c_{23} = 2^{-7}$ (cf[IW], Lemma V.10.6, p. 399).

For $m > 2$ we apply Hölder's inequality and use the result for $m = 2$ to obtain

$$P\left(\int_0^a |B(u)|^m du < \epsilon\right) \le P\left(\int_0^a |B(u)|^2 du < a^{1-\frac{2}{m}}\epsilon^{\frac{2}{m}}\right)$$
$$\le \sqrt{2}\exp(-c_{23}a^{1+\frac{2}{m}}\,\epsilon^{-\frac{2}{m}})$$

for every $\epsilon > 0$. This proves the proposition. $\qquad\square$

**Proposition 3.10.** Assume the notation and hypotheses of Lemma 3.5. Then for every $m \ge 2$ and $q > 0$, there exist positive constants $c_{24}$ and $c_{25}$ such that for all $\epsilon > 0$,

$$P\left(\int_0^\tau |y_u|^m du < \epsilon, \ \tau \ge \epsilon^q\right) < c_{24}\exp\left(-c_{25}\epsilon^{q+\frac{2(q-1)}{m}}\right).$$

The constants $c_{24}$ and $c_{25}$ depend only on $c_3$ (the bound for the drift and diffusion coefficients of $y$), $\delta$, $m$, and $q$. In particular, they are independent of $y_0$ and the characteristics of $\tau$.

Proof. Expressing (3.5) in components, it is sufficient to treat the case $d = 1$, $n \ge 1$. In this case, the process $y$ may be written in the form

$$y_t = B(\tau_4(t)) + \int_0^t b(u)\,du, \quad 0 \le t \le T,$$

where

$$\tau_4(t) \equiv \int_0^t |a(u)|^2\,du, \quad 0 \le t \le T,$$

and $B : [0, \infty) \times \varnothing \to \mathbf{R}$ is a one-dimensional adapted Brownian motion started at $B(0) = y_0$. By assumption, we may take $|a(u)| \ge \delta > 0$ for $0 \le u \le \tau$. Hence $\tau \ge \epsilon^q$ implies $\tau_4(\tau) \ge \delta^2\epsilon^q$. Furthermore, the function $\tau_4(t)$ is strictly increasing on $(0, \tau)$ and changes of the time variable yield

$$\int_0^\tau |y_u|^m du \ge c_{26}^{-1}\int_0^{\tau_4(\tau)} |y(\tau_4^{-1}(s))|^m\,ds$$
$$= c_{26}^{-1}\int_0^{\tau_4(\tau)} \left|B(s) + \int_0^s \frac{b(\tau_4^{-1}(u))}{|a(\tau_4^{-1}(u))|^2}\,du\right|^m ds,$$

where $c_{26} \equiv nc_3^2$. Thus

$$P\left(\int_0^\tau |y_u|^m \, du < \epsilon, \ \tau \geq \epsilon^q\right)$$

$$\leq P\left(\int_0^{\delta^2\epsilon^q} \left| B(s) + \int_0^s \frac{b(\tau_4^{-1}(u))}{|a(\tau_4^{-1}(u))|^2} \, du \right|^m ds < c_{26}\epsilon, \ \tau_4(\tau) \geq \delta^2\epsilon^q\right). \quad (3.27)$$

We now define a (bounded) process $h : [0, \infty) \times \emptyset \rightarrow \mathbf{R}$ by

$$h(u) \equiv \begin{cases} \dfrac{b(\tau_4^{-1}(u))}{|a(\tau_4^{-1}(u))|^2}, & u \in (0, \tau_4(\tau)) \\[2ex] \dfrac{b(\tau)}{|a(\tau)|^2}, & u \geq \tau_4(\tau). \end{cases}$$

and we denote by $B'$ the process

$$B'(s) \equiv B(s) + \int_0^s h(u) du, \quad 0 \leq s \leq \tau_4(T).$$

By the Girsanov theorem, $B'$ is a Brownian motion on $\emptyset$ with respect to the measure

$$dP' \equiv \left\{\exp\left(-\int_0^{\tau_4(T)} h(u) \, dB(u) - \frac{1}{2}\int_0^{\tau_4(T)} h^2(u) \, du\right)\right\} dP.$$

Denote by $\emptyset_\epsilon$ the event

$$\emptyset_\epsilon \equiv \left(\int_0^{\delta^2\epsilon^q} |B'(s)|^m \, ds < c_{26}\epsilon\right),$$

and by $G$ the Girsanov density

$$G \equiv \exp\left(-\int_0^{\tau_4(T)} h(u) dB(u) - \frac{1}{2}\int_0^{\tau_4(T)} h^2(u) \, du\right).$$

We now apply Hölder's inequality to (3.27) to obtain

$$P\left(\int_0^\tau |y_u|^m du < \epsilon, \ \tau \geq \epsilon^q\right) \leq P(\emptyset_\epsilon)$$

$$\leq \sqrt{E(G^{-2})P'(\emptyset_\epsilon)}.$$

By Proposition 3.9, we have

$$P'(\emptyset_\epsilon) \leq \sqrt{2}\exp\left(-2c_{25}\epsilon^{q+\frac{2(q-1)}{m}}\right), \quad (3.28)$$

where $c_{25} \equiv \frac{1}{2}c_{23}c_{26}^{-\frac{2}{m}}\delta^{2(1+\frac{2}{m})}$. The boundedness of $h$ and $\tau_4(T)$ imply the existence of a constant $c_{27}$, depending only on the bounds of the foregoing quantities, such that

$$G^{-2} \leq c_{27}\exp\left(2\int_0^{\tau_4(T)} h(u) \, dB(u) - 2\int_0^{\tau_4(T)} h^2(u) \, du\right). \quad (3.29)$$

The desired conclusion follows from (3.28) and (3.29), together with the fact that the exponential on the right hand side of (3.29) is a Girsanov density (note that it is obtained by replacing $h$ by $(-2h)$ in the relation defining $G$), and therefore has expectation equal to 1.     $\square$

*Proof of Lemma 3.2.* Note that for every $q > 0$, we may write

$$P\left( \int_0^{t \wedge \tau} |y_u|^m du < \epsilon \right) \le P_1 + P_2, \tag{3.30}$$

where

$$P_1 \equiv P\left( \int_0^{t \wedge \tau} |y_u|^m du < \epsilon, \ tau \wedge \tau \ge \epsilon^q \right)$$

and

$$P_2 \equiv P(t \wedge \tau < \epsilon^q).$$

By Proposition 3.10,

$$P_1 < c_{24} \exp\left( -c_{25}\epsilon^{q + \frac{2(q-1)}{m}} \right), \tag{3.31}$$

where $c_{24}$ and $c_{25}$ are independent of $t$. Now $\tau$ is exponentially positive; so if $T_0 > t > \epsilon^q$, then

$$P_2 < \exp(-c_{27}\epsilon^{-q}), \tag{3.32}$$

where $c_{27}$ and $T_0$ denote the characteristics of $\tau$. Combining (3.30)–(3.32), we obtain

$$P\left( \int_0^{t \wedge \tau} |y_u|^m du < \epsilon \right) \le c_{24} \exp\left( -c_{25}\epsilon^{q + \frac{2(q-1)}{m}} \right) + \exp(c_{27}\epsilon^{-q})$$

for $t \in (0, T_0)$ and $0 < \epsilon < t^{\frac{1}{q}}$. The lemma now follows by choosing as $q$ the value for which the two exponents $\left\{ q + \frac{2(q-1)}{m} \right\}$ and $-q$ coincide, namely $\frac{1}{m+1}$.     $\square$

*Proof of Lemma 3.6.* Choose and fix $m \ge \max\left\{ -\frac{p}{1+p}, 2 \right\}$ and set $q \equiv -\frac{m}{p(m+1)}$; so $q > 1$. Define a function $\psi : [0, \infty) \to [0, 1)$ by

$$\psi(z) \equiv \begin{cases} \exp(-z^{\frac{p}{m}}) & z > 0 \\ 0 & z = 0. \end{cases}$$

Note that

1. (i) $\psi$ is strictly increasing;
2. (ii) $\psi$ is convex in an interval $(0, c_{28})$, for some positive constant $c_{28}$.

Furthermore

$$P\left( \int_0^{t \wedge \tau} \exp(-|y_u|^p) du < \epsilon \right) = P\left( \int_0^{t \wedge \tau} \psi(|y_u|^m) du < \epsilon \right).$$

We break the proof of the lemma into two cases. Firstly, suppose that $|y_0| \geq c_{29}$ $(\equiv (\frac{1}{2}c_{28})^{\frac{1}{m}})$. Let $\tau_5 \equiv \inf\{s > 0 : |y_s - y_0| = \frac{1}{2}c_{29}\} \wedge \tau$. Then there exists a positive constant $c_{30}$, determined by $m$, $p$, and $\delta$, such that

$$P\left(\int_0^{t \wedge \tau} \exp(-|y_u|^p)du < \epsilon\right) \leq P(t \wedge \tau_5 \leq c_{30}\epsilon). \tag{3.33}$$

Applying Lemma 3.5 to the right hand side of (3.33), we deduce the existence of positive constants $c_{31}$ and $c_{32}$, such that if $t \in (0, c_{31})$ and $0 < \epsilon < \frac{t}{c_{30}}$, then

$$P\left(\int_0^{t \wedge \tau} \exp(-|y_u|^p)du < \epsilon\right) \leq \exp\left(-\frac{c_{32}}{\epsilon}\right).$$

The constants $c_{30}, c_{31}, c_{32}$ depend only on $m$, $p$, $c_3$, and the characteristics of $\tau$. Thus the conclusion of the lemma holds in this case.

On the other hand, suppose $|y_0| < c_{29}$. We now set

$$\tau_6 \equiv \inf\{s > 0 : |y_s|^m = c_{28}\} \wedge \tau.$$

Jensen's inequality yields

$$P\left(\int_0^{t \wedge \tau} \psi(|y_u|^m)du < \epsilon\right) \leq P\left(\int_0^{t \wedge \tau_6} |y_u|^m \, du \leq (t \wedge \tau_6)\psi^{-1}\left(\frac{\epsilon}{t \wedge \tau_6}\right)\right)$$
$$\leq P_1 + P_2$$

where

$$P_1 \equiv P\left(t \wedge \tau_6 \leq \frac{\epsilon}{c_{28}}\right)$$

and

$$P_2 \equiv P\left(\int_0^{t \wedge \tau} |y_u|^m du \leq (t \wedge \tau_6)\psi^{-1}\left(\frac{\epsilon}{t \wedge \tau_6}\right), \quad t \wedge \tau_6 > \frac{\epsilon}{c_{28}}\right).$$

Note that $P_1$ is of the same form as the probability on the right hand side of (3.33), and hence satisfies a similar estimate.

We now consider $P_2$. An elementary argument shows that the convexity of $\psi$ in the interval $(0, c_{28})$ implies that the function $\theta(u) \equiv u\psi^{-1}\left(\frac{\epsilon}{u}\right)$ is increasing for $u > \frac{\epsilon}{c_{28}}$. In particular, if $t \wedge \tau_6 > \frac{\epsilon}{c_{28}}$ then

$$(t \wedge \tau_6)\psi^{-1}\left(\frac{\epsilon}{t \wedge \tau_6}\right) \leq T\psi^{-1}\left(\frac{\epsilon}{T}\right)$$

where $T$ is any upper bound for $t$. This implies

$$P_2 \leq P\left(\int_0^{t \wedge \tau_6} |y_u|^m du\right) \leq T\psi^{-1}\left(\frac{\epsilon}{T}\right)\right). \tag{3.34}$$

We now apply Lemma 3.5 to estimate the right hand side of (3.34). Thus

$$P_2 < \exp\left(-c_5\left\{T\psi_r^{-1}\left(\frac{\epsilon}{T}\right)\right\}^{-\frac{1}{m+1}}\right)$$

$$\leq \exp(-c_7|\log\epsilon|^q), \tag{3.35}$$

for all $0 < t < c_{34}$ and $\epsilon < \exp\left(-c_{33}t^{-\frac{1}{q}}\right)$, where $c_7$, $c_{33}$, and $c_{34}$ are positive constants exhibiting the appropriate dependence. Then (3.35) gives an estimate for $P_2$ of the required form, and the proof of the lemma is complete. $\qquad\square$

# 4 A study of a class of degenerate functional stochastic differential equations

In the previous section we derived very sharp sufficient conditions for the existence of smooth densities for the class of Itô processes

$$d\xi_t = \sum_{i=1}^n X_i(\xi_t)dw_i + X_0(\xi_t)dt. \tag{4.1}$$

The arguments relied heavily on the existence of a reversible stochastic flow on $\mathbf{R}^d$ determined by the equation. In particular, a major role in the proofs was played by the inverse of the derivative of the stochastic flow (the process $Z$ in equation (2.17)). Furthermore, as is apparent from Theorem 2.7 and its proof, the Lie bracket condition of Hörmander appears as a result of the interaction between $Z$ and the vector fields $X_i$. The existence of a stochastic flow is the analytic counterpart of the probabilistic fact that the solutions to the classical Itô equation (4.1) is a Markov process. In this section (which comes form [BM2]) we study a class of SDEs with far less analytic and probabilistic structure.

In order to describe this class of equations, we introduce the following notation and assumptions. Denote by:

$\eta$ a continuous (deterministic) $\mathbf{R}^d$-valued path defined on a time interval $[-r, 0]$, for a fixed $r > 0$.

$g$, a bounded smooth map from $\mathbf{R}^d$ to $\mathbf{R}^d \otimes \mathbf{R}^n$ with bounded derivatives of all orders.

$C$ the space of continuous paths $\{x : [-r, \infty) \mapsto \mathbf{R}^d\}$, equipped with the uniform norm on compact time sets.

$B : C \mapsto C$ a smooth bounded map, with bounded derivatives of all orders. Assume also that $B$ is *non-anticipating*, i.e., for all $t \geq -r$ and $x \in C$, $B(t, x) \equiv B(x)_t$ depends only on $\{x(u) : u \leq t\}$.

$H_d[0, s]$, the $d$-dimensional Cameron–Martin space (cf. Section 2) with paths and norm defined on the time interval $[0, s]$. Suppose that for all $x \in C$, the map $DB(x)|H_d : H_d \mapsto H_d$ and that

$$\alpha \equiv \sup_{s \in [0,t], x \in C} \|DB(x)\|_{H_d[0,s]} < 1.$$

We consider the following stochastic functional differential equation

$$x_t = \eta(t), \quad -r \le t \le 0$$

$$x_t = \eta(0) + \int_0^t g(x_{s-r})dw_s + \int_0^t B(t,x)ds, \quad t \ge 0 \tag{4.2}$$

where, as before, $w$ is a standard Wiener process in $\mathbf{R}^n$. Owing to the past-state dependence of the coefficients in the equation, the solution $x$ will not in general be a Markov process.

We establish the existence of smooth densities for $x$ under hypotheses that allow degeneracy of the matrix-valued function $g$. The general scheme of Section 2 is used. As before, we construct a Malliavin covariance $\sigma$ for each $x_t$, $(t > 0)$ and deduce the existence of a smooth density for $x_t$ by demonstrating the strong invertibility of $\sigma$. However, the argument for doing this is, by necessity, quite different from that in Section 3. Indeed, because of the non-Markovian nature of the process $x$, it no longer generates a stochastic flow on $\mathbf{R}^d$. Thus the analysis used in Section 3 to study the non-degeneracy of $\sigma$ completely breaks down. We salvage the situation by proving and exploiting a striking property that seems peculiar to equations of the form (4.2); namely we show (cf. Lemma 4.5) that for all times $a \ge 0$ and $k \ge 1$, $x$ satisfies estimates of the form

$$P\left(\int_a^{a+r} |x_t|^2 dt < \varepsilon\right) = o(\varepsilon^k) \text{ as } \epsilon \to 0.$$

These probabilistic lower bounds on $x$ are shown to be propagated from the initial condition $\eta$ by the time delay $r$ in equation 4.2 (there is no reason to believe that similar estimates hold for classical diffusions). We use this property to deduce non-degeneracy of the Malliavin covariance matrices, under appropriate hypotheses on $g$ and $B$.

In addition to their theoretical interest, equations of the form (4.2) have important applications. For example, physical dynamical systems influenced by noise are often modeled by ordinary SDEs (4.1). However, this choice of model assumes that the evolution of the system at any given time depends *instantaneously* on the position of the system and the noise input. Since physical systems are always subject to some degree of inertia, this assumption is unrealistic. It would seem that equations of the form (4.2), where the drift coefficient $B$ is allowed to depend on the past history of the solution, are more realistic. The time-delayed form of the diffusion coefficient $g(x_{s-r})$ in equation (4.2) is similarly motivated ($r$ represents a "time lag" in the system).

The main result of this section is

**Theorem 4.1.** *Suppose there exist positive constants $\rho$, $\delta$, an integer $p \ge 2$ and a function $\phi : \mathbf{R}^d \to \mathbf{R}$ such that*

(i)   *the Lebesgue measure of the set $\{s \in [-r,0] : \phi(\eta(s)) = 0\}$ is zero.*
(ii)  *$g$ satisfies*

$$gg^*(x) \ge \begin{cases} |\phi(x)|^P I, & |\phi(x)| < \rho \\ \delta I, & |\phi(x)| \ge \rho. \end{cases} \tag{4.3}$$

*(iii) $\phi$ is $C^2$ with bounded first and second derivatives and there is a positive constant c such that*

$$\|\nabla\phi(x)\| \geq c, \text{ whenever } |\phi(x)| \leq \rho. \tag{4.4}$$

*Suppose $0 < \alpha\sqrt{2t}e^{\alpha t} < 1$. Then the random variable $x_t$ defined by (4.2) is absolutely continuous and has a $C^\infty$ density.*

We note that condition (ii) above allows degeneracy of $g$ on the hypersurface $D \subseteq \mathbf{R}^d$ where $\phi$ vanishes, while condition (iii) implies that $D$ does not contain any singular points.

Denote by $F$ the map $w \mapsto x_t$. In order to prove the theorem we must compute the Malliavin covariance matrix, defined formally by $\sigma \equiv DF(w)DF(w)^*$. This can be done as follows. Let $H$ denote the Cameron–Martin space and $P_m$ the sequence of piecewise linear projections defined in Chapter 2, we consider the natural restriction $\tilde{F}$ of the map $F$ to $H$, i.e., the map $h \in H \mapsto k_t$ where $k \equiv \eta$ on $[-r, 0]$ and

$$k'_s = g(k_{s-r})h_s + B(s, k), s > 0.$$

The matrix $\sigma$ is then obtained as the limit in probability of the matrix sequence

$$D\tilde{F}(P_m)D\tilde{F}(P_m)^*.$$

The result of this computation is:

**Lemma 4.2.** *The Malliavin covariance matrix $\sigma$ for the random variable $x_t$ is*

$$\sigma = \int_0^t Z_u g(x_{u-r})g(x_{u-r})^* Z_u^* du \tag{4.5}$$

*where the process $Z$ satisfies*

$$Z_s = I + \int_{(s+r)\wedge t}^t Dg(x_{u-r})(.,dw_u)^* Z_s + D_{H_d}B(x)^*\left(\int_0^{\cdot} Z_u du\right)_s. \tag{4.6}$$

It is at this point that the non-Markovian nature of the problem manifests itself analytically; in contrast to the situation in Section 3, there is now no reason to suppose that the matrix process $Z_s$ is non-degenerate for small values of $s$. We show in the next Lemma that it *is*, however, non-degenerate for values of $s$ close enough to $t$.

**Lemma 4.3.** *Under the hypotheses of Theorem 4.1, $Z_s \in GL(d)$ for $s \in [t - r, t]$ and there exists a deterministic constant $c < \infty$ such that $\|Z_s^{-1}\| \leq c$ for $s \in [t - r, t]$.*

*Proof.* For notational convenience, write the operator $D_{H_d}B(x)^*$ as $K(x)$. For $s \in [t - r, t]$ and any unit vector $e \in \mathbf{R}^d$, (4.6) gives

$$Z_s e = e + K(x)_s\left(\int_0^{\cdot} Z_u e\, du\right)_s. \tag{4.7}$$

Now

$$\left|K(x)_s\left(\int_0^{\cdot} Z_u e\ du\right)_s\right| \leq \left\|K(x)\left(\int_0^{\cdot} Z_u e\ du\right)_s\right\|_{H_d[0,s]}$$

$$\leq \alpha\left\|\int_0^{\cdot} Z_u e\ du\right\|_{H_d[0,s]} = \alpha\left(\int_0^s |Z_u e|^2\right)^{1/2}. \tag{4.8}$$

Substituting this back into (4.7), we deduce

$$\|Z_s\| \leq 1 + \alpha\left(\int_0^s \|Z_u\|^2 du\right)^{1/2}$$

which implies

$$\|Z_s\|^2 \leq 2 + 2\alpha\int_0^s \|Z_u\|^2 du.$$

Applying Gronwall's inequality to this yields

$$\|Z_s\| \leq \sqrt{2}e^{\alpha s} \leq \sqrt{2}e^{\alpha t}, \forall s \leq t.$$

Combining this with (4.8) gives

$$\left\|K(x)_s\left(\int_0^{\cdot} Z_u\ du\right)\right\| \leq \sqrt{2t}e^{\alpha t}\alpha < 1.$$

It follows from (4.7) that $Z_s \in GL(d)$,

$$Z_s^{-1} = \sum_{j=0}^{\infty}(-1)^j\left[K(x)_s\left(\int_0^{\cdot} Z_u\ du\right)_s\right]^j$$

and

$$\|Z_s^{-1}\| \leq (1 - \alpha\sqrt{2t}e^{\alpha t})^{-1}. \quad \square$$

Let $\lambda$ denote the smallest eigenvalue of $\sigma$ and set $\xi_s \equiv \phi(x_s)$, $s \geq -r$. It follows from (4.5) and Lemma 4.3 that

$$\lambda \geq \frac{1}{c}\int_{(t-r)\wedge 0}^t (|\xi_{u-r}|^p \wedge \delta)\ du. \tag{4.9}$$

We will need the following results to complete the proof of the above Theorem 4.1

**Lemma 4.4.** *Fix* $b > a > 0$. *Let* $y : [0, \infty) \times \emptyset \to \mathbf{R}$ *be a measurable stochastic process such that* $E \sup_{a \leq t \leq b} |y(t)|^p$ *is finite for every positive integer p. Suppose that y satisfies*

$$P\left(\int_a^b y(t)^2 dt < \varepsilon\right) = o(\varepsilon^k).$$

*Then for every positive constant* $\alpha$,

$$P\left(\int_a^b \{y(t)^2 \wedge \alpha\}\ dt < \varepsilon\right) = o(\varepsilon^k).$$

*Proof.* This result was originally proved as Lemma 3 in ([BM1], 91–94). We give a simplified proof of the result.

Let $A$ and $B$ denote, respectively, the sets $\{s \in [a, b] : y(s)^2 \leq \alpha\}$ and $\{s \in [a, b] : y(s)^2 > \alpha\}$. Then

$$
P\left(\int_a^b \{y(t)^2 \wedge \alpha\} dt < \varepsilon\right) = P\left(\int_A y(t)^2 dt + \alpha\lambda(B) < \varepsilon\right)
$$

$$
\leq P\left(\int_A y(t)^2 dt < \varepsilon, \ \alpha\lambda(B) < \varepsilon\right)
$$

$$
= P\left(\int_a^b y(t)^2 dt < \varepsilon + \int_B y(t)^2 dt, \ \lambda(B) < \varepsilon/\alpha\right)
$$

$$
\leq P_1 + P_2.
$$

Here

$$
P_1 := P\left(\int_a^b y(t)^2 dt < \varepsilon + \int_B y(t)^2 dt, \ \lambda(B) < \varepsilon/\alpha, \ \sup_{a \leq t \leq b} y(t)^2 \leq 1/\sqrt{\varepsilon}\right)
$$

and

$$
P_2 := P(\sup_{a \leq t \leq b} y(t)^2 > 1/\sqrt{\varepsilon}).
$$

Note that

$$
P_1 \leq P\left(\int_a^b y(t)^2 dt < \varepsilon + \sqrt{\varepsilon}/\alpha\right) = o(\varepsilon^k)
$$

by hypothesis.

Using the finite-moment hypothesis on $y$ and applying Markov–Chebyshev's inequality to probability $P_2$, we obtain $P_2 = o(\varepsilon^k)$. This completes the proof of the lemma. □

**Lemma 4.5 (Propagation lemma).** *Suppose that, for some* $-r < a < b$, $\xi$ *satisfies*

$$
P\left(\int_a^b \xi_s^2 ds < \varepsilon\right) = o(\varepsilon^k). \tag{4.10}
$$

*Then*

$$
P\left(\int_{a+r}^{b+r} \xi_s^2 ds < \varepsilon\right) = o(\varepsilon^k). \tag{4.11}
$$

*Proof.* Write $g = (g_1 \ldots g_n)$, where $g_i$, $1 \leq i \leq n$, are column $d$-vectors. Computing $\xi_s = \phi(x_s)$, $s > 0$, by Itô's formula gives

$$
d\xi_s = \sum_{i=1}^n \nabla\phi(x_s) \cdot g_i(x_{s-r}) dw_i(s) + G(s) ds, \quad s > 0 \tag{4.12}
$$

where $G$ is a bounded adapted real-valued process. We write

$$P\left(\int_{a+r}^{b+r} \xi_s^2 \, ds < \varepsilon\right) = P_1 + P_2$$

where

$$P_1 \equiv P\left(\int_{a+r}^{b+r} \xi_s^2 \, ds < \varepsilon, \sum_{i=1}^{n} \int_{a+r}^{b+r} [\nabla\phi(x_s) \cdot g_i(x_{s-r})]^2 \, ds \geq \varepsilon^{1/18}\right)$$

and

$$P_2 \equiv P\left(\int_{a+r}^{b+r} \xi_s^2 \, ds < \varepsilon, \sum_{i=1}^{n} \int_{a+r}^{b+r} [\nabla\phi(, x_s) \cdot g_i(x_{s-r})]^2 \, ds < \varepsilon^{1/18}\right).$$

In view of 4.12, an inequality of Kusuoka and Stroock (cf. [KS2], Lemma 6.5) implies that $P_1 = o(\varepsilon^k)$. Thus it is sufficient to show that $P_2$ also has this property.

Write

$$a(s) = \sum_{i=1}^{n} [\nabla\phi(x_s) \cdot g_i(x_{s-r})]^2.$$

Then by (4.3) and (4.4), it follows that $a(s) \geq c^2(|\xi_{s-r}|^p \wedge \delta)$ if $|\xi_s| \leq \rho$.
Define

$$A \equiv \{s \in [a+r, b+r] : |\xi_s| \leq \rho\} \text{ and } B \equiv \{s \in [a+r, b+r] : |\xi_s| > \rho\}.$$

Then

$$\begin{aligned}
P_2 &= P\left(\int_{a+r}^{b+r} \xi_s^2 \, ds < \varepsilon, \int_{a+r}^{b+r} a(s) \, ds < \varepsilon^{1/18}\right) \\
&\leq P\left(\int_{a+r}^{b+r} \xi_s^2 \, ds < \varepsilon, \int_A c^2(|\xi_{s-r}|^p \wedge \delta) \, ds < \varepsilon^{1/18}\right) \\
&= P\left(\int_{a+r}^{b+r} \xi_s^2 \, ds < \varepsilon, \int_{a+r}^{b+r} c^2(|\xi_{s-r}|^p \wedge \delta) \, ds < \varepsilon^{1/18} + \int_B c^2(|\xi_{s-r}|^p \wedge \delta) \, ds\right) \\
&\leq P\left(\int_{a+r}^{b+r} \xi_s^2 \, ds < \varepsilon, \int_{a+r}^{b+r} c^2(|\xi_{s-r}|^p \wedge \delta) \, ds < \varepsilon^{1/18} + c^2\delta\lambda(B)\right).
\end{aligned}$$

However $\int_{a+r}^{b+r} \xi_s^2 \, ds < \varepsilon$ implies $\lambda(B) < \varepsilon/\rho^2$. Thus the preceding probability is

$$\begin{aligned}
&\leq P\left(\int_{a+r}^{b+r} c^2(|\xi_{s-r}|^p \wedge \delta) \, ds < \varepsilon^{1/18} + c^2\delta\varepsilon/\rho^2\right) \\
&\leq P\left(\int_a^b (|\xi_s|^p \wedge \delta) \, ds < c'\varepsilon^{1/18}\right)
\end{aligned}$$

for some positive constant $c'$, and for small enough $\varepsilon$. Assumption 4.10, Lemma 4.4, and Jensen's inequality allow us to conclude that the probability on the right hand side of the above inequality is $o(\varepsilon^k)$. This implies that $P_2 = o(\varepsilon^k)$, and the proof of Lemma (4.2) is complete. □

We are now in a position to complete the proof of Theorem 4.1. Recall that we need to verify the condition

$$(\det \sigma)^{-1} \in \cap_{p \geq 1} L^p(\gamma). \tag{4.13}$$

As before let $\lambda$ denote the smallest eigenvalue of $\sigma$. Let $n$ denote the integer such that $t \in ((n-1)r, nr]$. First, suppose $n = 1$. Hypothesis (i) of Theorem 4.1 and (4.9) imply

$$\lambda \geq \frac{1}{c} \int_{-r}^{t-r} (|\xi_u|^p \wedge \delta) \, du = \frac{1}{c} \int_{-r}^{t-r} (|\eta_u|^p \wedge \delta) \, du > 0.$$

Since the second integral is deterministic, 4.13 trivially holds in this case.

On the other hand, suppose $n > 1$. Now $[t - (n+1)r, t - nr] \subset [-r, 0]$, and hypothesis (i) implies

$$P\left( \int_{t-(n+1)r}^{t-nr} \xi_s^2 \, ds < \varepsilon \right) = o(\varepsilon^k).$$

We now iterate Lemma 4.5 $n$ times on this estimate to obtain

$$P\left( \int_{t-r}^{t} \xi_s^2 \, ds < \varepsilon \right) = o(\varepsilon^k).$$

Applying Lemma 3.4 yields

$$P\left( \int_{t-r}^{t} \xi_s^2 \wedge \delta^{2/p} ds < \varepsilon \right) = o(\varepsilon^k) \tag{4.14}$$

Using Jensen's inequality in (4.14) gives

$$P\left( \int_{t-r}^{t} |\xi_s|^p \wedge \delta \, ds < \varepsilon \right) = o(\varepsilon^k).$$

Combining this with (4.9), we finally have

$$P(\lambda < \epsilon) = o(\epsilon^k).$$

This implies (4.13) and completes the proof of the theorem. □

## 5 Some open problems

In this section we describe a few open problems that we hope will serve as a stimulus to further research.

1. It would be interesting to generalize Theorem 4.1 to the class of fully hereditary functional differential equations

$$dx_t = A(t, x)dw + B(t, x)dt \tag{5.1}$$

where both $A(t, x)$ and $B(t, x)$ are allowed to depend on the whole history of the path $\{x_s : 0 \leq s \leq t\}$. Kusuoka and Stroock [KS1] addressed this problem under a strong ellipticity assumption, i.e., they have shown that $x_t$ has a smooth density for all positive $t$ if there exists $\delta > 0$ such that $A(t, x)A(t, x)^* \geq \delta I, \forall (t, x) \in [0, 1) \times C$. As far as I am aware, Theorem 4.1 is the only result establishing the existence of densities for a general class of non-Markov Itô processes under hypotheses that allow degeneracy of the diffusion coefficient. An analogous result in the fully general setting (5.1) would therefore be of considerable interest and importance.

2. Kusuoka–Stroock's discovery described in Section 3 of the non-hypoellipticity of the operators

$$\frac{\partial^2}{\partial x_1^2} + \exp(-|x_1|^{1/p})\frac{\partial^2}{\partial x_2^2} + \frac{\partial^2}{\partial x_3^2}, p \leq -1$$

is strikingly different when contrasted to a result proved by Fedii in 1971 [F]. He showed that the operator on $\mathbf{R}^2$

$$\frac{\partial^2}{\partial x^2} + \exp(-|x|^p)\frac{\partial^2}{\partial y^2}, p < 0$$

is hypoelliptic *for all* negative values of $p$. It would be interesting to gain a deeper understanding, either by classical or probabilistic means, of the role that dimension is playing in these results.

3. The probabilistic methods employed above can also be used to study *quasilinear* PDEs. For example, let $\psi$ denote a smooth non-linear and non-negative function defined on $[0, \infty) \times \mathbf{R}^d \times \mathbf{R}$ and consider the initial-value problem

$$\left. \begin{array}{l} \frac{\partial u}{\partial t} = Lu + \psi(t, x, u), \quad (t, x) \in (0, \infty) \times \mathbf{R}^d \\ u(0, x) = 0, \quad x \in \mathbf{R}^d \end{array} \right\} \tag{5.2}$$

where $L$ is defined in (1.1). In particular, a continuous weak solution to (5.2) is given by the probabilistic representation

$$u(t, x) = E\left[ \int_0^t \psi(s, \xi_{t-s}^x, u(s, \xi_{t-s}^x)) \, ds \right] \tag{5.3}$$

where $\xi^x$ denotes process $\xi$ in (1.2) with initial point $x \in \mathbf{R}^d$. It should be possible to use this probabilistic representation together with the ideas of Section 3 to show that, under suitable conditions, the solution $u$ to (5.2) is smooth for $t > 0$. We note, however, that the problem is considerably more difficult than for the linear case treated in Section 3, owing to the presence of $u$ in the right hand side of (5.3).

Quasilinear problems of this type have received a great deal of attention in recent years. Dynkin [D] and others have discovered a remarkable link between operators of the form $L + \psi$ and a collection of stochastic processes called *super processes*.

4. A related issue is the study of the quasilinear Dirichlet problem. Suppose $D$ is a bounded regular open subset of $\mathbf{R}^d$ with a $C^2$ boundary, $f$ is a smooth non-negative function defined on $\bar{D} \times \mathbf{R}$, and $g$ is a smooth non-negative function on $\partial D$

$$Lu = f(x, u), \quad x \in D \atop u(x) = g(x), \quad x \in \partial D \biggr\} . \tag{5.4}$$

Under mild further conditions, a weak continuous solution of equation (5.4) exists, given implicitly by

$$u(x) + E\left[ \int_0^\tau f(\xi^x(s), u(\xi^x(s))) \, ds \right] = E[g(\xi^x(\tau)] \tag{5.5}$$

where $\tau = \tau(x)$ is the first exit time of the diffusion $\xi^x$ from $D$. Again, one would hope to be able to study the regularity of the solution $u$ to (5.4) via the representation (5.5) by using the methods of Section 3. The goal would be to establish smoothness of $u$ on $\bar{D}$ under conditions that allow degeneracy of the operator $L$ (e.g., Hörmander's condition or even the superdegeneracy condition introduced in Theorem 3.1.).

# References

[Be1]   Bell, D. R., *Some properties of measures induced by solutions of stochastic differential equations*, Ph.D. Thesis, Univ. Warwick, 1982.

[Be2]   Bell, D. R., *Degenerate Stochastic Differential Equations and Hypoellipticity*, Pitman Monographs and Surveys in Pure and Applied Mathematics, Vol. 79, Longman, Essex, 1995.

[BM1]   Bell, D. R., and Mohammed, S.-E. A., The Malliavin calculus and stochastic delay equations, *JFunctÀnal.* **99**, no. 1 (1991) 75–99.

[BM2]   Bell, D. R., and Mohammed, S.-E. A., Smooth densities for degenerate stochastic delay equations with hereditary drift, *Ann. Prob.* **23** (1995), no. 4, 1875–1894.

[BM3]   Bell, D. R. and Mohammed, S.-E. A., An extension of Hörmander's theorem for infinitely degenerate second-order operators, *Duke Math J.* **78**, no. 3 (1995), 453–475.

[Bi1]   Bismut, J. M., Martingales, the Malliavin calculus and hypoellipticity under general Hörmander's conditions, *Z. Wahrsch. Verw. Gebiete* **56** (1981), 529–548.

[Bi2]   Bismut, J. M., *Large Deviations and the Malliavin Calculus*, Progress in Mathematics, vol. 45, Birkhäuser Boston, 1984.

[D]   Dynkin, E. B., Superdiffusions and parabolic nonlinear differential equations, *Ann. Prob.* **20**, no. 2 (1992), 942–962.

[F]   Fedii, V. S., On a criterion for hypoellipticity, *Math. USSR sb.* **14** (1971), 15–45.

[G]   L. Gross, Potential theory on Hilbert space, *J. Funct. Anal.* **1**, no. 1 (1967), 123–181.

[H]   Hörmander, L., Hypoelliptic second order differential equations, *Acta Math.* 119:3–4 (1967), 147–171.

[IW]   Ikeda, N., and Watanabe, S., *Stochastic Differential Equations and Diffusion Processes*, 2nd Edition, North-Holland-Kodansha, 1989.

[KS1]   Kusuoka, S., and Stroock, D., Applications of the Malliavin calculus, I, *Taniguchi Sympos. SA Katata* (1982), 271–306.

[KS2]   Kusuoka, S., and Stroock, D., Applications of the Malliavin calculus, Part II, *Journal of Faculty of Science, University of Tokyo*, Sec. 1A, Vol. 32, No. 1 (1985), 1–76.

[Ma1]   Malliavin, P., Stochastic calculus of variations and hypoelliptic operators, *Proceedings of the International Conference on Stochastic Differential Equations, Kyoto*, Kinokuniya, 1976, 195–263.

[Ma2]   Malliavin, P., $C^k$-hypoellipticity with degeneracy, part II, *Stochastic Analysis*, A. Friedman and M. Pinsky, eds., 1978, 327–340.

[Mi]    Michel, D., Régularité des lois conditionnelles en théorie du filtrage non-linéaire et calcul des variations stochastique, *J. Funct. Anal.* **41**, no. 1 (1981), 1–36.

[NP]    Nualart, D. and Pardoux, E., Stochastic calculus with anticipating integrands, *Probab. Theory Rel. Fields* **78** (1988), 535–581.

[O]     D. Ocone, Malliavin's calculus and stochastic integral representation of functionals of diffusion processes, *Stochastics* **12**, no. 3-4 (1984), 161–185.

# Curved Wiener Space Analysis

Bruce K. Driver

Department of Mathematics, University of California at San Diego, La Jolla, CA 92093-0112
driver@math.ucsd.edu

## 1 Introduction

The purpose of these notes is to first provide some basic background to Riemannian geometry and stochastic calculus on manifolds and then to cover some of the more recent developments pertaining to analysis on "curved Wiener spaces." Essentially no differential geometry is assumed, however, it is assumed that the reader is comfortable with stochastic calculus and differential equations on Euclidean spaces. Here is a brief description of what will be covered in the text below.

Section 2 is a basic introduction to differential geometry through imbedded submanifolds. Section 3 is an introduction to the Riemannian geometry that will be needed in the sequel. Section 4 records a number of results pertaining to flows of vector fields and "Cartan's rolling map." The stochastic version of these results will be important tools in the sequel. Section 5 is a rapid introduction to stochastic calculus on manifolds and related geometric constructions. Section 6 briefly gives applications of stochastic calculus on manifolds to representation formulas for derivatives of heat kernels. Section 7 is devoted to the study of the calculus and integral geometry associated with the path space of a Riemannian manifold equipped with a "Wiener measure." In particular, quasi-invariance, Poincaré and logarithmic Sobolev inequalities are developed for the Wiener measure on path spaces in this section. Section 8 is a short introduction to Malliavin's probabilistic methods for dealing with hypoelliptic diffusions. The appendix in section 9 records some basic martingale and stochastic differential equation estimates, which are mostly used in section 8.

Although the majority of these notes form a survey of known results, many proofs have been cleaned up and some proofs are new. Moreover, Section 8 is written using the geometric language introduced in these notes, which is not completely standard in the literature. I have also tried (without complete success) to give an overview of many of the major techniques that have been used to date in this subject. Although numerous references are given to the literature, the list is far from complete. I apologize in advance to anyone who feels cheated by not being included in the references. However, I do hope the list of references is sufficiently rich that the interested reader will be able to find additional information by looking at the related articles and the references that they contain.

## 2 Manifold primer

### Conventions

1. If $A$, $B$ are linear operators on some vector space, then $[A, B] := AB - BA$ is the *commutator* of $A$ and $B$.
2. If $X$ is a topological space we will write $A \subset_o X$, $A \sqsubset X$ and $A \sqsubset\sqsubset X$ to mean $A$ is an open, closed, and respectively a compact subset of $X$.
3. Given two sets $A$ and $B$, the notation $f : A \to B$ will mean that $f$ is a function from a subset $\mathcal{D}(f) \subset A$ to $B$. (We will allow $\mathcal{D}(f)$ to be the empty set.) The set $\mathcal{D}(f) \subset A$ is called the domain of $f$ and the subset $\mathcal{R}(f) := f(\mathcal{D}(f)) \subset B$ is called the range of $f$. If $f$ is injective, let $f^{-1} : B \to A$ denote the inverse function with domain $\mathcal{D}(f^{-1}) = \mathcal{R}(f)$ and range $\mathcal{R}(f^{-1}) = \mathcal{D}(f)$. If $f : A \to B$ and $g : B \to C$, then $g \circ f$ denotes the composite function from $A$ to $C$ with domain $\mathcal{D}(g \circ f) := f^{-1}(\mathcal{D}(g))$ and range $\mathcal{R}(g \circ f) := g \circ f(\mathcal{D}(g \circ f)) = g(\mathcal{R}(f) \cap \mathcal{D}(g))$.

**Notation 2.1.** Throughout these notes, let $E$ and $V$ denote finite dimensional vector spaces. A function $F : E \to V$ is said to be smooth if $\mathcal{D}(F)$ is open in $E$ ($\mathcal{D}(F) = \emptyset$ is allowed) and $F : \mathcal{D}(F) \to V$ is infinitely differentiable. Given a *smooth* function $F : E \to V$, let $F'(x)$ denote the differential of $F$ at $x \in \mathcal{D}(F)$. Explicitly, $F'(x) = DF(x)$ denotes the linear map from $E$ to $V$ determined by

$$DF(x)\, a = F'(x)a := \frac{d}{dt}|_0 F(x + ta) \; \forall \, a \in E. \tag{2.1}$$

We also let

$$F''(x)(v, w) = F''(x)(v, w) := (\partial_v \partial_w F)(x) = \frac{d}{dt}|_0 \frac{d}{ds}|_0 F(x + tv + sw). \tag{2.2}$$

### 2.1 Imbedded submanifolds

Rather than describe the most abstract setting for Riemannian geometry, for simplicity we choose to restrict our attention to imbedded submanifolds of a Euclidean space $E = \mathbb{R}^N$.[1] We will equip $\mathbb{R}^N$ with the standard inner product,

$$\langle a, b \rangle = \langle a, b \rangle_{\mathbb{R}^N} := \sum_{i=1}^N a_i b_i.$$

In general, we will denote inner products in these notes by $\langle \cdot, \cdot \rangle$.

---

[1] Because of the Whitney imbedding theorem (see for example Theorem 6-3 in Auslander and MacKenzie [9]), this is actually not a restriction.

**Definition 2.2.** A subset $M$ of $E$ (see Figure 1) is a *d-dimensional imbedded submanifold* (without boundary) of $E$ iff for all $m \in M$, there is a function $z : E \to \mathbb{R}^N$ such that:

1. $\mathcal{D}(z)$ is an open neighborhood of $E$ containing $m$,
2. $\mathcal{R}(z)$ is an open subset of $\mathbb{R}^N$,
3. $z : \mathcal{D}(z) \to \mathcal{R}(z)$ is a diffeomorphism (a smooth invertible map with smooth inverse), and
4. $z(M \cap \mathcal{D}(z)) = \mathcal{R}(z) \cap (\mathbb{R}^d \times \{0\}) \subset \mathbb{R}^N$.

(We write $M^d$ if we wish to emphasize that $M$ is a $d$-dimensional manifold.)

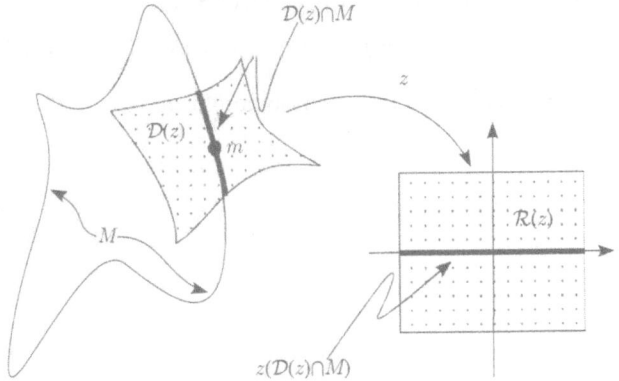

**Figure 1.** An imbedded one dimensional submanifold in $\mathbb{R}^2$.

**Notation 2.3.** Given an imbedded submanifold and diffeomorphism $z$ as in the above definition, we will write $z = (z_<, z_>)$ where $z_<$ is the first $d$ components of $z$ and $z_>$ consists of the last $N - d$ components of $z$. Also let $x : M \to \mathbb{R}^d$ denote the function defined by $\mathcal{D}(x) := M \cap \mathcal{D}(z)$ and $x := z_<|_{\mathcal{D}(x)}$. Notice that $\mathcal{R}(x) := x(\mathcal{D}(x))$ is an open subset of $\mathbb{R}^d$ and that $x^{-1} : \mathcal{R}(x) \to \mathcal{D}(x)$, thought of as a function taking values in $E$, is smooth. The bijection $x : \mathcal{D}(x) \to \mathcal{R}(x)$ is called a *chart* on $M$. Let $\mathcal{A} = \mathcal{A}(M)$ denote the collection of charts on $M$. The collection of charts $\mathcal{A} = \mathcal{A}(M)$ is often referred to as an *atlas* for $M$.

**Remark 2.4.** The imbedded submanifold $M$ is made into a topological space using the induced topology from $E$. With this topology, each chart $x \in \mathcal{A}(M)$ is a homeomorphism from $\mathcal{D}(x) \subset_o M$ to $\mathcal{R}(x) \subset_o \mathbb{R}^d$.

**Theorem 2.5 (A Basic Construction of Manifolds).** *Let* $F : E \to \mathbb{R}^{N-d}$ *be a smooth function and* $M := F^{-1}(\{0\}) \subset E$ *which we assume to be non-empty. Suppose that* $F'(m) : E \to \mathbb{R}^{N-d}$ *is surjective for all* $m \in M$. *Then* $M$ *is a* $d$-*dimensional imbedded submanifold of* $E$.

*Proof.* Let $m \in M$, we will begin by constructing a smooth function $G : E \to \mathbb{R}^d$ such that $(G, F)'(m) : E \to \mathbb{R}^N = \mathbb{R}^d \times \mathbb{R}^{N-d}$ is invertible. To do this, let $X = \text{Nul}(F'(m))$ and $Y$ be a complementary subspace so that $E = X \oplus Y$ and let $P : E \to X$ be the associated projection map (see Figure 2). Notice that $F'(m) : Y \to \mathbb{R}^{N-d}$ is a linear isomorphism of vector spaces and hence

$$\dim(X) = \dim(E) - \dim(Y) = N - (N - d) = d.$$

In particular, $X$ and $\mathbb{R}^d$ are isomorphic as vector spaces. Set $G(m) = APm$ where $A : X \to \mathbb{R}^d$ is an arbitrary but fixed linear isomorphism of vector spaces. Then for $x \in X$ and $y \in Y$,

$$(G, F)'(m)(x + y) = (G'(m)(x + y), F'(m)(x + y))$$
$$= (AP(x + y), F'(m)y) = (Ax, F'(m)y) \in \mathbb{R}^d \times \mathbb{R}^{N-d}$$

from which it follows that $(G, F)'(m)$ is an isomorphism.

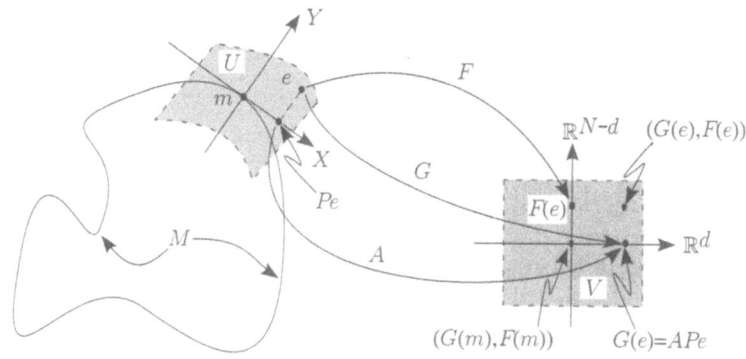

**Figure 2.** Constructing charts for $M$ using the inverse function theorem. For simplicity of the drawing, $m \in M$ is assumed to be the origin of $E = X \oplus Y$.

By the inverse function theorem, there exists a neighborhood $U \subset_o E$ of $m$ such that $V := (G, F)(U) \subset_o \mathbb{R}^N$ and $(G, F) : U \to V$ is a diffeomorphism. Let $z = (G, F)$ with $\mathcal{D}(z) = U$ and $\mathcal{R}(z) = V$, then $z$ is a chart of $E$ about $m$ satisfying the conditions of Definition 2.2. Indeed, items 1) – 3) are clear by construction. If $p \in M \cap \mathcal{D}(z)$ then $z(p) = (G(p), F(p)) = (G(p), 0) \in \mathcal{R}(z) \cap (\mathbb{R}^d \times \{0\})$. Conversely, if $p \in \mathcal{D}(z)$ is a point such that $z(p) = (G(p), F(p)) \in \mathcal{R}(z) \cap (\mathbb{R}^d \times \{0\})$, then $F(p) = 0$ and hence $p \in M \cap \mathcal{D}(z)$; so item 4) of Definition 2.2 is verified.  $\square$

**Example 2.6.** Let $gl(n, \mathbb{R})$ denote the set of all $n \times n$ real matrices. The following are examples of imbedded submanifolds.

1. Any open subset $M$ of $E$.

2. The graph,

$$\Gamma(f) := \left\{ (x, f(x)) \in \mathbb{R}^d \times \mathbb{R}^{N-d} : x \in \mathcal{D}(f) \right\} \subset \mathcal{D}(f) \times \mathbb{R}^{N-d} \subset \mathbb{R}^N,$$

of any smooth function $f : \mathbb{R}^d \to \mathbb{R}^{N-d}$ as can be seen by applying Theorem 2.5 with $F(x, y) := y - f(x)$. In this case it would be a good idea for the reader to produce an explicit chart $z$ as in Definition 2.2 such that $\mathcal{D}(z) = \mathcal{R}(z) = \mathcal{D}(f) \times \mathbb{R}^{N-d}$.

3. The unit sphere, $S^{N-1} := \{x \in \mathbb{R}^N : \langle x, x \rangle_{\mathbb{R}^N} = 1\}$, as is seen by applying Theorem 2.5 with $E = \mathbb{R}^N$ and $F(x) := \langle x, x \rangle_{\mathbb{R}^N} - 1$. Alternatively, express $S^{N-1}$ locally as the graph of smooth functions and then use item 2.

4. $GL(n, \mathbb{R}) := \{g \in gl(n, \mathbb{R}) | \det(g) \neq 0\}$, see item 1.

5. $SL(n, \mathbb{R}) := \{g \in gl(n, \mathbb{R}) | \det(g) = 1\}$ as is seen by taking $E = gl(n, \mathbb{R})$ and $F(g) := \det(g)$ and then applying Theorem 2.5 with the aid of Lemma 2.7 below.

6. $O(n) := \{g \in gl(n, \mathbb{R}) | g^{tr} g = I\}$ where $g^{tr}$ denotes the transpose of $g$. In this case take $F(g) := g^{tr} g - I$ thought of as a function from $E = gl(n, \mathbb{R})$ to $S(n)$, where

$$S(n) := \left\{ A \in gl(n, \mathbb{R}) : A^{tr} = A \right\}$$

is the subspace of symmetric matrices. To show $F'(g)$ is surjective, show

$$F'(g)(gB) = B + B^{tr} \text{ for all } g \in O(n) \text{ and } B \in gl(n, \mathbb{R}).$$

7. $SO(n) := \{g \in O(n) | \det(g) = 1\}$, an open subset of $O(n)$.

8. $M \times N \subset E \times V$, where $M$ and $N$ are imbedded submanifolds of $E$ and $V$ respectively. The reader should verify this by constructing appropriate charts for $E \times V$ by taking "tensor" products of the charts for $E$ and $V$ associated to $M$ and $N$ respectively.

9. The $n$-dimensional torus,

$$T^n := \{z \in \mathbb{C}^n : |z_i| = 1 \text{ for } i = 1, 2, \ldots, n\} = (S^1)^n,$$

where $z = (z_1, \ldots, z_n)$ and $|z_i| = \sqrt{z_i \bar{z}_i}$. This follows by induction using items 3. and 8. Alternatively apply Theorem 2.5 with $F(z) := \left( |z_1|^2 - 1, \ldots, |z_n|^2 - 1 \right)$.

**Lemma 2.7.** *Suppose $g \in GL(n, \mathbb{R})$ and $A \in gl(n, \mathbb{R})$, then*

$$\det'(g)A = \det(g)\mathrm{tr}(g^{-1}A). \tag{2.3}$$

*Proof.* By definition we have

$$\det{}'(g)A = \frac{d}{dt}|_0 \det(g + tA) = \det(g)\frac{d}{dt}|_0 \det(I + tg^{-1}A).$$

So it suffices to prove $\frac{d}{dt}|_0 \det(I + tB) = \mathrm{tr}(B)$ for all matrices $B$. If $B$ is upper triangular, then $\det(I + tB) = \prod_{i=1}^n (1 + tB_{ii})$ and hence by the product rule,

$$\frac{d}{dt}|_0 \det(I + tB) = \sum_{i=1}^{n} B_{ii} = \operatorname{tr}(B).$$

This completes the proof because 1) every matrix can be put into upper triangular form by a similarity transformation, and 2) "det" and "tr" are invariant under similarity transformations. □

**Definition 2.8.** Let $E$ and $V$ be two finite dimensional vector spaces and $M^d \subset E$ and $N^k \subset V$ be two imbedded submanifolds. A function $f : M \rightarrow N$ is said to be *smooth* if for all charts $x \in \mathcal{A}(M)$ and $y \in \mathcal{A}(N)$ the function $y \circ f \circ x^{-1} : \mathbb{R}^d \rightarrow \mathbb{R}^k$ is smooth.

**Exercise 2.9.** Let $M^d \subset E$ and $N^k \subset V$ be two imbedded submanifolds as in Definition 2.8.

1. Show that a function $f : \mathbb{R}^k \rightarrow M$ is smooth iff $f$ is smooth when thought of as a function from $\mathbb{R}^k$ to $E$.
2. If $F : E \rightarrow V$ is a smooth function such that $F(M \cap \mathcal{D}(F)) \subset N$, show that $f := F|_M : M \rightarrow N$ is smooth.
3. Show the composition of smooth maps between imbedded submanifolds is smooth.

**Proposition 2.10.** *Assuming the notation in Definition 2.8, a function $f : M \rightarrow N$ is smooth iff there is a smooth function $F : E \rightarrow V$ such that $f = F|_M$.*

*Proof.* (Sketch.) Suppose that $f : M \rightarrow N$ is smooth, $m \in M$ and $n = f(m)$. Let $z$ be as in Definition 2.2 and $w$ be a chart on $N$ such that $n \in \mathcal{D}(w)$, by shrinking the domain of $z$ if necessary, we may assume that $\mathcal{R}(z) = U \times W$ where $U \subset_o \mathbb{R}^d$ and $W \subset_o \mathbb{R}^{N-d}$ in which case $z(M \cap \mathcal{D}(z)) = U \times \{0\}$. For $\xi \in \mathcal{D}(z)$, let $F(\xi) := f(z^{-1}(z_<(\xi), 0))$ with $z = (z_<, z_>)$ as in Notation 2.3. Then $F : \mathcal{D}(z) \rightarrow N$ is a smooth function such that $F|_{M \cap \mathcal{D}(z)} = f|_{M \cap \mathcal{D}(z)}$. The function $F$ is smooth. Indeed, letting $x = z_<|_{\mathcal{D}(z) \cap M}$,

$$w_< \circ F = w_< \circ f(z^{-1}(z_<(\xi), 0)) = w_< \circ f \circ x^{-1} \circ (z_<(\cdot), 0)$$

which, being the composition of the smooth maps $w_< \circ f \circ x^{-1}$ (smooth by assumption) and $\xi \rightarrow (z_<(\xi), 0)$, is smooth as well. Hence by definition, $F$ is smooth as claimed. Using a standard partition of unity argument (which we omit), it is possible to piece this local argument together to construct a globally defined smooth function $F : E \rightarrow V$ such that $f = F|_M$. □

**Definition 2.11.** A function $f : M \rightarrow N$ is a *diffeomorphism* if $f$ is smooth and has a smooth inverse. The set of diffeomorphisms $f : M \rightarrow M$ is a group under composition which will be denoted by $\operatorname{Diff}(M)$.

## 2.2 Tangent planes and spaces

**Definition 2.12.** Given an imbedded submanifold $M \subset E$ and $m \in M$, let $\tau_m M \subset E$ denote the collection of all vectors $v \in E$ such that there exists a smooth path $\sigma : (-\varepsilon, \varepsilon) \rightarrow M$ with $\sigma(0) = m$ and $v = \frac{d}{ds}|_0 \sigma(s)$. The subset $\tau_m M$ is called the *tangent plane* to $M$ at $m$ and $v \in \tau_m M$ is called a *tangent vector*, see Figure 3.

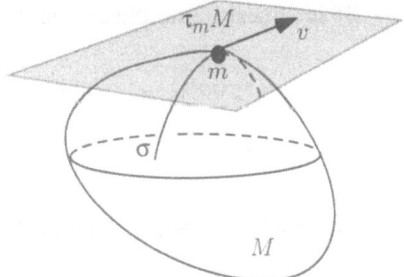

**Figure 3.** Tangent plane, $\tau_m M$, to $M$ at $m$ and a vector, $v$, in $\tau_m M$.

**Theorem 2.13.** *For each $m \in M$, $\tau_m M$ is a d-dimensional subspace of E. If $z : E \to \mathbb{R}^N$ is as in Definition 2.2, then $\tau_m M = \mathrm{Nul}(z'_>(m))$. If $x$ is a chart on $M$ such that $m \in \mathcal{D}(x)$, then*

$$\{\frac{d}{ds}|_0 x^{-1}(x(m) + se_i)\}_{i=1}^d$$

*is a basis for $\tau_m M$, where $\{e_i\}_{i=1}^d$ is the standard basis for $\mathbb{R}^d$.*

*Proof.* Let $\sigma : (-\varepsilon, \varepsilon) \to M$ be a smooth path with $\sigma(0) = m$ and $v = \frac{d}{ds}|_0 \sigma(s)$ and $z$ be a chart (for $E$) around $m$ as in Definition 2.2 such that $x = z_<$. Then $z_>(\sigma(s)) = 0$ for all $s$ and therefore,

$$0 = \frac{d}{ds}|_0 z_>(\sigma(s)) = z'_>(m)v$$

which shows that $v \in \mathrm{Nul}(z'_>(m))$, i.e., $\tau_m M \subset \mathrm{Nul}(z'_>(m))$.

Conversely, suppose that $v \in \mathrm{Nul}(z'_>(m))$. Let $w = z'_<(m)v \in \mathbb{R}^d$ and $\sigma(s) := x^{-1}(z_<(m) + sw) \in M$ defined for $s$ near 0. Differentiating the identity $z^{-1} \circ z = id$ at $m$ shows

$$\left(z^{-1}\right)'(z(m))z'(m) = I.$$

Therefore,

$$\sigma'(0) = \frac{d}{ds}|_0 x^{-1}(z_<(m) + sw) = \frac{d}{ds}|_0 z^{-1}(z_<(m) + sw, 0)$$

$$= \left(z^{-1}\right)'((z_<(m), 0))(z'_<(m)v, 0)$$

$$= \left(z^{-1}\right)'((z_<(m), 0))(z'_<(m)v, z'_>(m)v)$$

$$= \left(z^{-1}\right)'(z(m))z'(m)v = v,$$

and so by definition $v = \sigma'(0) \in \tau_m M$. We have now shown $\mathrm{Nul}(z'_>(m)) \subset \tau_m M$ which completes the proof that $\tau_m M = \mathrm{Nul}(z'_>(m))$.

Since $z'_<(m) : \tau_m M \to \mathbb{R}^d$ is a linear isomorphism, the above argument also shows

$$\frac{d}{ds}|_0 x^{-1}(x(m) + sw) = \left(z'_<(m)|_{\tau_m M}\right)^{-1} w \in \tau_m M \ \forall \ w \in \mathbb{R}^d.$$

In particular it follows that

$$\{\frac{d}{ds}|_0 x^{-1}(x(m) + se_i)\}_{i=1}^d = \{\left(z'_<(m)|_{\tau_m M}\right)^{-1} e_i\}_{i=1}^d$$

is a basis for $\tau_m M$ (see Figure 4 below).                                          □

The following proposition is an easy consequence of Theorem 2.13 and the proof of Theorem 2.5.

**Proposition 2.14.** *Suppose that M is an imbedded submanifold constructed as in Theorem 2.5. Then* $\tau_m M = \mathrm{Nul}\big(F'(m)\big)$.

**Exercise 2.15.** Show:

1. $\tau_m M = E$, if $M$ is an open subset of $E$.
2. $\tau_g GL(n, \mathbb{R}) = gl(n, \mathbb{R})$, for all $g \in GL(n, \mathbb{R})$.
3. $\tau_m S^{N-1} = \{m\}^\perp$ for all $m \in S^{N-1}$.
4. Let $sl(n, \mathbb{R})$ be the traceless (zero trace) matrices,

$$sl(n, \mathbb{R}) := \{A \in gl(n, \mathbb{R})| \, \mathrm{tr}(A) = 0\}. \tag{2.4}$$

Then

$$\tau_g SL(n, \mathbb{R}) = \{A \in gl(n, \mathbb{R})|g^{-1}A \in sl(n, \mathbb{R})\}$$

and in particular $\tau_I SL(n, \mathbb{R}) = sl(n, \mathbb{R})$.
5. Let $so\,(n, \mathbb{R})$ be the skew symmetric matrices,

$$so\,(n, \mathbb{R}) := \{A \in gl(n, \mathbb{R})|A = -A^{\mathrm{tr}}\}.$$

Then

$$\tau_g O(n) = \{A \in gl(n, \mathbb{R})|g^{-1}A \in so\,(n, \mathbb{R})\}$$

and in particular $\tau_I O\,(n) = so\,(n, \mathbb{R})$. *Hint:* $g^{-1} = g^{\mathrm{tr}}$ for all $g \in O(n)$.
6. If $M \subset E$ and $N \subset V$ are imbedded submanifolds then

$$\tau_{(m,n)}(M \times N) = \tau_m M \times \tau_n N \subset E \times V.$$

It is quite possible that $\tau_m M = \tau_{m'} M$ for some $m \neq m'$, with $m$ and $m'$ in $M$ (think of the sphere). Because of this, it is helpful to label each of the tangent planes with their base point.

**Definition 2.16.** The *tangent space* $(T_m M)$ to $M$ at $m$ is given by

$$T_m M := \{m\} \times \tau_m M \subset M \times E.$$

Let

$$TM := \cup_{m \in M} T_m M,$$

and call $TM$ the *tangent space* (or *tangent bundle*) of $M$. A *tangent vector* is a point $v_m := (m, v) \in TM$ and we let $\pi : TM \to M$ denote the *canonical projection* by $\pi(v_m) = m$. Each tangent space is made into a vector space with the vector space operations being defined by: $c(v_m) := (cv)_m$ and $v_m + w_m := (v + w)_m$.

**Exercise 2.17.** Prove that $TM$ is an imbedded submanifold of $E \times E$. *Hint:* suppose that $z : E \to \mathbb{R}^N$ is a function as in the Definition 2.2. Define $\mathcal{D}(Z) := \mathcal{D}(z) \times E$ and $Z : \mathcal{D}(Z) \to \mathbb{R}^N \times \mathbb{R}^N$ by $Z(x, a) := (z(x), z'(x)a)$. Use $Z$'s of this type to check $TM$ satisfies Definition 2.2.

**Notation 2.18.** In the sequel, given a smooth path $\sigma : (-\varepsilon, \varepsilon) \to M$, we will abuse notation and write $\sigma'(0)$ for either

$$\frac{d}{ds}|_0 \sigma(s) \in \tau_{\sigma(0)} M$$

or for

$$(\sigma(0), \frac{d}{ds}|_0 \sigma(s)) \in T_{\sigma(0)} M = \{\sigma(0)\} \times \tau_{\sigma(0)} M.$$

Also given a chart $x = (x^1, x^2, \ldots, x^d)$ on $M$ and $m \in \mathcal{D}(x)$, let $\partial/\partial x^i|_m$ denote the element $T_m M$ determined by $\partial/\partial x^i|_m = \sigma'(0)$, where $\sigma(s) := x^{-1}(x(m) + se_i)$, i.e.,

$$\frac{\partial}{\partial x^i}|_m = (m, \frac{d}{ds}|_0 x^{-1}(x(m) + se_i)), \tag{2.5}$$

see Figure 4.

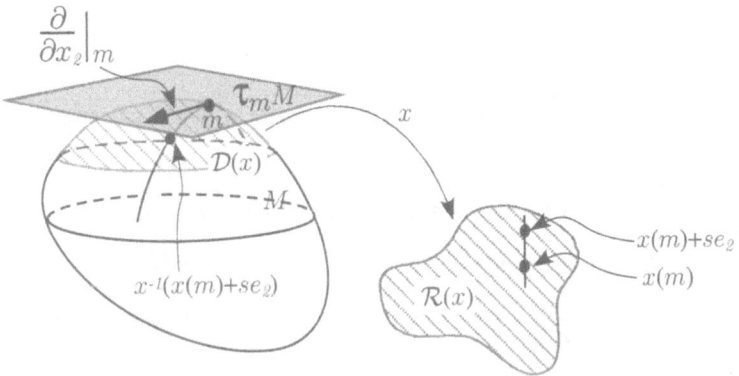

**Figure 4.** Forming a basis of tangent vectors.

The reason for the strange notation in Eq. (2.5) will be explained after Notation 2.20. By definition, every element of $T_m M$ is of the form $\sigma'(0)$ where $\sigma$ is a smooth path into $M$ such that $\sigma(0) = m$. Moreover by Theorem 2.13, $\{\partial/\partial x^i|_m\}_{i=1}^d$ is a basis for $T_m M$.

**Definition 2.19.** Suppose that $f : M \to V$ is a smooth function, $m \in \mathcal{D}(f)$ and $v_m \in T_m M$. Write

$$v_m f = df(v_m) := \frac{d}{ds}|_0 f(\sigma(s)),$$

where $\sigma$ is any smooth path in $M$ such that $\sigma'(0) = v_m$. The function $df : TM \to V$ will be called the *differential of* $f$.

**Notation 2.20.** If $M$ and $N$ are two manifolds $f : M \times N \to V$ is a smooth function, we will write $d_M f(\cdot, n)$ to indicate that we are computing the differential of the function $m \in M \to f(m, n) \in V$ for fixed $n \in N$.

To understand the notation in (2.5), suppose that $f = F \circ x = F(x^1, x^2, \ldots, x^d)$ where $F : \mathbb{R}^d \to \mathbb{R}$ is a smooth function and $x$ is a chart on $M$. Then

$$\frac{\partial f(m)}{\partial x^i} := \frac{\partial}{\partial x^i}|_m f = (D_i F)(x(m)),$$

where $D_i$ denotes the $i^{\text{th}}$-partial derivative of $F$. Also notice that $dx^j \left( \frac{\partial}{\partial x^i}|_m \right) = \delta_{ij}$ so that $\{dx^i|_{T_m M}\}_{i=1}^d$ is the dual basis of $\{\partial/\partial x^i|_m\}_{i=1}^d$ and therefore if $v_m \in T_m M$ then

$$v_m = \sum_{i=1}^d dx^i(v_m) \frac{\partial}{\partial x^i}|_m. \tag{2.6}$$

This explicitly exhibits $v_m$ as a first order differential operator acting on "germs" of smooth functions defined near $m \in M$.

**Remark 2.21 (Product Rule).** Suppose that $f : M \to V$ and $g : M \to \text{End}(V)$ are smooth functions, then

$$v_m(gf) = \frac{d}{ds}|_0 [g(\sigma(s)) f(\sigma(s))] = v_m g \cdot f(m) + g(m) v_m f$$

or equivalently

$$d(gf)(v_m) = dg(v_m) f(m) + g(m) df(v_m).$$

This last equation will be abbreviated as $d(gf) = dg \cdot f + g df$.

**Definition 2.22.** Let $f : M \to N$ be a smooth map of imbedded submanifolds. Define the *differential*, $f_*$, of $f$ by

$$f_* v_m = (f \circ \sigma)'(0) \in T_{f(m)} N,$$

where $v_m = \sigma'(0) \in T_m M$, and $m \in \mathcal{D}(f)$.

**Lemma 2.23.** *The differentials defined in Definitions 2.19 and 2.22 are well defined linear maps on $T_m M$ for each $m \in \mathcal{D}(f)$.*

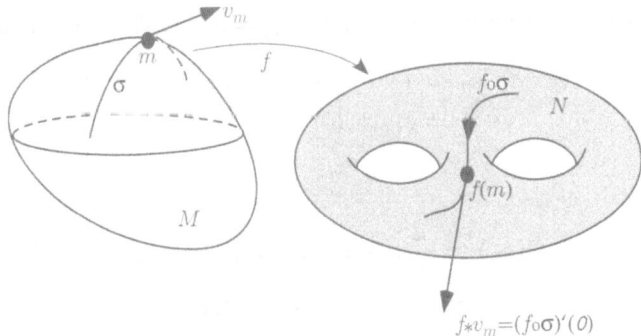

**Figure 5.** The differential of $f$.

*Proof.* I will only prove that $f_*$ is well defined, since the case of $df$ is similar. By Proposition 2.10, there is a smooth function $F : E \to V$, such that $f = F|_M$. Therefore by the chain rule

$$f_* v_m = (f \circ \sigma)'(0) := \left[ \frac{d}{ds} \Big|_0 f(\sigma(s)) \right]_{f(\sigma(0))} = \left[ F'(m)v \right]_{f(m)}, \qquad (2.7)$$

where $\sigma$ is a smooth path in $M$ such that $\sigma'(0) = v_m$. It follows from (2.7) that $f_* v_m$ does not depend on the choice of the path $\sigma$. It is also clear from (2.7), that $f_*$ is linear on $T_m M$. $\qquad \square$

**Remark 2.24.** Suppose that $F : E \to V$ is a smooth function and that $f := F|_M$. Then as in the proof of Lemma 2.23

$$df(v_m) = F'(m)v \qquad (2.8)$$

for all $v_m \in T_m M$, and $m \in \mathcal{D}(f)$. Incidentally, since the left hand sides of (2.7) and (2.8) are defined "intrinsically," the right members of (2.7) and (2.8) are independent of the possible choices of functions $F$ which extend $f$.

**Lemma 2.25 (Chain Rules).** *Suppose that $M$, $N$, and $P$ are imbedded submanifolds and $V$ is a finite dimensional vector space. Let $f : M \to N$, $g : N \to P$, and $h : N \to V$ be smooth functions. Then:*

$$(g \circ f)_* v_m = g_*(f_* v_m), \qquad \forall\, v_m \in TM \qquad (2.9)$$

*and*

$$d(h \circ f)(v_m) = dh(f_* v_m), \qquad \forall\, v_m \in TM. \qquad (2.10)$$

*These equations will be written more concisely as $(g \circ f)_* = g_* f_*$ and $d(h \circ f) = dh f_*$ respectively.*

*Proof.* Let $\sigma$ be a smooth path in $M$ such that $v_m = \sigma'(0)$. Then, see Figure 6,

$$
\begin{aligned}
(g \circ f)_* v_m &:= (g \circ f \circ \sigma)'(0) = g_*(f \circ \sigma)'(0) \\
&= g_* f_* \sigma'(0) = g_* f_* v_m.
\end{aligned}
$$

Similarly,

$$
\begin{aligned}
d(h \circ f)(v_m) &:= \frac{d}{ds}\big|_0 (h \circ f \circ \sigma)(s) = dh((f \circ \sigma)'(0)) \\
&= dh(f_* \sigma'(0)) = dh(f_* v_m).
\end{aligned}
$$

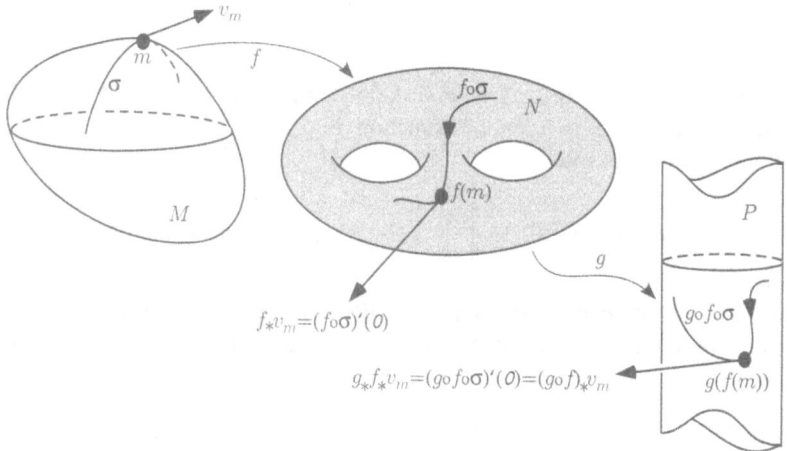

**Figure 6.** The chain rule.

If $f : M \to V$ is a smooth function, $x$ is a chart on $M$, and $m \in \mathcal{D}(f) \cap \mathcal{D}(x)$, we will write $\partial f(m)/\partial x^i$ for $df\left(\partial/\partial x^i|_m\right)$. Combining this notation with Eq. (2.6) leads to the pleasing formula,

$$
df = \sum_{i=1}^{d} \frac{\partial f}{\partial x^i} dx^i, \tag{2.11}
$$

by which we mean

$$
df(v_m) = \sum_{i=1}^{d} \frac{\partial f(m)}{\partial x^i} dx^i(v_m).
$$

Suppose that $f : M^d \to N^k$ is a smooth map of imbedded submanifolds, $m \in M$, $x$ is a chart on $M$ such that $m \in \mathcal{D}(x)$, and $y$ is a chart on $N$ such that $f(m) \in \mathcal{D}(y)$. Then the matrix of

$$
f_{*m} := f_*|_{T_m M} : T_m M \to T_{f(m)} N
$$

relative to the bases $\{\partial/\partial x^i|_m\}_{i=1}^d$ of $T_m M$ and $\{\partial/\partial y^j|_{f(m)}\}_{j=1}^k$ of $T_{f(m)} N$ is $(\partial(y^j \circ f)(m)/\partial x^i)$. Indeed, if $v_m = \sum_{i=1}^d v^i \partial/\partial x^i|_m$, then

$$f_* v_m = \sum_{j=1}^k dy^j (f_* v_m) \partial/\partial y^j|_{f(m)}$$

$$= \sum_{j=1}^k d(y^j \circ f)(v_m) \partial/\partial y^j|_{f(m)} \qquad \text{(by Eq. (2.10))}$$

$$= \sum_{j=1}^k \sum_{i=1}^d \frac{\partial(y^j \circ f)(m)}{\partial x^i} \cdot dx^i(v_m) \partial/\partial y^j|_{f(m)} \qquad \text{(by Eq. (2.11))}$$

$$= \sum_{j=1}^k \sum_{i=1}^d \frac{\partial(y^j \circ f)(m)}{\partial x^i} v^i \partial/\partial y^j|_{f(m)}.$$

**Example 2.26.** Let $M = O(n)$, $k \in O(n)$, and $f : O(n) \to O(n)$ be defined by $f(g) := kg$. Then $f$ is a smooth function on $O(n)$ because it is the restriction of a smooth function on $gl(n, \mathbb{R})$. Given $A_g \in T_g O(n)$, by Eq. (2.7),

$$f_* A_g = (kg, kA) = (kA)_{kg}$$

(In the future we denote $f$ by $L_k$; $L_k$ is **left translation** by $k \in O(n)$.)

**Definition 2.27.** A *Lie group* is a manifold, $G$, which is also a group such that the group operations are smooth functions. The tangent space, $\mathfrak{g} := \text{Lie}(G) := T_e G$, to $G$ at the identity $e \in G$ is called the *Lie algebra* of $G$.

**Exercise 2.28.** Verify that $GL(n, \mathbb{R})$, $SL(n, \mathbb{R})$, $O(n)$, $SO(n)$ and $T^n$ (see Example 2.6) are all Lie groups and

$$\text{Lie}(GL(n, \mathbb{R})) \cong gl(n, \mathbb{R}),$$
$$\text{Lie}(SL(n, \mathbb{R}))) \cong sl(n, \mathbb{R})$$
$$\text{Lie}(O(n))) = \text{Lie}(SO(n))) \cong so(n, \mathbb{R}) \text{ and }$$
$$\text{Lie}(T^n)) \cong (i\mathbb{R})^n \subset \mathbb{C}^n.$$

See Exercise 2.15 for the notation being used here.

**Exercise 2.29 (Continuation of Exercise 2.17).** Show for each chart $x$ on $M$ that the function

$$\phi(v_m) := (x(m), dx(v_m)) = x_* v_m$$

is a chart on $TM$. Note that $\mathcal{D}(\phi) := \cup_{m \in \mathcal{D}(x)} T_m M$.

The following lemma gives an important example of a smooth function on $M$ which will be needed when we consider $M$ as a "Riemannian manifold."

**Lemma 2.30.** *Suppose that $(E, \langle \cdot, \cdot \rangle)$ is an inner product space and the $M \subset E$ is an imbedded submanifold. For each $m \in M$, let $P(m)$ denote the orthogonal projection of $E$ onto $\tau_m M$ and $Q(m) := I - P(m)$ denote the orthogonal projection onto $\tau_m M^\perp$. Then $P$ and $Q$ are smooth functions from $M$ to $gl(E)$, where $gl(E)$ denotes the vector space of linear maps from $E$ to $E$.*

*Proof.* Let $z : E \to \mathbb{R}^N$ be as in Definition 2.2. To simplify notation, let $F(p) := z_>(p)$ for all $p \in \mathcal{D}(z)$, so that $\tau_m M = \mathrm{Nul}\left(F'(m)\right)$ for $m \in \mathcal{D}(x) = \mathcal{D}(z) \cap M$. Since $F'(m) : E \to \mathbb{R}^{N-d}$ is surjective, an elementary exercise in linear algebra shows

$$(F'(m)F'(m)^*) : \mathbb{R}^{N-d} \to \mathbb{R}^{N-d}$$

is invertible for all $m \in \mathcal{D}(x)$. The orthogonal projection $Q(m)$ may be expressed as;

$$Q(m) = F'(m)^*(F'(m)F'(m)^*)^{-1}F'(m). \tag{2.12}$$

Since being invertible is an open condition, $(F'(\cdot)F'(\cdot)^*)$ is invertible in an open neighborhood $\mathcal{N} \subset E$ of $\mathcal{D}(x)$. Hence $Q$ has a smooth extension $\tilde{Q}$ to $\mathcal{N}$ given by

$$\tilde{Q}(x) := F'(x)^*(F'(x)F'(x)^*)^{-1}F'(x).$$

Since $Q|_{\mathcal{D}(x)} = \tilde{Q}|_{\mathcal{D}(x)}$ and $\tilde{Q}$ is smooth on $\mathcal{N}$, $Q|_{\mathcal{D}(x)}$ is also smooth. Since $z$, as in Definition 2.2, was arbitrary and smoothness is a local property, it follows that $Q$ is smooth on $M$. Clearly, $P := I - Q$ is also a smooth function on $M$. $\qquad\square$

**Definition 2.31.** A *local vector field* $Y$ on $M$ is a smooth function $Y : M \to TM$ such that $Y(m) \in T_m M$ for all $m \in \mathcal{D}(Y)$, where $\mathcal{D}(Y)$ is assumed to be an open subset of $M$. Let $\Gamma(TM)$ denote the collection of globally defined (i.e., $\mathcal{D}(Y) = M$) smooth vector fields $Y$ on $M$.

Note that $\partial/\partial x^i$ are local vector fields on $M$ for each chart $x \in \mathcal{A}(M)$ and $i = 1, 2, \ldots, d$. The next exercise asserts that these vector fields are smooth.

**Exercise 2.32.** Let $Y$ be a vector field on $M$, $x \in \mathcal{A}(M)$ be a chart on $M$, and $Y^i := dx^i(Y)$, then

$$Y(m) := \sum_{i=1}^{d} Y^i(m) \, \partial/\partial x^i|_m \ \forall \, m \in \mathcal{D}(x),$$

which we abbreviate as $Y = \sum_{i=1}^{d} Y^i \partial/\partial x^i$. Show the condition that $Y$ is smooth translates into the statement that each of the functions $Y^i$ is smooth.

**Exercise 2.33.** Let $Y : M \to TM$, be a vector field. Then

$$Y(m) = (m, y(m)) = y(m)_m$$

for some function $y : M \to E$ such that $y(m) \in \tau_m M$ for all $m \in \mathcal{D}(Y) = \mathcal{D}(y)$. Show that $Y$ is smooth iff $y : M \to E$ is smooth.

**Example 2.34.** Let $M = SL(n, \mathbb{R})$ and $A \in sl(n, \mathbb{R}) = \tau_I SL(n, \mathbb{R})$, i.e. $A$ is a $n \times n$ real matrix such that tr $(A) = 0$. Then $\tilde{A}(g) := L_{g*}A_e = (g, gA)$ for $g \in M$ is a smooth vector field on $M$.

**Example 2.35.** Keep the notation of Lemma 2.30. Let $y : M \to E$ be any smooth function. Then $Y(m) := (m, P(m)y(m))$ for all $m \in M$ is a smooth vector field on $M$.

**Definition 2.36.** Given $Y \in \Gamma(TM)$ and $f \in C^\infty(M)$, let $Yf \in C^\infty(M)$ be defined by $(Yf)(m) := df(Y(m))$, for all $m \in \mathcal{D}(f) \cap \mathcal{D}(Y)$. In this way the vector field $Y$ may be viewed as a first order differential operator on $C^\infty(M)$.

**Notation 2.37.** The *Lie bracket* of two smooth vector fields, $Y$ and $W$, on $M$ is the vector field $[Y, W]$ which acts on $C^\infty(M)$ by the formula

$$[Y, W]f := Y(Wf) - W(Yf), \quad \forall f \in C^\infty(M). \tag{2.13}$$

(In general one might suspect that $[Y, W]$ is a second order differential operator, however this is not the case, see Exercise 2.38.) Sometimes it will be convenient to write $L_Y W$ for $[Y, W]$.

**Exercise 2.38.** Show that $[Y, W]$ is again a first order differential operator on $C^\infty(M)$ coming from a vector field. In particular, if $x$ is a chart on $M$, $Y = \sum_{i=1}^d Y^i \partial/\partial x^i$ and $W = \sum_{i=1}^d W^i \partial/\partial x^i$, then on $\mathcal{D}(x)$,

$$[Y, W] = \sum_{i=1}^d (YW^i - WY^i)\partial/\partial x^i. \tag{2.14}$$

**Proposition 2.39.** *If $Y(m) = (m, y(m))$ and $W(m) = (m, w(m))$ and $y, w : M \to E$ are smooth functions such that $y(m), w(m) \in \tau_m M$, then we may express the Lie bracket, $[Y, W](m)$, as*

$$[Y, W](m) = (m, (Yw - Wy)(m)) = (m, dw(Y(m)) - dy(W(m))). \tag{2.15}$$

*Proof.* Let $f$ be a smooth function $M$ which we may take, by Proposition 2.10, to be the restriction of a smooth function on $E$. Similarly we may assume that $y$ and $w$ are smooth functions on $E$ such that $y(m), w(m) \in \tau_m M$ for all $m \in M$. Then

$$\begin{aligned}
(YW - WY)f &= Y[f'w] - W[f'y] \\
&= f''(y, w) - f''(w, y) + f'(Yw) - f'(Wy) \\
&= f'(Yw - Wy)
\end{aligned} \tag{2.16}$$

wherein the last equality we have used the fact that mixed partial derivatives commute to conclude

$$f''(u, v) - f''(v, u) := (\partial_u \partial_v - \partial_v \partial_u) f = 0 \,\forall\, u, v \in E.$$

Taking $f = z_>$ in Eq. (2.16) with $z = (z_<, z_>)$ being a chart on $E$ as in Definition 2.2, shows

$$0 = (YW - WY)z_> (m) = z'_> (dw(Y(m)) - dy(W(m)))$$

and thus $(m, dw(Y(m)) - dy(W(m))) \in T_m M$. With this observation, we then have

$$f'(Yw - Wy) = df((m, dw(Y(m)) - dy(W(m))))$$

which combined with Eq. (2.16) verifies Eq. (2.15).     □

**Exercise 2.40.** Let $M = SL(n, \mathbb{R})$ and $A, B \in sl(n, \mathbb{R})$ and $\tilde{A}$ and $\tilde{B}$ be the associated left invariant vector fields on $M$ as introduced in Example 2.34. Show $\left[\tilde{A}, \tilde{B}\right] = \widetilde{[A, B]}$ where $[A, B] := AB - BA$ is the matrix commutator of $A$ and $B$.

## 2.3 More references

The reader wishing to learn about manifolds is referred to [1, 9, 19, 41, 42, 95, 111, 112, 113, 114, 115, 164]. The texts by Kobayashi and Nomizu are very thorough while the books by Klingenberg give an idea of why differential geometers are interested in loop spaces. There is a vast literature on Lie groups and their representations. Here are just two books that I have found very useful, [24, 178].

# 3 Riemannian geometry primer

This section introduces the following objects: 1) Riemannian metrics, 2) Riemannian volume forms, 3) gradients, 4) divergences, 5) Laplacians, 6) covariant derivatives, 7) parallel translations, and 8) curvatures.

## 3.1 Riemannian metrics

**Definition 3.1.** A *Riemannian metric*, $\langle \cdot, \cdot \rangle$ (also denoted by $g$), on $M$ is a smoothly varying choice of inner product, $g_m = \langle \cdot, \cdot \rangle_m$, on each of the tangent spaces $T_m M$, $m \in M$. The smoothness condition is the requirement that the function $m \in M \to \langle X(m), Y(m) \rangle_m \in \mathbb{R}$ is smooth for all smooth vector fields $X$ and $Y$ on $M$.

It is customary to write $ds^2$ for the function on $TM$ defined by

$$ds^2(v_m) := \langle v_m, v_m \rangle_m = g_m(v_m, v_m). \tag{3.1}$$

By polarization, the Riemannian metric $\langle \cdot, \cdot \rangle$ is uniquely determined by the function $ds^2$. Given a chart $x$ on $M$ and $v \in T_m M$, by equations (3.1) and (2.6) we have

$$ds^2(v_m) = \sum_{i,j=1}^{d} \langle \partial/\partial x^i |_m, \partial/\partial x^j |_m \rangle_m dx^i(v_m) dx^j(v_m). \tag{3.2}$$

We will abbreviate this equation in the future by writing

$$ds^2 = \sum_{i,j=1}^{d} g_{ij}^x dx^i dx^j \tag{3.3}$$

where

$$g_{i,j}^x(m) := \langle \partial/\partial x^i |_m, \partial/\partial x^j |_m \rangle_m = g\left( \partial/\partial x^i |_m, \partial/\partial x^j |_m \right).$$

Typically $g_{i,j}^x$ will be abbreviated by $g_{ij}$ if no confusion is likely to arise.

**Example 3.2.** Let $M = \mathbb{R}^N$ and let $x = (x^1, x^2, \ldots, x^N)$ denote the standard chart on $M$, i.e., $x(m) = m$ for all $m \in M$. The standard Riemannian metric on $\mathbb{R}^N$ is determined by

$$ds^2 = \sum_{i=1}^{N} (dx^i)^2 = \sum_{i=1}^{N} dx^i \cdot dx^i,$$

and so $g^x$ is the identity matrix. The general Riemannian metric on $\mathbb{R}^N$ is determined by $ds^2 = \sum_{i,j=1}^{N} g_{ij} dx^i dx^j$, where $g = (g_{ij})$ is a smooth $gl(N, \mathbb{R})$-valued function on $\mathbb{R}^N$ such that $g(m)$ is positive definite matrix for all $m \in \mathbb{R}^N$.

Let $M$ be an imbedded submanifold of a finite dimensional inner product space $(E, \langle \cdot, \cdot \rangle)$. The manifold $M$ *inherits* a metric from $E$ determined by

$$ds^2(v_m) = \langle v, v \rangle \ \forall \ v_m \in TM.$$

It is a well known deep fact that *all* finite dimensional Riemannian manifolds may be constructed in this way, see Nash [143] and Moser [138, 139, 140]. To simplify the exposition, in the sequel we will usually assume that $(E, \langle \cdot, \cdot \rangle)$ is an inner product space, $M^d \subset E$ is an imbedded submanifold, and the Riemannian metric on $M$ is determined in this way, i.e.,

$$\langle v_m, w_m \rangle = \langle v, w \rangle_{\mathbb{R}^N}, \quad \forall \ v_m, w_m \in T_m M \text{ and } m \in M.$$

In this setting the components $g_{i,j}^x$ of the metric $ds^2$ relative to a chart $x$ may be computed as $g_{i,j}^x(m) = \langle \phi_{;i}(x(m)), \phi_{;j}(x(m)) \rangle$, where $\{e_i\}_{i=1}^{d}$ is the standard basis for $\mathbb{R}^d$,

$$\phi := x^{-1} \text{ and } \phi_{;i}(a) := \frac{d}{dt} |_0 \phi(a + te_i).$$

**Example 3.3.** Let $M = G := SL(n, \mathbb{R})$ and $A_g \in T_g M$.

1. Then

$$ds^2(A_g) := \text{tr}(A^*A) \tag{3.4}$$

   defines a Riemannian metric on $G$; this metric is the inherited metric from the inner product space $E = gl(n, \mathbb{R})$ with inner product $\langle A, B \rangle := \text{tr}(A^*B)$.
2. A more "natural" choice of a metric on $G$ is

$$ds^2(A_g) := \text{tr}((g^{-1}A)^* g^{-1}A). \tag{3.5}$$

This metric is invariant under left translations, i.e., $ds^2(L_{k*}A_g) = ds^2(A_g)$, for all $k \in G$ and $A_g \in TG$. According to the imbedding theorem of Nash and Moser, it would be possible to find another imbedding of $G$ into a Euclidean space, $E$, so that the metric in (3.5) is inherited from an inner product on $E$.

**Example 3.4.** Let $M = \mathbb{R}^3$ be equipped with the standard Riemannian metric and $(r, \varphi, \theta)$ be spherical coordinates on $M$ (see Figure 7). Here $r$, $\varphi$, and $\theta$ are taken

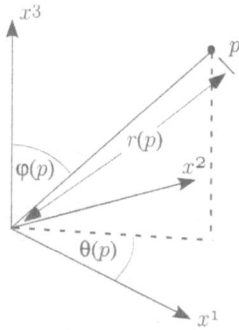

**Figure 7.** Defining the spherical coordinates, $(r, \theta, \phi)$ on $\mathbb{R}^3$.

to be functions on $\mathbb{R}^3 \setminus \{p \in \mathbb{R}^3 : p_2 = 0 \text{ and } p_1 > 0\}$ defined by $r(p) = |p|$, $\varphi(p) = \cos^{-1}(p_3/|p|) \in (0, \pi)$, and $\theta(p) \in (0, 2\pi)$ is given by $\theta(p) = \tan^{-1}(p_2/p_1)$ if $p_1 > 0$ and $p_2 > 0$ with similar formulas for $(p_1, p_2)$ in the other three quadrants of $\mathbb{R}^2$. Since $x^1 = r \sin \varphi \cos \theta$, $x^2 = r \sin \varphi \sin \theta$, and $x^3 = r \cos \varphi$, it follows using (2.11) that,

$$dx^1 = \frac{\partial x^1}{\partial r} dr + \frac{\partial x^1}{\partial \varphi} d\varphi + \frac{\partial x^1}{\partial \theta} d\theta$$

$$= \sin \varphi \cos \theta dr + r \cos \varphi \cos \theta d\varphi - r \sin \varphi \sin \theta d\theta,$$

$$dx^2 = \sin \varphi \sin \theta dr + r \cos \varphi \sin \theta d\varphi + r \sin \varphi \cos \theta d\theta,$$

and

$$dx^3 = \cos \varphi dr - r \sin \varphi d\varphi.$$

An elementary calculation now shows that

$$ds^2 = \sum_{i=1}^{3} (dx^i)^2 = dr^2 + r^2 d\varphi^2 + r^2 \sin^2 \varphi d\theta^2. \tag{3.6}$$

From this last equation, we see that

$$g^{(r,\varphi,\theta)} = \begin{bmatrix} 1 & 0 & 0 \\ 0 & r^2 & 0 \\ 0 & 0 & r^2 \sin^2 \varphi \end{bmatrix}. \tag{3.7}$$

**Exercise 3.5.** Let $M := \{m \in \mathbb{R}^3 : |m|^2 = \rho^2\}$, so that $M$ is a sphere of radius $\rho$ in $\mathbb{R}^3$. Since $r = \rho$ on $M$ and $dr\,(v) = 0$ for all $v \in T_m M$, it follows from (3.6) that the induced metric $ds^2$ on $M$ is given by

$$ds^2 = \rho^2 d\varphi^2 + \rho^2 \sin^2 \varphi d\theta^2, \tag{3.8}$$

and hence

$$g^{(\varphi,\theta)} = \begin{bmatrix} \rho^2 & 0 \\ 0 & \rho^2 \sin^2 \varphi \end{bmatrix}. \tag{3.9}$$

### 3.2 Integration and the volume measure

**Definition 3.6.** Let $f \in C_c^\infty(M)$ (the smooth functions on $M^d$ with compact support) and assume the support of $f$ is contained in $\mathcal{D}(x)$, where $x$ is some chart on $M$. Set

$$\int_M f dx = \int_{\mathcal{R}(x)} f \circ x^{-1}(a) da,$$

where $da$ denotes Lebesgue measure on $\mathbb{R}^d$.

The problem with this notion of integration is that (as the notation indicates) $\int_M f dx$ depends on the choice of chart $x$. To remedy this, consider a small cube $C(\delta)$ of side $\delta$ contained in $\mathcal{R}(x)$, see Figure 8. We wish to estimate "the volume" of $\phi(C(\delta))$ where $\phi := x^{-1} : \mathcal{R}(x) \to \mathcal{D}(x)$. Heuristically, we expect the volume of $\phi(C(\delta))$ to be approximately equal to the volume of the parallelepiped, $\tilde{C}(\delta)$, in the tangent space $T_m M$ determined by

$$\tilde{C}(\delta) := \left\{ \sum_{i=1}^{d} s_i \delta \cdot \phi_{;i}(x\,(m)) | 0 \le s_i \le 1, \text{ for } i = 1, 2, \ldots, d \right\}, \tag{3.10}$$

where we are using the notation proceeding Example 3.3, see Figure 8.

Since $T_m M$ is an inner product space, the volume of $\tilde{C}(\delta)$ is well defined. For example choose an isometry $\theta : T_m M \to \mathbb{R}^d$ and define the volume of $\tilde{C}(\delta)$ to be $m\left(\theta(\tilde{C}(\delta))\right)$ where $m$ is Lebesgue measure on $\mathbb{R}^d$. The next elementary lemma will be used to give a formula for the volume of $\tilde{C}\,(\delta)$.

**Lemma 3.7.** If $V$ is a finite dimensional inner product space, $\{v_i\}_{i=1}^{\dim V}$ is any basis for $V$ and $A : V \to V$ is a linear transformation, then

$$\det(A) = \frac{\det\left[\langle Av_i, v_j \rangle\right]}{\det\left[\langle v_i, v_j \rangle\right]}, \tag{3.11}$$

where $\det\left[\langle Av_i, v_j \rangle\right]$ is the determinant of the matrix with $i$-$j^{th}$-entry being $\langle Av_i, v_j \rangle$. Moreover if

$$\tilde{C}(\delta) := \left\{ \sum_{i=1}^{d} \delta s_i \cdot v_i : 0 \le s_i \le 1, \text{ for } i = 1, 2, \ldots, d \right\}$$

then the volume of $\tilde{C}\,(\delta)$ is $\delta^d \sqrt{\det\left[\langle v_i, v_j \rangle\right]}$.

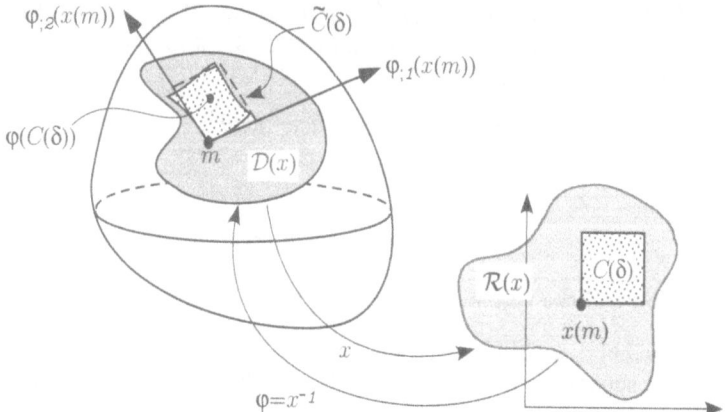

**Figure 8.** Defining the Riemannian "volume element."

*Proof.* Let $\{e_i\}_{i=1}^{\dim V}$ be an orthonormal basis for $V$, then

$$\langle Av_i, v_j \rangle = \sum_{l,k} \langle v_i, e_l \rangle \langle Ae_l, e_k \rangle \langle e_k, v_j \rangle$$

and therefore by the multiplicative property of the determinant,

$$\det\left[\langle Av_i, v_j \rangle\right] = \det\left[\langle v_i, e_l \rangle\right] \det\left[\langle Ae_l, e_k \rangle\right] \det\left[\langle e_k, v_j \rangle\right]$$
$$= \det(A) \det\left[\langle v_i, e_l \rangle\right] \cdot \det\left[\langle e_k, v_j \rangle\right]. \tag{3.12}$$

Taking $A = I$ in this equation then shows that the

$$\det\left[\langle v_i, v_j \rangle\right] = \det\left[\langle v_i, e_l \rangle\right] \cdot \det\left[\langle e_k, v_j \rangle\right]. \tag{3.13}$$

Dividing (3.13) into (3.12) proves (3.11).

For the second assertion, it suffices to assume $V = \mathbb{R}^d$ with the usual inner-product. Define $T : \mathbb{R}^d \to \mathbb{R}^d$ so that $T e_i = v_i$ where $\{e_i\}_{i=1}^d$ is the standard basis for $\mathbb{R}^d$, then $\tilde{C}(\delta) = T\left([0, \delta]^d\right)$ and hence

$$m\left(\tilde{C}(\delta)\right) = |\det T| \, m\left([0, \delta]^d\right) = \delta^d \, |\det T| = \delta^d \sqrt{\det(T^{\mathrm{tr}} T)}$$
$$= \delta^d \sqrt{\det\left[\langle T^{\mathrm{tr}} T e_i, e_j \rangle\right]} = \delta^d \sqrt{\det\left[\langle T e_i, T e_j \rangle\right]} = \delta^d \sqrt{\det\left[\langle v_i, v_j \rangle\right]}.$$

Using the second assertion in Lemma 3.7, the volume of $\tilde{C}(\delta)$ in Eq. (3.10) is $\delta^d \sqrt{\det g^x(m)}$, where $g_{ij}^x(m) = \langle \phi_{;i}(x(m)), \phi_{;j}(x(m)) \rangle_m$. Because of the above computations, it is reasonable to try to define a new integral on $\mathcal{D}(x) \subset M$ by

$$\int_{\mathcal{D}(x)} f \, d\lambda_{\mathcal{D}(x)} := \int_{\mathcal{D}(x)} f \sqrt{g^x} dx,$$

letting $\lambda_{\mathcal{D}(x)}$ be the measure satisfying

$$d\lambda_{\mathcal{D}(x)} = \sqrt{g^x} dx \tag{3.14}$$

where $\sqrt{g^x}$ is shorthand for $\sqrt{\det g^x}$.

**Lemma 3.8.** *Suppose that $y$ and $x$ are two charts on $M$, then*

$$g_{l,k}^y = \sum_{i,j=1}^{d} g_{i,j}^x \frac{\partial x^i}{\partial y^k} \frac{\partial x^j}{\partial y^l}. \tag{3.15}$$

*Proof.* Inserting the identities

$$dx^l = \sum_{k=1}^{d} \frac{\partial x^i}{\partial y^k} dy^k \text{ and } dx^j = \sum_{l=1}^{d} \frac{\partial x^j}{\partial y^l} dy^l$$

and into the formula $ds^2 = \sum_{i,j=1}^{d} g_{i,j}^x dx^i dx^j$ gives

$$ds^2 = \sum_{i,j,k,l=1}^{d} g_{i,j}^x \frac{\partial x^i}{\partial y^k} \frac{\partial x^j}{\partial y^l} dy^l dy^k$$

from which (3.15) follows.                                                     □

**Exercise 3.9.** Suppose that $x$ and $y$ are two charts on $M$ and $f \in C_c^\infty(M)$ such that the support of $f$ is contained in $\mathcal{D}(x) \cap \mathcal{D}(y)$. Using Lemma 3.8 and the change of variable formula show,

$$\int_{\mathcal{D}(x) \cap \mathcal{D}(y)} f \sqrt{g^x} dx = \int_{\mathcal{D}(x) \cap \mathcal{D}(y)} f \sqrt{g^y} dy.$$

**Theorem 3.10 (Riemann Volume Measure).** *There exists a unique measure, $\lambda_M$ on the Borel $\sigma$-algebra of $M$ such that for any chart $x$ on $M$,*

$$d\lambda_M(x) = d\lambda_{\mathcal{D}(x)} = \sqrt{g^x} dx \text{ on } \mathcal{D}(x). \tag{3.16}$$

*Proof.* Choose a countable collection of charts, $\{x_i\}_{i=1}^\infty$ such that $M = \cup_{i=1}^\infty \mathcal{D}(x_i)$ and let $U_1 := \mathcal{D}(x_1)$ and $U_i := \mathcal{D}(x_i) \setminus (\cup_{j=1}^{i-1} \mathcal{D}(x_j))$ for $i \geq 1$. Then if $B \subset X$ is a Borel set, define the measure $\lambda_M(B)$ by

$$\lambda_M(B) := \sum_{i=1}^\infty \lambda_{\mathcal{D}(x_i)}(B \cap U_i). \tag{3.17}$$

If $x$ is any chart on $M$ and $B \subset \mathcal{D}(x)$, then $B \cap U_i \subset \mathcal{D}(x_i) \cap \mathcal{D}(x)$ and so by Exercise 3.9, $\lambda_{\mathcal{D}(x_i)}(B \cap U_i) = \lambda_{\mathcal{D}(x)}(B)$. Using this identity in (3.17) implies

$$\lambda_M(B) := \sum_{i=1}^{\infty} \lambda_{\mathcal{D}(x)}(B \cap U_i) = \lambda_{\mathcal{D}(x)}(B)$$

and hence we have proved the existence of $\lambda_M$. The uniqueness assertion is easy and will be left to the reader. $\qquad\square$

**Example 3.11.** Let $M = \mathbb{R}^3$ with the standard Riemannian metric, and let $x$ denote the standard coordinates on $M$ determined by $x(m) = m$ for all $m \in M$. Then $\lambda_{\mathbb{R}^3}$ is Lebesgue measure which in spherical coordinates may be written as

$$d\lambda_{\mathbb{R}^3} = r^2 \sin \varphi dr d\varphi d\theta$$

because $\sqrt{g^{(r,\varphi,\theta)}} = r^2 \sin \varphi$ by (3.7). Similarly using (3.9),

$$d\lambda_M = \rho^2 \sin \varphi d\varphi d\theta$$

when $M \subset \mathbb{R}^3$ is the sphere of radius $\rho$ centered at $0 \in \mathbb{R}^3$.

**Exercise 3.12.** Compute the "volume element," $d\lambda_{\mathbb{R}^3}$, for $\mathbb{R}^3$ in cylindrical coordinates.

**Theorem 3.13 (Change of Variables Formula).** *Let $(M, \langle \cdot, \cdot \rangle_M)$ and $(N, \langle \cdot, \cdot \rangle_N)$ be two Riemannian manifolds, $\psi : M \to N$ be a diffeomorphism and $\rho \in C^{\infty}(M, (0, \infty))$ be determined by the equation*

$$\rho(m) = \sqrt{\det\left[\psi_{*m}^{\mathrm{tr}} \psi_{*m}\right]} \text{ for all } m \in M,$$

*where $\psi_{*m}^{\mathrm{tr}}$ denotes the adjoint of $\psi_{*m}$ relative to Riemannian inner products on $T_m M$ and $T_{\psi(m)} N$. If $f : N \to \mathbb{R}_+$ is a positive Borel measurable function, then*

$$\int_N f d\lambda_N = \int_M \rho \cdot (f \circ \psi) d\lambda_M.$$

*In particular if $\psi$ is an isometry, i.e., $\psi_{*m} : T_m M \to T_{\psi(m)} N$ is orthogonal for all $m$, then*

$$\int_N f d\lambda_N = \int_M f \circ \psi \, d\lambda_M.$$

*Proof.* By a partition of unity argument (see the proof of Theorem 3.10), it suffices to consider the case where $f$ has "small" support, i.e., we may assume that the support of $f \circ \psi$ is contained in $\mathcal{D}(x)$ for some chart $x$ on $M$. Letting $\phi := x^{-1}$, by (3.11) of Lemma 3.7,

$$\frac{\det\left[\langle \partial_i (\psi \circ \phi)(t), \partial_j (\psi \circ \phi)(t) \rangle_N\right]}{\det\left[\langle \partial_i \phi(t), \partial_j \phi(t) \rangle_M\right]}$$

$$= \frac{\det\left[\langle \psi_* \partial_i \phi(t), \psi_* \partial_j \phi(t) \rangle_N\right]}{\det\left[\langle \partial_i \phi(t), \partial_j \phi(t) \rangle_M\right]} = \frac{\det\left[\langle \psi_*^{\mathrm{tr}} \psi_* \partial_i \phi(t), \partial_j \phi(t) \rangle_M\right]}{\det\left[\langle \partial_i \phi(t), \partial_j \phi(t) \rangle_M\right]}$$

$$= \det\left[\psi_{*\phi(t)}^{\mathrm{tr}} \psi_{*\phi(t)}\right] = \rho^2(\phi(t)).$$

This implies

$$\int_N f d\lambda_N = \int_{\mathcal{R}(x)} f \circ (\psi \circ \phi) (t) \sqrt{\det \left[ \langle \partial_i (\psi \circ \phi) (t), \partial_j (\psi \circ \phi) (t) \rangle_N \right]} dt$$

$$= \int_{\mathcal{R}(x)} (f \circ \psi) \circ \phi(t) \cdot \rho (\phi (t)) \sqrt{\det \left[ \langle \partial_i \phi (t), \partial_j \phi (t) \rangle_M \right]} dt$$

$$= \int_{\mathcal{D}(x)} (f \circ \psi) \cdot \rho \cdot \sqrt{g^x} dx = \int_M \rho \cdot f \circ \psi \, d\lambda_M.$$

**Example 3.14.** Let $M = SL(n, \mathbb{R})$ as in Example 3.3 and let $\langle \cdot, \cdot \rangle_M$ be the metric given by (3.5). Because $L_g : M \to M$ is an isometry, Theorem 3.13 implies

$$\int_{SL(n,\mathbb{R})} f (gx) d\lambda_G (x) = \int_{SL(n,\mathbb{R})} f (x) d\lambda_G (x) \text{ for all } g \in G.$$

Thus $\lambda_G$ is invariant under left translations by elements of $G$ and such an invariant left invariant measure is called a *"left Haar"* measure on $G$.

Similarly if $G = O(n)$ with Riemannian metric determined by (3.5), then, since $g \in G$ is orthogonal, we have

$$ds^2 (A_g) := \mathrm{tr}((g^{-1}A)^* g^{-1}A) = \mathrm{tr}((g^*A)^* g^{-1}A) = \mathrm{tr}(A^* gg^{-1}A) = \mathrm{tr}(A^*A)$$

and

$$\mathrm{tr}((Ag^{-1})^* Ag^{-1}) = \mathrm{tr}(gA^* Ag^{-1}) = \mathrm{tr}(A^* Ag^{-1}g) = \mathrm{tr}(A^*A).$$

Therefore, both left and right translations by element $g \in G$ are isometries for this Riemannian metric on $O(m)$ and so by Theorem 3.13,

$$\int_{O(n)} f (gx) d\lambda_G (x) = \int_{O(n)} f (x) d\lambda_G (x) = \int_{O(n)} f (xg) d\lambda_G (x)$$

for all $g \in G$.

## 3.3 Gradients, divergence, and Laplacians

In the sequel, let $M$ be a Riemannian manifold, $x$ be a chart on $M$, $g_{ij} := \langle \partial/\partial x^i, \partial/\partial x^j \rangle$, and $ds^2 = \sum_{i,j=1}^d g_{ij} dx^i dx^j$.

**Definition 3.15.** Let $g^{ij}$ denote the $i$-$j^{\text{th}}$ – matrix element for the inverse matrix to the matrix, $(g_{ij})$.

Given $f \in C^\infty(M)$ and $m \in M$, $df_m := df|_{T_m M}$ is a linear functional on $T_m M$. Hence there is a unique vector $v_m \in T_m M$ such that $df_m = \langle v_m, \cdot \rangle_m$.

**Definition 3.16.** The vector $v_m$ above is called the *gradient* of $f$ at $m$ and will be denoted by either grad $f(m)$ or $\nabla f(m)$.

**Exercise 3.17.** If $x$ is a chart on $M$ and $m \in \mathcal{D}(x)$ then

$$\nabla f(m) = \operatorname{grad} f(m) = \sum_{i,j=1}^{d} g^{ij}(m) \frac{\partial f(m)}{\partial x^i} \frac{\partial}{\partial x^j} \big|_m, \qquad (3.18)$$

where as usual, $g_{ij} = g_{ij}^x$ and $g^{ij} = (g_{ij})^{-1}$. Notice from Eq. (3.18) that $\nabla f$ is a smooth vector field on $M$.

**Exercise 3.18.** Suppose $M \subset \mathbb{R}^N$ is an imbedded submanifold with the induced Riemannian structure. Let $F : \mathbb{R}^N \to \mathbb{R}$ be a smooth function and set $f := F|_M$. Then $\operatorname{grad} f(m) = (P(m)\nabla F(m))_m$, where $\nabla F(m)$ denotes the usual gradient on $\mathbb{R}^N$, and $P(m)$ denotes orthogonal projection of $\mathbb{R}^N$ onto $\tau_m M$.

We now introduce the divergence of a vector field $Y$ on $M$.

**Lemma 3.19 (Divergence).** *To every smooth vector field $Y$ on $M$ there is a unique smooth function, $\nabla \cdot Y = \operatorname{div} Y$, on $M$ such that*

$$\int_M Yf \, d\lambda_M = -\int_M \operatorname{div} Y \cdot f \, d\lambda_M, \qquad \forall f \in C_c^\infty(M). \qquad (3.19)$$

*(The function, $\nabla \cdot Y = \operatorname{div} Y$, is called the divergence of $Y$.) Moreover if $x$ is a chart on $M$, then on its domain, $\mathcal{D}(x)$,*

$$\nabla \cdot Y = \operatorname{div} Y = \sum_{i=1}^{d} \frac{1}{\sqrt{g}} \frac{\partial(\sqrt{g}Y^i)}{\partial x^i} = \sum_{i=1}^{d} \left\{ \frac{\partial Y^i}{\partial x^i} + \frac{\partial \log \sqrt{g}}{\partial x^i} Y^i \right\} \qquad (3.20)$$

*where $Y^i := dx^i(Y)$ and $\sqrt{g} = \sqrt{g^x} = \sqrt{(\det(g_{ij}^x))}$.*

*Proof.* (Sketch) Suppose that $f \in C_c^\infty(M)$ such that the support of $f$ is contained in $\mathcal{D}(x)$. Because $Yf = \sum_{i=1}^{d} Y^i \partial f / \partial x^i$,

$$\int_M Yf \, d\lambda_M = \int_M \sum_{i=1}^{d} Y^i \partial f / \partial x^i \cdot \sqrt{g} dx = -\int_M \sum_{i=1}^{d} f \frac{\partial(\sqrt{g}\,Y^i)}{\partial x^i} dx$$

$$= -\int_M f \sum_{i=1}^{d} \frac{1}{\sqrt{g}} \frac{\partial(\sqrt{g}Y^i)}{\partial x^i} d\lambda_M,$$

where the second equality follows from an integration by parts. This shows that if $\operatorname{div} Y$ exists it must be given on $\mathcal{D}(x)$ by (3.20). This proves the uniqueness assertion. Using what we have already proven, it is easy to conclude that the formula for $\operatorname{div} Y$ is chart independent. Hence, we may define a smooth function $\operatorname{div} Y$ on $M$ using (3.20) in each coordinate chart $x$ on $M$. It is then possible to show (again using a smooth partition of unity argument) that this function satisfies (3.19). $\square$

**Remark 3.20.** We may write (3.19) as

$$\int_M \langle Y, \text{grad } f \rangle \, d\lambda_M = - \int_M \text{div} Y \cdot f \, d\lambda_M, \quad \forall f \in C_c^\infty(M), \tag{3.21}$$

so that "div" is the negative of the formal adjoint of "grad."

**Exercise 3.21 (Product Rule).** If $f \in C^\infty(M)$ and $Y \in \Gamma(TM)$ then

$$\nabla \cdot (fY) = \langle \nabla f, Y \rangle + f \, \nabla \cdot Y.$$

**Lemma 3.22 (Integration by Parts).** *Suppose that* $Y \in \Gamma(TM)$, $f \in C_c^\infty(M)$, *and* $h \in C^\infty(M)$, *then*

$$\int_M Yf \cdot h \, d\lambda_M = \int_M f\{-Yh - h \cdot \text{div} Y\} \, d\lambda_M.$$

*Proof.* By the definition of div $Y$ and the product rule,

$$\int_M fh \, \text{div} Y \, d\lambda_M = - \int_M Y(fh) \, d\lambda_M = - \int_M \{hYf + fYh\} \, d\lambda_M.$$

**Definition 3.23.** The *Laplacian* on $M$ is the second order differential operator, $\Delta : C^\infty(M) \to C^\infty(M)$, defined by

$$\Delta f := \text{div}(\text{grad } f) = \nabla \cdot \nabla f. \tag{3.22}$$

In local coordinates,

$$\Delta f = \frac{1}{\sqrt{g}} \sum_{i,j=1}^d \partial_i \{ \sqrt{g} g^{ij} \partial_j f \}, \tag{3.23}$$

where $\partial_i = \partial/\partial x^i$, $g = g^x$, $\sqrt{g} = \sqrt{\det g}$, and $(g^{ij}) = (g^x_{ij})^{-1}$.

**Remark 3.24.** The Laplacian, $\Delta f$, may be characterized by the equation:

$$\int_M \Delta f \cdot h \, d\lambda_M = - \int_M \langle \nabla f, \nabla h \rangle \, d\lambda_M,$$

which is to hold for all $f \in C^\infty(M)$ and $h \in C_c^\infty(M)$.

**Example 3.25.** Suppose that $M = \mathbb{R}^N$ with the standard Riemannian metric $ds^2 = \sum_{i=1}^N (dx^i)^2$, then the standard formulas:

$$\text{grad } f = \sum_{i=1}^N \partial f/\partial x^i \cdot \partial/\partial x^i, \quad \text{div} Y = \sum_{i=1}^N \partial Y^i/\partial x^i \text{ and } \Delta f = \sum_{i=1}^N \frac{\partial^2 f}{(\partial x^i)^2}$$

are easily verified, where $f$ is a smooth function on $\mathbb{R}^N$ and $Y = \sum_{i=1}^N Y^i \partial/\partial x^i$ is a smooth vector field.

**Exercise 3.26.** Let $M = \mathbb{R}^3$, $(r, \varphi, \theta)$ be spherical coordinates on $\mathbb{R}^3$, $\partial_r = \partial/\partial r$, $\partial_\varphi = \partial/\partial\varphi$, and $\partial_\theta = \partial/\partial\theta$. Given a smooth function $f$ and a vector field $Y = Y_r \partial_r + Y_\varphi \partial_\varphi + Y_\theta \partial_\theta$ on $\mathbb{R}^3$ verify:

$$\text{grad } f = (\partial_r f)\partial_r + \frac{1}{r^2}(\partial_\varphi f)\partial_\varphi + \frac{1}{r^2 \sin^2 \varphi}(\partial_\theta f)\partial_\theta,$$

$$\text{div} Y = \frac{1}{r^2 \sin \varphi}\{\partial_r(r^2 \sin \varphi Y_r) + \partial_\varphi(r^2 \sin \varphi Y_\varphi) + r^2 \sin \varphi \partial_\theta Y_\theta\}$$

$$= \frac{1}{r^2}\partial_r(r^2 Y_r) + \frac{1}{\sin \varphi}\partial_\varphi(\sin \varphi Y_\varphi) + \partial_\theta Y_\theta,$$

and

$$\Delta f = \frac{1}{r^2}\partial_r(r^2 \partial_r f) + \frac{1}{r^2 \sin \varphi}\partial_\varphi(\sin \varphi \partial_\varphi f) + \frac{1}{r^2 \sin^2 \varphi}\partial_\theta^2 f.$$

**Example 3.27.** Let $M = G = O(n)$ with Riemannian metric determined by (3.5) and for $A \in \mathfrak{g} := T_e G$ let $\tilde{A} \in \Gamma(TG)$ be the left invariant vector field,

$$\tilde{A}(x) := L_{x*}A = \frac{d}{dt}\Big|_0 x e^{tA}$$

as was done for $SL(n, \mathbb{R})$ in Example 2.34. Using the invariance of $d\lambda_G$ under right translations established in Example 3.14, we find for $f, h \in C^1(G)$ that

$$\int_G \tilde{A}f(x) \cdot h(x) d\lambda_G(x) = \int_G \frac{d}{dt}\Big|_0 f\left(xe^{tA}\right) \cdot h(x) d\lambda_G(x)$$

$$= \frac{d}{dt}\Big|_0 \int_G f\left(xe^{tA}\right) \cdot h(x) d\lambda_G(x)$$

$$= \frac{d}{dt}\Big|_0 \int_G f(x) \cdot h\left(xe^{-tA}\right) d\lambda_G(x)$$

$$= \int_G f(x) \cdot \frac{d}{dt}\Big|_0 h\left(xe^{-tA}\right) d\lambda_G(x)$$

$$= -\int_G f(x) \cdot \tilde{A}h(x) d\lambda_G(x).$$

Taking $h \equiv 1$ implies

$$0 = \int_G \tilde{A}f(x) d\lambda_G(x) = \int_G \left\langle \tilde{A}(x), \nabla f(x)\right\rangle d\lambda_G(x)$$

$$= -\int_G \nabla \cdot \tilde{A}(x) \cdot f(x) d\lambda_G(x)$$

from which we learn $\nabla \cdot \tilde{A} = 0$.

Now letting $S_0 \subset \mathfrak{g}$ be an orthonormal basis for $\mathfrak{g}$, because $L_{g*}$ is an isometry, $\{\tilde{A}(g) : A \in S_0\}$ is an orthonormal basis for $T_g G$ for all $g \in G$. Hence

$$\nabla f\,(g) = \sum_{A \in S_0} \left\langle \nabla f\,(g), \tilde{A}\,(g) \right\rangle \tilde{A}\,(g) = \sum_{A \in S_0} \left( \tilde{A} f \right) (g)\,\tilde{A}\,(g)\,.$$

and, by the product rule and $\nabla \cdot \tilde{A} = 0$,

$$\Delta f = \nabla \cdot \nabla f = \sum_{A \in S_0} \nabla \cdot \left[ \left( \tilde{A} f \right) \tilde{A} \right] = \sum_{A \in S_0} \left\langle \nabla \tilde{A} f, \tilde{A} \right\rangle = \sum_{A \in S_0} \tilde{A}^2 f.$$

## 3.4 Covariant derivatives and curvature

**Definition 3.28.** We say a smooth path $s \to V(s)$ in $TM$ is a *vector field along a smooth path* $s \to \sigma(s)$ in $M$ if $\pi \circ V(s) = \sigma(s)$, i.e. $V(s) \in T_{\sigma(s)}M$ for all $s$. (Recall that $\pi$ is the canonical projection defined in Definition 2.16.)

Note: if $V$ is a smooth path in $TM$ then $V$ is a vector field along $\sigma := \pi \circ V$. This section is motivated by the desire to have the notion of the derivative of a smooth path $V(s) \in TM$. On one hand, since $TM$ is a manifold, we may write $V'(s)$ as an element of $TTM$. However, this is not what we will want for later purposes. We would like the derivative of $V$ to again be a path back in $TM$, not in $TTM$. In order to define such a derivative, we will need to use more than just the manifold structure of $M$, see Definition 3.31 below.

**Notation 3.29.** In the sequel, we assume that $M^d$ is an imbedded submanifold of an inner product space $(E = \mathbb{R}^N, \langle \cdot, \cdot \rangle)$, and that $M$ is equipped with the inherited Riemannian metric. Also let $P(m)$ denote orthogonal projection of $E$ onto $\tau_m M$ for all $m \in M$ and $Q(m) := I - P(m)$ be orthogonal projection onto $(\tau_m M)^{\perp}$.

The following elementary lemma will be used throughout the sequel.

**Lemma 3.30.** *The differentials of the orthogonal projection operators, $P$ and $Q$, satisfy*

$$0 = dP + dQ,$$
$$P dQ = -dPQ = dQQ \text{ and}$$
$$Q dP = -dQP = dPP.$$

*In particular,*
$$Q dPQ = Q dQQ = P dPP = P dQP = 0.$$

*Proof.* The first equality comes from differentiating the identity, $I = P + Q$, the second from differentiating $0 = PQ$ and the third from differentiating $0 = QP$.     □

**Definition 3.31 (Levi-Civita Covariant Derivative).** Let $V(s) = (\sigma(s), v(s)) = v(s)_{\sigma(s)}$ be a smooth path in $TM$ (see Figure 9), then the *covariant derivative,* $\nabla V(s)/ds$, is the vector field along $\sigma$ defined by

$$\frac{\nabla V(s)}{ds} := (\sigma(s), P(\sigma(s)) \frac{d}{ds} v(s)). \tag{3.24}$$

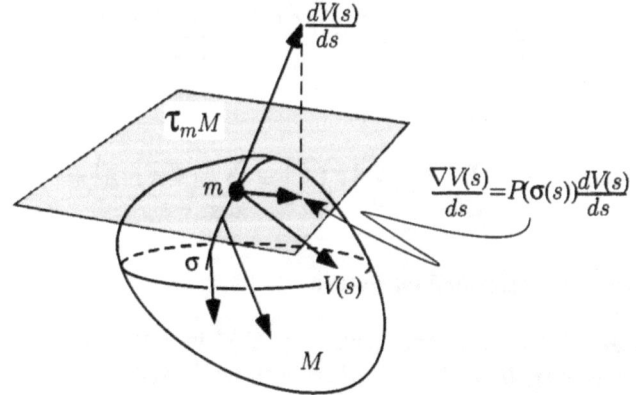

**Figure 9.** The Levi-Civita covariant derivative.

**Proposition 3.32 (Properties of $\nabla/ds$).** *Let $W(s) = (\sigma(s), w(s))$ and $V(s) = (\sigma(s), v(s))$ be two smooth vector fields along a path $\sigma$ in $M$. Then:*

*1. $\nabla W(s)/ds$ may be computed as:*

$$\frac{\nabla W(s)}{ds} := (\sigma(s), \frac{d}{ds}w(s) + (dQ(\sigma'(s)))w(s)). \tag{3.25}$$

*2. $\nabla$ is metric compatible, i.e.,*

$$\frac{d}{ds}\langle W(s), V(s)\rangle = \langle \frac{\nabla W(s)}{ds}, V(s)\rangle + \langle W(s), \frac{\nabla V(s)}{ds}\rangle. \tag{3.26}$$

*Now suppose that $(s,t) \to \sigma(s,t)$ is a smooth function into $M$, $W(s,t) = (\sigma(s,t), w(s,t))$ is a smooth function into $TM$, $\sigma'(s,t) := (\sigma(s,t), \frac{d}{ds}\sigma(s,t))$ and $\dot\sigma(s,t) = (\sigma(s,t), \frac{d}{dt}\sigma(s,t))$. (Notice by assumption that $w(s,t) \in T_{\sigma(s,t)}M$ for all $(s,t)$.)*

*3. $\nabla$ has zero torsion, i.e.,*

$$\frac{\nabla\sigma'}{dt} = \frac{\nabla\dot\sigma}{ds}. \tag{3.27}$$

*4. If $R$ is the curvature tensor of $\nabla$ defined by*

$$R(u_m, v_m)w_m = (m, [dQ(u_m), dQ(v_m)]w), \tag{3.28}$$

*then*

$$\left[\frac{\nabla}{dt}, \frac{\nabla}{ds}\right]W := (\frac{\nabla}{dt}\frac{\nabla}{ds} - \frac{\nabla}{ds}\frac{\nabla}{dt})W = R(\dot\sigma, \sigma')W. \tag{3.29}$$

*Proof.* Differentiate the identity, $P(\sigma(s))w(s) = w(s)$, relative to $s$ implies

$$(dP(\sigma'(s)))w(s) + P(\sigma(s))\frac{d}{ds}w(s) = \frac{d}{ds}w(s)$$

from which (3.25) follows.

For (3.26) just compute:

$$
\begin{aligned}
\frac{d}{ds}\langle W(s), V(s)\rangle &= \frac{d}{ds}\langle w(s), v(s)\rangle \\
&= \left\langle \frac{d}{ds}w(s), v(s)\right\rangle + \left\langle w(s), \frac{d}{ds}v(s)\right\rangle \\
&= \left\langle \frac{d}{ds}w(s), P(\sigma(s))v(s)\right\rangle + \left\langle P(\sigma(s))w(s), \frac{d}{ds}v(s)\right\rangle \\
&= \left\langle P(\sigma(s))\frac{d}{ds}w(s), v(s)\right\rangle + \left\langle w(s), P(\sigma(s))\frac{d}{ds}v(s)\right\rangle \\
&= \left\langle \frac{\nabla W(s)}{ds}, V(s)\right\rangle + \left\langle W(s), \frac{\nabla V(s)}{ds}\right\rangle,
\end{aligned}
$$

where the third equality relies on $v(s)$ and $w(s)$ being in $\tau_{\sigma(s)}M$ and the fourth equality relies on $P(\sigma(s))$ being an orthogonal projection.

From the definitions of $\sigma'$, $\dot{\sigma}$, $\nabla/dt$, $\nabla/ds$ and the fact that mixed partial derivatives commute,

$$
\begin{aligned}
\frac{\nabla\sigma'(s,t)}{dt} &= \frac{\nabla}{dt}(\sigma(t,s), \sigma'(s,t)) = (\sigma(t,s), P(\sigma(s,t))\frac{d}{dt}\frac{d}{ds}\sigma(t,s)) \\
&= (\sigma(t,s), P(\sigma(s,t))\frac{d}{ds}\frac{d}{dt}\sigma(t,s)) = \nabla\dot{\sigma}(s,t)/ds,
\end{aligned}
$$

which proves (3.27).

For (3.29) we observe,

$$
\begin{aligned}
\frac{\nabla}{dt}\frac{\nabla}{ds}W(s,t) &= \frac{\nabla}{dt}(\sigma(s,t), \frac{d}{ds}w(s,t) + dQ(\sigma'(s,t))w(s,t)) \\
&= (\sigma(s,t), \eta_+(s,t))
\end{aligned}
$$

where (with the arguments $(s,t)$ suppressed from the notation)

$$
\begin{aligned}
\eta_+ &= \frac{d}{dt}\left[\frac{d}{ds}w + dQ(\sigma')w\right] + dQ(\dot{\sigma})\left[\frac{d}{ds}w + dQ(\sigma')w\right] \\
&= \frac{d}{dt}\frac{d}{ds}w + \left(\frac{d}{dt}[dQ(\sigma')]\right)w + dQ(\sigma')\frac{d}{dt}w + dQ(\dot{\sigma})\frac{d}{ds}w + dQ(\dot{\sigma})dQ(\sigma')w.
\end{aligned}
$$

Therefore

$$
\left[\frac{\nabla}{dt}, \frac{\nabla}{ds}\right]W = (\sigma, \eta_+ - \eta_-),
$$

where $\eta_-$ is defined the same as $\eta_+$ with all $s$ and $t$ derivatives interchanged. Hence, it follows (using again $\frac{d}{dt}\frac{d}{ds}w = \frac{d}{ds}\frac{d}{dt}w$) that

$$
\left[\frac{\nabla}{dt}, \frac{\nabla}{ds}\right]W = (\sigma, [\frac{d}{dt}(dQ(\sigma'))]w - [\frac{d}{ds}(dQ(\dot{\sigma}))]w + [dQ(\dot{\sigma}), dQ(\sigma')]w).
$$

The proof of (3.28) is finished because

$$\frac{d}{dt}(dQ(\sigma')) - \frac{d}{ds}(dQ(\dot{\sigma})) = \frac{d}{dt}\frac{d}{ds}(Q \circ \sigma) - \frac{d}{ds}\frac{d}{dt}(Q \circ \sigma) = 0.$$

**Example 3.33.** Let $M = \{m \in \mathbb{R}^N : |m| = \rho\}$ be the sphere of radius $\rho$. In this case $Q(m) = \frac{1}{\rho^2}mm^{\mathrm{tr}}$ for all $m \in M$. Therefore

$$dQ(v_m) = \frac{1}{\rho^2}\{vm^{\mathrm{tr}} + mv^{\mathrm{tr}}\} \ \forall \ v_m \in T_m M$$

and hence

$$dQ(u_m)dQ(v_m) = \frac{1}{\rho^4}\{um^{\mathrm{tr}} + mu^{\mathrm{tr}}\}\{vm^{\mathrm{tr}} + mv^{\mathrm{tr}}\}$$

$$= \frac{1}{\rho^4}\{\rho^2 uv^{\mathrm{tr}} + \langle u, v\rangle Q(m)\}.$$

So the curvature tensor is given by

$$R(u_m, v_m)w_m = (m, \frac{1}{\rho^2}\{uv^{\mathrm{tr}} - vu^{\mathrm{tr}}\}w) = (m, \frac{1}{\rho^2}\{\langle v, w\rangle u - \langle u, w\rangle v\}).$$

**Exercise 3.34.** Show the curvature tensor of the cylinder

$$M = \{(x, y, z) \in \mathbb{R}^3 : x^2 + y^2 = 1\}$$

is zero.

**Definition 3.35 (Covariant Derivative on $\Gamma(TM)$).** Suppose that $Y$ is a vector field on $M$ and $v_m \in T_m M$. Define $\nabla_{v_m} Y \in T_m M$ by

$$\nabla_{v_m} Y := \frac{\nabla Y(\sigma(s))}{ds}\Big|_{s=0},$$

where $\sigma$ is any smooth path in $M$ such that $\sigma'(0) = v_m$.

If $Y(m) = (m, y(m))$, then

$$\nabla_{v_m} Y = (m, P(m)dy(v_m)) = (m, dy(v_m) + dQ(v_m)y(m)),$$

from which it follows $\nabla_{v_m} Y$ is well defined, i.e. $\nabla_{v_m} Y$ is independent of the choice of $\sigma$ such that $\sigma'(0) = v_m$. The following proposition relates curvature and torsion to the covariant derivative $\nabla$ on vector fields.

**Proposition 3.36.** *Let $m \in M$, $v \in T_m M$, $X, Y, Z \in \Gamma(TM)$, and $f \in C^\infty(M)$, then the following relations hold.*

1. Product Rule $\nabla_v(f \cdot X) = df(v) \cdot X(m) + f(m) \cdot \nabla_v X.$
2. Zero Torsion $\nabla_X Y - \nabla_Y X - [X, Y] = 0.$
3. Zero Torsion *For all $v_m, w_m \in T_m M$, $dQ(v_m)w_m = dQ(w_m)v_m$.*

4. Curvature Tensor $R(X, Y)Z = [\nabla_X, \nabla_Y]Z - \nabla_{[X,Y]}Z$, where

$$[\nabla_X, \nabla_Y]Z := \nabla_X(\nabla_Y Z) - \nabla_Y(\nabla_X Z).$$

*Moreover if $u, v, w, z \in T_m M$, then $R$ has the following symmetries*

a: $R(u_m, v_m) = -R(v_m, u_m)$
b: $[R(u_m, v_m)]^{\text{tr}} = -R(u_m, v_m)$ *and*
c: *if $z_m \in \tau_m M$, then*

$$\langle R(u_m, v_m)w_m, z_m \rangle = \langle R(w_m, z_m)u_m, v_m \rangle. \tag{3.30}$$

5. Ricci Curvature Tensor *For each $m \in M$, let $\text{Ric}_m : T_m M \to T_m M$ be defined by*

$$\text{Ric}_m v_m := \sum_{a \in S} R(v_m, a)a, \tag{3.31}$$

*where $S \subset T_m M$ is an orthonormal basis. Then $\text{Ric}_m^{\text{tr}} = \text{Ric}_m$ and $\text{Ric}_m$ may be computed as*

$$\langle \text{Ric}_m u, v \rangle = \text{tr}(dQ(dQ(u)v) - dQ(v)dQ(u)) \text{ for all } u, v \in T_m M. \tag{3.32}$$

*Proof.* The product rule is easily checked and may be left to the reader. For the second and third items, write $X(m) = (m, x(m))$, $Y(m) = (m, y(m))$, and $Z(m) = (m, z(m))$ where $x, y, z : M \to \mathbb{R}^N$ are smooth functions such that $x(m)$, $y(m)$, and $z(m)$ are in $\tau_m M$ for all $m \in M$. Then using (2.15), we have

$$(\nabla_X Y - \nabla_Y X)(m) = (m, P(m)(dy(X(m)) - dx(Y(m))))$$
$$= (m, (dy(X(m)) - dx(Y(m)))) = [X, Y](m), \tag{3.33}$$

which proves the second item. Since $(\nabla_X Y)(m)$ may also be written as

$$(\nabla_X Y)(m) = (m, dy(X(m)) + dQ(X(m))y(m)),$$

(3.33) may be expressed as $dQ(X(m))y(m) = dQ(Y(m))x(m)$ which implies the third item.

Similarly for the fourth item:

$$\nabla_X \nabla_Y Z = \nabla_X(\cdot, Yz + (YQ)z)$$
$$= (\cdot, XYz + (XYQ)z + (YQ)Xz + (XQ)(Yz + (YQ)z)),$$

where $YQ := dQ(Y)$ and $Yz := dz(Y)$. Interchanging $X$ and $Y$ in this last expression and then subtracting gives:

$$[\nabla_X, \nabla_Y]Z = (\cdot, [X, Y]z + ([X, Y]Q)z + [XQ, YQ]z)$$
$$= \nabla_{[X,Y]}Z + R(X, Y)Z.$$

The anti-symmetry properties in items 4a) and 4b) follow easily from (3.28). For example for 4b), $dQ(u_m)$ and $dQ(v_m)$ are symmetric operators and hence

$$[R(u_m, v_m)]^{\text{tr}} = [dQ(u_m), dQ(v_m)]^{\text{tr}} = [dQ(v_m)^{\text{tr}}, dQ(u_m)^{\text{tr}}]$$
$$= [dQ(v_m), dQ(u_m)] = -[dQ(u_m), dQ(v_m)] = -R(u_m, v_m).$$

To prove (3.30) we make use of the zero-torsion condition $dQ(v_m)w_m = dQ(w_m)v_m$ and the fact that $dQ(u_m)$ is symmetric to learn

$$\langle R(u_m, v_m)w, z\rangle = \langle[dQ(u_m), dQ(v_m)]w, z\rangle$$
$$= \langle[dQ(u_m)dQ(v_m) - dQ(v_m)dQ(u_m)]w, z\rangle$$
$$= \langle dQ(v_m)w, dQ(u_m)z\rangle - \langle dQ(u_m)w, dQ(v_m)z\rangle$$
$$= \langle dQ(w)v, dQ(z)u\rangle - \langle dQ(w)u, dQ(z)v\rangle \qquad (3.34)$$
$$= \langle[dQ(z), dQ(w)]v, u\rangle = \langle R(z, w)v, u\rangle = \langle R(w, z)u, v\rangle$$

where we have used the anti-symmetry properties in 4a. and 4b. By (3.34) with $v = w = a$,

$$\langle \text{Ric}\, u, z\rangle = \sum_{a \in S} \langle R(u, a)a, z\rangle$$
$$= \sum_{a \in S} [\langle dQ(a)a, dQ(u)z\rangle - \langle dQ(u)a, dQ(a)z\rangle]$$
$$= \sum_{a \in S} [\langle a, dQ(a)dQ(u)z\rangle - \langle dQ(u)a, dQ(z)a\rangle]$$
$$= \sum_{a \in S} [\langle a, dQ(dQ(u)z)a\rangle - \langle dQ(z)dQ(u)a, a\rangle]$$
$$= \text{tr}(dQ(dQ(u)z) - dQ(z)dQ(u))$$

which proves (3.32). The assertion that $\text{Ric}_m : T_m M \to T_m M$ is a symmetric operator follows easily from this formula and item 3. $\qquad \square$

**Notation 3.37.** To each $v \in \mathbb{R}^N$, let $\partial_v$ denote the vector field on $\mathbb{R}^N$ defined by

$$\partial_v(\text{at } x) = v_x = \frac{d}{dt}|_0(x + tv).$$

So if $F \in C^\infty(\mathbb{R}^N)$, then

$$(\partial_v F)(x) := \frac{d}{dt}|_0 F(x + tv) = F'(x)\, v$$

and

$$(\partial_v \partial_w F)(x) = F''(x)\,(v, w),$$

see Notation 2.1.

Notice that if $w : \mathbb{R}^N \to \mathbb{R}^N$ is a function and $v \in \mathbb{R}^N$, then

$$(\partial_v \partial_w F)(x) = \partial_v \left[F'(\cdot)\, w(\cdot)\right](x) = F'(x)\,\partial_v w(x) + F''(x)\,(v, w(x)).$$

The following variant of item 4. of Proposition 3.36 will be useful in proving the key Bochner-Weitenböck identity in Theorem 3.49 below.

**Proposition 3.38.** *Suppose that* $Z \in \Gamma(TM)$, $v, w \in T_m M$ *and let* $X, Y \in \Gamma(TM)$ *such that* $X(m) = v$ *and* $Y(m) = w$. *Then*

1. $\nabla^2_{v \otimes w} Z$ *defined by*

$$\nabla^2_{v \otimes w} Z := \left( \nabla_X \nabla_Y Z - \nabla_{\nabla_X Y} Z \right)(m) \tag{3.35}$$

   *is well defined, independent of the possible choices for* $X$ *and* $Y$.
2. *If* $Z(m) = (m, z(m))$ *with* $z : \mathbb{R}^N \to \mathbb{R}^N$ *a smooth function such that* $z(m) \in \tau_m M$ *for all* $m \in M$, *then*

$$\nabla^2_{v \otimes w} Z = dQ(v) dQ(w) z(m) + P(m) z''(m)(v, w) - P(m) z'(m) [dQ(v) w]. \tag{3.36}$$

3. *The curvature tensor* $R(v, w)$ *may be computed as*

$$\nabla^2_{v \otimes w} Z - \nabla^2_{w \otimes v} Z = R(v, w) Z(m). \tag{3.37}$$

4. *If* $V$ *is a smooth vector field along a path* $\sigma(s)$ *in* $M$, *then the following product rule holds,*

$$\frac{\nabla}{ds} \left( \nabla_{V(s)} Z \right) = \left( \nabla_{\frac{\nabla}{ds} V(s)} Z \right) + \nabla^2_{\sigma'(s) \otimes V(s)} Z. \tag{3.38}$$

*Proof.* We will prove items 1 and 2 by showing the right sides of (3.35) and Eq. (3.36) are equal. To do this write $X(m) = (m, x(m))$, $Y(m) = (m, y(m))$, and $Z(m) = (m, z(m))$ where $x, y, z : \mathbb{R}^N \to \mathbb{R}^N$ are smooth functions such that $x(m)$, $y(m)$, and $z(m)$ are in $\tau_m M$ for all $m \in M$. Then, suppressing $m$ from the notation,

$$\begin{aligned}
\nabla_X \nabla_Y Z - \nabla_{\nabla_X Y} Z &= P \partial_x \left[ P \partial_y z \right] - P \partial_{P \partial_x y} z \\
&= P(\partial_x P) \partial_y z + P \partial_x \partial_y z - P \partial_{P \partial_x y} z \\
&= P(\partial_x P) \partial_y z + P z''(x, y) + P z' [\partial_x y - P \partial_x y] \\
&= (\partial_x P) Q \partial_y z + P z''(x, y) + P z' [Q \partial_x y].
\end{aligned}$$

Differentiating the identity, $Qy = 0$ on $M$ shows $Q \partial_x y = - (\partial_x Q) y$ which combined with the previous equation gives

$$\begin{aligned}
\nabla_X \nabla_Y Z - \nabla_{\nabla_X Y} Z &= (\partial_x P) Q \partial_y z + P z''(x, y) - P z' [(\partial_x Q) Y] \tag{3.39} \\
&= - (\partial_x P)(\partial_y Q) z + P z''(X, Y) - P z' [(\partial_x Q) Y].
\end{aligned}$$

Evaluating this expression at $m$ proves the right side of (3.36).

Equation (3.37) now follows from (3.36) and (3.28), item 3 of Proposition 3.36 and the fact the $z''(v, w) = z''(w, v)$ because mixed partial derivatives commute.

We give two proofs of (3.38). For the first proof, choose local vector fields $\{E_i\}_{i=1}^d$ defined in a neighborhood of $\sigma(s)$ such that $\{E_i(\sigma(s))\}_{i=1}^d$ is a basis for $T_{\sigma(s)} M$ for each $s$. We may then write $V(s) = \sum_{i=1}^d V_i(s) E_i(\sigma(s))$ and therefore,

$$\frac{\nabla}{ds} V(s) = \sum_{i=1}^d \left\{ V_i'(s) E_i(\sigma(s)) + V_i(s) \nabla_{\sigma'(s)} E_i \right\} \tag{3.40}$$

and

$$\frac{\nabla}{ds}\left(\nabla_{V(s)}Z\right) = \frac{\nabla}{ds}\left(\sum_{i=1}^{d} V_i\,(s)\,\left(\nabla_{E_i}Z\right)(\sigma\,(s))\right)$$

$$= \sum_{i=1}^{d} V_i'\,(s)\,\left(\nabla_{E_i}Z\right)(\sigma\,(s)) + \sum_{i=1}^{d} V_i\,(s)\,\nabla_{\sigma'(s)}\left(\nabla_{E_i}Z\right).$$

Using (3.35),

$$\nabla_{\sigma'(s)}\left(\nabla_{E_i}Z\right) = \nabla^2_{\sigma'(s)\otimes E_i(\sigma(s))}Z + \left(\nabla_{\nabla_{\sigma'(s)}E_i}Z\right)$$

and using this in the previous equation along with (3.40) shows

$$\frac{\nabla}{ds}\left(\nabla_{V(s)}Z\right) = \nabla_{\sum_{i=1}^{d}\left\{V_i'(s)E_i(\sigma(s))+V_i(s)\nabla_{\sigma'(s)}E_i\right\}}Z + \sum_{i=1}^{d} V_i\,(s)\,\nabla^2_{\sigma'(s)\otimes E_i(\sigma(s))}Z$$

$$= \left(\nabla_{\frac{\nabla}{ds}V(s)}Z\right) + \nabla^2_{\sigma'(s)\otimes V(s)}Z.$$

For the second proof, write $V\,(s) = (\sigma\,(s), v\,(s)) = v\,(s)_{\sigma(s)}$ and $p\,(s) := P\,(\sigma\,(s))$, then

$$\frac{\nabla}{ds}\left(\nabla_V Z\right) - \left(\nabla_{\frac{\nabla}{ds}V}Z\right) = p\frac{d}{ds}\left(pz'\,(v)\right) - pz'\,(pv')$$

$$= p\left[p'z'\,(v) + pz''\,(\sigma', v) + pz'\,(v')\right] - pz'\,(pv')$$

$$= pp'z'\,(v) + pz''\,(\sigma', v) + pz'\,(qv')$$

$$= p'qz'\,(v) + pz''\,(\sigma', v) - pz'\,(q'v)$$

$$= \nabla^2_{\sigma'(s)\otimes V(s)}Z$$

wherein the last equation we have made use of (3.39).    □

## 3.5 Formulas for the divergence and the Laplacian

**Theorem 3.39.** *Let $Y$ be a vector field on $M$, then*

$$\mathrm{div}\,Y = \mathrm{tr}(\nabla Y). \tag{3.41}$$

*(Note: $(v_m \to \nabla_{v_m}Y) \in \mathrm{End}(T_m M)$ for each $m \in M$, so it makes sense to take the trace.) Consequently, if $f$ is a smooth function on $M$, then*

$$\Delta f = \mathrm{tr}(\nabla\,\mathrm{grad}\,f). \tag{3.42}$$

*Proof.* Let $x$ be a chart on $M$, $\partial_i := \partial/\partial x^i$, $\nabla_i := \nabla_{\partial_i}$, and $Y^i := dx^i\,(Y)$. Then by the product rule and the fact that $\nabla$ is torsion free (item 2 of Proposition 3.36),

$$\nabla_i Y = \sum_{j=1}^{d} \nabla_i (Y^j \partial_j) = \sum_{j=1}^{d} (\partial_i Y^j \partial_j + Y^j \nabla_i \partial_j),$$

and $\nabla_i \partial_j = \nabla_j \partial_i$. Hence,

$$\operatorname{tr}(\nabla Y) = \sum_{i=1}^{d} dx^i (\nabla_i Y) = \sum_{i=1}^{d} \partial_i Y^i + \sum_{i,j=1}^{d} dx^i (Y^j \nabla_i \partial_j)$$

$$= \sum_{i=1}^{d} \partial_i Y^i + \sum_{i,j=1}^{d} dx^i (Y^j \nabla_j \partial_i).$$

Therefore, according to (3.20), to finish the proof it suffices to show that

$$\sum_{i=1}^{d} dx^i (\nabla_j \partial_i) = \partial_j \log \sqrt{g}.$$

From Lemma 2.7,

$$\partial_j \log \sqrt{g} = \frac{1}{2} \partial_j \log(\det g) = \frac{1}{2} \operatorname{tr}(g^{-1} \partial_j g) = \frac{1}{2} \sum_{k,l=1}^{d} g^{kl} \partial_j g_{kl},$$

and using (3.26) we have

$$\partial_j g_{kl} = \partial_j \langle \partial_k, \partial_l \rangle = \langle \nabla_j \partial_k, \partial_l \rangle + \langle \partial_k, \nabla_j \partial_l \rangle.$$

Combining the last two equations along with the symmetry of $g^{kl}$ implies

$$\partial_j \log \sqrt{g} = \sum_{k,l=1}^{d} g^{kl} \langle \nabla_j \partial_k, \partial_l \rangle = \sum_{k=1}^{d} dx^k (\nabla_j \partial_k),$$

where we have used

$$\sum_{k=1}^{d} g^{kl} \langle \cdot, \partial_l \rangle = dx^k.$$

This last equality is easily verified by applying both sides of this equation to $\partial_i$ for $i = 1, 2, \ldots, n$. $\qquad \square$

**Definition 3.40 (One forms).** A *one form* $\omega$ on $M$ is a smooth function $\omega : TM \to \mathbb{R}$ such that $\omega_m := \omega|_{T_m M}$ is linear for all $m \in M$. Note: if $x$ is a chart of $M$ with $m \in \mathcal{D}(x)$, then

$$\omega_m = \sum_{i=1}^{d} \omega_i(m) dx^i|_{T_m M},$$

where $\omega_i := \omega(\partial/\partial x^i)$. The condition that $\omega$ is smooth is equivalent to the condition that each of the functions $\omega_i$ is smooth on $M$. Let $\Omega^1(M)$ denote the smooth one-forms on $M$.

Given a one-form, $\omega \in \Omega^1(M)$, there is a unique vector field $X$ on $M$ such that $\omega_m = \langle X(m), \cdot \rangle_m$ for all $m \in M$. Using this observation, we may extend the definition of $\nabla$ to one-forms by requiring

$$\nabla_{v_m} \omega := \langle \nabla_{v_m} X, \cdot \rangle \in T_m^* M := (T_m M)^*. \tag{3.43}$$

**Lemma 3.41 (Product Rule).** *Keep the notation of the above paragraph. Let $Y \in \Gamma(TM)$, then*

$$v_m[\omega(Y)] = (\nabla_{v_m} \omega)(Y(m)) + \omega(\nabla_{v_m} Y). \tag{3.44}$$

*Moreover, if $\theta : M \to (\mathbb{R}^N)^*$ is a smooth function and*

$$\omega(v_m) := \theta(m)v$$

*for all $v_m \in TM$, then*

$$(\nabla_{v_m} \omega)(w_m) = d\theta(v_m)w - \theta(m)dQ(v_m)w = (d(\theta P)(v_m))w, \tag{3.45}$$

*where $(\theta P)(m) := \theta(m)P(m) \in (\mathbb{R}^N)^*$.*

*Proof.* Using the metric compatibility of $\nabla$,

$$v_m(\omega(Y)) = v_m(\langle X, Y \rangle) = \langle \nabla_{v_m} X, Y(m) \rangle + \langle X(m), \nabla_{v_m} Y \rangle$$
$$= (\nabla_{v_m} \omega)(Y(m)) + \omega(\nabla_{v_m} Y).$$

Writing $Y(m) = (m, y(m)) = y(m)_m$ and using (3.44), it follows that

$$(\nabla_{v_m} \omega)(Y(m)) = v_m(\omega(Y)) - \omega(\nabla_{v_m} Y)$$
$$= v_m(\theta(\cdot)y(\cdot)) - \theta(m)(dy(v_m) + dQ(v_m)y(m))$$
$$= (d\theta(v_m))y(m) - \theta(m)(dQ(v_m))y(m).$$

Choosing $Y$ such that $Y(m) = w_m$ proves the first equality in (3.45). The second equality in (3.45) is a simple consequence of the formula

$$d(\theta P) = d\theta(\cdot)P + \theta dP = d\theta(\cdot)P - \theta dQ.$$

Before continuing, let us record the following useful corollary of the previous proof.

**Corollary 3.42.** *To every one-form $\omega$ on $M$, there exists $f_i, g_i \in C^\infty(M)$ for $i = 1, 2, \ldots, N$ such that*

$$\omega = \sum_{i=1}^{N} f_i dg_i. \tag{3.46}$$

*Proof.* Let $f_i(m) := \theta(m)P(m)e_i$ and $g_i(m) = x^i(m) = \langle m, e_i \rangle_{\mathbb{R}^N}$ where $\{e_i\}_{i=1}^N$ is the standard basis for $\mathbb{R}^N$ and $P(m)$ is orthogonal projection of $\mathbb{R}^N$ onto $\tau_m M$ for each $m \in M$.                                               $\square$

**Definition 3.43.** For $f \in C^\infty(M)$ and $v_m$, $w_m$ in $T_m M$, let

$$\nabla df(v_m, w_m) := (\nabla_{v_m} df)(w_m),$$

so that

$$\nabla df : \cup_{m \in M}(T_m M \times T_m M) \to \mathbb{R}.$$

We call $\nabla df$ the *Hessian* of $f$.

**Lemma 3.44.** Let $f \in C^\infty(M)$, $F \in C^\infty(\mathbb{R}^N)$ such that $f = F|_M$, $X, Y \in \Gamma(TM)$ and $v_m$, $w_m \in T_m M$. Then:

1. $\nabla df(X, Y) = XYf - df(\nabla_X Y)$.
2. $\nabla df(v_m, w_m) = F''(m)(v, w) - F'(m)dQ(v_m)w$.
3. $\nabla df(v_m, w_m) = \nabla df(w_m, v_m)$, *this is another manifestation of zero torsion.*

*Proof.* Using the product rule (see (3.44)):

$$XYf = X(df(Y)) = (\nabla_X df)(Y) + df(\nabla_X Y),$$

and hence

$$\nabla df(X, Y) = (\nabla_X df)(Y) = XYf - df(\nabla_X Y).$$

This proves item 1. From this last equation and Proposition 3.36 ($\nabla$ has zero torsion), it follows that

$$\nabla df(X, Y) - \nabla df(Y, X) = [X, Y]f - df(\nabla_X Y - \nabla_Y X) = 0.$$

This proves the third item upon choosing $X$ and $Y$ such that $X(m) = v_m$ and $Y(m) = w_m$. Item 2 follows easily from Lemma 3.41 applied with $\theta := F'$. □

**Definition 3.45.** Given a point $m \in M$, a *local orthonormal frame* $\{E_i\}_{i=1}^d$ at $m$ is a collection of local vector fields defined near $m$ such that $\{E_i(p)\}_{i=1}^d$ is an orthonormal basis for $T_p M$ for all $p$ near $m$.

**Corollary 3.46.** Suppose that $F \in C^\infty(\mathbb{R}^N)$, $f := F|_M$, and $m \in M$. Let $\{e_i\}_{i=1}^d$ be an orthonormal basis for $\tau_m M$ and let $\{E_i\}_{i=1}^d$ be an orthonormal frame near $m \in M$. Then

$$\Delta f(m) = \sum_{i=1}^d \nabla df(E_i(m), E_i(m)), \tag{3.47}$$

$$\Delta f(m) = \sum_{i=1}^d \{(E_i E_i f)(m) - df(\nabla_{E_i(m)} E_i)\}, \tag{3.48}$$

and

$$\Delta f(m) = \sum_{i=1}^d F''(m)(e_i, e_i) - F'(m)(dQ(E_i(m))e_i) \tag{3.49}$$

where $E_i(m) := (m, e_i)$.

*Proof.* By Theorem 3.39, $\Delta f = \sum_{i=1}^{d} \langle \nabla_{E_i} \operatorname{grad} f, E_i \rangle$ and by (3.43), $\nabla_{E_i} df = \langle \nabla_{E_i} \operatorname{grad} f, \cdot \rangle$. Therefore

$$\Delta f = \sum_{i=1}^{d} (\nabla_{E_i} df)(E_i) = \sum_{i=1}^{d} \nabla df(E_i, E_i),$$

which proves (3.47). Equations (3.48) and (3.49) follows from (3.47) and Lemma 3.44.

$\square$

**Notation 3.47.** Let $\{e_i\}_{i=1}^{N}$ be the standard basis on $\mathbb{R}^N$ and define $X_i(m) := P(m) e_i$ for all $m \in M$ and $i = 1, 2, \ldots, N$.

In the next proposition we will express the gradient, divergence and the Laplacian in terms of the vector fields, $\{X_i\}_{i=1}^{N}$. These formulas will prove very useful when we start discussing Brownian motion on $M$.

**Proposition 3.48.** *Let* $f \in C^\infty(M)$ *and* $Y \in \Gamma(TM)$ *then*

1. $v_m = \sum_{i=1}^{N} \langle v_m, X_i(m) \rangle X_i(m)$ *for all* $v_m \in T_m M$.
2. $\nabla f = \operatorname{grad} f = \sum_{i=1}^{N} X_i f \cdot X_i$
3. $\nabla \cdot Y = \operatorname{div}(Y) = \sum_{i=1}^{N} \langle \nabla_{X_i} Y, X_i \rangle$
4. $\sum_{i=1}^{N} \nabla_{X_i} X_i = 0$
5. $\Delta f = \sum_{i=1}^{N} X_i^2 f$.

*Proof.* 1. The main point is to show

$$\sum_{i=1}^{N} X_i(m) \otimes X_i(m) = \sum_{i=1}^{d} u_i \otimes u_i \qquad (3.50)$$

where $\{u_i\}_{i=1}^{d}$ is an orthonormal basis for $T_m M$. But this is easily proved since

$$\sum_{i=1}^{N} X_i(m) \otimes X_i(m) = \sum_{i=1}^{N} P(m) e_i \otimes P(m) e_i$$

and the latter expression is independent of the choice of orthonormal basis $\{e_i\}_{i=1}^{N}$ for $\mathbb{R}^N$. Hence if we choose $\{e_i\}_{i=1}^{N}$ so that $e_i = u_i$ for $i = 1, \ldots, d$, then

$$\sum_{i=1}^{N} P(m) e_i \otimes P(m) e_i = \sum_{i=1}^{d} u_i \otimes u_i$$

as desired. Since $\sum_{i=1}^{N} \langle v_m, X_i(m) \rangle X_i(m)$ is quadratic in $X_i$, it now follows that

$$\sum_{i=1}^{N} \langle v_m, X_i(m) \rangle X_i(m) = \sum_{i=1}^{d} \langle v_m, u_i \rangle u_i = v_m.$$

2. This is an immediate consequence of item 1:

$$\text{grad } f\,(m) = \sum_{i=1}^{N} \langle \text{grad } f\,(m)\,,\,X_i\,(m)\rangle X_i\,(m) = \sum_{i=1}^{N} X_i f\,(m) \cdot X_i\,(m)\,.$$

3. Again $\sum_{i=1}^{N} \langle \nabla_{X_i} Y, X_i\rangle\,(m)$ is quadratic in $X_i$ and so by (3.50) and Theorem 3.39,

$$\sum_{i=1}^{N} \langle \nabla_{X_i} Y, X_i\rangle\,(m) = \sum_{i=1}^{d} \langle \nabla_{u_i} Y, u_i\rangle\,(m) = \text{div}(Y)\,.$$

4. By definition of $X_i$ and $\nabla$ and using Lemma 3.30,

$$\sum_{i=1}^{N} \left(\nabla_{X_i} X_i\right)(m) = \sum_{i=1}^{N} P\,(m)\,dP\,(X_i\,(m))\,e_i = \sum_{i=1}^{N} dP\,(P\,(m)\,e_i)\,Q\,(m)\,e_i.\quad (3.51)$$

The latter expression is independent of the choice of orthonormal basis $\{e_i\}_{i=1}^{N}$ for $\mathbb{R}^N$. So again we may choose $\{e_i\}_{i=1}^{N}$ so that $e_i = u_i$ for $i = 1,\ldots,d$, in which case $P\,(m)\,e_j = 0$ for $j > d$ and so each summand in the right member of (3.51) is zero.

5. To compute $\Delta f$, use items 2–4, the definition of $\nabla f$ and the product rule to find

$$\Delta f = \nabla \cdot (\nabla f) = \sum_{i=1}^{N} \langle \nabla_{X_i} \nabla f, X_i\rangle$$

$$= \sum_{i=1}^{N} X_i \langle \nabla f, X_i\rangle - \sum_{i=1}^{N} \langle \nabla f, \nabla_{X_i} X_i\rangle = \sum_{i=1}^{N} X_i X_i f.$$

The following commutation formulas are at the heart of many of the results to appear in the latter sections of these note.

**Theorem 3.49 (The Bochner–Weitenböck Identity).** *Let $f \in C^{\infty}\,(M)$ and $a, b, c \in T_m M$, then*

$$\langle \nabla^2_{a \otimes b} \nabla f, c\rangle = \langle \nabla^2_{a \otimes c} \nabla f, b\rangle \tag{3.52}$$

*and if $S \subset T_m M$ is an orthonormal basis, then*

$$\sum_{a \in S} \nabla^2_{a \otimes a} \nabla f = (\text{grad } \Delta f)\,(m) + \text{Ric } \nabla f\,(m)\,. \tag{3.53}$$

This result is the first indication that the Ricci tensor is going to play an important role in later developments. The proof will be given after the next technical lemma, which will be helpful in simplifying the proof of the theorem.

**Lemma 3.50.** *Given $m \in M$ and $v \in T_m M$ there exists $V \in \Gamma\,(TM)$ such that $V\,(m) = v$ and $\nabla_w V = 0$ for all $w \in T_m M$. Moreover if $\{e_i\}_{i=1}^{d}$ is an orthonormal basis for $T_m M$, there exists a local orthonormal frame $\{E_i\}_{i=1}^{d}$ near $m$ such that $\nabla_w E_i = 0$ for all $w \in T_m M$.*

*Proof.* In the proof to follow it is assumed that $V$, $Q$ and $P$ have all been extended off $M$ to smooth function on the ambient space. If $V$ is to exist, we must have

$$0 = \nabla_w V = V'(m)\, w + \partial_w Q\,(m)\, v,$$

such that

$$V'(m)\, w = -\partial_w Q\,(m)\, v \text{ for all } w \in T_m M.$$

This helps to motivate defining $V$ by

$$V(x) := P(x)\,(v - (\partial_{x-m} Q)\,(m)\, v) \in T_x M \text{ for all } x \in M.$$

By construction, $V(m) = v$ and making use of the identities in Lemma 3.30,

$$\begin{aligned}
\nabla_w V &= \partial_w \left[ P(x)\,(v - (\partial_{x-m} Q)\,(m)\, v) \right]|_{x=m} + (\partial_w Q)\,(m)\, v \\
&= (\partial_w P)\,(m)\, v - P(m)\,(\partial_w Q)\,(m)\, v + (\partial_w Q)\,(m)\, v \\
&= (\partial_w P)\,(m)\, v + Q(m)\,(\partial_w Q)\,(m)\, v = (\partial_w P)\,(m)\, v + (\partial_w Q)\,(m)\, v = 0
\end{aligned}$$

as desired.

For the second assertion, choose a local frame $\{V_i\}_{i=1}^d$ such that $V_i(m) = e_i$ and $\nabla_w V_i = 0$ for all $i$ and $w \in T_m M$. The desired frame $\{E_i\}_{i=1}^d$ is now constructed by performing Gram-Schmidt orthogonalization on $\{V_i\}_{i=1}^d$. The resulting orthonormal frame, $\{E_i\}_{i=1}^d$, still satisfies $\nabla_w E_i = 0$ for all $w \in T_m M$. For example, $E_1 = \langle V_1, V_1 \rangle^{-1/2} V_1$ and since

$$w \langle V_1, V_1 \rangle = 2 \langle \nabla_w V_1, V_1(m) \rangle = 0$$

it follows that

$$\nabla_w E_1 = w\left( \langle V_1, V_1 \rangle^{-1/2} \right) \cdot V_1(m) + \langle V_1, V_1 \rangle^{-1/2}(m)\, \nabla_w V_1(m) = 0.$$

The similar verifications that $\nabla_w E_j = 0$ for $j = 2, \ldots, d$ will be left to the reader.  ☐

*Proof of Theorem 3.49.* Let $a, b, c \in T_m M$ and suppose $A, B, C \in \Gamma(TM)$ have been chosen as in Lemma 3.50, so that $A(m) = a$, $B(m) = b$ and $C(m) = c$ with $\nabla_w A = \nabla_w B = \nabla_w C = 0$ for all $w \in T_m M$. Then

$$\begin{aligned}
ABCf &= AB \langle \nabla f, C \rangle = A \langle \nabla_B \nabla f, C \rangle + A \langle \nabla f, \nabla_B C \rangle \\
&= \langle \nabla_A \nabla_B \nabla f, C \rangle + \langle \nabla_B \nabla f, \nabla_A C \rangle + A \langle \nabla f, \nabla_B C \rangle
\end{aligned}$$

which evaluated at $m$ gives

$$\begin{aligned}
(ABCf)(m) &= (\langle \nabla_A \nabla_B \nabla f, C \rangle + A \langle \nabla f, \nabla_B C \rangle)(m) \\
&= \langle \nabla^2_{a \otimes b} \nabla f, c \rangle + (A \langle \nabla f, \nabla_B C \rangle)(m)
\end{aligned}$$

wherein the last equality we have used $(\nabla_A B)(m) = 0$. Interchanging $B$ and $C$ in this equation and subtracting then implies

$$(A\,[B,\,C]\,f)\,(m) = \langle \nabla^2_{a\otimes b} \nabla f, c\rangle - \langle \nabla^2_{a\otimes c} \nabla f, b\rangle + (A\langle \nabla f, \nabla_B C - \nabla_C B\rangle)\,(m)$$
$$= \langle \nabla^2_{a\otimes b} \nabla f, c\rangle - \langle \nabla^2_{a\otimes c} \nabla f, b\rangle + (A\langle \nabla f, [B, C]\rangle)\,(m)$$
$$= \langle \nabla^2_{a\otimes b} \nabla f, c\rangle - \langle \nabla^2_{a\otimes c} \nabla f, b\rangle + (A[B, C]f)\,(m)$$

and this equation implies (3.52).

Now suppose that $\{E_i\}_{i=1}^d \subset T_m M$ is an orthonormal frame as in Lemma 3.50 and $e_i = E_i\,(m)$. Then, using Proposition 3.38,

$$\sum_{i=1}^d \langle \nabla^2_{e_i\otimes e_i} \nabla f, c\rangle = \sum_{i=1}^d \langle \nabla^2_{e_i\otimes c} \nabla f, e_i\rangle = \sum_{i=1}^d \langle \nabla^2_{c\otimes e_i} \nabla f + R\,(e_i, c)\,\nabla f\,(m), e_i\rangle.$$
$$(3.54)$$

Since

$$\sum_{i=1}^d \langle \nabla^2_{c\otimes e_i} \nabla f, e_i\rangle = \sum_{i=1}^d \left(\langle \nabla_C \nabla_{E_i} \nabla f, E_i\rangle\right)(m) = \sum_{i=1}^d \left(C\langle \nabla_{E_i} \nabla f, E_i\rangle\right)(m)$$
$$= (C\Delta f)\,(m) = \langle (\nabla\Delta f)\,(m), c\rangle$$

and (using $R\,(e_i, c)^{\mathrm{tr}} = R\,(c, e_i)$)

$$\sum_{i=1}^d \langle R\,(e_i, c)\,\nabla f\,(m), e_i\rangle = \sum_{i=1}^d \langle \nabla f\,(m), R\,(c, e_i)\,e_i\rangle$$
$$= \langle \nabla f\,(m), \mathrm{Ric}\,c\rangle = \langle \mathrm{Ric}\,\nabla f\,(m), c\rangle,$$

(3.54) implies

$$\sum_{i=1}^d \langle \nabla^2_{e_i\otimes e_i} \nabla f, c\rangle = \langle (\nabla\Delta f)\,(m) + \mathrm{Ric}\,\nabla f\,(m), c\rangle$$

which proves (3.53) since $c \in T_m M$ was arbitrary.

## 3.6 Parallel translation

**Definition 3.51.** Let $V$ be a smooth path in $TM$. $V$ is said to be *parallel* or *covariantly constant* if $\nabla V(s)/ds \equiv 0$.

**Theorem 3.52.** *Let $\sigma$ be a smooth path in $M$ and $(v_0)_{\sigma(0)} \in T_{\sigma(0)}M$. Then there exists a unique smooth vector field $V$ along $\sigma$ such that $V$ is parallel and $V(0) = (v_0)_{\sigma(0)}$. Moreover if $V(s)$ and $W(s)$ are parallel along $\sigma$, then $\langle V(s), W(s)\rangle = \langle V(0), W(0)\rangle$ for all $s$.*

*Proof.* If $V$ and $W$ are parallel, then

$$\frac{d}{ds}\langle V(s), W(s)\rangle = \left\langle \frac{\nabla}{ds}V(s), W(s)\right\rangle + \left\langle V(s), \frac{\nabla}{ds}W(s)\right\rangle = 0$$

which proves the last assertion of the theorem. If a parallel vector field $V(s) = (\sigma(s), v(s))$ along $\sigma(s)$ is to exist, then

$$dv(s)/ds + dQ(\sigma'(s))v(s) = 0 \quad \text{and} \quad v(0) = v_0. \tag{3.55}$$

By existence and uniqueness of solutions to ordinary differential equations, there is exactly one solution to (3.55). Hence, if $V$ exists it is unique.

Now let $v$ be the unique solution to (3.55) and set $V(s) := (\sigma(s), v(s))$. To finish the proof it suffices to show that $v(s) \in \tau_{\sigma(s)}M$. Equivalently, we must show that $w(s) := q(s)v(s)$ is identically zero, where $q(s) := Q(\sigma(s))$. Letting $v'(s) = dv(s)/ds$ and $p(s) = P(\sigma(s))$, then (3.55) states $v' = -q'v$ and from Lemma 3.30 we have $pq' = q'q$. Thus the function $w$ satisfies

$$w' = q'v + qv' = q'v - qq'v = pq'v = q'qv = q'w$$

with $w(0) = 0$. But this linear ordinary differential equation has $w \equiv 0$ as its unique solution. $\qquad\square$

**Definition 3.53 (Parallel Translation).** Given a smooth path $\sigma$, let $//_s(\sigma) : T_{\sigma(0)}M \to T_{\sigma(s)}M$ be defined by $//_s(\sigma)(v_0)_{\sigma(0)} = V(s)$, where $V$ is the unique parallel vector field along $\sigma$ such that $V(0) = (v_0)_{\sigma(0)}$. We call $//_s(\sigma)$ *parallel translation* along $\sigma$ up to time $s$.

**Remark 3.54.** Notice that $//_s(\sigma)v_{\sigma(0)} = (u(s)v)_{\sigma(0)}$, where $s \to u(s) \in \mathrm{Hom}(\tau_{\sigma(0)}M, \mathbb{R}^N)$ is the unique solution to the differential equation

$$u'(s) + dQ(\sigma'(s))u(s) = 0 \quad \text{with} \quad u(0) = P(\sigma(0)). \tag{3.56}$$

Because of Theorem 3.52, $u(s) : \tau_{\sigma(0)}M \to \mathbb{R}^N$ is an isometry for all $s$ and the range of $u(s)$ is $\tau_{\sigma(s)}M$. Moreover, if we let $\bar{u}(s)$ denote the solution to

$$\bar{u}'(s) - \bar{u}(s)dQ(\sigma'(s)) = 0 \text{ with } \bar{u}(0) = P(\sigma(0)), \tag{3.57}$$

then

$$\frac{d}{ds}[\bar{u}(s)u(s)] = \bar{u}'(s)u(s) + \bar{u}(s)u'(s)$$

$$= \bar{u}(s)dQ(\sigma'(s))u(s) - \bar{u}(s)dQ(\sigma'(s))u(s) = 0.$$

Hence $\bar{u}(s)u(s) = P(\sigma(0))$ for all $s$ and therefore $\bar{u}(s)$ is the inverse to $u(s)$ thought of as a linear operator from $\tau_{\sigma(0)}M$ to $\tau_{\sigma(s)}M$. See also Lemma 3.57 below.

The following techniques for computing covariant derivatives will be useful in the sequel.

**Lemma 3.55.** *Suppose* $Y \in \Gamma(TM)$, $\sigma(s)$ *is a path in* $M$, $W(s) = (\sigma(s), w(s))$ *is a vector field along* $\sigma$ *and let* $//_s = //_s(\sigma)$ *be parallel translation along* $\sigma$. *Then*

1. $\frac{\nabla}{ds}W(s) = //_s \frac{d}{ds}\left[//_s^{-1}W(s)\right].$

*2. For any $v \in T_{\sigma(0)}M$,*

$$\frac{\nabla}{ds}\nabla_{//_s v}Y = \nabla^2_{\sigma'(s)\otimes//_s v}Y. \tag{3.58}$$

*where $\nabla^2_{\sigma'(s)\otimes//_s v}Y$ was defined in Proposition 3.38.*

*Proof.* Let $\bar{u}$ be as in (3.57). From (3.25),

$$\frac{\nabla W(s)}{ds} = \left(\frac{d}{ds}w(s) + dQ(\sigma'(s)))w(s)\right)_{\sigma(s)}$$

while, using Remark 3.54,

$$\begin{aligned}
\frac{d}{ds}\left[//_s^{-1}W(s)\right] &= \left(\frac{d}{ds}[\bar{u}(s)w(s)]\right)_{\sigma(s)} \\
&\doteq \left(\bar{u}'(s)W(s) + \bar{u}(s)w'(s)\right)_{\sigma(s)} \\
&= \left(\bar{u}(s)dQ(\sigma'(s))w(s) + \bar{u}(s)w'(s)\right)_{\sigma(s)} \\
&= //_s^{-1}\frac{\nabla W(s)}{ds}.
\end{aligned}$$

This proves the first item. We will give two proofs of the second item, the first proof being extrinsic while the second will be intrinsic. In each of these proofs there will be an implied sum on repeated indices.

*First proof.* Let $\{X_i\}_{i=1}^N \subset \Gamma(TM)$ be as in Notation 3.47, then by Proposition 3.48,

$$//_s v = \langle //_s v, X_i(\sigma(s))\rangle X_i(\sigma(s)) = \langle v, //_s^{-1}X_i(\sigma(s))\rangle X_i(\sigma(s)) \tag{3.59}$$

and therefore,

$$\begin{aligned}
\frac{\nabla}{ds}\nabla_{//_s v}Y &= \frac{\nabla}{ds}\left[\langle //_s v, X_i(\sigma(s))\rangle \cdot (\nabla_{X_i}Y)(\sigma(s))\right] \\
&= \langle //_s v, X_i(\sigma(s))\rangle \cdot \nabla_{\sigma'(s)}(\nabla_{X_i}Y) + \langle //_s v, \nabla_{\sigma'(s)}X_i\rangle \cdot (\nabla_{X_i}Y)(\sigma(s)). \tag{3.60}
\end{aligned}$$

Now

$$\nabla_{\sigma'(s)}(\nabla_{X_i}Y) = \nabla^2_{\sigma'(s)\otimes X_i}Y + \nabla_{\sigma'(s)X_i}Y$$

and so again using Proposition 3.48,

$$\langle //_s v, X_i(\sigma(s))\rangle \cdot \nabla_{\sigma'(s)}(\nabla_{X_i}Y) = \nabla^2_{\sigma'(s)\otimes//_s v}Y + \langle //_s v, X_i(\sigma(s))\rangle \cdot \nabla_{\sigma'(s)X_i}Y. \tag{3.61}$$

Taking $\nabla/ds$ of (3.59) shows

$$0 = \langle //_s v, \nabla_{\sigma'(s)}X_i\rangle X_i(\sigma(s)) + \langle //_s v, X_i(\sigma(s))\rangle \nabla_{\sigma'(s)}X_i.$$

and so

$$\langle //_s v, X_i \left( \sigma \left( s \right) \right) \rangle \cdot \nabla_{\sigma'(s) X_i} Y = -\langle //_s v, \nabla_{\sigma'(s)} X_i \rangle \cdot \left( \nabla_{X_i} Y \right) \left( \sigma \right) \left( s \right). \qquad (3.62)$$

Assembling (3.59), (3.61) and (3.62) proves (3.58).

*Second proof.* Let $\{E_i\}_{i=1}^d$ be an orthonormal frame near $\sigma \left( s \right)$, then

$$\frac{\nabla}{ds} \nabla_{//_s v} Y = \frac{\nabla}{ds} \left[ \langle //_s v, E_i \left( \sigma \left( s \right) \right) \rangle \cdot \left( \nabla_{E_i} Y \right) \left( \sigma \left( s \right) \right) \right]$$
$$= \langle //_s v, \nabla_{\sigma'(s)} E_i \rangle \cdot \left( \nabla_{E_i} Y \right) \left( \sigma \left( s \right) \right) + \langle //_s v, E_i \left( \sigma \left( s \right) \right) \rangle \cdot \nabla_{\sigma'(s)} \nabla_{E_i} Y.$$
$$(3.63)$$

Working as in the first proof,

$$\langle //_s v, E_i \left( \sigma \left( s \right) \right) \rangle \cdot \nabla_{\sigma'(s)} \nabla_{E_i} Y = \langle //_s v, E_i \left( \sigma \left( s \right) \right) \rangle \cdot \left( \nabla^2_{\sigma'(s) \otimes E_i} Y + \nabla_{\nabla_{\sigma'(s)} E_i} Y \right)$$
$$= \nabla^2_{\sigma'(s) \otimes //_s v} Y + \nabla_{\langle //_s v, E_i(\sigma(s)) \rangle \nabla_{\sigma'(s)} E_i} Y$$

and using

$$0 = \frac{\nabla}{ds} //_s v = \langle //_s v, \nabla_{\sigma'(s)} E_i \rangle \cdot E_i \left( \sigma \left( s \right) \right) + \langle //_s v, E_i \left( \sigma \left( s \right) \right) \rangle \cdot \nabla_{\sigma'(s)} E_i$$

we find

$$\langle //_s v, E_i \left( \sigma \left( s \right) \right) \rangle \cdot \nabla_{\sigma'(s)} \nabla_{E_i} Y = \nabla^2_{\sigma'(s) \otimes //_s v} Y - \langle //_s v, \nabla_{\sigma'(s)} E_i \rangle \cdot \left( \nabla_{E_i} Y \right) \left( \sigma \left( s \right) \right).$$

This equation combined with (3.63) again proves (3.58). □

The remainder of this section discusses a covariant derivative on $M \times \mathbb{R}^N$ which "extends" $\nabla$ defined above. This will be needed in Section 5, where it will be convenient to have a covariant derivative on the normal bundle:

$$N(M) := \cup_{m \in M} \left( \{m\} \times \tau_m M^\perp \right) \subset M \times \mathbb{R}^N.$$

Analogous to the definition of $\nabla$ on $TM$, it is reasonable to extend $\nabla$ to the normal bundle $N(M)$ by setting

$$\frac{\nabla V(s)}{ds} = \left( \sigma(s), Q(\sigma(s)) v'(s) \right) = \left( \sigma(s), v'(s) + dP(\sigma'(s)) v(s) \right),$$

for all smooth paths $s \rightarrow V(s) = \left( \sigma(s), v(s) \right)$ in $N(M)$. Then this covariant derivative on the normal bundle satisfies analogous properties to $\nabla$ on the tangent bundle $TM$. The covariant derivatives on $TM$ and $N(M)$ can be put together to make a covariant derivative on $M \times \mathbb{R}^N$. Explicitly, if $V(s) = \left( \sigma(s), v(s) \right)$ is a smooth path in $M \times \mathbb{R}^N$, let $p(s) := P(\sigma(s))$, $q(s) := Q(\sigma(s))$ and then define

$$\frac{\nabla V(s)}{ds} := \left( \sigma(s), p(s) \frac{d}{ds} \{p(s) v(s)\} + q(s) \frac{d}{ds} \{q(s) v(s)\} \right).$$

Since

$$\frac{\nabla V(s)}{ds} = (\sigma(s), \frac{d}{ds}\{p(s)v(s)\} + q'(s)p(s)v(s)$$

$$+ \frac{d}{ds}\{q(s)v(s)\} + p'(s)q(s)v(s))$$

$$= (\sigma(s), v'(s) + q'(s)p(s)v(s) + p'(s)q(s)v(s))$$

$$= (\sigma(s), v'(s) + dQ(\sigma'(s))P(\sigma(s))v(s) + dP(\sigma'(s))Q(\sigma(s))v(s))$$

we may write $\nabla V(s)/ds$ as

$$\frac{\nabla V(s)}{ds} = (\sigma(s), v'(s) + \Gamma(\sigma'(s))v(s)) \tag{3.64}$$

where

$$\Gamma(w_m)v := dQ(w_m)P(m)v + dP(w_m)Q(m)v \tag{3.65}$$

for all $w_m \subset TM$ and $v \in \mathbb{R}^N$.

It should be clear from the above computation that the covariant derivative defined in (3.64) agrees with those already defined on $TM$ and $N(M)$. Many of the properties of the covariant derivative on $TM$ follow quite naturally from this fact and (3.64).

**Lemma 3.56.** *For each* $w_m \in TM$, $\Gamma(w_m)$ *is a skew symmetric* $N \times N$-*matrix. Hence, if* $u(s)$ *is the solution to the differential equation*

$$u'(s) + \Gamma(\sigma'(s))u(s) = 0 \quad with \quad u(0) = I, \tag{3.66}$$

*then* $u$ *is an orthogonal matrix for all* $s$.

*Proof.* Since $\Gamma = dQP + dPQ$ and $P$ and $Q$ are orthogonal projections and hence symmetric, the adjoint $\Gamma^{tr}$ of $\Gamma$ is given by

$$\Gamma^{tr} = PdQ + QdP = -dPQ - dQP = -\Gamma.$$

where Lemma 3.30 was used in the second equality. Hence $\Gamma$ is a skew-symmetric valued one form. Now let $u$ denote the solution to (3.66) and $A(s) := \Gamma(\sigma'(s))$. Then

$$\frac{d}{ds}u^{tr}u = (-Au)^{tr}u + u^{tr}(-Au) = u^{tr}(A - A)u = 0,$$

which shows that $u^{tr}(s)u(s) = u^{tr}(0)u(0) = I$. □

**Lemma 3.57.** *Let* $u$ *be the solution to (3.66). Then*

$$u(s)(\tau_\sigma(0)M) = \tau_\sigma(s)M \tag{3.67}$$

*and*

$$u(s)(\tau_\sigma(0)M)^\perp = \tau_\sigma(s)M^\perp. \tag{3.68}$$

*In particular, if* $v \in \tau_{\sigma(0)}M$ ($v \in \tau_{\sigma(0)}M^\perp$) *then* $V(s) := (\sigma(s), u(s)v)$ *is the parallel vector field along* $\sigma$ *in* $TM$ ($N(M)$) *such that* $V(0) = v_{\sigma(0)}$.

*Proof.* By the product rule,

$$\frac{d}{ds}\{u^{\text{tr}} P(\sigma) u\} = u^{\text{tr}}\{\Gamma(\sigma') P(\sigma) + dP(\sigma') - P(\sigma)\Gamma(\sigma')\}u. \qquad (3.69)$$

Moreover, making use of Lemma 3.30,

$$\begin{aligned}
\Gamma(\sigma') P(\sigma) &- P(\sigma)\Gamma(\sigma') + dP(\sigma') \\
&= dP(\sigma') + [dQ(\sigma')P(\sigma) + dP(\sigma')Q(\sigma)] P(\sigma) \\
&\quad - P(\sigma)[dQ(\sigma')P(\sigma) + dP(\sigma')Q(\sigma)] \\
&= dP(\sigma') + dQ(\sigma')P(\sigma) - dP(\sigma')Q(\sigma) \\
&= dP(\sigma') + dQ(\sigma') = 0,
\end{aligned}$$

which combined with (3.69) shows $\frac{d}{ds}\{u^{\text{tr}} P(\sigma) u\} = 0$. Therefore,

$$u^{\text{tr}}(s) P(\sigma(s)) u(s) = P(\sigma(0))$$

for all $s$. Combining this with Lemma 3.56, shows

$$P(\sigma(s)) u(s) = u(s) P(\sigma(0)).$$

This last equation is equivalent to (3.67). (3.68) has completely analogous proof or can be seen easily from the fact that $P + Q = I$.                                      □

## 3.7  More references

I recommend [86] and [42] for more details on Riemannian geometry. The references, [1, 19, 41, 42, 86, 95, 111, 112, 113, 114, 115, 149] and the complete five volume set of Spivak's books on differential geometry starting with [164] are also very useful.

# 4  Flows and Cartan's development map

The results of this section will serve as a warm-up for their stochastic counter parts. These types of theorems will be crucial for the path space analysis results to be developed in Sections 7 and 8 below.

## 4.1  Time-dependent smooth flows

**Notation 4.1.** Given a smooth *time dependent vector* field, $(t, m) \to X_t(m) \in T_m M$ on a manifold $M$, let $T_t^X(m)$ denote the solution to the ordinary differential equation,

$$\frac{d}{dt} T_t^X(m) = X_t \circ T_t^X(m) \text{ with } T_0^X(m) = m.$$

If $X$ is *time independent* we will write $e^{tX}(m)$ for $T_t^X(m)$. We call $T^X$ the *flow* of $X$. See Figure 10.

**Figure 10.** Going with the flow. Here we suppose that $X$ is a time independent vector field which is indicated by the arrows in the picture and the curve is the corresponding flow line starting at $m \in M$.

**Theorem 4.2 (Flow Theorem).** *Suppose that $X_t$ is a smooth time dependent vector field on $M$. Then for each $m \in M$, there exists a maximal open interval $J_m \subset \mathbb{R}$ such that $0 \in J_m$ and $t \to T_t^X (m)$ exists for $t \in J_m$. Moreover the set $\mathcal{D} (X) :=$ $\cup_m (J_m \times \{m\}) \subset \mathbb{R} \times M$ is open and the map $(t, m) \in \mathcal{D} (X) \to T_t^X (m) \in M$ is a smooth map.*

*Proof.* Let $Y_t$ be a smooth extension of $X_t$ to a vector field on $E$ where $E$ is the Euclidean space in which $M$ is imbedded. The stated results with $X$ replaced by $Y$ follow from the standard theory of ordinary differential equations on Euclidean spaces. Let $T_t^Y$ denote the flow of $Y$ on $E$. We will construct $T^X$ by setting $T_t^X (m) := T_t^Y (m)$ for all $m \in M$ and $t \in J_m$. In order for this to work we must show that $T_t^Y (m) \in M$ whenever $m \in M$.

To verify this last assertion, let $x$ be a chart on $M$ such that $m \in \mathcal{D} (x)$, then $\sigma (t)$ solves $\dot{\sigma} (t) = X_t (\sigma (t))$ with $\sigma (0) = m$ iff

$$\frac{d}{dt} [x \circ \sigma (t)] = dx (\dot{\sigma} (t)) = dx (X_t (\sigma (t))) = dx \left( X_t \circ x^{-1} (x \circ \sigma (t)) \right)$$

with $x \circ \sigma (0) = m$. Since this is a differential equation for $x \circ \sigma (t) \in \mathcal{R} (z)$ and $\mathcal{R} (z)$ is an open subset $\mathbb{R}^d$, the standard local existence theorem for ordinary differential equations implies $x \circ \sigma (t)$ exists for a small time. This then implies $\sigma (t) \in M$ exists for small $t$ and satisfies

$$\dot{\sigma} (t) = X_t (\sigma (t)) = Y_t (\sigma (t)) \text{ with } \sigma (0) = m.$$

By uniqueness of solutions to ordinary differential equations, we must have $T_t^Y (m) = \sigma (t)$ for small $t$ and in particular $T_t^Y (m) \in M$ for small $t$. Let

$$\tau := \sup \left\{ t \in J_m : T_s^Y (m) \in M \text{ for } 0 \leq s \leq t \right\}$$

and for the sake of contradiction suppose that $[0, \tau] \subset J_m$. Then by continuity, $T_\tau^Y (m) \in M$ and by repeating the above argument using a chart $x$ on $M$ centered at $T_\tau^Y (m)$, we would find that $T_t^Y(m) \in M$ for $t$ in a neighborhood of $\tau$. This contradicts the definition of $\tau$ and hence we may conclude that $\tau$ is the right end point of $J_m$. A similar argument works for $t \in J_m$ with $t < 0$ and hence $T_t^Y(m) \in M$ for all $t \in J_m$.     □

**Assumption 1 (Completeness).** *For simplicity in these notes it will always be assumed that $X$ is complete, i.e., $J_m = \mathbb{R}$ for all $m \in M$ and hence $\mathcal{D}(X) = \mathbb{R} \times M$. This will be the case if, for example, $M$ is compact or $M$ is imbedded in $\mathbb{R}^N$ and the vector field $X$ satisfies a Lipschitz condition. (Later we will restrict to the compact case.)*

**Notation 4.3.** For $g, h \in \mathrm{Diff}(M)$ let $Ad_g h := g \circ h \circ g^{-1}$. We will also write $Ad_g$ for the linear transformation on $\Gamma(TM)$ defined by

$$Ad_g Y = \frac{d}{ds}\Big|_0 Ad_g e^{sY} = \frac{d}{ds}\Big|_0 g \circ e^{sY} \circ g^{-1} = g_*\left(Y \circ g^{-1}\right)$$

for all $Y \in \Gamma(TM)$. (The vector space $\Gamma(TM)$ should be interpreted as the Lie algebra of the diffeomorphism group, $\mathrm{Diff}(M)$.)

In order to verify $T_t^X$ is invertible, let $T_{t,s}^X$ denote the solution to

$$\frac{d}{dt} T_{t,s}^X = X_t \circ T_{t,s}^X \text{ with } T_{s,s}^X = id.$$

**Lemma 4.4.** *Suppose that $X_t$ is a complete time dependent vector field on $M$, then $T_t^X \in \mathrm{Diff}(M)$ for all $t$ and*

$$\left(T_t^X\right)^{-1} = T_{0,t}^X = T_t^{-Ad_{(T^X)^{-1}}X}, \tag{4.1}$$

*where*

$$\left(Ad_{(T^X)^{-1}}X\right)_t := Ad_{(T_t^X)^{-1}} X_t.$$

*Proof.* If $s, t, u \in \mathbb{R}$, then $S_t := T_{t,s}^X \circ T_{s,u}^X$ solves

$$\dot{S}_t = X_t \circ S_t \text{ with } S_s = T_{s,u}^X$$

which is the same equation that $t \to T_{t,u}^X$ solves and therefore $T_{t,u}^X = T_{t,s}^X \circ T_{s,u}^X$. In particular, $T_{0,t}^X$ is the inverse to $T_t^X$. Moreover if we let $T_t := T_t^X$ and $S_t := T_t^{-1}$ then

$$0 = \frac{d}{dt} id = \frac{d}{dt}[T_t \circ S_t] = X_t \circ T_t \circ S_t + T_{t*}\dot{S}_t.$$

So it follows that $S_t$ solves

$$\dot{S}_t = -T_{t*}^{-1} X_t \circ T_t \circ S_t = -\left(Ad_{T_t^{-1}} X_t\right) \circ S_t$$

which proves the second equality in (4.1).     □

## 4.2 Differentials of $T_t^X$

In the later sections of this article, we will make heavy use of the stochastic analogues of the following two differentiation theorems.

**Theorem 4.5 (Differentiating** $m \to T_t^X(m)$**).** *Suppose $\nabla$ is the Levi-Civita[2] covariant derivative on $TM$ and $T_t = T_t^X$ as above, then*

$$\frac{\nabla}{dt} T_{t*} v = \nabla_{T_{t*}v} X_t \text{ for all } v \in TM. \tag{4.2}$$

*If we further let $m \in M$, $//_t = //_t (\tau \to T_\tau(m))$ be parallel translation relative to $\nabla$ along the flow line $\tau \to T_\tau(m)$ and $z_t := //_t^{-1} T_{t*m}$, then*

$$\frac{d}{dt} z_t v = //_t^{-1} \nabla_{//_t z_t v} X_t \text{ for all } v \in T_m M. \tag{4.3}$$

*(This is a linear differential equation for $z_t \in \text{End}(T_m M)$.)*

*Proof.* Let $\sigma(s)$ be a smooth path in $M$ such that $\sigma'(0) = v$, then

$$\frac{\nabla}{dt} T_{t*} v = \frac{\nabla}{dt}\frac{d}{ds}|_0 T_t(\sigma(s)) = \frac{\nabla}{ds}|_0 \frac{d}{dt} T_t(\sigma(s))$$

$$= \frac{\nabla}{ds}|_0 X_t(T_t(\sigma(s))) = \nabla_{T_{t*}v} X_t$$

wherein the second equality we have used $\nabla$ has zero torsion. Equation (4.3) follows directly from (4.2) using $\frac{\nabla}{dt} = //_t \frac{d}{dt}//_t^{-1}$, see Lemma 3.55.     □

**Remark 4.6.** As a warm up for writing the stochastic version of (4.3) in Itô form let us pause to compute $\frac{\nabla}{dt}(\nabla_{T_{t*}v} Y)$ for $Y \in \Gamma(TM)$. Using (3.38), (3.37) and (3.35) of Proposition 3.38,

$$\frac{\nabla}{dt} \nabla_{T_{t*}v} Y = \nabla^2_{\dot{T}_t(m) \otimes T_{t*}v} Y + \nabla_{\frac{\nabla}{dt} T_{t*}v} Y = \nabla^2_{X_t(T_t(m)) \otimes T_{t*}v} Y + \nabla_{\nabla_{T_{t*}v} X_t} Y$$

$$= \nabla^2_{T_{t*}v \otimes X_t(T_t(m))} Y + R^\nabla(X_t(T_t(m)), T_{t*}v) Y(T_t(m)) + \nabla_{\nabla_{T_{t*}v} X_t} Y$$

$$= R^\nabla(X_t(T_t(m)), T_{t*}v) Y(T_t(m)) + \nabla_{T_{t*}v}(\nabla_{X_t} Y). \tag{4.4}$$

**Theorem 4.7 (Differentiating** $T_t^X$ **in** $X$**).** *Suppose $(t, m) \to X_t(m)$ and $(t, m) \to Y_t(m)$ are smooth time dependent vector fields on $M$ and let*

$$\partial_Y T_t^X := \frac{d}{ds}|_0 T_t^{X+sY}. \tag{4.5}$$

*Then*

$$\partial_Y T_t^X = T_{t*}^X \int_0^t (T_{\tau*}^X)^{-1} Y_\tau \circ T_\tau^X d\tau = T_{t*}^X \int_0^t Ad_{T_\tau^X}^{-1} Y_\tau d\tau. \tag{4.6}$$

*This formula may also be written as*

$$\partial_Y T_t^X = \left(\int_0^t Ad_{T_{t,\tau}^X} Y_\tau d\tau\right) \circ T_t^X = \left(\int_0^t Ad_{T_t^X \circ (T_\tau^X)^{-1}} Y_\tau d\tau\right) \circ T_t^X. \tag{4.7}$$

---

[2] Actually, for those in the know, any torsion zero covariant derivative could be used here.

*Proof.* To simplify notation, let $T_t := T_t^X$ and define $V_t := \left(T_{t*}^X\right)^{-1} \partial_Y T_t^X$. Then $V_0 = 0$ and $\partial_Y T_t^X = T_{t*}^X V_t$ or equivalently, for all $f \in C^\infty(M)$,

$$\frac{d}{ds}\big|_0 f \circ T_t^{X+sY} = \left(T_{t*}^X V_t\right) f = V_t\left(f \circ T_t^X\right).$$

Given $f \in C^\infty(M)$, on one hand we have

$$\frac{d}{dt}\frac{d}{ds}\big|_0 f \circ T_t^{X+sY} = \frac{d}{dt}\left[V_t(f \circ T_t^X)\right] = \dot{V}_t(f \circ T_t^X) + V_t(X_t f \circ T_t^X)$$
$$= \left(T_{t*}^X \dot{V}_t\right) f + V_t(X_t f \circ T_t^X)$$

while on the other hand

$$\frac{d}{ds}\big|_0 \frac{d}{dt} f \circ T_t^{X+sY} = \frac{d}{ds}\big|_0 \left[((X_t + sY_t) f) \circ T_t^{X+sY}\right]$$
$$= (Y_t f) \circ T_t^X + V_t\left(X_t f \circ T_t^X\right)$$
$$= \left(Y_t \circ T_t^X\right) f + V_t\left(X_t f \circ T_t^X\right).$$

Since $\left[\frac{d}{dt}, \frac{d}{ds}\big|_0\right] = 0$, the previous two displayed equations imply $\left(T_{t*}^X \dot{V}_t\right) f = \left(Y_t \circ T_t^X\right) f$ and because this holds for all $f \in C^\infty(M)$,

$$T_{t*}^X \dot{V}_t = Y_t \circ T_t^X. \tag{4.8}$$

Solving (4.8) for $\dot{V}_t$ and then integrating on $t$ shows

$$V_t = \int_0^t \left(T_{\tau*}^X\right)^{-1} Y_\tau \circ T_\tau^X d\tau.$$

which along with the relation, $\partial_Y T_t^X = T_{t*}^X V_t$, implies (4.6).

We may now rewrite the formula in (4.6) as

$$\partial_Y T_t^X = T_{t*}^X \left(\int_0^t Ad_{T_\tau^X}^{-1} Y_\tau d\tau\right) \circ \left(T_t^X\right)^{-1} \circ T_t^X = Ad_{T_t^X}\left(\int_0^t Ad_{T_\tau^X}^{-1} Y_\tau d\tau\right) \circ T_t^X$$
$$= \left(\int_0^t Ad_{T_t^X} Ad_{T_\tau^X}^{-1} Y_\tau d\tau\right) \circ T_t^X = \left(\int_0^t Ad_{T_t^X \circ (T_\tau^X)^{-1}} Y_\tau d\tau\right) \circ T_t^X$$
$$= \left(\int_0^t Ad_{T_{t,\tau}^X} Y_\tau d\tau\right) \circ T_t^X$$

which gives (4.7).                                                                 □

**Example 4.8.** Suppose that $G$ is a Lie group, $\mathfrak{g} := \text{Lie}(G)$, $A_t$ and $B_t$ are two smooth $\mathfrak{g}$-valued functions and $g_t^A \in G$ solves the equation

$$\frac{d}{dt} g_t^A = \tilde{A}_t\left(g_t^A\right) \text{ with } g_0^A = e \in G$$

where $\tilde{A}_t(x) := L_{x*}A_t$ is the *left invariant* vector field on $G$ associated to $A_t \in \mathfrak{g}$, see Examples 2.34 and 3.27. Then

$$\partial_B g_t^A = R_{g_t^A*}\int_0^t Ad_{g_\tau^A}B_\tau d\tau$$

where

$$Ad_g A = R_{g^{-1}*}L_{g*}A \text{ for all } g \in G \text{ and } A \in \mathfrak{g}.$$

*Proof.* Let $T_t^A$ denote the flow of $A_t$. Because $A_t$ is left invariant,

$$T_t^A(x) = xg_t^A = R_{g_t^A}x$$

as the reader should verify. Thus

$$\partial_B g_t^A = \partial_B T_t^A(e) = R_{g_t^A*}\int_0^t \left(R_{g_\tau^A*}\right)^{-1}\tilde{B}_\tau \circ R_{g_\tau^A}(e)\,d\tau$$

$$= R_{g_t^A*}\int_0^t \left(R_{g_\tau^A*}\right)^{-1}\tilde{B}_\tau\left(g_\tau^A\right)d\tau = R_{g_t^A*}\int_0^t \left(R_{g_\tau^A*}\right)^{-1}L_{g_\tau^A*}B_\tau d\tau$$

$$= R_{g_t^A*}\int_0^t Ad_{g_\tau^A}B_\tau d\tau. \quad \square$$

The next theorem expresses $[X_t, Y]$ using the flow $T^X$. The stochastic analog of this theorem is a key ingredient in the "Malliavin calculus," see Proposition 8.14 below.

**Theorem 4.9.** *If $X_t$ and $T_t^X$ are as above and $Y \in \Gamma(TM)$, then*

$$\frac{d}{dt}\left[\left(T_{t*}^X\right)^{-1}Y \circ T_t^X\right] = \left(T_{t*}^X\right)^{-1}[X_t, Y] \circ T_t^X \tag{4.9}$$

*or equivalently put*

$$\frac{d}{dt}Ad_{T_t^X}^{-1} = Ad_{T_t^X}^{-1}L_{X_t} \tag{4.10}$$

*where $L_X Y := [X, Y]$.*

*Proof.* Let $V_t := \left(T_{t*}^X\right)^{-1}Y \circ T_t^X$ which is equivalent to $T_{t*}^X V_t = Y \circ T_t^X$, or more explicitly to

$$Yf \circ T_t^X = \left(Y \circ T_t^X\right)f = \left(T_{t*}^X V_t\right)f = V_t\left(f \circ T_t^X\right) \text{ for all } f \in C^\infty(M).$$

Differentiating this equation in $t$ then shows

$$(X_t Yf) \circ T_t^X = \dot{V}_t\left(f \circ T_t^X\right) + V_t\left(X_t f \circ T_t^X\right)$$

$$= \left(T_{t*}^X \dot{V}_t\right)f + \left(T_{t*}^X V_t\right)X_t f$$

$$= \left(T_{t*}^X \dot{V}_t\right)f + \left(Y \circ T_t^X\right)X_t f$$

$$= \left(T_{t*}^X \dot{V}_t\right)f + (YX_t f) \circ T_t^X.$$

Therefore

$$\left(T_{t*}^{X}\dot{V}_{t}\right)f = ([X_{t},Y]\,f)\circ T_{t}^{X}$$

from which we conclude $T_{t*}^{X}\dot{V}_{t} = [X_{t},Y]\circ T_{t}^{X}$ and therefore

$$\dot{V}_{t} = \left(T_{t*}^{X}\right)^{-1}[X_{t},Y]\circ T_{t}^{X}. \qquad \square$$

## 4.3 Cartan's development map

For this section assume that $M$ is compact[3] Riemannian manifold and let $W^{\infty}(T_{0}M)$ be the collection of piecewise smooth paths, $b : [0, 1] \to T_{o}M$ such that $b\,(0) = 0_{o} \in T_{o}M$ and let $W_{o}^{\infty}(M)$ be the collection of piecewise smooth paths, $\sigma : [0, 1] \to M$ such that $\sigma\,(0) = o \in M$.

**Theorem 4.10 (Development Map).** *To each $b \in W^{\infty}(T_{0}M)$ there is a unique $\sigma \in W_{o}^{\infty}(M)$ such that*

$$\sigma'(s) := (\sigma(s), d\sigma(s)/ds) = //_{s}(\sigma)b'(s) \quad and \quad \sigma(0) = o, \tag{4.11}$$

*where $//_{s}(\sigma)$ denotes parallel translation along $\sigma$.*

*Proof.* Suppose that $\sigma$ is a solution to (4.11) and $//_{s}(\sigma)v_{o} = (o, u(s)v)$, where $u(s) : \tau_{o}M \to \mathbb{R}^{N}$. Then $u$ satisfies the differential equation

$$u'(s) + dQ(\sigma'(s))u(s) = 0 \quad \text{with} \quad u(0) = u_{0}, \tag{4.12}$$

where $u_{0}v := v$ for all $v \in \tau_{o}M$, see Remark 3.54. Hence (4.11) is equivalent to the following pair of coupled ordinary differential equations

$$\sigma'(s) = u(s)b'(s) \quad \text{with} \quad \sigma(0) = o, \tag{4.13}$$

and

$$u'(s) + dQ((\sigma(s), u(s)b'(s)))u(s) = 0 \quad \text{with} \quad u(0) = u_{0}. \tag{4.14}$$

Therefore the uniqueness assertion follows from standard uniqueness theorems for ordinary differential equations. The slickest proof of existence to (4.11) is to first introduce the orthogonal frame bundle, $O\,(M)$, on $M$ defined by $O\,(M) := \cup_{m\in M}O_{m}(M)$ where $O_{m}(M)$ is the set of all isometries, $u : T_{o}M \to T_{m}M$. It is then possible to show that $O\,(M)$ is an imbedded submanifold in $\mathbb{R}^{N} \times \text{Hom}\,(\tau_{o}M, \mathbb{R}^{N})$ and that coupled pair of ordinary differential equations (4.13) and (4.14) may be viewed as a flow equation on $O(M)$. Hence the existence of solutions may be deduced from the Theorem 4.2, see, for example, [47] for details of this method. Here I will sketch a proof which does not require us to develop the frame bundle formalism in detail.

---

[3] It would actually be sufficient to assume that $M$ is a *complete* Riemannian manifold for this section.

Looking at the proof of Lemma 2.30, $Q$ has an extension to a neighborhood in $\mathbb{R}^N$ of $m \in M$ in such a way that $Q(x)$ is still an orthogonal projection onto $\text{Nul}(F'(x))$, where $F(x) = z_>(x)$ is as in Lemma 2.30. Hence for small $s$, we may define $\sigma$ and $u$ to be the unique solutions to (4.13) and (4.14) with values in $\mathbb{R}^N$ and $\text{Hom}(\tau_o M, \mathbb{R}^N)$ respectively. The key point now is to show that $\sigma(s) \in M$ and that the range of $u(s)$ is $\tau_{\sigma(s)} M$.

Using the same proof as in Theorem 3.52, $w(s) := Q(\sigma(s))u(s)$ satisfies,

$$w' = dQ\left(\sigma'\right)u + Q\left(\sigma\right)u' = dQ\left(\sigma'\right)u - Q\left(\sigma\right)dQ(\sigma')u$$
$$= P\left(\sigma\right)dQ\left(\sigma'\right)u = dQ\left(\sigma'\right)Q\left(\sigma\right)u = dQ\left(\sigma'\right)w,$$

where Lemma 3.30 was used in the last equality. Since $w(0) = 0$, it follows by uniqueness of solutions to linear ordinary differential equations that $w \equiv 0$ and hence

$$\text{Ran}\,[u(s)] \subset \text{Nul}\,[Q(\sigma(s))] = \text{Nul}\,\left[F'(\sigma(s))\right].$$

Consequently

$$dF(\sigma(s))/ds = F'(\sigma(s))d\sigma(s)/ds = F'(\sigma(s))u(s)b'(s) = 0$$

for small $s$ and since $F(\sigma(0)) = F(o) = 0$, it follows that $F(\sigma(s)) = 0$, i.e., $\sigma(s) \in M$. So we have shown that there is a solution $(\sigma, u)$ to (4.13) and (4.14) for small $s$ such that $\sigma$ stays in $M$ and $u(s)$ is parallel translation along $s$. By standard ordinary differential equation methods, there is a maximal solution $(\sigma, u)$ with these properties. Notice that $(\sigma, u)$ is a path in $M \times \text{Iso}(T_oM, \mathbb{R}^N)$, where $\text{Iso}(T_oM, \mathbb{R}^N)$ is the set of isometries

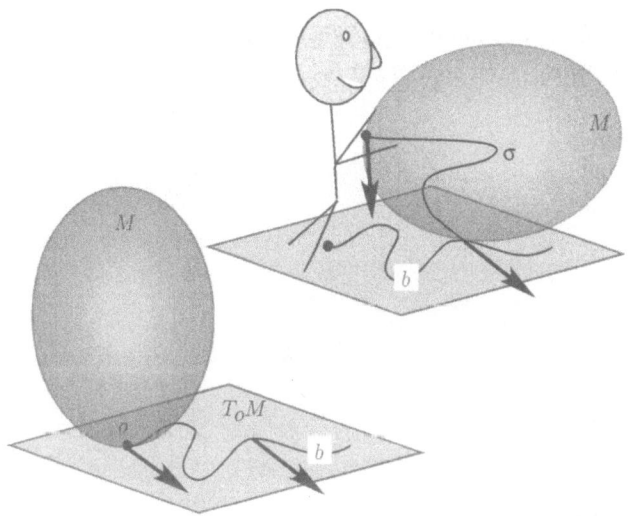

**Figure 11.** Monsieur Cartan is shown here rolling, without "slipping," a manifold $M$ along a curve, $b$, in $T_oM$ to produce a curve, $\sigma$, on $M$.

from $T_oM$ to $\mathbb{R}^N$. Since $M \times \mathrm{Iso}(T_oM, \mathbb{R}^N)$ is a compact space, $(\sigma, u)$ can not explode. Therefore $(\sigma, u)$ is defined on the same interval where $b$ is defined.                                                        □

The geometric interpretation of Cartan's map is to roll the manifold $M$ along a freshly painted curve $b$ in $T_oM$ to produce a curve $\sigma$ on $M$ (see Figure 11).

**Notation 4.11.** Let $\phi : W^\infty(T_0M) \to W_o^\infty(M)$ be the map $b \to \sigma$, where $\sigma$ is the solution to (4.11). It is easy to construct the inverse map $\Psi := \phi^{-1}$. Namely, $\Psi(\sigma) = b$, where

$$\Psi_s(\sigma) = b(s) := \int_0^s //_r(\sigma)^{-1}\sigma'(r)dr.$$

We now conclude this section by computing the differentials of $\Psi$ and $\phi$. For more details on computations of this nature the reader is referred to [46, 47] and the references therein.

**Theorem 4.12 (Differential of $\Psi$).** *Let $(t, s) \to \Sigma(t, s)$ be a smooth map into $M$ such that $\Sigma(t, \cdot) \in W_o^\infty(M)$ for all $t$. Let*

$$H(s) := \dot{\Sigma}(0, s) := (\Sigma(0, s), d\Sigma(t, s)/dt|_{t=0}),$$

*so that $H$ is a vector field along $\sigma := \Sigma(0, \cdot)$. One should view $H$ as an element of the "tangent space" to $W_o^\infty(M)$ at $\sigma$, see Figure 12. Let $u(s) := //_s(\sigma)$, $h(s) := //_s(\sigma)^{-1}H(s)$ $b := \Psi_s(\sigma)$ and, for all $a, c \in T_oM$, let*

$$(R_u(a, c))(s) := u(s)^{-1}R(u(s)a, u(s)c)u(s). \tag{4.15}$$

*Then*

$$d\Psi(H) = d\Psi(\Sigma(t, \cdot))/dt|_{t=0} = h + \int_0^s \left(\int_0^s R_u(h, \delta b)\right)\delta b, \tag{4.16}$$

*where $\delta b(s)$ is short hand notation for $b'(s)ds$, and $\int_0 f\delta b$ denotes the function $s \to \int_0^s f(r)b'(r)dr$ when $f$ is a path of matrices.*

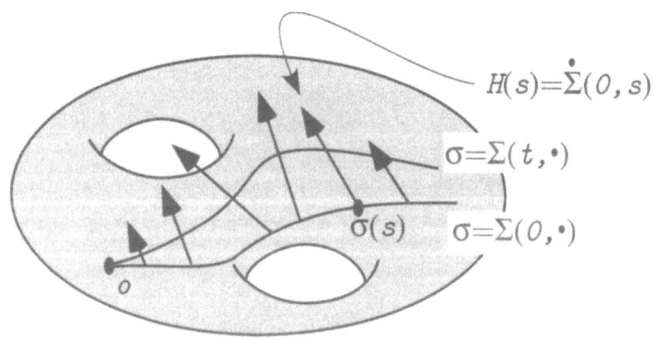

**Figure 12.** A variation of $\sigma$ giving rise to a vector field along $\sigma$.

*Proof.* To simplify notation let " $\cdot$ "= $\frac{d}{dt}|_0$, " $'$ "= $\frac{d}{ds}$, $B(t,s) := \Psi(\Sigma(t,\cdot))(s)$, $U(t,s) := //_s(\Sigma(t,\cdot))$, $u(s) := //_s(\sigma) = U(0,s)$ and

$$\dot{b}(s) := (d\Psi(H))(s) := dB(t,s)/dt|_{t=0}.$$

I will also suppress $(t,s)$ from the notation when possible. With this notation

$$\Sigma' = UB', \quad \dot{\Sigma} = H = uh, \tag{4.17}$$

and

$$\frac{\nabla U}{ds} = 0. \tag{4.18}$$

In (4.18), $\frac{\nabla U}{ds} : T_o M \to T_\Sigma M$ is defined by $\frac{\nabla U}{ds} = P(\Sigma) U'$ or equivalently by

$$\frac{\nabla U}{ds} a := \frac{\nabla (Ua)}{ds} \text{ for all } a \in T_o M.$$

Taking $\nabla/dt$ of (4.17) at $t = 0$ gives, with the aid of Proposition 3.32,

$$\frac{\nabla U}{dt}|_{t=0} b' + u\dot{b}' = \nabla \Sigma'/dt|_{t=0} = \nabla \dot{\Sigma}/ds = uh'.$$

Therefore,

$$\dot{b}' = h' + Ab', \tag{4.19}$$

where $A := -U^{-1} \frac{\nabla U}{dt}|_{t=0}$, i.e.,

$$\frac{\nabla U}{dt}(0,\cdot) = -uA.$$

Taking $\nabla/ds$ of this last equation and using $\nabla u/ds = 0$ along with Proposition 3.32 gives

$$-uA' = \frac{\nabla}{ds}\frac{\nabla}{dt}U\bigg|_{t=0} = \left[\frac{\nabla}{ds},\frac{\nabla}{dt}\right]U\bigg|_{t=0} = R(\sigma',H)u$$

and hence $A' = R_u(h,b')$. By integrating this identity using $A(0) = 0$ ($\nabla U(t,0)/dt = 0$ since $U(t,0) := //_0(\Sigma(t,\cdot)) = I$ is independent of $t$) shows

$$A = \int_0 R_u(h,\delta b) \tag{4.20}$$

The theorem now follows by integrating (4.19) relative to $s$ making use of (4.20) and the fact that $\dot{b}(0) = 0$. □

**Theorem 4.13 (Differential of $\phi$).** *Let* $b,k \in W^\infty(T_o M)$ *and* $(t,s) \to B(t,s)$ *be a smooth map into* $T_o M$ *such that* $B(t,\cdot) \in W^\infty(T_o M)$ , $B(0,s) = b(s)$, *and* $\dot{B}(0,s) = k(s)$. *(For example take* $B(t,s) = b(s) + tk(s)$.) *Then*

$$\phi_*(k_b) := \frac{d}{dt}|_0 \phi(B(t,\cdot)) = //.(\sigma)h,$$

*where $\sigma := \phi(b)$ and $h$ is the first component in the solution $(h, A)$ to the pair of coupled differential equations:*

$$k' = h' + Ab', \quad with \quad h(0) = 0 \tag{4.21}$$

*and*

$$A' = R_u(h, b') \quad with \quad A(0) = 0. \tag{4.22}$$

*Proof.* This theorem has an analogous proof to that of Theorem 4.12. We can also deduce the result from Theorem 4.12 by defining $\Sigma$ by $\Sigma(t, s) := \phi_s(B(t, \cdot))$. We now assume the same notation used in Theorem 4.12 and its proof. Then $B(t, \cdot) = \Psi(\Sigma(t, \cdot))$ and hence by Theorem 4.12

$$k = \frac{d}{dt}|_0 \Psi(\Sigma(t, \cdot)) = d\Psi(H) = h + \int_0^{\cdot} (\int_0^{\cdot} R_u(h, \delta b)) \delta b.$$

Therefore, defining $A := \int_0 R_u(h, \delta b)$ and differentiating this last equation relative to $s$, it follows that $A$ solves (4.22) and that $h$ solves (4.21). $\quad\square$

The following theorem is a mild extension of Theorem 4.12 to include the possibility that $\Sigma(t, \cdot) \notin W_o^{\infty}(M)$ when $t \neq 0$, i.e., the base point may change.

**Theorem 4.14.** *Let $(t, s) \to \Sigma(t, s)$ be a smooth map into $M$ such that $\sigma := \Sigma(0, \cdot) \in W_o^{\infty}(M)$. Define $H(s) := d\Sigma(t, s)/dt|_{t=0}$, $\sigma := \Sigma(0, \cdot)$, and $h(s) := //_s(\sigma)^{-1}H(s)$. (Note: $H(0)$ and $h(0)$ are no longer necessarily equal to zero.) Let*

$$U(t, s) := //_s(\Sigma(t, \cdot))//_t(\Sigma(\cdot, 0)) : T_oM \to T_{\Sigma(t,s)}M,$$

*so that $\nabla U(t, 0)/dt = 0$ and $\nabla U(t, s)/ds \equiv 0$. Set $B(t, s) := \int_0^s U(t, r)^{-1} \Sigma'(t, r) dr$, then*

$$\dot{b}(s) := \frac{d}{dt}|_0 B(t, s) = h_s + \int_0^s \left( \int_0^{\cdot} R_u(h, \delta b) \right) \delta b, \tag{4.23}$$

*where as before $b := \Psi(\sigma)$.*

*Proof.* The proof is almost identical to the proof of Theorem 4.12 and hence will be omitted. $\quad\square$

# 5 Stochastic calculus on manifolds

In this section and the rest of the chapter the reader is assumed to be well versed in stochastic calculus in the Euclidean context.

**Notation 5.1.** In the sequel we will always assume there is any underlying filtered probability space $(\Omega, \{\mathcal{F}_s\}_{s\geq 0}, \mathcal{F}, \mu)$ satisfying the "usual hypothesis." Namely, $\mathcal{F}$ is $\mu$-complete, $\mathcal{F}_s$ contains all of the null sets in $\mathcal{F}$, and $\mathcal{F}_s$ is right continuous. As usual $\mathbb{E}$ will be used to denote the expectation relative to the probability measure $\mu$.

**Definition 5.2.** For simplicity, we will call a function $\Sigma : \mathbb{R}_+ \times \Omega \to V$ ($V$ a vector space) a *process* if $\Sigma_s = \Sigma(s) := \Sigma(s, \cdot)$ is $\mathcal{F}_s$-measurable for all $s \in \mathbb{R}_+ := [0, \infty)$, i.e., a process will mean an adapted process unless otherwise stated. As above, we will always assume that $M$ is an imbedded submanifold of $\mathbb{R}^N$ with the induced Riemannian structure. An *M-valued semi-martingale* is a *continuous* $\mathbb{R}^N$-valued semi-martingale ($\Sigma$) such that $\Sigma(s, \omega) \in M$ for all $(s, \omega) \in \mathbb{R}_+ \times \Omega$. It will be convenient to let $\lambda$ be the distinguished process: $\lambda(s) = \lambda_s := s$.

Since $f \in C^\infty(M)$ is the restriction of a smooth function $F$ on $\mathbb{R}^N$, it follows by Itô's lemma that $f \circ \Sigma = F \circ \Sigma$ is a real-valued semi-martingale if $\Sigma$ is an $M$-valued semi-martingale. Conversely, if $\Sigma$ is an $M$-valued process and $f \circ \Sigma$ is a real-valued semi-martingale for all $f \in C^\infty(M)$ then $\Sigma$ is an $M$-valued semi-martingale. Indeed, let $x = (x^1, \dots, x^N)$ be the standard coordinates on $\mathbb{R}^N$, then $\Sigma^i := x^i \circ \Sigma$ is a real semi-martingale for each $i$, which implies that $\Sigma$ is a $\mathbb{R}^N$-valued semi-martingale.

**Notation 5.3 (Fisk–Stratonovich Integral).** Suppose $V$ is a finite dimensional vector space and

$$\pi = \{0 = s_0 < s_1 < s_2 < \cdots\}$$

is a partition of $\mathbb{R}_+$ with $\lim_{n \to \infty} s_n = \infty$. To such a partition $\pi$, let $|\pi| := \sup_i |s_{i+1} - s_i|$ be the *mesh size* of $\pi$ and $s \wedge s_i := \min\{s, s_i\}$. To each $\mathrm{Hom}\left(\mathbb{R}^N, V\right)$ – valued semi-martingale $Z_t$ and each $M$-valued semi-martingale $\Sigma_t$, the *Fisk–Stratonovich integral* of $Z$ relative to $\Sigma$ is defined by

$$\int_0^s Z\delta\Sigma = \lim_{|\pi| \to 0} \sum_{i=0}^\infty \frac{1}{2}\left(Z_{s \wedge s_i} + Z_{s \wedge s_{i+1}}\right)\left(\Sigma_{s \wedge s_{i+1}} - \Sigma_{s \wedge s_i}\right)$$

$$= \int_0^s Z d\Sigma + \frac{1}{2}\int_0^s dZ d\Sigma \in V$$

where

$$\int_0^s Z d\Sigma = \lim_{|\pi| \to 0} \sum_{i=0}^\infty Z_{s \wedge s_i}\left(\Sigma_{s \wedge s_{i+1}} - \Sigma_{s \wedge s_i}\right) \in V$$

is the *Itô integral* and

$$[Z, \Sigma]_s = \int_0^s dZ d\Sigma := \lim_{|\pi| \to 0} \sum_{i=0}^\infty \left(Z_{s \wedge s_i} - Z_{s \wedge s_{i+1}}\right)\left(\Sigma_{s \wedge s_{i+1}} - \Sigma_{s \wedge s_i}\right) \in V$$

is the *mutual variation (or co-variation)* of $Z$ and $\Sigma$. (All limits may be taken in the sense of uniform convergence on compact subsets of $\mathbb{R}_+$ in probability.)

## 5.1 Stochastic differential equations on manifolds

**Notation 5.4.** Suppose that $\{X_i\}_{i=0}^n \subset \Gamma(TM)$ are vector fields on $M$. For $a \in \mathbb{R}^n$ let

$$X_a(m) := \mathbf{X}(m)a := \sum_{i=1}^n a_i X_i(m)$$

With this notation, $X(m) : \mathbb{R}^n \to T_m M$ is a linear map for each $m \in M$.

**Definition 5.5.** Given an $\mathbb{R}^n$-valued semi-martingale, $\beta_s$, we say an $M$-valued semi-martingale $\Sigma_s$ solves the Fisk–Stratonovich stochastic differential equation

$$\delta \Sigma_s = \mathbf{X}(\Sigma_s)\,\delta\beta_s + X_0(\Sigma_s)\,ds := \sum_{i=1}^{n} X_i(\Sigma_s)\,\delta\beta_s^i + X_0(\Sigma_s)\,ds \qquad (5.1)$$

if for all $f \in C^\infty(M)$,

$$\delta f(\Sigma_s) = \sum_{i=1}^{n} (X_i f)(\Sigma_s)\,\delta\beta_s^i + X_0 f(\Sigma_s)\,ds,$$

i.e., if

$$f(\Sigma_s) = f(\Sigma_0) + \sum_{i=1}^{n} \int_0^s (X_i f)(\Sigma_r)\,\delta\beta_r^i + \int_0^s X_0 f(\Sigma_r)\,dr.$$

**Lemma 5.6 (Itô Form of (5.1)).** *Suppose that* $\beta = B$ *is an* $\mathbb{R}^n$-*valued Brownian motion and let* $L := \frac{1}{2}\sum_{i=1}^{n} X_i^2 + X_0$. *Then an* $M$-*valued semi-martingale* $\Sigma_s$ *solves (5.1) iff*

$$f(\Sigma_s) = f(\Sigma_0) + \sum_{i=1}^{n} \int_0^s (X_i f)(\Sigma_r)\,dB_r^i + \int_0^s Lf(\Sigma_r)\,dr \qquad (5.2)$$

*for all* $f \in C^\infty(M)$.

*Proof.* Suppose that $\Sigma_s$ solves (5.1), then

$$d[(X_i f)(\Sigma_r)] = \sum_{j=1}^{n} (X_j X_i f)(\Sigma_r)\,\delta B_s^j + X_0 X_i f(\Sigma_s)\,ds$$

$$= \sum_{j=1}^{n} (X_j X_i f)(\Sigma_r)\,dB_s^j + d\,(BV)$$

where $BV$ denotes a process of bounded variation. Hence

$$\int_0^s (X_i f)(\Sigma_r)\,\delta B_r^i = \sum_{i=1}^{n} \int_0^s (X_i f)(\Sigma_r)\,dB_r^i + \frac{1}{2}\int_0^s d[(X_i f)(\Sigma_r)]\,dB_r^i$$

$$= \sum_{i=1}^{n} \int_0^s (X_i f)(\Sigma_r)\,dB_r^i + \frac{1}{2}\sum_{i,j=1}^{n} \int_0^s (X_j X_i f)(\Sigma_r)\,dB_s^j dB_r^i$$

$$= \sum_{i=1}^{n} \int_0^s (X_i f)(\Sigma_r)\,dB_r^i + \frac{1}{2}\int_0^s \sum_{i=1}^{n} X_i^2 f(\Sigma_r)\,dr.$$

Similarly if (5.2) holds for all $f \in C^\infty(M)$ we have

$$d[(X_i f)(\Sigma_r)] = (X_j X_i f)(\Sigma_r)\,dB_s^j + LX_i f(\Sigma_s)\,ds$$

and so as above

$$\int_0^s (X_i f)(\Sigma_r)\, \delta B_r^i = \sum_{i=1}^n \int_0^s (X_i f)(\Sigma_r)\, dB_r^i + \frac{1}{2}\int_0^s \sum_{i=1}^n X_i^2 f(\Sigma_r)\, dr.$$

Solving for $\int_0^s (X_i f)(\Sigma_r)\, dB_r^i$ and putting the result into (5.2) shows

$$f(\Sigma_s) = f(\Sigma_0) + \sum_{i=1}^n \int_0^s (X_i f)(\Sigma_r)\, \delta B_r^i$$

$$-\frac{1}{2}\int_0^s \sum_{i=1}^n X_i^2 f(\Sigma_r)\, dr + \int_0^s Lf(\Sigma_r)\, dr$$

$$= f(\Sigma_0) + \sum_{i=1}^n \int_0^s (X_i f)(\Sigma_r)\, \delta B_r^i + \int_0^s X_0 f(\Sigma_r)\, dr.$$

To avoid technical problems with possible explosions of stochastic differential equations in the sequel, we make the following assumption.

**Assumption 2.** *Unless otherwise stated, in the remainder of these notes, M will be a compact manifold imbedded in $E := \mathbb{R}^N$.*

To shortcut the development of a number of issues here it is useful to recall the following Wong and Zakai type approximation theorem for solutions to Fisk–Stratonovich stochastic differential equations.

**Notation 5.7.** Let $\{B_s\}_{s\in[0,T]}$ be a standard $\mathbb{R}^n$-valued Brownian motion. Given a partition

$$\pi = \{0 = s_0 < s_1 < s_2 < \dots < s_k = T\}$$

of $[0, T]$, let

$$|\pi| = \max\{s_i - s_{i-1} : i = 1, 2, \dots, k\}$$

and

$$B_\pi(s) = B(s_{i-1}) + (s - s_{i-1})\frac{\Delta_i B}{\Delta_i s} \text{ if } s \in (s_{i-1}, s_i],$$

where $\Delta_i B := B(s_i) - B(s_{i-1})$ and $\Delta_i s := s_i - s_{i-1}$. Notice that $B_\pi(s)$ is a continuous piecewise linear path in $\mathbb{R}^n$.

**Theorem 5.8 (Wong–Zakai type approximation theorem).** *Let $a \in \mathbb{R}^N$,*

$$f : \mathbb{R}^n \times \mathbb{R}^N \to \mathrm{Hom}(\mathbb{R}^n, \mathbb{R}^N) \text{ and } f_0 : \mathbb{R}^n \times \mathbb{R}^N \to \mathbb{R}^N$$

*be twice differentiable functions with bounded continuous derivatives. Let $\pi$ and $B_\pi$ be as in Notation 5.7 and $\xi_\pi(s)$ denote the solution to the ordinary differential equation:*

$$\xi'_\pi(s) = f(B_\pi(s), \xi_\pi(s))B'_\pi(s) + f_0(B_\pi(s), \xi_\pi(s)), \qquad \xi_\pi(0) = a \qquad (5.3)$$

*and $\xi$ denote the solution to the Fisk–Stratonovich SDE,*

$$d\xi_s = f(B_s, \xi_s)\delta B_s + f_0(B_s, \xi_s)ds, \qquad \xi_0 = a. \tag{5.4}$$

*Then, for any $\gamma \in (0, \frac{1}{2})$ and $p \in [1, \infty)$, there is a constant $C(p, \gamma) < \infty$ such that*

$$\lim_{|\pi| \to 0} \mathbb{E} \left[ \sup_{s \le T} |\xi_\pi(s) - \xi_s|^p \right] \le C(p, \gamma)|\pi|^{\gamma p}. \tag{5.5}$$

This theorem is a special case of Theorem 5.7.3 and Example 5.7.4 in Kunita [116]. Theorems of this type have a long history starting with Wong and Zakai [180, 181]. The reader may also find this and related results in the following *partial* list of references: [7, 10, 11, 20, 22, 44, 68, 94, 103, 107, 108, 118, 117, 126, 129, 132, 134, 135, 141, 142, 151, 166, 174, 167, 175, 177]. Also see [8, 53] and the references therein for more of the geometry associated to the Wong and Zakai approximation scheme.

**Remark 5.9 (Transfer Principle).** Theorem 5.8 is a manifestation of the *transfer principle* (coined by Malliavin) which loosely states: to get a correct stochastic formula one should take the corresponding deterministic smooth formula and replace all derivatives by Fisk–Stratonovich differentials. We will see examples of this principle over and over again in the sequel.

**Theorem 5.10.** *Given a point $m \in M$ there exists a unique $M$-valued semi martingale $\Sigma$ which solves (5.1) with the initial condition, $\Sigma_0 = m$. We will write $T_s(m)$ for $\Sigma_s$ if we wish to emphasize the dependence of the solution on the initial starting point $m \in M$.*

(*Proof of Existence.*) If for the moment we assumed that the Brownian motion $B_s$ were differentiable in $s$, (5.1) could be written as

$$\Sigma'_s = X_s(\Sigma_s) \text{ with } \Sigma_0 = m$$

where

$$X_s(m) := \sum_{i=1}^n X_i(m) \left(B^i\right)'(s) + X_0(m)$$

and the existence of $\Sigma_s$ could be deduced from Theorem 4.2. We will make this rigorous with an application of Theorem 5.8.

Let $\{Y_i\}_{i=0}^n$ be smooth vector fields on $E$ with compact support such that $Y_i = X_i$ on $M$ for each $i$ and let $B_\pi(s)$ be as in Notation 5.7 and define

$$X^\pi_s(m) := \sum_{i=1}^n X_i(m) \left(B^i_\pi\right)'(s) + X_0(m) \text{ and}$$

$$Y^\pi_s(m) := \sum_{i=1}^n Y_i(m) \left(B^i_\pi\right)'(s) + Y_0(m).$$

Then by Theorem 4.2 we may use $X^\pi$ and $Y^\pi$ to generate (random) flows $T^\pi := T^{X^\pi}$ on $M$ and $\tilde{T}^\pi := T^{Y^\pi}$ on $E$ respectively. Moreover, as in the proof of Theorem 4.2 we know $T_s^\pi(m) = \tilde{T}_s^\pi(m)$ for all $m \in M$. An application of Theorem 5.8 now shows that $\Sigma_s := \tilde{T}_s(m) := \lim_{|\pi| \to 0} \tilde{T}_s^\pi(m) = \lim_{|\pi| \to 0} T_s^\pi(m) \in M$ exists[4] and satisfies the Fisk–Stratonovich differential equation *on* $E$,

$$d\Sigma_s = \sum_{i=1}^{n} Y_i(\Sigma_s) \, \delta B_s^i + Y_0(\Sigma_s) \, ds \text{ with } \Sigma_0 = m. \tag{5.6}$$

Given $f \in C^\infty(M)$, let $F \in C^\infty(E)$ be chosen so that $f = F|_M$. Then (5.6) implies

$$d[F(\Sigma_s)] = \sum_{i=1}^{n} Y_i F(\Sigma_s) \, \delta B_s^i + Y_0 F(\Sigma_s) \, ds. \tag{5.7}$$

Since we have already seen $\Sigma_s \in M$ and by construction $Y_i = X_i$ on $M$, we have $F(\Sigma_s) = f(\Sigma_s)$ and $Y_i F(\Sigma_s) = X_i f(\Sigma_s)$. Therefore (5.7) implies

$$d[f(\Sigma_s)] = \sum_{i=1}^{n} X_i f(\Sigma_s) \, \delta B_s^i + Y_0 F(\Sigma_s) \, ds,$$

i.e., $\Sigma_s$ solves (5.1) as desired.

*Proof of uniqueness.* If $\Sigma$ is a solution to (5.1), then for $F \in C^\infty(E)$, we have

$$\begin{aligned} dF(\Sigma_s) &= \sum_{i=1}^{n} X_i F(\Sigma_s) \, \delta B_s^i + X_0 F(\Sigma_s) \, ds \\ &= \sum_{i=1}^{n} Y_i F(\Sigma_s) \, \delta B_s^i + Y_0 F(\Sigma_s) \, ds \end{aligned}$$

which shows, by taking $F$ to be the standard linear coordinates on $E$, $\Sigma_s$ also solves (5.6). But this is a stochastic differential equation on a Euclidean space $E$ with smooth compactly supported coefficients and therefore has a *unique* solution. ☐

## 5.2 Line integrals

For $a, b \in \mathbb{R}^N$, let $\langle a, b \rangle_{\mathbb{R}^N} := \sum_{i=1}^{N} a_i b_i$ denote the standard inner product on $\mathbb{R}^N$. Also let $\mathfrak{gl}(N) = \mathfrak{gl}(N, \mathbb{R})$ be the set of $N \times N$ real matrices. (It is not necessary to assume $M$ is compact for most of the results in this section.)

**Theorem 5.11.** *As above, for $m \in M$, let $P(m)$ and $Q(m)$ denote orthogonal projection or $\mathbb{R}^N$ onto $\tau_m M$ and $\tau_m M^\perp$ respectively. Then for any $M$-valued semi-martingale $\Sigma$,*

$$0 = Q(\Sigma)\delta\Sigma \text{ and } d\Sigma = P(\Sigma)\delta\Sigma,$$

*i.e.,*

$$\Sigma_s - \Sigma_0 = \int_0^s P(\Sigma_r)\delta\Sigma_r.$$

---

[4] Here we have used the fact that $M$ is a closed subset of $\mathbb{R}^N$.

*Proof.* We will first assume that $M$ is the level set of a function $F$ as in Theorem 2.5. Then we may assume that

$$Q(x) = \phi(x)F'(x)^*(F'(x)F'(x)^*)^{-1}F'(x),$$

where $\phi$ is smooth function on $\mathbb{R}^N$ such that $\phi := 1$ in a neighborhood of $M$ and the support of $\phi$ is contained in the set: $\{x \in \mathbb{R}^N | F'(x)$ is surjective$\}$. By Itô's lemma

$$0 = d0 = d(F(\Sigma)) = F'(\Sigma)\delta\Sigma.$$

The lemma follows in this special case by multiplying the above equation through by $\phi(\Sigma)F'(\Sigma)^*(F'(\Sigma)F'(\Sigma)^*)^{-1}$ (see the proof of Lemma 2.30).

For the general case, choose two open covers $\{V_i\}$ and $\{U_i\}$ of $M$ such that each $\bar{V}_i$ is compactly contained in $U_i$, there is a smooth function $F_i \in C_c^\infty(U_i \to \mathbb{R}^{N-d})$ such that $V_i \cap M = V_i \cap \{F_i^{-1}(\{0\})\}$ and $F_i$ has a surjective differential on $V_i \cap M$. Choose $\phi_i \in C_c^\infty(\mathbb{R}^N)$ such that the support of $\phi_i$ is contained in $V_i$ and $\sum \phi_i = 1$ on $M$, with the sum being locally finite. (For the existence of such covers and functions, see the discussion of partitions of unity in any reasonable book about manifolds.) Notice that $\phi_i \cdot F_i \equiv 0$ and that $F_i \cdot \phi_i' \equiv 0$ on $M$ so that

$$0 = d\{\phi_i(\Sigma)F_i(\Sigma)\} = (\phi_i'(\Sigma)\delta\Sigma)F_i(\Sigma) + \phi_i(\Sigma)F_i'(\Sigma)\delta\Sigma$$
$$= \phi_i(\Sigma)F_i'(\Sigma)\delta\Sigma.$$

Multiplying this equation by $\Psi_i(\Sigma)F_i'(\Sigma)^*(F_i'(\Sigma)F_i'(\Sigma)^*)^{-1}$, where each $\Psi_i$ is a smooth function on $\mathbb{R}^N$ such that $\Psi_i \equiv 1$ on the support of $\phi_i$ and the support of $\Psi_i$ is contained in the set where $F_i'$ is surjective, we learn that

$$0 = \phi_i(\Sigma)F_i'(\Sigma)^*(F_i'(\Sigma)F_i'(\Sigma)^*)^{-1}F_i'(\Sigma)\delta\Sigma = \phi_i(\Sigma)Q(\Sigma)\delta\Sigma \qquad (5.8)$$

for all $i$. By a stopping time argument we may assume that $\Sigma$ never leaves a compact set, and therefore we may choose a finite subset $I$ of the indices $\{i\}$ such that $\sum_{i \in I} \phi_i(\Sigma)Q(\Sigma) = Q(\Sigma)$. Hence, summing over $i \in I$ in (5.8) shows that $0 = Q(\Sigma)\delta\Sigma$. Since $Q + P = I$, it follows that

$$d\Sigma = I\delta\Sigma = [Q(\Sigma) + P(\Sigma)]\delta\Sigma = P(\Sigma)\delta\Sigma. \qquad \square$$

The following notation will be needed to define line integrals along a semimartingale $\Sigma$.

**Notation 5.12.** Let $P(m)$ be orthogonal projection of $\mathbb{R}^N$ onto $\tau_m M$ as above.

1. Given a one-form $\alpha$ on $M$ let $\tilde{\alpha} : M \to (\mathbb{R}^N)^*$ be defined by

$$\tilde{\alpha}(m)v := \alpha((P(m)v)_m) \qquad (5.9)$$

for all $m \in M$ and $v \in \mathbb{R}^N$.

2. Let $\Gamma(T^*M \otimes T^*M)$ denote the set of functions $\rho : \bigcup_{m \in M} T_m M \otimes T_m M \to \mathbb{R}$ such that $\rho_m := \rho|_{T_m M \otimes T_m M}$ is linear, and $m \to \rho(X(m) \otimes Y(m))$ is a smooth function on $M$ for all smooth vector fields $X, Y \in \Gamma(TM)$. (Riemannian metrics and Hessians of smooth functions are examples of elements of $\Gamma(T^*M \otimes T^*M)$.)

3. For $\rho \in \Gamma(T^*M \otimes T^*M)$, let $\tilde{\rho} : M \to (\mathbb{R}^N \otimes \mathbb{R}^N)^*$ be defined by

$$\tilde{\rho}(m)(v \otimes w) := \rho((P(m)v)_m \otimes (P(m)w)_m). \tag{5.10}$$

**Definition 5.13.** Let $\alpha$ be a one form on $M$, $\rho \in \Gamma(T^*M \otimes T^*M)$, and $\Sigma$ be an $M$-valued semi-martingale. Then the *Fisk–Stratonovich* integral of $\alpha$ along $\Sigma$ is:

$$\int_0^{\cdot} \alpha(\delta\Sigma) := \int_0^{\cdot} \tilde{\alpha}(\Sigma)\delta\Sigma, \tag{5.11}$$

and the *Itô* integral is given by:

$$\int_0^{\cdot} \alpha(\bar{d}\Sigma) := \int_0^{\cdot} \tilde{\alpha}(\Sigma)d\Sigma, \tag{5.12}$$

where the stochastic integrals on the right hand sides of (5.11) and (5.12) are Fisk–Stratonovich and Itô integrals respectively. Formally, $\bar{d}\Sigma := P(\Sigma)d\Sigma$. We also define a *quadratic integral*:

$$\int_0^{\cdot} \rho(d\Sigma \otimes d\Sigma) := \int_0^{\cdot} \tilde{\rho}(\Sigma)(d\Sigma \otimes d\Sigma) := \sum_{i,j=1}^{N} \int_0^{\cdot} \tilde{\rho}(\Sigma)(e_i \otimes e_j)d[\Sigma^i, \Sigma^j], \tag{5.13}$$

where $\{e_i\}_{i=1}^{N}$ is an orthonormal basis for $\mathbb{R}^N$, $\Sigma^i := \langle e_i, \Sigma \rangle$, and $d[\Sigma^i, \Sigma^j]$ is the differential of the mutual quadratic variation of $\Sigma^i$ and $\Sigma^j$.

So as not to confuse $[\Sigma^i, \Sigma^j]$ with a commutator or a Lie bracket, in the sequel we will write $d\Sigma^i d\Sigma^j$ for $d[\Sigma^i, \Sigma^j]$.

**Remark 5.14.** The above definitions may be generalized as follows. Suppose that $\alpha$ is now a $T^*M$-valued semi-martingale and $\Sigma$ is the $M$-valued semi-martingale such that $\alpha_s \in T^*_{\Sigma_s} M$ for all $s$. Then we may define

$$\tilde{\alpha}_s v := \alpha_s((P(\Sigma_s)v)_{\Sigma_s}),$$

$$\int_0^{\cdot} \alpha(\delta\Sigma) := \int_0^{\cdot} \tilde{\alpha}\delta\Sigma, \tag{5.14}$$

and

$$\int_0^{\cdot} \alpha(\bar{d}\Sigma) := \int_0^{\cdot} \tilde{\alpha}d\Sigma. \tag{5.15}$$

Similarly, if $\rho$ is a process in $T^*M \otimes T^*M$ such that $\rho_s \in T^*_{\Sigma_s} M \otimes T^*_{\Sigma_s} M$, let

$$\int_0^{\cdot} \rho(d\Sigma \otimes d\Sigma) = \int_0^{\cdot} \tilde{\rho}(d\Sigma \otimes d\Sigma), \tag{5.16}$$

where

$$\tilde{\rho}_s(v \otimes w) := \rho_s((P(\Sigma_s)v)_{\Sigma_s} \otimes (P(\Sigma_s)v)_{\Sigma_s})$$

and

$$d\Sigma \otimes d\Sigma = \sum_{i,j=1}^{N} e_i \otimes e_j d\Sigma^i d\Sigma^j \tag{5.17}$$

as in (5.13).

**Lemma 5.15.** *Suppose that* $\alpha = f dg$ *for some functions* $f, g \in C^\infty(M)$, *then*

$$\int_0^{\cdot} \alpha(\delta\Sigma) = \int_0^{\cdot} f(\Sigma)\delta[g(\Sigma)].$$

*Since, by Corollary 3.42, any one-form* $\alpha$ *on* $M$ *may be written as* $\alpha = \sum_{i=1}^{N} f_i dg_i$ *with* $f_i, g_i \in C^\infty(M)$, *it follows that the Fisk–Stratonovich integral is intrinsically defined independent of how* $M$ *is imbedded into a Euclidean space.*

*Proof.* Let $G$ be a smooth function on $\mathbb{R}^N$ such that $g = G|_M$. Then $\tilde{\alpha}(m) = f(m)G'(m)P(m)$, so that

$$\int_0^{\cdot} \alpha(\delta\Sigma) = \int_0^{\cdot} f(\Sigma)G'(\Sigma)P(\Sigma)\delta\Sigma$$

$$= \int_0^{\cdot} f(\Sigma)G'(\Sigma)\delta\Sigma \qquad \text{(by Theorem 5.11)}$$

$$= \int_0^{\cdot} f(\Sigma)\delta[G(\Sigma)] \qquad \text{(by Itô's Lemma)}$$

$$= \int_0^{\cdot} f(\Sigma)\delta[g(\Sigma)]. \qquad (g(\Sigma) = G(\Sigma))$$

**Lemma 5.16.** *Suppose that* $\rho = f dh \otimes dg$, *where* $f, g, h \in C^\infty(M)$, *then*

$$\int_0^{\cdot} \rho(d\Sigma \otimes d\Sigma) = \int_0^{\cdot} f(\Sigma)d[h(\Sigma), g(\Sigma)] =: \int_0^{\cdot} f(\Sigma)d[h(\Sigma)]d[g(\Sigma)].$$

*Since, by an argument similar to that in Corollary 3.42, any* $\rho \in \Gamma(T^*M \otimes T^*M)$ *may be written as a finite linear combination* $\rho = \sum_i f_i dh_i \otimes dg_i$ *with* $f_i, h_i, g_i \in C^\infty(M)$, *it follows that the quadratic integral is intrinsically defined independent of the imbedding.*

*Proof.* By Theorem 5.11, $\delta\Sigma = P(\Sigma)\delta\Sigma$, so that

$$\Sigma_s^i = \Sigma_0^i + \int_0^{\cdot} (e_i, P(\Sigma)d\Sigma) + B.V.$$

$$= \Sigma_0^i + \sum_k \int_0^{\cdot} (e_i, P(\Sigma)e_k)d\Sigma^k + B.V.,$$

where $B.V.$ denotes a process of bounded variation. Therefore

$$d[\Sigma^i, \Sigma^j] = \sum_{k,l} (e_i, P(\Sigma)e_k)(e_i, P(\Sigma)e_l)d\Sigma^k d\Sigma^l. \tag{5.18}$$

Now let $H$ and $G$ be in $C^\infty(\mathbb{R}^N)$ such that $h = H|_M$ and $g = G|_M$. By Itô's lemma and (5.18),

$$\begin{aligned}
d[h(\Sigma), g(\Sigma)] &= \sum_{i,j} (H'(\Sigma)e_i)(G'(\Sigma)e_j)d[\Sigma^i, \Sigma^j] \\
&= \sum_{i,j,k,l} (H'(\Sigma)e_i)(G'(\Sigma)e_j)(e_i, P(\Sigma)e_k)(e_i, P(\Sigma)e_l)d\Sigma^k d\Sigma^l \\
&= \sum_{k,l} (H'(\Sigma)P(\Sigma)e_k)(G'(\Sigma)P(\Sigma)e_l)d\Sigma^k d\Sigma^l.
\end{aligned}$$

Since

$$\tilde\rho(m) = f(m) \cdot (H'(m)P(m)) \otimes (G'(m)P(m)),$$

it follows from (5.13) and the two above displayed equations that

$$\begin{aligned}
\int_0^\cdot f(\Sigma)d[h(\Sigma), g(\Sigma)] &:= \int_0^\cdot \sum_{k,l} f(\Sigma)(H'(\Sigma)P(\Sigma)e_k)(G'(\Sigma)P(\Sigma)e_l)d\Sigma^k d\Sigma^l \\
&= \int_0^\cdot \tilde\rho(\Sigma)(d\Sigma \otimes d\Sigma) =: \int_0^\cdot \rho(d\Sigma \otimes d\Sigma).
\end{aligned}$$

**Theorem 5.17.** *Let $\alpha$ be a one form on $M$, and $\Sigma$ be an $M$-valued semi-martingale. Then*

$$\int_0^\cdot \alpha(\delta\Sigma) = \int_0^\cdot \alpha(\bar{d}\Sigma) + \frac{1}{2}\int_0^\cdot \nabla\alpha(d\Sigma \otimes d\Sigma), \tag{5.19}$$

*where $\nabla\alpha(v_m \otimes w_m) := (\nabla_{v_m}\alpha)(w_m)$ and $\nabla\alpha$ is defined in Definition 3.40, also see Lemma 3.41. (This shows that the Itô integral depends not only on the manifold structure of $M$ but on the geometry of $M$ as reflected in the Levi-Civita covariant derivative $\nabla$.)*

*Proof.* Let $\tilde\alpha$ be as in (5.9). For the purposes of the proof, suppose that $\tilde\alpha : M \to (\mathbb{R}^N)^*$ has been extended to a smooth function from $\mathbb{R}^N \to (\mathbb{R}^N)^*$. We still denote this extension by $\tilde\alpha$. Then using (5.18),

$$\begin{aligned}
\int_0^\cdot \alpha(\delta\Sigma) &:= \int_0^\cdot \tilde\alpha(\Sigma)\delta\Sigma \\
&= \int_0^\cdot \tilde\alpha(\Sigma)d\Sigma + \frac{1}{2}\int_0^\cdot \tilde\alpha'(\Sigma)(d\Sigma)d\Sigma \\
&= \int_0^\cdot \alpha(\bar{d}\Sigma) + \frac{1}{2}\sum_{i,j,k,l} \int_0^\cdot \tilde\alpha'(\Sigma)(e_i)e_j(e_i, P(\Sigma)e_k)(e_i, P(\Sigma)e_l)d\Sigma^k d\Sigma^l
\end{aligned}$$

$$= \int_0^{\cdot} \alpha(\bar{d}\Sigma) + \frac{1}{2} \sum_{k,l} \int_0^{\cdot} \tilde{\alpha}'(\Sigma)(P(\Sigma)e_k)P(\Sigma)e_l d\Sigma^k d\Sigma^l$$

$$= \int_0^{\cdot} \alpha(\bar{d}\Sigma) + \frac{1}{2} \sum_{k,l} \int_0^{\cdot} d\tilde{\alpha}((P(\Sigma)e_k)_\Sigma)P(\Sigma)e_l d\Sigma^k d\Sigma^l.$$

But by (3.45), we know for all $v_m, w_m \in TM$ that

$$\nabla\alpha(v_m \otimes w_m) = d\tilde{\alpha}(v_m)w$$

which combined with the previous equation implies

$$\int_0^{\cdot} \alpha(\delta\Sigma) = \int_0^{\cdot} \alpha(\bar{d}\Sigma) + \frac{1}{2} \sum_{k,l} \int_0^{\cdot} \nabla\alpha((P(\Sigma)e_k)_\Sigma \otimes (P(\Sigma)e_l)_\Sigma)d\Sigma^k d\Sigma^l$$

$$= \int_0^{\cdot} \alpha(\bar{d}\Sigma) + \frac{1}{2} \sum_{k,l} \int_0^{\cdot} \nabla\alpha(d\Sigma \otimes d\Sigma).$$

**Corollary 5.18 (Itô's Lemma for Manifolds).** *If* $u \in C^\infty((0, T) \times M)$ *and* $\Sigma$ *is an M-valued semi-martingale, then*

$$d[u(s, \Sigma_s)] = (\partial_s u)(s, \Sigma_s)\, ds$$

$$+ d_M[u(s, \cdot)](\bar{d}\Sigma_s) + \frac{1}{2}(\nabla d_M u(s, \cdot))(d\Sigma_s \otimes d\Sigma_s), \qquad (5.20)$$

*where, as in Notation 2.20,* $d_M u(s, \cdot)$ *is being used to denote the differential of the map:* $m \in M \to u(s, m)$.

*Proof.* Let $U \in C^\infty((0, T) \times \mathbb{R}^N)$ such that $u(s, \cdot) = U(s, \cdot)|_M$. Then by Itô's lemma and Theorem 5.11,

$$d[u(s, \Sigma_s)] = d[U(s, \Sigma_s)] = (\partial_s U)(s, \Sigma_s)\, ds + D_\Sigma U(s, \Sigma_s)\delta\Sigma_s$$
$$= (\partial_s U)(s, \Sigma_s)\, ds + D_\Sigma U(s, \Sigma_s)P(\Sigma_s)\delta\Sigma_s$$
$$= (\partial_s u)(s, \Sigma_s)\, ds + d_M[u(s, \cdot)](\delta\Sigma_s)$$
$$= (\partial_s u)(s, \Sigma_s)\, ds + d_M[u(s, \cdot)](\bar{d}\Sigma_s)$$
$$+ \frac{1}{2}(\nabla d_M u(s, \cdot))(d\Sigma_s \otimes d\Sigma_s),$$

wherein the last equality is a consequence of Theorem 5.17.    $\square$

### 5.3 M-valued Martingales and Brownian motions

**Definition 5.19.** An $M$-valued semi-martingale $\Sigma$ is said to be a (local) *martingale* (more precisely a $\nabla$-martingale) if

$$\int_0^{\cdot} df(\bar{d}\Sigma) = f(\Sigma) - f(\Sigma_0) - \frac{1}{2}\int_0^{\cdot} \nabla df(d\Sigma \otimes d\Sigma) \qquad (5.21)$$

is a (local) martingale for all $f \in C^\infty(M)$. (See Theorem 5.17 for the truth of the equality in Eq. (5.21).) The process $\Sigma$ is said to be a *Brownian motion* if

$$f(\Sigma) - f(\Sigma_0) - \frac{1}{2} \int_0^{\cdot} \Delta f(\Sigma) d\lambda \tag{5.22}$$

is a local martingale for all $f \in C^\infty(M)$, where $\lambda(s) := s$ and $\int_0^{\cdot} \Delta f(\Sigma) d\lambda$ denotes the process $s \to \int_0^s \Delta f(\Sigma) d\lambda$.

**Theorem 5.20 (Projection Construction of Brownian Motion).** *Suppose that $B = \left(B^1, B^2, \ldots, B^N\right)$ is an $N$-dimensional Brownian motion. Then there is a unique $M$-valued semi-martingale $\Sigma$ which solves the Fisk–Stratonovich stochastic differential equation,*

$$\delta\Sigma = P(\Sigma)\delta B \quad with \quad \Sigma_0 = o \in M, \tag{5.23}$$

*see Figure 13. Moreover, $\Sigma$ is an $M$-valued Brownian motion.*

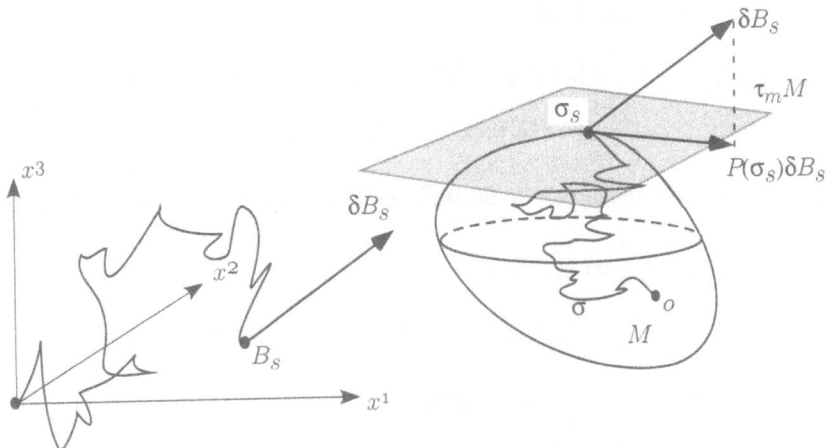

**Figure 13.** Projection construction of Brownian motion on $M$.

*Proof.* Let $\{e_i\}_{i=1}^N$ be the standard basis for $\mathbb{R}^N$ and $X_i(m) := P(m)e_i \in T_m M$ for each $i = 1, 2, \ldots, N$ and $m \in M$. Then (5.23) is equivalent to the SDE,

$$\delta\Sigma = \sum_{i=1}^N X_i(\Sigma)\delta B^i \quad with \quad \Sigma_0 = o \in M$$

which has a unique solution by Theorem 5.10. Using Lemma 5.6, this equation may be rewritten in the Itô form as

$$d[f(\Sigma)] = \sum_{i=1}^{N} X_i f(\Sigma) dB^i + \frac{1}{2} \sum_{i=1}^{N} X_i^2 f(\Sigma) ds \text{ for all } f \in C^{\infty}(M).$$

This completes the proof since $\sum_{i=1}^{N} X_i^2 = \Delta$ by Proposition 3.48.    □

**Lemma 5.21 (Lévy's Criteria).** *For each $m \in M$, let $\mathcal{I}(m) := \sum_{i=1}^{d} E_i \otimes E_i$, where $\{E_i\}_{i=1}^{d}$ is an orthonormal basis for $T_m M$. An $M$-valued semi-martingale, $\Sigma$, is a Brownian motion iff $\Sigma$ is a martingale and*

$$d\Sigma \otimes d\Sigma = \mathcal{I}(\Sigma)d\lambda. \tag{5.24}$$

*More precisely, this last condition is to be interpreted as:*

$$\int_0^{\cdot} \rho(d\Sigma \otimes d\Sigma) = \int_0^{\cdot} \rho(\mathcal{I}(\Sigma))d\lambda \; \forall \; \rho \in \Gamma(T^*M \otimes T^*M). \tag{5.25}$$

*Proof.* ($\Rightarrow$) Suppose that $\Sigma$ is a Brownian motion on $M$ (so (5.22) holds) and $f, g \in C^{\infty}(M)$. Then on the one hand

$$d(f(\Sigma)g(\Sigma)) = d[f(\Sigma)] \cdot g(\Sigma) + f(\Sigma)d[g(\Sigma)] + d[f(\Sigma), g(\Sigma)]$$

$$\cong \frac{1}{2}\{\Delta f(\Sigma)g(\Sigma) + f(\Sigma)\Delta g(\Sigma)\}d\lambda + d[f(\Sigma), g(\Sigma)],$$

where "$\cong$" denotes equality up to the differential of a martingale. On the other hand,

$$d(f(\Sigma)g(\Sigma)) \cong \frac{1}{2}\Delta(fg)(\Sigma)d\lambda$$

$$= \frac{1}{2}\{\Delta f(\Sigma)g(\Sigma) + f(\Sigma)\Delta g(\Sigma) + 2\langle \text{grad } f, \text{grad} g \rangle (\Sigma)\}d\lambda.$$

Comparing the above two equations implies that

$$d[f(\Sigma), g(\Sigma)] = \langle \text{grad } f, \text{grad} g \rangle (\Sigma)d\lambda = df \otimes dg(\mathcal{I}(\Sigma))d\lambda.$$

Therefore by Lemma 5.16, if $\rho = h \cdot df \otimes dg$ then

$$\int_0^{\cdot} \rho(d\Sigma \otimes d\Sigma) = \int_0^{\cdot} h(\Sigma)d[f(\Sigma), g(\Sigma)]$$

$$= \int_0^{\cdot} h(\Sigma)(df \otimes dg)(\mathcal{I}(\Sigma))d\lambda = \int_0^{\cdot} \rho(\mathcal{I}(\Sigma))d\lambda.$$

Since the general element $\rho$ of $\Gamma(T^*M \otimes T^*M)$ is a finite linear combination of expressions of the form $hdf \otimes dg$, it follows that (5.24) holds. Moreover, (5.24) implies

$$(\nabla df)(d\Sigma \otimes d\Sigma) = (\nabla df)(\mathcal{I}(\Sigma))d\lambda = \Delta f(\Sigma)d\lambda \tag{5.26}$$

and therefore,

$$f(\Sigma) - f(\Sigma_0) - \frac{1}{2}\int_0^{\cdot} \nabla df(d\Sigma \otimes d\Sigma)$$

$$= f(\Sigma) - f(\Sigma_0) - \frac{1}{2}\int_0^{\cdot} \Delta f(\Sigma)d\lambda \tag{5.27}$$

is a martingale and so by definition $\Sigma$ is a martingale.

Conversely assume $\Sigma$ is a martingale and (5.24) holds. Then (5.26) and (5.27) hold and they imply $\Sigma$ is a Brownian motion, see Definition 5.19. $\qquad\square$

**Definition 5.22** ($\delta^{\nabla}V := P\delta V$)**.** Suppose $\alpha$ is a one form on $M$ and $V$ is a $TM$-valued semi-martingale, i.e., $V_s = (\Sigma_s, v_s)$, where $\Sigma$ is an $M$-valued semi-martingale and $v$ is an $\mathbb{R}^N$-valued semi-martingale such that $v_s \in \tau_{\Sigma_s} M$ for all $s$. Then we define:

$$\int_0^{\cdot} \alpha(\delta^{\nabla}V) := \int_0^{\cdot} \tilde{\alpha}(\Sigma)\delta v = \int_0^{\cdot} \alpha(\Sigma)\,(P\,(\Sigma)\,\delta v)\,. \tag{5.28}$$

**Remark 5.23.** Suppose that $\alpha(v_m) = \theta(m)v$, where $\theta : M \to (\mathbb{R}^N)^*$ is a smooth function. Then

$$\int_0^{\cdot} \alpha(\delta^{\nabla}V) := \int_0^{\cdot} \theta(\Sigma)P(\Sigma)\delta v = \int_0^{\cdot} \theta(\Sigma)\{\delta v + dQ(\delta\Sigma)v\},$$

where we have used the identity:

$$\delta^{\nabla}V = P(\Sigma)\delta v = \delta v + dQ(\delta\Sigma)v. \tag{5.29}$$

This last identity follows by taking the differential of the identity, $v = P(\Sigma)v$, as in the proof of Proposition 3.32.

**Proposition 5.24 (Product Rule).** *Keeping the notation of above, we have*

$$\delta(\alpha(V)) = \nabla\alpha(\delta\Sigma \otimes V) + \alpha(\delta^{\nabla}V), \tag{5.30}$$

*where* $\nabla\alpha(\delta\Sigma \otimes V) := \gamma(\delta\Sigma)$ *and* $\gamma$ *is the* $T^*M$-valued semi-martingale defined by

$$\gamma_s\,(w) := \nabla\alpha(w \otimes V_s) = (\nabla_w \alpha)\,(V_s)\,\text{for any } w \in T_{\Sigma_s}M.$$

*Proof.* Let $\theta : \mathbb{R}^N \to (\mathbb{R}^N)^*$ be a smooth map such that $\tilde{\alpha}(m) = \theta(m)|_{\tau_m M}$ for all $m \in M$. By Lemma 5.15, $\delta(\theta(\Sigma)P(\Sigma)) = d(\theta P)(\delta\Sigma)$ and hence by Lemma 3.41, $\delta(\theta(\Sigma)P(\Sigma))v = \nabla\alpha(\delta\Sigma \otimes V)$, where $\nabla\alpha(v_m \otimes w_m) := (\nabla_{v_m}\alpha)(w_m)$ for all $v_m, w_m \in TM$. Therefore:

$$\delta(\alpha(V)) = \delta(\theta(\Sigma)v) = \delta(\theta(\Sigma)P(\Sigma)v) = (d(\theta P)(\delta\Sigma))v + \theta(\Sigma)P(\Sigma)\delta v$$

$$= (d(\theta P)(\delta\Sigma))v + \tilde{\alpha}(\Sigma)\delta v = \nabla\alpha(\delta\Sigma \otimes V) + \alpha(\delta^{\nabla}V). \qquad\square$$

## 5.4 Stochastic parallel translation and development maps

**Definition 5.25.** A $TM$-valued semi-martingale $V$ is said to be *parallel* if $\delta^\nabla V \equiv 0$, i.e., $\int_0^{\cdot} \alpha(\delta^\nabla V) \equiv 0$ for all one-forms $\alpha$ on $M$.

**Proposition 5.26.** *A $TM$-valued semi-martingale $V = (\Sigma, v)$ is parallel iff*

$$\int_0^{\cdot} P(\Sigma)\delta v = \int_0^{\cdot} \{\delta v + dQ(\delta\Sigma)v\} \equiv 0. \tag{5.31}$$

*Proof.* Let $x = (x^1, \ldots, x^N)$ denote the standard coordinates on $\mathbb{R}^N$. If $V$ is parallel then,

$$0 \equiv \int_0^{\cdot} dx^i (\delta^\nabla V) = \int_0^{\cdot} \langle e_i, P(\Sigma)\delta v \rangle$$

for each $i$ which implies (5.31). The converse follows from Remark 5.23.    □

In the following theorem, $V_0$ is said to be a measurable vector field on $M$ if $V_0(m) = (m, v(m))$ with $v : M \to \mathbb{R}^N$ being a (Borel) measurable function such that $v(m) \in \tau_m M$ for all $m \in M$.

**Theorem 5.27 (Stochastic Parallel Translation on $M \times \mathbb{R}^N$).** *Let $\Sigma$ be an $M$-valued semi-martingale, and $V_0(m) = (m, v(m))$ be a measurable vector field on $M$, then there is a unique parallel $TM$-valued semi-martingale $V$ such that $V_0 = V_0(\Sigma_0)$ and $V_s \in T_{\Sigma_s} M$ for all $s$. Moreover, if $u$ denotes the solution to the SDE*

$$\delta u + \Gamma(\delta\Sigma)u = 0 \quad \text{with} \quad u_0 = I \in O(N), \tag{5.32}$$

*(where $O(N)$ is as in Example 2.6 and $\Gamma$ is as in (3.65)) then $V_s = (\Sigma_s, u_s v(\Sigma_0))$. The process $u$ defined in (5.32) is orthogonal for all $s$ and satisfies $P(\Sigma_s)u_s = u_s P(\Sigma_0)$. Moreover if $\Sigma_0 = o \in M$ a.e. and $v \in \tau_o M$ and $w \perp \tau_o M$, then $u_s v$ and $u_s w$ satisfy*

$$\delta [u_s v] + dQ(\delta\Sigma) u_s v = P(\Sigma) \delta [u_s v] = 0 \tag{5.33}$$

*and*

$$\delta [u_s w] + dP(\delta\Sigma) u_s v = Q(\Sigma) \delta [u_s v] = 0. \tag{5.34}$$

*Proof.* The assertions prior to (5.33) are the stochastic analogs of Lemmas 3.56 and 3.57. The proof may be given by replacing $\frac{d}{ds}$ everywhere in the proofs of Lemmas 3.56 and 3.57 by $\delta_s$ to get a proof in this stochastic setting. (5.33) and (5.34) are now easily verified, for example using $P(\Sigma) uv = uv$, we have

$$\delta [uv] = \delta [P(\Sigma) uv] = P(\delta\Sigma) uv + P(\Sigma) \delta [uv]$$

which proves the first equality in (5.33). For the second equality in (5.33),

$$\begin{aligned} P(\Sigma) \delta [uv] &= -P(\Sigma) \Gamma(\delta\Sigma) [uv] \\ &= -P(\Sigma) [dQ(\delta\Sigma)P(\Sigma) + dP(\delta\Sigma)Q(\Sigma)] [uv] \\ &= -dQ(\delta\Sigma)Q(\Sigma) P(\Sigma)\delta [uv] = 0 \end{aligned}$$

where Lemma 3.30 was used in the third equality. The proof of (5.34) is completely analogous. The skeptical reader is referred to Section 3 of Driver [47] for more details.    □

**Definition 5.28 (Stochastic Parallel Translation).** Given $v \in \mathbb{R}^N$ and an $M$-valued semi-martingale $\Sigma$, let $//_s(\Sigma)v_{\Sigma_0} = (\Sigma_s, u_s v)$, where $u$ solves (5.32). (Note: $V_s = //_s(\Sigma)V_0$.)

In the remainder of these notes, I will often abuse notation and write $u_s$ instead of $//_s := //_s(\Sigma)$ and $v_s$ rather than $V_s = (\Sigma_s, v_s)$. For example, the reader should sometimes interpret $u_s v$ as $//_s(\Sigma)v_{\Sigma_0}$ depending on the context. Essentially, we will be identifying $\tau_m M$ with $T_m M$ when no particular confusion will arise.

*Convention.* Let us now fix a *base point* $o \in M$ and unless otherwise noted, we will assume that all $M$-valued semi-martingales, $\Sigma$, start at $o \in M$, i.e., $\Sigma_0 = o$ a.e.

To each $M$-valued semi-martingale, $\Sigma$, let $\Psi(\Sigma) := b$ where

$$b := \int_0^{\cdot} //^{-1}\delta\Sigma = \int_0^{\cdot} u^{-1}\delta\Sigma = \int_0^{\cdot} u^{\mathrm{tr}}\delta\Sigma.$$

Then $b = \Psi(\Sigma)$ is a $T_o M$-valued semi-martingale such that $b_0 = 0_o \in T_o M$. The converse holds as well.

**Theorem 5.29 (Stochastic Development Map).** *Suppose that $o \in M$ is given and $b$ is a $T_o M$-valued semi-martingale. Then there exists a unique $M$-valued semi-martingale $\Sigma$ such that*

$$\delta\Sigma_s = //_s \delta b_s = u_s \delta b_s \quad with \quad \Sigma_0 = o \tag{5.35}$$

*where $u$ solves (5.32).*

*Proof.* This theorem is a stochastic analog of Theorem 4.10 and the reader is again referred to Figure 11. To prove the existence and uniqueness, we may follow the method in the proof of Theorem 4.10. Namely, the pair $(\Sigma, u) \in M \times O(N)$ solves a SDE of the form

$$\delta\Sigma = u\delta b \quad with \quad \Sigma_0 = o$$
$$\delta u = -\Gamma(\delta\Sigma)u = -\Gamma(u\delta b)u \quad with \quad u_0 = I \in O(N)$$

which after a little effort can be expressed in a form for which Theorem 5.10 may be applied. The details will be left to the reader, or see (for example) Section 3 of Driver [47]. $\square$

**Notation 5.30.** As in the smooth case, define $\Sigma = \phi(b)$, so that

$$\Psi(\Sigma) := \phi^{-1}(b) = \int_0^{\cdot} //_r(\Sigma)^{-1}\delta\Sigma_r.$$

In what follows, we will assume that $b_s$, $u_s$ (or equivalently $//_s(\Sigma)$), and $\Sigma_s$ are related by Equations (5.35) and (5.32), i.e., $\Sigma = \phi(b)$ and $u = // = //(\Sigma)$. Recall that $\bar{d}\Sigma = P(\Sigma)d\Sigma$ is the Itô differential of $\Sigma$, see Definition 5.13.

**Proposition 5.31.** *Let $\Sigma = \phi(b)$, then*

$$\bar{d}\Sigma = P(\Sigma)d\Sigma = udb. \tag{5.36}$$

*Also*

$$d\Sigma \otimes d\Sigma = udb \otimes udb := \sum_{i,j=1}^{d} ue_i \otimes ue_j db^i db^j, \tag{5.37}$$

*where $\{e_i\}_{i=1}^{d}$ is an orthonormal basis for $T_oM$ and $b = \sum_{i=1}^{d} b^i e_i$. More precisely*

$$\int_0^{\cdot} \rho(d\Sigma \otimes d\Sigma) = \int_0^{\cdot} \sum_{i,j=1}^{d} \rho(ue_i \otimes ue_j) db^i db^j,$$

*for all $\rho \in \Gamma(T^*M \otimes T^*M)$.*

*Proof.* Consider the identity:

$$d\Sigma = u\delta b = udb + \frac{1}{2} dudb$$

$$= udb - \frac{1}{2}\Gamma(\delta\Sigma)udb = udb - \frac{1}{2}\Gamma(udb)udb$$

where $\Gamma$ is as defined in (3.65). Hence

$$\bar{d}\Sigma = P(\Sigma)d\Sigma = udb - \frac{1}{2}\sum_{i,j=1}^{d} P(\Sigma)\Gamma((ue_i)_\Sigma)ue_j db^i db^j.$$

The proof of (5.36) is finished upon observing,

$$P\Gamma P = P\{dQP + dPQ\}P = PdQP = PQdQ = 0.$$

The proof of (5.37) is easy and will be left for the reader.     □

**Fact 5.32.** If $(M, g)$ is a complete Riemannian manifold and the Ricci curvature tensor is bounded from below[5], then $\Delta = \Delta_g$ acting on $C_c^\infty(M)$ is essentially self-adjoint, i.e., the closure $\bar{\Delta}$ of $\Delta$ is an unbounded self-adjoint operator on $L^2(M, d\lambda)$. (Here $d\lambda = \sqrt{g}dx^1 \dots dx^n$ is being used to denote the Riemann volume measure on $M$.) Moreover, the semi-group $e^{t\bar{\Delta}/2}$ has a smooth integral kernel, $p_t(x, y)$, such that

$$p_t(x, y) \geq 0 \text{ for all } x, y \in M$$

$$\int_M p_t(x, y)d\lambda(y) = 1 \text{ for all } x \in M \text{ and}$$

$$\left(e^{t\bar{\Delta}/2}f\right)(x) = \int_M p_t(x, y)f(y)d\lambda(y) \text{ for all } f \in L^2(M).$$

If $f \in C_c^\infty(M)$, the function $u(t, x) := e^{t\bar{\Delta}/2}f(x)$ is smooth for $t > 0$ and $x \in M$ and $Le^{t\bar{\Delta}/2}f(x)$ is continuous for $t \geq 0$ and $x \in M$ for any smooth linear differential operator $L$ on $C^\infty(M)$. For these results, see for example Strichartz [165], Dodziuk [43] and Davies [41].

---

[5] These assumptions are always satisfied when $M$ is compact.

**Theorem 5.33 (Stochastic Rolling Constructions).** *Assume M is compact and let* $\Sigma$, $u_s = //_s$, *and b be as in Theorem 5.29, then:*

1. $\Sigma$ *is a martingale iff b is a $T_oM$-valued martingale.*
2. $\Sigma$ *is a Brownian motion iff b is a $T_oM$-valued Brownian motion.*

*Furthermore if $\Sigma$ is a Brownian motion, $T \in (0, \infty)$ and $f \in C^\infty(M)$, then*

$$M_s := \left( e^{(T-s)\tilde{\Delta}/2} f \right) (\Sigma_s)$$

*is a martingale for $s \in [0, T]$ and*

$$dM_s = \left( de^{(T-s)\tilde{\Delta}/2} f \right) (u_s db_s)_{\Sigma_s} = \left( de^{(T-s)\tilde{\Delta}/2} f \right) (//_s db_s). \qquad (5.38)$$

*Proof.* Keep the same notation as in Proposition 5.31 and let $f \in C^\infty(M)$. By Proposition 5.31, if $b$ is a martingale, then $\int_0^\cdot df(\bar{d}\Sigma) = \int_0^\cdot df(udb)$ is also a martingale and hence $\Sigma$ is a martingale, see Definition 5.19. Combining this with Corollary 5.18 and Proposition 5.31,

$$d[f(\Sigma)] = df(\bar{d}\Sigma) + \frac{1}{2}\nabla df(d\Sigma \otimes d\Sigma)$$

$$= df(udb) + \frac{1}{2}\nabla df(udb \otimes udb).$$

Since $u$ is an isometry and if $b$ is a Brownian motion, then $udb \otimes udb = \mathcal{I}(\Sigma)d\lambda$. Hence

$$d[f(\Sigma)] = df(udb) + \frac{1}{2}\Delta f(\Sigma)d\lambda$$

from which it follows that $\Sigma$ is a Brownian motion.

Conversely, if $\Sigma$ is an $M$-valued martingale, then

$$N := \sum_{i=1}^N \int_0^\cdot dx^i(\bar{d}\Sigma)e_i = \sum_{i=1}^N \int_0^\cdot \langle e_i, udb\rangle e_i = \int_0^\cdot udb \qquad (5.39)$$

is a martingale, where $x = (x^1, \ldots, x^N)$ are standard coordinates on $\mathbb{R}^N$ and $\{e_i\}_{i=1}^N$ is the standard basis for $\mathbb{R}^N$. From (5.39), it follows that $b = \int_0^\cdot u^{-1}dN$ is also a martingale.

Now suppose that $\Sigma$ is an $M$-valued Brownian motion, then we have already proved that $b$ is a martingale. To finish the proof it suffices by Lévy's criteria (Lemma 5.21) to show that $db \otimes db = \mathcal{I}(o)d\lambda$. But $\Sigma = N + $ (bounded variation) and hence

$$db \otimes db = u^{-1}d\Sigma \otimes u^{-1}d\Sigma = u^{-1}dN \otimes u^{-1}dN$$

$$= (u^{-1} \otimes u^{-1})(d\Sigma \otimes d\Sigma)$$

$$= (u^{-1} \otimes u^{-1})\mathcal{I}(\Sigma)d\lambda = \mathcal{I}(o)d\lambda,$$

wherein (5.24) was used in the fourth equality and the orthogonality of $u$ was used in the last equality.

To prove (5.38), let $M_s = u(s, \Sigma_s)$ where $u(s, x) := \left(e^{(T-s)\bar{\Delta}/2} f\right)(x)$ which satisfies

$$\partial_s u(s, x) + \frac{1}{2}\Delta u(s, x) = 0 \text{ with } u(T, x) = f(x)$$

By Itô's Lemma (see Corollary 5.18) along with Lemma 5.21 and Proposition 5.31,

$$dM_s = \partial_s u(s, \Sigma_s)\, ds + d_M[u(s, \cdot)](\bar{d}\Sigma_s) + \frac{1}{2}\nabla d_M[u(s, \cdot)](d\Sigma_s \otimes d\Sigma_s)$$

$$= \partial_s u(s, \Sigma_s)\, ds + \frac{1}{2}\Delta u(s, \Sigma_s)\, ds + \left(d_M e^{(T-s)\bar{\Delta}/2} f\right)((u_s db_s)_{\Sigma_s})$$

$$= \left(d_M e^{(T-s)\bar{\Delta}/2} f\right)((u_s db_s)_{\Sigma_s}). \qquad \square$$

The rolling construction of Brownian motion seems to have first been discovered by Eells and Elworthy [63] who used ideas of Gangolli [87]. The relationship of the stochastic development map to stochastic differential equations on the orthogonal frame bundle $O(M)$ of $M$ is pointed out in Elworthy [66, 67, 68]. The frame bundle point of view has also been extensively developed by Malliavin, see for example [130, 129, 131]. For a more detailed history of the stochastic development map, see Elworthy [68], pp. 156–157. The reader may also wish to consult [74, 103, 116, 132, 171, 101].

**Corollary 5.34.** *If $\Sigma$ is a Brownian motion on $M$,*

$$\pi = \{0 = s_0 < s_1 < \cdots < s_n = T\}$$

*is a partition of $[0, T]$ and $f \in C^\infty(M^n)$, then*

$$\mathbb{E} f(\Sigma_{s_1}, \ldots, \Sigma_{s_n}) = \int_{M^n} f(x_1, x_2, \ldots, x_n) \prod_{i=1}^{n} p_{\Delta_i s}(x_{i-1}, x_i)\, d\lambda(x_i) \qquad (5.40)$$

*where $\Delta_i s := s_i - s_{i-1}$, $x_0 := o$ and $\lambda := \lambda_M$. In particular $\Sigma$ is a Markov process relative to the filtration, $\{\mathcal{F}_s\}$ where $\mathcal{F}_s$ is the $\sigma$-algebra generated by $\{\Sigma_\tau : \tau \le s\}$.*

*Proof.* By standard measure theoretic arguments, it suffices to prove (5.40) when $f$ is a product function of the form $f(x_1, x_2, \ldots, x_n) = \prod_{i=1}^{n} f_i(x_i)$ with $f_i \in C^\infty(M)$. By Theorem 5.33, $M_s := e^{(T-s)\bar{\Delta}/2} f_n(\Sigma_s)$ is a martingale for $s \le T$ and therefore

$$\mathbb{E}\left[f(\Sigma_{s_1}, \ldots, \Sigma_{s_n})\right] = \mathbb{E}\left[\prod_{i=1}^{n-1} f_i(\Sigma_{s_i}) \cdot M_T\right] = \mathbb{E}\left[\prod_{i=1}^{n-1} f_i(\Sigma_{s_i}) \cdot M_{s_{n-1}}\right]$$

$$= \mathbb{E}\left[\prod_{i=1}^{n-1} f_i(\Sigma_{s_i}) \cdot (P_{\Delta_n s} f_n)(\Sigma_{s_{n-1}})\right]. \qquad (5.41)$$

In particular if $n = 1$, it follows that

$$\mathbb{E}\left[f_1\left(\Sigma_T\right)\right] = \mathbb{E}\left[\left(e^{T\tilde{\Delta}/2}f_1\right)\left(\Sigma_0\right)\right] = \int_M p_T\left(o, x_1\right)f_1\left(x_1\right)d\lambda\left(x_1\right).$$

Now assume we have proved (5.40) with $n$ replaced by $n - 1$ and to simplify notation let $g\left(x_1, x_2, \ldots, x_{n-1}\right) := \prod_{i=1}^{n-1} f_i\left(x_i\right)$. It would then follow from (5.41) that

$$\mathbb{E}\left[f\left(\Sigma_{s_1}, \ldots, \Sigma_{s_n}\right)\right]$$

$$= \int_{M^{n-1}} g\left(x_1, x_2, \ldots, x_{n-1}\right)\left(e^{\frac{s_n - s_{n-1}}{2}\tilde{\Delta}}f_n\right)\left(x_{n-1}\right)\prod_{i=1}^{n-1} p_{\Delta_i s}\left(x_{i-1}, x_i\right)d\lambda\left(x_i\right)$$

$$= \int_{M^{n-1}} g\left(x_1, x_2, \ldots, x_{n-1}\right)\left[\int_M f_n\left(x_n\right)p_{\Delta_n s}\left(x_{n-1}, x_n\right)d\lambda\left(x_n\right)\right]$$

$$\times \prod_{i=1}^{n-1} p_{\Delta_i s}\left(x_{i-1}, x_i\right)d\lambda\left(x_i\right)$$

$$= \int_{M^n} f\left(x_1, x_2, \ldots, x_n\right)\prod_{i=1}^{n} p_{\Delta_i s}\left(x_{i-1}, x_i\right)d\lambda\left(x_i\right).$$

This completes the induction step and hence also the proof of the theorem.    □

## 5.5 More constructions of semi-Martingales and Brownian motions

Let $\Gamma$ be the one form on $M$ with values in the skew symmetric $N \times N$ matrices defined by $\Gamma = dQP + dPQ$ as in (3.65). Given an $M$-valued semi-martingale $\Sigma$, let $u$ denote parallel translation along $\Sigma$ as defined in (5.32) of Theorem 5.27.

**Lemma 5.35 (Orthogonality Lemma).** *Suppose that $B$ is an $\mathbb{R}^N$-valued semi-martingale and $\Sigma$ is the solution to*

$$\delta\Sigma = P(\Sigma)\delta B \quad \text{with} \quad \Sigma_0 = o \in M. \tag{5.42}$$

*Let $\{e_i\}_{i=1}^{N}$ be any orthonormal basis for $\mathbb{R}^N$ and define $B^i := \langle e_i, B\rangle$ then*

$$P(\Sigma)dB \otimes Q(\Sigma)dB := \sum_{i,j=1}^{N} P(\Sigma)e_i \otimes Q(\Sigma)e_j\left(dB^i dB^j\right) = 0.$$

*Proof.* Suppose $\{v_i\}_{i=1}^{N}$ is another orthonormal basis for $\mathbb{R}^N$. Using the bilinearity of the joint quadratic variation,

$$[\langle e_i, B\rangle, \langle e_j, B\rangle] = \sum_{k,l}[\langle e_i, v_k\rangle\langle v_k, B\rangle, \langle e_j, v_l\rangle\langle v_l, B\rangle]$$

$$= \sum_{k,l}\langle e_i, v_k\rangle\langle e_j, v_l\rangle[\langle v_k, B\rangle, \langle v_l, B\rangle].$$

Therefore,

$$\sum_{i,j=1}^{N} P(\Sigma)e_i \otimes Q(\Sigma)e_j \cdot d\left[B^i, B^j\right]$$

$$= \sum_{i,j,k,l=1}^{N} \left[P(\Sigma)e_i \otimes Q(\Sigma)e_j\right]\langle e_i, v_k\rangle\langle e_j, v_l\rangle d[\langle v_k, B\rangle, \langle v_l, B\rangle]$$

$$= \sum_{k,l=1}^{N} \left[P(\Sigma)v_k \otimes Q(\Sigma)v_l\right] d[\langle v_k, B\rangle, \langle v_l, B\rangle]$$

which shows $P(\Sigma)dB \otimes Q(\Sigma)dB$ is well defined.

Now define

$$\tilde{B} := \int_0^{\cdot} u^{-1}dB \text{ and } \tilde{B}^i := \langle e_i, \tilde{B}\rangle = \int_0^{\cdot} \langle ue_i, dB\rangle$$

where $u$ is parallel translation along $\Sigma$ in $M \times \mathbb{R}^N$ as defined in (5.32). Then

$$P(\Sigma)dB \otimes Q(\Sigma)dB = \sum_{i,j,k,l=1}^{N} P(\Sigma)ue_k \otimes Q(\Sigma)ue_l\langle e_i, ue_k\rangle\langle e_j, ue_l\rangle \left(dB^i dB^j\right)$$

$$= \sum_{k,l=1}^{N} P(\Sigma)ue_k \otimes Q(\Sigma)ue_l \left(d\tilde{B}^k d\tilde{B}^l\right)$$

$$= \sum_{k,l=1}^{N} uP(o)e_k \otimes uQ(o)e_l \left(d\tilde{B}^k d\tilde{B}^l\right)$$

wherein we have used $P(\Sigma)u = uP(o)$ and $Q(\Sigma)u = uQ(o)$, see Theorem 5.27. This last expression is easily seen to be zero by choosing $\{e_i\}$ such that $P(o)e_i = e_i$ for $i = 1, 2, \ldots, d$ and $Q(o)e_j = e_j$ for $j = d+1, \ldots, N$.     □

The next proposition is a stochastic analogue of Lemma 3.55 and the proof is very similar to that of Lemma 3.55.

**Proposition 5.36.** *Suppose that $V$ is a $TM$-valued semi-martingale, $\Sigma = \pi(V)$ so that $\Sigma$ is an $M$-valued semi-martingale and $V_s \in T_{\Sigma_s}M$ for all $s \geq 0$. Then*

$$//_s \delta_s \left[//_s^{-1}V_s\right] = \delta_s^{\nabla} V_s =: P(\Sigma_s)\delta V_s \tag{5.43}$$

*where $//_s$ is stochastic parallel translation along $\Sigma$. If $Y_s \in \Gamma(TM)$ is a time dependent vector field, then*

$$\delta_s \left[//_s^{-1}Y_s(\Sigma_s)\right] = //_s^{-1}\left(\frac{d}{ds}Y_s\right)(\Sigma_s)\,ds + //_s^{-1}\nabla_{\delta\Sigma_s}Y_s \tag{5.44}$$

*and for $w \in T_oM$,*

$$//_s^{-1}\delta_s^\nabla\left[\nabla_{//_sw}Y_s\right] = \delta_s\left[//_s^{-1}\nabla_{//_sw}Y_s\right]$$

$$= //_s^{-1}\nabla_{\delta\Sigma_s\otimes//_sw}^2 Y_s + //_s^{-1}\left[\nabla_{//_sw}\left(\frac{d}{ds}Y_s\right)\right]ds. \qquad (5.45)$$

*Furthermore if $\Sigma_s$ is a Brownian motion, then*

$$d\left[//_s^{-1}Y_s(\Sigma_s)\right] = //_s^{-1}\nabla_{//_sdb_s}Y_s + //_s^{-1}\left(\frac{d}{ds}Y_s\right)(\Sigma_s)\,ds$$

$$+ \frac{1}{2}\sum_{i=1}^d //_s^{-1}\nabla_{//_se_i\otimes//_se_i}^2 Y_s\,ds \qquad (5.46)$$

*where $\{e_i\}_{i=1}^d$ is an orthonormal basis for $T_oM$.*

*Proof.* We will use the convention of summing on repeated indices and write $u_s$ for stochastic parallel translation $//_s$, in $TM$ along $\Sigma$. Recall that $u_s$ solves

$$\delta u_s + dQ\,(\delta\Sigma_s)\,u_s = 0 \text{ with } u_0 = I_{T_oM}.$$

Define $\bar{u}_s$ as the solution to:

$$\delta\bar{u}_s = \bar{u}_s dQ\,(\delta\Sigma_s) \text{ with } \bar{u}_0 = I_{T_oM}.$$

Then

$$\delta\,(\bar{u}_s u_s) = -\bar{u}_s dQ\,(\delta\Sigma_s)\,u_s + \bar{u}_s dQ\,(\delta\Sigma_s)\,u_s = 0$$

from which it follows that $\bar{u}_s u_s = I$ for all $s$ and hence $\bar{u}_s = u_s^{-1}$. This proves (5.43) since

$$u_s\delta_s\left[u_s^{-1}V_s\right] = u_s\left[u_s^{-1}dQ\,(\delta\Sigma_s)\,V_s + u_s^{-1}\delta V_s\right]$$

$$= dQ\,(\delta\Sigma_s)\,V_s + \delta V_s = \delta^\nabla V_s,$$

where the last equality comes from (5.29).

Applying (5.43) to $V_s := Y_s(\Sigma_s)$ gives

$$\delta_s\left[//_s^{-1}Y_s(\Sigma_s)\right] = //_s^{-1}P(\Sigma_s)\,\delta_s\,[Y_s(\Sigma_s)]$$

$$= //_s^{-1}P(\Sigma_s)\left(\frac{d}{ds}Y_s\right)(\Sigma_s)\,ds + //_s^{-1}P(\Sigma_s)\,Y_s'(\Sigma_s)\,\delta_s\Sigma_s$$

$$= //_s^{-1}\left(\frac{d}{ds}Y_s\right)(\Sigma_s)\,ds + //_s^{-1}\nabla_{\delta_s\Sigma_s}Y_s,$$

which proves (5.44).

To prove (5.45), let $X_i(m) = P(m)\,e_i$ for $i = 1, 2, \ldots, N$. By Proposition 3.48,

$$\nabla_{//_sw}Y_s = \langle//_sw, X_i(\Sigma_s)\rangle\,(\nabla_{X_i}Y_s)(\Sigma_s) \qquad (5.47)$$

$$= \langle w, //_s^{-1}X_i(\Sigma_s)\rangle\,(\nabla_{X_i}Y_s)(\Sigma_s)$$

and

$$//_s w = \langle //_s w, X_i(\Sigma_s)\rangle X_i(\Sigma_s) = \langle w, //_s^{-1} X_i(\Sigma_s)\rangle X_i(\Sigma_s)$$

or equivalently,

$$w = \langle w, //_s^{-1} X_i(\Sigma_s)\rangle //_s^{-1} X_i(\Sigma_s). \tag{5.48}$$

Taking the covariant differential of (5.47), making use of (5.44), gives

$$\delta_s^\nabla \left[\nabla_{//_s w} Y_s\right]$$

$$= \langle //_s w, \nabla_{\delta_s \Sigma_s} X_i\rangle \left(\nabla_{X_i} Y_s\right)(\Sigma_s) + \langle //_s w, X_i(\Sigma_s)\rangle \nabla_{\delta_s \Sigma_s} \nabla_{X_i} Y_s$$

$$+ \langle //_s w, X_i(\Sigma_s)\rangle \left(\nabla_{X_i}\left(\frac{d}{ds} Y_s\right)\right)(\Sigma_s)\, ds$$

$$= \langle //_s w, \nabla_{\delta_s \Sigma_s} X_i\rangle \left(\nabla_{X_i} Y_s\right)(\Sigma_s) + \langle //_s w, X_i(\Sigma_s)\rangle \nabla^2_{\delta_s \Sigma_s \otimes X_i} Y_s$$

$$+ \langle //_s w, X_i(\Sigma_s)\rangle \nabla_{\nabla_{\delta_s \Sigma_s} X_i} Y_s + \left(\nabla_{//_s w}\left(\frac{d}{ds} Y_s\right)\right)(\Sigma_s)\, ds$$

$$= \left(\nabla_{\langle //_s w, \nabla_{\delta_s \Sigma_s} X_i\rangle X_i(\Sigma_s) + \langle //_s w, X_i(\Sigma_s)\rangle \nabla_{\delta_s \Sigma_s} X_i} Y_s\right)(\Sigma_s)$$

$$+ \nabla^2_{\delta_s \Sigma_s \otimes //_s w} Y_s + \left(\nabla_{//_s w}\left(\frac{d}{ds} Y_s\right)\right)(\Sigma_s)\, ds, \tag{5.49}$$

Taking the differential of (5.48) implies

$$0 = \delta w = \langle w, //_s^{-1}\nabla_{\delta_s \Sigma_s} X_i\rangle //_s^{-1} X_i(\Sigma_s) + \langle w, //_s^{-1} X_i(\Sigma_s)\rangle //_s^{-1}\nabla_{\delta_s \Sigma_s} X_i$$

which upon multiplying by $//_s$ shows

$$\langle //_s w, \nabla_{\delta_s \Sigma_s} X_i\rangle X_i(\Sigma_s) + \langle //_s w, X_i(\Sigma_s)\rangle \nabla_{\delta_s \Sigma_s} X_i = 0.$$

Using this identity in (5.49) completes the proof of (5.45).

Now suppose that $\Sigma_s$ is a Brownian motion and $b_s = \Psi_s(\Sigma)$ is the anti-developed $T_oM$-valued Brownian motion associated to $\Sigma$. Then by (5.44),

$$d\left[//_s^{-1} Y_s(\Sigma_s)\right] = //_s^{-1}\left(\frac{d}{ds} Y_s\right)(\Sigma_s)\, ds + //_s^{-1}\nabla_{//_s \delta b_s} Y_s$$

$$= //_s^{-1}\left(\frac{d}{ds} Y_s\right)(\Sigma_s)\, ds + \left(//_s^{-1}\nabla_{//_s e_i} Y_s\right)\delta b_s^i.$$

Using (5.45),

$$\left(//_s^{-1}\nabla_{//_s e_i} Y_s\right)\delta b_s^i = \left(//_s^{-1}\nabla_{//_s e_i} Y_s\right) db_s^i + \frac{1}{2} d\left(//_s^{-1}\nabla_{//_s e_i} Y_s\right) db_s^i$$

$$= //_s^{-1}\nabla_{//_s db_s} Y_s + \frac{1}{2}//_s^{-1}\nabla^2_{\delta \Sigma_s \otimes //_s e_i} Y_s db_s^i$$

$$= //_s^{-1}\nabla_{//_s db_s} Y_s + \frac{1}{2}//_s^{-1}\nabla^2_{//_s e_j \otimes //_s e_i} Y_s db_s^i db_s^j$$

$$= //_s^{-1}\nabla_{//_s db_s} Y_s + \frac{1}{2}//_s^{-1}\nabla^2_{//_s e_i \otimes //_s e_i} Y_s ds.$$

Combining the last two equations proves (5.46).     □

**Theorem 5.37.** *Let $\Sigma_s$ denote the solution to (5.1) with $\Sigma_0 = o \in M$, $\beta = B$ and $b_s = \Psi_s(\Sigma) \in T_oM$. Then*

$$b_s = \int_0^s //_r^{-1}(\Sigma)\left[\mathbf{X}(\Sigma_r)\,\delta B_r + X_0(\Sigma_r)\,dr\right]$$

$$= \int_0^s //_r^{-1}(\Sigma)\,\mathbf{X}(\Sigma_r)\,dB_r$$

$$+ \int_0^s //_r^{-1}\left[\frac{1}{2}\sum_{i,j=1}^n \left(\nabla_{X_i}X_j\right)(\Sigma_r)\,dB_r^i dB_r^j + X_0(\Sigma_r)\,dr\right]. \tag{5.50}$$

*Hence if $B$ is a Brownian motion, then*

$$b_s = \int_0^s //_r^{-1}(\Sigma)\,\mathbf{X}(\Sigma_r)\,dB_r$$

$$+ \int_0^s //_r^{-1}\left[\frac{1}{2}\sum_{i=1}^n \left(\nabla_{X_i}X_i\right)(\Sigma_r) + X_0(\Sigma_r)\right]dr. \tag{5.51}$$

*Proof.* By the definition of $b$,

$$db_s = //_s^{-1}(\Sigma)\left[\mathbf{X}(\Sigma_s)\,\delta B_s + X_0(\Sigma_s)\,ds\right]$$

$$= //_s^{-1}(\Sigma)\left[\mathbf{X}(\Sigma_s)\,dB_s + X_0(\Sigma_s)\,ds\right] + \frac{1}{2}d\left[//_s^{-1}(\Sigma)\,\mathbf{X}(\Sigma_s)\right]dB_s$$

$$= //_s^{-1}(\Sigma)\left[\mathbf{X}(\Sigma_s)\,dB_s + X_0(\Sigma_s)\,ds\right] + \frac{1}{2}\left[//_s^{-1}(\Sigma)\,\nabla_{\mathbf{X}(\Sigma_s)dB_s}\mathbf{X}\right]dB_s$$

$$= //_s^{-1}(\Sigma)\left[\mathbf{X}(\Sigma_s)\,dB_s + ds\right] + \frac{1}{2}//_s^{-1}(\Sigma)\sum_{i,j=1}^n\left(\nabla_{X_i}X_j\right)(\Sigma_s)\,dB_s^i dB_s^j$$

which combined with the identity,

$$d\left[//_s^{-1}(\Sigma)\,\mathbf{X}(\Sigma_s)\right]dB_s = \left[//_s^{-1}(\Sigma)\,\nabla_{d\Sigma_s}\mathbf{X}\right]dB_s = \left[//_s^{-1}(\Sigma)\,\nabla_{\mathbf{X}(\Sigma_s)dB_s}\mathbf{X}\right]dB_s$$

$$= \sum_{i,j=1}^n\left(\nabla_{X_i}X_j\right)(\Sigma_s)\,dB_s^i dB_s^j$$

proves (5.50). $\qquad\qquad\square$

**Corollary 5.38.** *Suppose $B_s$ is an $\mathbb{R}^n$-valued Brownian motion, $\Sigma_s$ is the solution to (5.1) with $\beta = B$ and $\frac{1}{2}\sum_{k=1}^n\left(\nabla_{X_k}X_k\right) + X_0 = 0$, then $\Sigma$ is an $M$-valued martingale with quadratic variation,*

$$d\Sigma_s \otimes d\Sigma_s = \sum_{k=1}^n X_k(\Sigma_s) \otimes X_k(\Sigma_s)\,ds. \tag{5.52}$$

*Proof.* By (5.51) and Theorem 5.33, $\Sigma$ is a martingale and from (5.1),

$$d\Sigma^i d\Sigma^j = \sum_{k,l=1}^{n} X_k^i(\Sigma) X_l^j(\Sigma) dB^k dB^l = \sum_{k=1}^{n} X_k^i(\Sigma) X_k^j(\Sigma) ds$$

where $\{e_i\}_{i=1}^{N}$ is the standard basis for $\mathbb{R}^N$, $\Sigma^i := \langle \Sigma, e_i \rangle$ and $X_k^i(\Sigma) = \langle X_k(\Sigma), e_i \rangle$. Using this identity in Eq. (5.17), shows

$$d\Sigma_s \otimes d\Sigma_s = \sum_{i,j=1}^{N} \sum_{k=1}^{n} e_i \otimes e_j X_k^i(\Sigma) X_k^j(\Sigma) ds = \sum_{k=1}^{n} X_k(\Sigma_s) \otimes X_k(\Sigma_s) ds. \quad \square$$

**Corollary 5.39.** *Suppose now that $B_s$ is an $\mathbb{R}^N$-valued semi-martingale and $\Sigma_s$ is the solution to (5.42) in Lemma 5.35. If $B$ is a martingale, then $\Sigma$ is a martingale and if $B$ is a Brownian motion, then $\Sigma$ is a Brownian motion.*

*Proof.* Solving (5.42) is the same as solving (5.1) with $n = N$, $\beta = B$, $X_0 \equiv 0$ and $X_i(m) = P(m) e_i$ for all $i = 1, 2, \ldots, N$. Since

$$\nabla_{X_i} X_j = P dP(X_i) e_j = dP(X_i) Q e_j = dP(P e_i) Q e_j,$$

it follows from orthogonality Lemma 5.35 that

$$\sum_{i,j=1}^{n} \left( \nabla_{X_i} X_j \right)(\Sigma_r) dB_r^i dB_r^j = 0.$$

Therefore from (5.50), $b_s := \int_0^s //_r^{-1} \delta \Sigma_r$ is a $T_o M$-martingale, which is equivalent to $\Sigma_s$ being an $M$-valued martingale. Finally if $B$ is a Brownian motion, then from (5.52), $\Sigma$ has quadratic variation given by

$$d\Sigma_s \otimes d\Sigma_s = \sum_{i=1}^{N} P(\Sigma_s) e_i \otimes P(\Sigma_s) e_i ds \tag{5.53}$$

Since $\sum_{i=1}^{N} P(m) e_i \otimes P(m) e_i$ is independent of the choice of orthonormal basis for $\mathbb{R}^N$, we may choose $\{e_i\}$ such that $\{e_i\}_{i=1}^{d}$ is an orthonormal basis for $\tau_m M$ to learn

$$\sum_{i=1}^{N} P(m) e_i \otimes P(m) e_i = \mathcal{I}(m).$$

Using this in (5.53) we learn that $d\Sigma_s \otimes d\Sigma_s = \mathcal{I}(\Sigma_s) ds$ and hence $\Sigma$ is a Brownian motion on $M$ by the Lévy criteria, see Lemma 5.21. $\qquad \square$

**Theorem 5.40.** *Let B be any $\mathbb{R}^N$-valued semi-martingale, $\Sigma$ be the solution to (5.42),*

$$b := \int_0^{\cdot} u^{-1}\delta\Sigma = \int_0^{\cdot} u^{-1}P(\Sigma)\delta B \qquad (5.54)$$

*be the anti-development of $\Sigma$ and*

$$\beta := \int_0^{\cdot} u^{-1}Q(\Sigma)dB = Q(o)\int_0^{\cdot} u^{-1}dB \qquad (5.55)$$

*be the "normal" process. Then*

$$b = \int_0^{\cdot} u^{-1}P(\Sigma)dB = P(o)\int_0^{\cdot} u^{-1}dB, \qquad (5.56)$$

*i.e., the Fisk–Stratonovich integral may be replaced by the Itô integral. Moreover if B is a standard $\mathbb{R}^N$-valued Brownian motion then $(b, \beta)$ is also a standard $\mathbb{R}^N$-valued Brownian and the processes, $b_{\dot{s}}$, $\Sigma_s$ and $//_s$ are all independent of $\beta$.*

*Proof.* Let $p = P(\Sigma)$ and $u$ be parallel translation on $M \times \mathbb{R}^N$ (see (5.32)), then

$$d(u^{-1}P(\Sigma)) \cdot dB = u^{-1}[\Gamma(\delta\Sigma)P(\Sigma)dB + dP(\delta\Sigma)dB]$$
$$= u^{-1}[(dQ(\delta\Sigma)P(\Sigma) + dP(\delta\Sigma)Q(\Sigma))P(\Sigma)dB + dP(\delta\Sigma)dB]$$
$$= u^{-1}[dQ(\delta\Sigma)P(\Sigma)dB - dQ(\delta\Sigma)dB]$$
$$= -u^{-1}dQ(\delta\Sigma)Q(\Sigma)dB = -u^{-1}dQ(P(\Sigma)dB)Q(\Sigma)dB = 0$$

where we have again used $P(\Sigma)dB \otimes Q(\Sigma)dB = 0$. This proves (5.56).

Now suppose that $B$ is a Brownian motion. Since $(b, \beta) = \int_0^{\cdot} u^{-1}dB$ and $u$ is an orthogonal process, it easily follows using Lévy's criteria that $(b, \beta)$ is a standard Brownian motion and in particular, $\beta$ is independent of $b$. Since $(\Sigma, u)$ satisfies the coupled pair of SDEs

$$d\Sigma = u\delta b \text{ and } du + \Gamma(u\delta b)u = 0 \text{ with}$$
$$\Sigma_0 = o \text{ and } u_0 = I \in \text{End}(\mathbb{R}^N),$$

it follows that $(\Sigma, u)$ is a functional of $b$ and hence the process $(\Sigma, u)$ are independent of $\beta$. □

## 5.6 The differential in the starting point of a stochastic flow

In this section let $B_s$ be an $\mathbb{R}^n$-valued Brownian motion and for each $m \in M$ let $T_s(m) = \cdot \Sigma_s$ where $\Sigma_s$ is the solution to (5.1) with $\Sigma_0 = m$. It is well known, see Kunita [116] that there is a version of $T_s(m)$ which is continuous in $s$ and smooth in $m$, moreover the differential of $T_s(m)$ relative to $m$ solves the SDE found by differentiating (5.1). Let

$$Z_s := T_{s*o} \text{ and } z_s := //_s^{-1}Z_s \in \text{End}(T_oM) \qquad (5.57)$$

where $//_s$ is stochastic parallel translation along $\Sigma_s := T_s(o)$.

**Theorem 5.41.** *For all* $v \in T_oM$

$$\delta_s^\nabla Z_s v = \left(\nabla_{Z_s v}\mathbf{X}\right)\delta B_s + \left(\nabla_{Z_s v}X_0\right)ds \text{ with } Z_0 v = v. \tag{5.58}$$

*Alternatively* $z_s$ *satisfies*

$$dz_s v = //_s^{-1}\left(\nabla_{//_s z_s v}\mathbf{X}\right)\delta B_s + //_s^{-1}\left(\nabla_{//_s z_s v}X_0\right)ds. \tag{5.59}$$

*Proof.* Equations (5.58) and (5.59) are the formal analogues (4.2) and (4.3) respectively. Because of Proposition 5.36, (5.58) is equivalent to (5.59). To prove (5.58), differentiate (5.1) in $m$ in the direction $v \in T_oM$ to find

$$\delta_s Z_s v = DX_i\left(\Sigma_s\right)Z_s v \circ \delta B_s^i + DX_0\left(\Sigma_s\right)Z_s v ds \text{ with } Z_0 v = v.$$

Multiplying this equation through by $P\left(\Sigma_s\right)$ on the left then gives (5.58).  □

**Notation 5.42.** The pull back, $\mathrm{Ric}_{//_s}$, of the Ricci tensor by parallel translation is defined by

$$\mathrm{Ric}_{//_s} := //_s^{-1}\,\mathrm{Ric}_{\Sigma_s}\,//_s. \tag{5.60}$$

**Theorem 5.43 (Itô form of (5.59)).** *The Itô form of (5.59) is*

$$dz_s v = //_s^{-1}\left(\nabla_{//_s z_s v}\mathbf{X}\right)dB_s + \alpha_s ds \tag{5.61}$$

*where*

$$\alpha_s := //_s^{-1}\left[\nabla_{//_s z_s v}\left(\sum_{i=1}^n \nabla_{X_i}X_i + X_0\right) - \frac{1}{2}\sum_{i=1}^n R^\nabla\left(//_s z_s v, X_i\left(\Sigma_s\right)\right)X_i\left(\Sigma_s\right)\right]ds. \tag{5.62}$$

*If we further assume that* $n = N$ *and* $X_i\left(m\right) = P\left(m\right)e_i$ *(so that (5.1) is equivalent to (5.42) if* $X_0 \equiv 0$*), then* $\alpha_s = -\frac{1}{2}\mathrm{Ric}_{//_s}z_s v ds$*, i.e., (5.59) is equivalent to*

$$dz_s v = //_s^{-1}P\left(\Sigma_s\right)dP\left(//_s z_s v\right)dB_s + \left[//_s^{-1}\nabla_{//_s z_s v}X_0 - \frac{1}{2}\mathrm{Ric}_{//_s}z_s v\right]ds. \tag{5.63}$$

*Proof.* In this proof there will always be an implied sum on repeated indices. Using Proposition 5.36,

$$
\begin{aligned}
d\left[//_s^{-1}\left(\nabla_{//_s z_s v}\mathbf{X}\right)\right]dB_s &= //_s^{-1}\left[\nabla_{\mathbf{X}(\Sigma_s)dB_s \otimes //_s z_s v}^2\mathbf{X} + \nabla_{//_s dz_s v}\mathbf{X}\right]dB_s \\
&= //_s^{-1}\left[\nabla_{\mathbf{X}(\Sigma_s)dB_s \otimes //_s z_s v}^2\mathbf{X} + \nabla_{(\nabla_{//_s z_s v}\mathbf{X})dB_s}\mathbf{X}\right]dB_s \\
&= //_s^{-1}\left[\nabla_{X_i(\Sigma_s)\otimes //_s z_s v}^2 X_i + \nabla_{(\nabla_{//_s z_s v}X_i)}X_i\right]ds. \tag{5.64}
\end{aligned}
$$

Now by Proposition 3.38,

$$\nabla^2_{X_i(\Sigma_s)\otimes//_s z_s v} X_i = \nabla^2_{//_s z_s v\otimes X_i(\Sigma_s)} X_i ds + R^\nabla (X_i(\Sigma_s), //_s z_s v) X_i(\Sigma_s)$$
$$= \nabla^2_{//_s z_s v\otimes X_i(\Sigma_s)} X_i ds - R^\nabla (//_s z_s v, X_i(\Sigma_s)) X_i(\Sigma_s)$$
$$= \left[\nabla_{//_s z_s v}\nabla_{X_i} X_i - \nabla_{\nabla_{//_s z_s v} X_i} X_i\right]$$
$$- R^\nabla (//_s z_s v, X_i(\Sigma_s)) X_i(\Sigma_s)$$

which combined with (5.64) implies

$$d\left[//_s^{-1}\left(\nabla_{//_s z_s v}\mathbf{X}\right)\right] dB_s = //_s^{-1}\left[\nabla_{//_s z_s v}\nabla_{X_i} X_i - R^\nabla (//_s z_s v, X_i(\Sigma_s)) X_i(\Sigma_s)\right] ds. \tag{5.65}$$

Equation (5.61) now follows directly from this equation and (5.59).

If we further assume $n = N$, $X_i(m) = P(m) e_i$ and $X_0(m) = 0$, then

$$\left(\nabla_{//_s z_s v}\mathbf{X}\right) dB_s = //_s^{-1} P(\Sigma_s) dP(//_s z_s v) dB_s. \tag{5.66}$$

Moreover, from the definition of the Ricci tensor in (3.31) and making use of (3.50) in the proof of Proposition 3.48 we have

$$R^\nabla (//_s z_s v, X_i(\Sigma_s)) X_i(\Sigma_s) = \text{Ric}_{//_s} //_s z_s v. \tag{5.67}$$

Combining (5.66) and (5.67) along with $\nabla_{X_i} X_i = 0$ (from Proposition 3.48) with (5.61) and (5.62) implies (5.63).     □

In the next result, we will filter out the "redundant noise" in (5.63). This is useful for deducing intrinsic formula from their extrinsic cousins, see, for example, Corollary 6.4 and Theorem 7.39 below.

**Theorem 5.44 (Filtering out the Redundant Noise).** *Keep the same setup in Theorem 5.43 with $n = N$ and $X_i(m) = P(m) e_i$. Further let $\mathcal{M}$ be the $\sigma$-algebra generated by the solution $\Sigma = \{\Sigma_s : s \geq 0\}$. Then there is a version, $\bar{z}_s$, of $\mathbb{E}[z_s|\mathcal{M}]$ such that $s \to \bar{z}_s$ is continuous and $\bar{z}$ satisfies,*

$$\bar{z}_s v = v + \int_0^s \left[//_r^{-1}\left(\nabla_{//_r \bar{z}_r v} X_0\right) - \frac{1}{2}\text{Ric}_{//_r} \bar{z}_r v\right] dr. \tag{5.68}$$

*In particular if $X_0 = 0$, then*

$$\frac{d}{ds}\bar{z}_s = -\frac{1}{2}\text{Ric}_{//_s} \bar{z}_s \text{ with } \bar{z}_0 = id, \tag{5.69}$$

*Proof.* In this proof, we let $b_s$ be the martingale part of the anti-development map, $\Psi_s(\Sigma)$, i.e.,

$$b_s := \int_0^s //_r^{-1} P(\Sigma_r) \delta B_r = \int_0^s //_r^{-1} P(\Sigma_r) dB_r.$$

Since $(\Sigma_s, u_s)$ solves the SDE

$$\delta \Sigma_s = u_s \delta b_s + X_0 (\Sigma_s) \, ds \text{ with } \Sigma_0 = o$$
$$\delta u = -\Gamma (\delta \Sigma) u = -\Gamma (u \delta b) u \text{ with } u_0 = I \in O(N)$$

it follows that $(\Sigma, u)$ may be expressed as a function of the Brownian motion, $b$. Therefore by the martingale representation property, see Corollary 7.20 below, any measurable function, $f(\Sigma)$, of $\Sigma$ may be expressed as

$$f(\Sigma) = f_0 + \int_0^1 \langle a_r, db_r \rangle = f_0 + \int_0^1 \langle a_r, //_r^{-1} [P(\Sigma_r) \, dB_r] \rangle.$$

Hence, using $P \, dP = dP \, Q$, the previous equation and the isometry property of the Itô integral,

$$\mathbb{E} \left\{ \left[ \int_0^s [P(\Sigma_r) \, dP \, (//_r z_r v) \, dB_r] \, f(\Sigma) \right\} \right.$$
$$= \mathbb{E} \left\{ \left[ \int_0^s [dP \, (//_r z_r v) \, Q(\Sigma_r) \, dB_r] \int_0^1 \langle P(\Sigma_r) \, //_r a_r, dB_r \rangle \right\} \right.$$
$$= \mathbb{E} \left\{ \left[ \int_0^s [dP \, (//_r z_r v) \, Q(\Sigma_r) \, P(\Sigma_r) \, //_r a_r] \, dr \right\} = 0. \right.$$

This shows that

$$\mathbb{E} \left[ \int_0^s P(\Sigma_r) \, dP \, (//_r z_r v) \, dB_r | \mathcal{M} \right] = 0$$

and hence taking the conditional expectation, $\mathbb{E} [\cdot | \mathcal{M}]$, of the integrated version of (5.63) implies (5.68). In performing this operation we have used the fact that $(\Sigma, //)$ is $\mathcal{M}$-measurable and that $z_s$ appears linearly in (5.63). I have also glossed over the technicality of passing the conditional expectation past the integrals involving a $ds$ term. For this detail and a much more general presentation of these ideas the reader is referred to Elworthy, Li and Le Jan [71].                    □

### 5.7 More references

For more details on the sorts of results in this section, the books by Elworthy [69], Emery [74], and Ikeda and Watanabe [104], Malliavin [132], Stroock [171], and Hsu [101] are highly recommended. The following articles and books are also relevant, [14, 20, 21, 40, 64, 63, 65, 110, 129, 137, 144, 154, 155, 156, 179].

## 6 Heat kernel derivative formula

In this short section we will illustrate how to derive Bismut type formulas for derivatives of heat kernels. For more details and more general formulae see, Driver and Thalmaier [58], Elworthy, Le Jan and Li [71], Stroock and Turetsky [173, 172] and Hsu [99] and the references therein. Throughout this section $\Sigma_s$ will be an $M$-valued semi-martingale, $//_s$ will be stochastic parallel translation along $\Sigma$ and

$$b_s = \Psi_s(\Sigma) := \int_0^s //_r^{-1} \delta \Sigma_r.$$

Furthermore, let $Q_s$ denote the unique solution to the differential equation:

$$\frac{dQ_s}{ds} = -\frac{1}{2} Q_s \text{Ric}_{//_s} \text{ with } Q_0 = I. \tag{6.1}$$

See (5.60) for the definition of $\text{Ric}_{//_s}$.

**Lemma 6.1.** *Let* $f : M \to \mathbb{R}$ *be a smooth function, $t > 0$ and for $s \in [0, t]$ let*

$$F(s, m) := (e^{(t-s)\bar{\Delta}/2} f)(m). \tag{6.2}$$

*If $\Sigma_s$ is an $M$-valued Brownian motion, then the process $s \in [0, t] \to Q_s //_s^{-1} \nabla F(s, \Sigma_s)$ is a martingale and*

$$d\left[ Q_s //_s^{-1} \nabla F(s, \Sigma_s) \right] = Q_s //_s^{-1} \nabla_{//_s db_s} \nabla F(s, \cdot). \tag{6.3}$$

*Proof.* Let $W_s := //_s^{-1} \nabla F(s, \Sigma_s)$. Then by Proposition 5.36 and Theorem 3.49,

$$dW_s = \left[ //_s^{-1} \nabla \partial_s F(s, \Sigma_s) + \frac{1}{2} //_s^{-1} \nabla^2_{//_s e_i \otimes //_s e_i} \nabla F(s, \cdot) \right] ds$$
$$+ //_s^{-1} \nabla_{//_s e_i} \nabla F(s, \cdot) db_s^i$$
$$= \frac{1}{2} //_s^{-1} \left[ \nabla^2_{//_s e_i \otimes //_s e_i} \nabla F(s, \cdot) - (\nabla \Delta F(s, \cdot))(\Sigma_s) \right] ds$$
$$+ //_s^{-1} \nabla_{//_s e_i} \nabla F(s, \cdot) db_s^i$$
$$= \frac{1}{2} //_s^{-1} \text{Ric} \nabla F(s, \Sigma_s) ds + //_s^{-1} \nabla_{//_s e_i} \nabla F(s, \cdot) db_s^i$$
$$= \frac{1}{2} \text{Ric}_{//_s} W_s ds + //_s^{-1} \nabla_{//_s e_i} \nabla F(s, \cdot) db_s^i$$

where $\{e_i\}_{i=1}^d$ is an orthonormal basis for $T_o M$ and there is an implied sum on repeated indices. Hence if $Q$ solves (6.1), then

$$d[Q_s W_s] = -\frac{1}{2} Q_s \text{Ric}_{//_s} W_s ds + Q_s \left[ \frac{1}{2} \text{Ric}_{//_s} W_s ds + //_s^{-1} \nabla_{//_s e_i} \nabla F(s, \cdot) db_s^i \right]$$
$$= Q_s //_s^{-1} \nabla_{//_s e_i} \nabla F(s, \cdot) db_s^i$$

which proves (6.3) and shows that $Q_s W_s$ is a martingale as desired. $\qquad\square$

**Theorem 6.2 (Bismut).** *Let* $f : M \to \mathbb{R}$ *be a smooth function and $\Sigma$ be an $M$-valued Brownian motion with $\Sigma_0 = o$, then for $0 < t_0 \le t < \infty$,*

$$\nabla(e^{t\Delta/2} f)(o) = \frac{1}{t_0} E\left[ \left( \int_0^{t_0} Q_r db_r \right) f(\Sigma_t) \right]. \tag{6.4}$$

*Proof.* The proof given here is modelled on Remark 6 on p. 84 in Bismut [21] and the proof of Theorem 2.1 in Elworthy and Li [72]. Also see Norris [145, 144, 146]. For $(s, m) \in [0, t] \times M$ let $F$ be defined as in (6.2). We wish to compute the differential of $k_s := \left( \int_0^s Q_r db_r \right) F(s, \Sigma_s)$. By (5.38), $d[F(s, \Sigma_s)] = \langle \nabla (F(s, \cdot))(\Sigma_s), //_s db_s \rangle$ and therefore

$$dk_s = F(s, \Sigma_s) Q_s db_s + \left( \int_0^s Q_r db_r \right) \langle \nabla (F(s, \cdot))(\Sigma_s), //_s db_s \rangle$$

$$+ \sum_{i=1}^d \langle \nabla (F(s, \cdot))(\Sigma_s), //_s e_i \rangle Q_s e_i \, ds.$$

From this we conclude that

$$\mathbb{E}[k_{t_0}] = \mathbb{E}[k_0] + \mathbb{E} \int_0^{t_0} \sum_{i=1}^d \langle //_s^{-1} \nabla (F(s, \cdot))(\Sigma_s), e_i \rangle Q_s e_i \, ds$$

$$= \int_0^{t_0} \mathbb{E} \left[ Q_s //_s^{-1} \nabla (F(s, \cdot))(\Sigma_s) \right] ds$$

$$= \int_0^{t_0} \mathbb{E} \left[ Q_0 //_0^{-1} \nabla (F(0, \cdot))(\Sigma_0) \right] ds = t_0 \nabla (e^{t\Delta/2} f)(o)$$

wherein the the third equality we have used (by Lemma 6.1) that $s \to Q_s //_s^{-1} \nabla (F(s, \cdot))(\Sigma_s)$ is a martingale. Hence

$$\nabla (e^{t\Delta/2} f)(o) = \frac{1}{t_0} \mathbb{E} \left[ \left( \int_0^{t_0} Q_s db_s \right) (e^{(t-t_0)\Delta/2} f)(\Sigma_{t_0}) \right]$$

from which (6.4) follows using either the Markov property of $\Sigma_s$ or the fact that $s \to \left( e^{(t-s)\Delta/2} f \right)(\Sigma_s)$ is a martingale.    □

The following theorem is an non-intrinsic form of Theorem 6.2. In this theorem we will be using the notation introduced before Theorem 5.41. Namely, let $\{X_i\}_{i=0}^n \subset \Gamma(TM)$ be as in Notation 5.4, $B_s$ be an $\mathbb{R}^n$-valued Brownian motion, and $T_s(m) = \Sigma_s$ where $\Sigma_s$ is the solution to (5.1) with $\Sigma_s = m \in M$ and $\beta = B$.

**Theorem 6.3 (Elworthy–Li).** *Assume that* $X(m) : \mathbb{R}^n \to T_m M$ *(recall* $X(m) a := \sum_{i=1}^n X_i(m) a_i)$ *is surjective for all* $m \in M$ *and let*

$$X(m)^\# = \left[ X(m) |_{\text{Nul}(X(m))^\perp} \right]^{-1} : T_m M \to \mathbb{R}^n, \tag{6.5}$$

*where the orthogonal complement is taken relative to the standard inner product on* $\mathbb{R}^n$. *(See Lemma 7.38 below for more on* $X(m)^\#$.) *Then for all* $v \in T_o M$, $0 < t_o < t < \infty$ *and* $f \in C(M)$ *we have*

$$v \left( e^{tL/2} f \right) = \frac{1}{t_0} \mathbb{E} \left[ f(\Sigma_t) \int_0^{t_0} \langle X(\Sigma_s)^\# Z_s v, dB_s \rangle \right] \tag{6.6}$$

*where* $Z_s = T_{s*o}$ *as in (5.57).*

*Proof.* Let $L = \sum_{i=1}^{n} X_i^2 + 2X_0$ be the generator of the diffusion, $\{T_s(m)\}_{s \geq 0}$. Since $X(m) : \mathbb{R}^n \to T_m M$ is surjective for all $m \in M$, $L$ is an elliptic operator on $C^{\infty}(M)$. So, using results similar to those in Fact 5.32, it makes sense to define $F_s(m) := \left(e^{(t-s)L/2} f\right)(m)$ and $N_s^m = F_s(T_s(m))$. Then

$$\partial_s F_s + \frac{1}{2} L F_s = 0 \text{ with } F_t = f$$

and by Itô's lemma,

$$dN_s^m = d[F_s(T_s(m))] = \sum_{i=1}^{n} (X_i F_s)(T_s(m))dB_s^i. \tag{6.7}$$

This shows $N_s^m$ is a martingale for all $m \in M$ and, upon integrating (6.7) on $s$, that

$$f(T_t(m)) = e^{tL/2} f(m) + \sum_{i=1}^{n} \int_0^t (X_i F_s)(T_s(m))dB_s^i.$$

Hence if $a_s \in \mathbb{R}^n$ is a predictable process such that $\mathbb{E} \int_0^t |a_s|^2 ds < \infty$, then by the Itô isometry property,

$$\mathbb{E} \left[ f(T_t(m)) \int_0^t \langle a, dB \rangle \right] = \int_0^t \mathbb{E}[(X_i F_s)(T_s(m))a_i(s)] ds$$

$$= \int_0^t \mathbb{E}[(d_M F_s)(X(T_s(m))a_s)] ds. \tag{6.8}$$

Suppose that $\ell_s \in \mathbb{R}$ is a continuous piecewise differentiable function and let $a_s := \ell_s' X(\Sigma_s)^{\#} Z_s v$. Then from (6.8) we have

$$\mathbb{E} \left[ f(\Sigma_t) \int_0^t \langle \ell_s' X(\Sigma_s)^{\#} Z_s v, dB_s \rangle \right] = \int_0^t \ell_s' \mathbb{E}[(d_M F_s)(Z_s v)] ds. \tag{6.9}$$

Since $N_s^m = F_s(T_s(m))$ is a martingale for all $m$, we may deduce that

$$v(m \to N_s^m) = d_M F_s(T_{s*o} v) = d_M F_s(Z_s v) \tag{6.10}$$

is a martingale as well for any $v \in T_o M$. In particular, $s \in [0, t] \to \mathbb{E}[(d_M F_s)(Z_s v)]$ is constant and evaluating this expression at $s = 0$ and $s = t$ implies

$$\mathbb{E}[(d_M F_s)(Z_s v)] = v\left(e^{tL/2} f\right) = \mathbb{E}[(d_M f)(Z_t v)]. \tag{6.11}$$

Using (6.11) in (6.9) then shows

$$\mathbb{E} \left[ f(\Sigma_t) \int_0^t \langle \ell_s' X(\Sigma_s)^{\#} Z_s v, dB_s \rangle \right] = (\ell_t - \ell_0) v\left(e^{tL/2} f\right)$$

which, by taking $\ell_s = s \wedge t_0$, implies (6.6).     $\square$

**Corollary 6.4.** *Theorem 6.3 may be used to deduce Theorem 6.2.*

*Proof.* Apply Theorem 6.3 with $n = N$, $X_0 \equiv 0$ and $X_i(m) = P(m)e_i$ for $i = 1, \ldots, N$ to learn

$$v\left(e^{t\Delta/2}f\right) = \frac{1}{t_0}\mathbb{E}\left[f(\Sigma_t)\int_0^{t_0}\langle Z_s v, dB_s\rangle\right] = \frac{1}{t_0}\mathbb{E}\left[f(\Sigma_t)\int_0^{t_0}\langle //_s z_s v, dB_s\rangle\right] \tag{6.12}$$

where we have used $L = \Delta$ (see Proposition 3.48) and $\mathbf{X}(m)^\# = P(m)$ in this setting. By Theorem 5.40,

$$\int_0^{t_0}\langle //_s z_s v, dB_s\rangle = \int_0^{t_0}\langle //_s z_s v, P(\Sigma_s)\,dB_s\rangle$$

$$= \int_0^{t_0}\langle z_s v, //_s^{-1}P(\Sigma_s)\,dB_s\rangle = \int_0^{t_0}\langle z_s v, db_s\rangle$$

and therefore (6.12) may be written as

$$v\left(e^{t\Delta/2}f\right) = \frac{1}{t_0}\mathbb{E}\left[f(\Sigma_t)\int_0^{t_0}\langle z_s v, db_s\rangle\right].$$

Using Theorem 5.44 to factor out the redundant noise, this may also be expressed as

$$v\left(e^{t\Delta/2}f\right) = \frac{1}{t_0}\mathbb{E}\left[f(\Sigma_t)\int_0^{t_0}\langle \bar{z}_s v, db_s\rangle\right] = \frac{1}{t_0}\mathbb{E}\left[f(\Sigma_t)\int_0^{t_0}\langle v, \bar{z}_s^{\text{tr}}db_s\rangle\right] \tag{6.13}$$

where $\bar{z}_s$ solves (5.69). By taking transposes of (5.69) it follows that $\bar{z}_s^{\text{tr}}$ satisfies (6.1) and hence $\bar{z}_s^{\text{tr}} = Q_s$. Since $v \in T_o M$ was arbitrary, Equation (6.4) is now an easy consequence of (6.13) and the definition of $\nabla(e^{t\Delta/2}f)(o)$. $\qquad\square$

# 7 Calculus on $W(M)$

In this section, $(M, o)$ is assumed to be either a compact Riemannian manifold equipped with a fixed point $o \in M$ or $M = \mathbb{R}^d$ with $o = 0$.

**Notation 7.1.** We will be interested in the following path spaces:

$$W(T_o M) := \{\omega \in C([0, 1] \to T_o M)|\omega(0) = 0_o \in T_o M\},$$

$$H(T_o M) := \{h \in W(T_o M) : h(0) = 0, \ \& \ \langle h, h\rangle_H := \int_0^1 |h'(s)|_{T_o M}^2 ds < \infty\}$$

and

$$W(M) := \{\sigma \in C([0, 1] \to M) : \sigma(0) = 0 \in M\}.$$

(By convention $\langle h, h\rangle_H = \infty$ if $h \in W(T_o M)$ is not absolutely continuous.) We refer to $W(T_o M)$ as *Wiener space*, $W(M)$ as *curved Wiener space* and $H(T_o M)$ or $H\left(\mathbb{R}^d\right)$ as the *Cameron–Martin Hilbert space*.

**Definition 7.2.** Let $\mu$ and $\mu_{W(M)}$ denote the Wiener measures on $W\,(T_oM)$ and $W\,(M)$ respectively, i.e., $\mu = \mathrm{Law}\,(b)$ and $\mu_{W(M)} = \mathrm{Law}\,(\Sigma)$ where $b$ and $\Sigma$ are Brownian motions on $T_oM$ and $M$ starting at $0 \in T_oM$ and $o \in M$ respectively.

**Notation 7.3.** The probability space in this section will often be $\left(W\,(M)\,,\mathcal{F},\mu_{W(M)}\right)$, where $\mathcal{F}$ is the completion of the $\sigma$-algebra generated by the projection maps, $\Sigma_s$ : $W\,(M) \to M$ defined by $\Sigma_s\,(\sigma) = \sigma_s$ for $s \in [0, 1]$. We make this into a filtered probability space by taking $\mathcal{F}_s$ to be the $\sigma$-algebra generated by $\{\Sigma_r : r \le s\}$ and the null sets in $\mathcal{F}_s$. Also let $//_s$ be a stochastic parallel translation along $\Sigma$.

**Definition 7.4.** A function $F : W(M) \to \mathbb{R}$ is called a $C^k$-*cylinder function* if there exists a partition

$$\pi := \{0 = s_0 < s_1 < s_2 \cdots < s_n = 1\} \tag{7.1}$$

of $[0, 1]$ and $f \in C^k(M^n)$ such that

$$F(\sigma) = f(\sigma_{s_1}, \ldots, \sigma_{s_n}) \text{ for all } \sigma \in W\,(M)\,. \tag{7.2}$$

If $M = \mathbb{R}^d$, we further require that $f$ and all of its derivatives up to order $k$ have at most polynomial growth at infinity. The collection of $C^k$-cylinder functions will be denoted by $\mathcal{F}C^k\,(W\,(M))\,.$

**Definition 7.5.** The *continuous tangent space* to $W(M)$ at $\sigma \in W(M)$ is the set $CT_\sigma W(M)$ of continuous vector-fields along $\sigma$ which are zero at $s = 0$ :

$$CT_\sigma W(M) = \{X \in C([0, 1], TM)|X_s \in T_{\sigma_s}M \; \forall \; s \in [0, 1] \text{ and } X(0) = 0\}. \tag{7.3}$$

To motivate the above definition, consider a differentiable path in $\gamma \in W(M)$ going through $\sigma$ at $t = 0$. Writing $\gamma\,(t)\,(s)$ as $\gamma\,(t, s)\,$, the derivative $X_s := \frac{d}{dt}|_0\gamma(t, s) \in T_{\sigma(s)}M$ of such a path should, by definition, be a tangent vector to $W(M)$ at $\sigma$.

We now wish to define a "Riemannian metric" on $W(M)$. It turns out that the continuous tangent space $CT_\sigma W(M)$ is too large for our purposes, see for example the Cameron–Martin Theorem 7.13 below. To remedy this we will introduce a Riemannian structure on an a.e. defined "sub-bundle" of $CTW\,(M)\,.$

**Definition 7.6.** A *Cameron–Martin process*, $h$, is a $T_oM$-valued process on $W\,(M)$ such that $s \to h(s)$ is in $H$, $\mu_{W(M)}$ a.e. Contrary to our earlier assumptions, we do *not* assume that $h$ is adapted unless explicitly stated.

**Definition 7.7.** Suppose that $X$ is a $TM$-valued process on $\left(W\,(M)\,,\mu_{W(M)}\right)$ such that the process $\pi\,(X_s) = \Sigma_s \in M$. We will say $X$ is a *Cameron–Martin vector field* if

$$h_s := //_s^{-1} X_s \tag{7.4}$$

is a Cameron–Martin valued process and

$$\langle X, X \rangle_\mathcal{X} := \mathbb{E}[\langle h, h \rangle_H] < \infty. \tag{7.5}$$

A Cameron–Martin vector field $X$ is said to be adapted if $h := //^{-1}X$ is adapted. The set of Cameron–Martin vector fields will be denoted by $\mathcal{X}$ and those which are adapted will be denoted by $\mathcal{X}_a$.

**Remark 7.8.** Notice that $\mathcal{X}$ is a Hilbert space with the inner product determined by $\langle \cdot, \cdot \rangle_{\mathcal{X}}$ in (7.5). Furthermore, $\mathcal{X}_a$ is a Hilbert-subspace of $\mathcal{X}$.

**Notation 7.9.** Given a Cameron–Martin process $h$, let $X^h := //h$. In this way we may identify Cameron–Martin processes with Cameron–Martin vector fields.

We define a "metric", $G$,[6] on $\mathcal{X}$ by

$$G(X^h, X^h) = \langle h, h \rangle_H. \tag{7.6}$$

With this notation we have $\langle X, X \rangle_{\mathcal{X}} = \mathbb{E}[G(X, X)]$.

**Remark 7.10.** Notice, if $\sigma$ is a smooth path then the expression in (7.6) could be written as

$$G(X, X) = \int_0^1 g\left(\frac{\nabla}{ds}X(s), \frac{\nabla}{ds}X(s)\right) ds,$$

where $\frac{\nabla}{ds}$ denotes the covariant derivative along the path $\sigma$ which is induced from the covariant derivative $\nabla$. This is a typical metric used by differential geometers on path and loop spaces.

**Notation 7.11.** Given a Cameron-Martin vector field $X$ on $\left(W(M), \mu_{W(M)}\right)$ and a cylinder function $F \in \mathcal{F}C^1(W(M))$ as in Eq. (7.2), let $XF$ denote the random variable

$$XF(\sigma) := \sum_{i=1}^n (\text{grad}_i F(\sigma), X_{s_i}(\sigma)), \tag{7.7}$$

where

$$\text{grad}_i F(\sigma) := (\text{grad}_i f)(\sigma_{s_1}, \dots, \sigma_{s_n}) \tag{7.8}$$

and $(\text{grad}_i f)$ denotes the gradient of $f$ relative to the $i^{\text{th}}$ variable.

**Notation 7.12.** The *gradient*, $DF$, of a smooth cylinder function, $F$, on $W(M)$ is the unique Cameron–Martin process such that $G(DF, X) = XF$ for all $X \in \mathcal{X}$. The explicit formula for $D$, as the reader should verify, is

$$(DF)_s = //_s \left(\sum_{i=1}^n s \wedge s_i //_{s_i}^{-1} \text{grad}_i F(\sigma)\right). \tag{7.9}$$

The formula in (7.9) defines a densely defined operator, $D: L^2(\mu) \to \mathcal{X}$ with $\mathcal{D}(D) = \mathcal{F}C^1(W(M))$ as its domain.

---

[6] The function $G$ is to be loosely interpreted as a Riemannian metric on $W(M)$.

## 7.1 Classical Wiener space calculus

In this subsection (which is a warm up for the sequel) we will specialize to the case where $M = \mathbb{R}^d$, $o = 0 \in \mathbb{R}^d$. To simplify notation let $W := W(\mathbb{R}^d)$, $H := H(\mathbb{R}^d)$, $\mu = \mu_{W(\mathbb{R}^d)}$, $b_s(\omega) = \omega_s$ for all $s \in [0, 1]$ and $\omega \in W$. Recall that $\{\mathcal{F}_s : s \in [0, 1]\}$ is the filtration on $W$ as explained in Notation 7.3 where we are now writing $b$ for $\Sigma$. Cameron and Martin [25, 26, 27] and Cameron [28] began the study of calculus on this classical Wiener space. They proved the following two results, see [26], Theorem 2, p. 387 and [28], Theorem II, p. 919, respectively. (There have been many extensions of these results partly initiated by Gross' work in [90, 91].)

**Theorem 7.13 (Cameron & Martin 1944).** *Let* $(W, \mathcal{F}, \mu)$ *be the classical Wiener space described above and for* $h \in W$, *define* $T_h : W \to W$ *by* $T_h(\omega) = \omega + h$ *for all* $\omega \in W$. *If* $h$ *is* $C^1$, *then* $\mu T_h^{-1}$ *is absolutely continuous relative to* $\mu$.

This theorem was extended by Maruyama [133] and Girsanov [88] to allow the same conclusion for $h \in H$ and more general Cameron–Martin processes. Moreover it is now well known $\mu T_h^{-1} \lambda \mu$ iff $h \in H$. From the Cameron and Martin theorem one may prove Cameron's integration by parts formula.

**Theorem 7.14 (Cameron 1951).** *Let* $h \in H$ *and* $F, G \in L^{\infty-}(\mu) := \cap_{1 \le p < \infty} L^p(\mu)$ *such that* $\partial_h F := \frac{d}{d\varepsilon} F \circ T_{\varepsilon h}|_{\varepsilon=0}$ *and* $\partial_h G := \frac{d}{d\varepsilon} G \circ T_{\varepsilon h}|_{\varepsilon=0}$ *where the derivatives are supposed to exist[7] in* $L^p(\mu)$ *for all* $1 \le p < \infty$. *Then*

$$\int_W \partial_h F \cdot G \, d\mu = \int_W F \partial_h^* G \, d\mu,$$

*where* $\partial_h^* G = -\partial_h G + z_h G$ *and* $z_h := \int_0^1 \langle h'(s), db_s \rangle_{\mathbb{R}^d}$.

In this flat setting parallel translation is trivial, i.e., $//_s = id$ for all $s$. Hence, the gradient operator $D$ in (7.9) reduces to the equation,

$$(DF)_s(\omega) = \left( \sum_{i=1}^n s \wedge s_i \operatorname{grad}_i F(\omega_s) \right).$$

Similarly the association of a Cameron–Martin vector field $X$ on $W(\mathbb{R}^d)$ with a Cameron–Martin valued process $h$ in (7.4) is simply that $X = h$.

We will now recall that adapted Cameron–Martin vector fields, $X = h$, are in the domain of $D^*$. From this fact it will easily follow that $D^*$ is densely defined.

**Theorem 7.15.** *Let* $h$ *be an adapted Cameron Martin process (vector field) on* $W$. *Then* $h \in \mathcal{D}(D^*)$ *and*

$$D^* h = \int_0^1 \langle h', db \rangle.$$

---

[7] The notion of derivative stated here is weaker than the notion given in [28]. Nevertheless Cameron's proof covers this case without any essential change.

*Proof.* We start by proving the theorem under the additional assumption that

$$\sup_{s \in [0,1]} |h'_s| \leq C, \tag{7.10}$$

where $C$ is a non-random constant. For each $t \in \mathbb{R}$ let $b(t, s) = b_s(t) = b_s + t h_s$. By Girsanov's theorem, $s \to b_s(t)$ (for fixed $t$) is a Brownian motion relative to $Z_t \cdot \mu$, where

$$Z_t := \exp\left(-\int_0^1 t \langle h'_s, db_s \rangle - \frac{1}{2} t^2 \int_0^1 \langle h'_s, h'_s \rangle ds \right).$$

Hence if $F$ is a smooth cylinder function on $W$,

$$\mathbb{E}\left[F\left(b(t, \cdot)\right) \cdot Z_t\right] = \mathbb{E}\left[F(b)\right].$$

Differentiating this equation in $t$ at $t = 0$, using

$$\langle DF, h \rangle_H = \frac{d}{dt}|_0 F\left(b\left(t, \cdot\right)\right) \text{ and } \frac{d}{dt}|_0 Z_t = -\int_0^1 \langle h', db \rangle,$$

shows

$$\mathbb{E}\left[\langle DF, h \rangle_H\right] - \mathbb{E}\left[F \int_0^1 \langle h', db \rangle\right] = 0.$$

From this equation it follows that $h \in \mathcal{D}(D^*)$ and $D^* h = \int_0^1 \langle h', db \rangle$. So it now only remains to remove the restriction placed on $h$ in (7.10).

Let $h$ be a general adapted Cameron–Martin vector field and for each $n \in \mathbb{N}$, let

$$h_n(s) := \int_0^s h'(r) \cdot 1_{|h'(r)| \leq n} dr. \tag{7.11}$$

(Notice that $h_n$ is still adapted.) By the special case above we know that $h_n \in \mathcal{D}(D^*)$ and $D^* h_n = \int_0^1 \langle h'_n, db \rangle$. Therefore,

$$\mathbb{E}\left|D^*(h_m - h_n)\right|^2 = \mathbb{E}\int_0^1 |h'_m - h'_n|^2 ds \to 0 \text{ as } m, n \to \infty$$

from which it follows that $D^* h_n$ is convergent. Because $D^*$ is a closed operator, $h \in \mathcal{D}(D^*)$ and

$$D^* h = \lim_{n \to \infty} D^* h_n = \lim_{n \to \infty} \int_0^1 \langle h'_n, db \rangle = \int_0^1 \langle h', db \rangle. \qquad \square$$

**Corollary 7.16.** *The operator $D^*$ is densely defined and hence $D$ is closable. (Let $\bar{D}$ denote the closure of $D$.)*

*Proof.* Let $h \in H$ and $F$ and $K$ be smooth cylinder functions. Then, by the product rule,

$$\langle DF, Kh \rangle_{\mathcal{X}} = \mathbb{E}[\langle KDF, h \rangle_H] = \mathbb{E}[\langle D(KF) - FDK, h \rangle_H]$$
$$= \mathbb{E}[F \cdot KD^*h - F \langle DK, h \rangle_H].$$

Therefore $Kh \in \mathcal{D}(D^*)$ ($\mathcal{D}(D^*)$ is the domain of $D^*$) and

$$D^*(Kh) = KD^*h - \langle DK, h \rangle_H.$$

Since the subspace,

$$\{Kh | h \in H \text{ and } K \text{ is a smooth cylinder function}\},$$

is a dense subspace of $\mathcal{X}$, $D^*$ is densely defined.                    □

### 7.1.1  Martingale representation property and the Clark-Ocone formula

**Lemma 7.17.** *Let* $F(b) = f(b_{s_1}, \ldots, b_{s_n})$ *be the smooth cylinder function on* $W$ *as in Definition 7.4, then*

$$F = \mathbb{E}F + \int_0^1 \langle a_s, db_s \rangle, \tag{7.12}$$

*where* $a_s$ *is a bounded, piecewise-continuous (in* $s$*) and predictable process. Furthermore, the jumps points of* $a_s$ *are contained in the set* $\{s_1, \ldots, s_n\}$ *and* $a_s \equiv 0$ *if* $s \geq s_n$.

*Proof.* The proof will be by induction on $n$. First assume that $n = 1$, so that $F(b) = f(b_t)$ for some $0 < t \leq 1$. Let $H(s, m) := (e^{(t-s)\Delta/2} f)(m)$ for $0 \leq s \leq t$ and $m \in \mathbb{R}^d$. Then, by Itô's formula (or see (5.38)),

$$dH(s, b_s) = \langle \text{grad } H(s, b_s), db_s \rangle$$

which upon integrating on $s \in [0, t]$ gives

$$F(b) = (e^{t\Delta/2} f)(o) + \int_0^t \langle \text{grad} H(s, b_s), db_s \rangle = \mathbb{E}F + \int_0^1 \langle a_s, db_s \rangle,$$

where $a_s = 1_{s \leq t} //_s^{-1} \text{ grad } H(s, b_s)$. This proves the $n = 1$ case. To finish the proof it suffices to show that we may reduce the assertion of the lemma at the level $n$ to the assertion at the level $n - 1$.

Let $F(b) = f(b_{s_1}, \ldots, b_{s_n})$,

$$(\Delta_n f)(x_1, x_2, \ldots, x_n) = (\Delta g)(x_n) \text{ and}$$
$$(\text{grad}_n f)(x_1, x_2, \ldots, x_n) = \nabla g(x_n)$$

where $g(x) := f(x_1, x_2, \ldots, x_{n-1}, x)$. (So $\Delta_n f$ and $\text{grad}_n f$ is the Laplacian and the gradient of $f$ in the $n^{\text{th}}$-variable.) Itô's lemma applied to the process,

$$s \in [s_{n-1}, s_n] \rightarrow H(s, b) := (e^{(s_n - s)\Delta_n/2} f)(b_{s_1}, \ldots, b_{s_{n-1}}, b_s)$$

gives

$$dH(s, b) = \langle \mathrm{grad}_n e^{(s_n - s)\Delta_n/2} f)(b_{s_1}, \ldots, b_{s_{n-1}}, b_s, db_s \rangle$$

and hence

$$F(b) = (e^{(s_n - s_{n-1})\Delta_n/2} f)(b_{s_1}, \ldots, b_{s_{n-1}}, b_{s_{n-1}})$$
$$+ \int_{s_{n-1}}^{s_n} \langle \mathrm{grad}_n e^{(s_n - s)\Delta_n/2} f)(b_{s_1}, \ldots, b_{s_{n-1}}, b_s, db_s \rangle$$
$$= (e^{(s_n - s_{n-1})\Delta_n/2} f)(b_{s_1}, \ldots, b_{s_{n-1}}, b_{s_{n-1}}) + \int_{s_{n-1}}^{s_n} \langle \alpha_s, db_s \rangle, \qquad (7.13)$$

where $\alpha_s := (\mathrm{grad}_n e^{(s_n - s)\Delta_n/2} f)(b_{s_1}, \ldots, b_{s_{n-1}}, b_s)$ for $s \in (s_{n-1}, s_n)$. By induction we know that the smooth cylinder function

$$(e^{(s_n - s_{n-1})\Delta_n/2} f)(b_{s_1}, \ldots, b_{s_{n-1}}, b_{s_{n-1}})$$

may be written as a constant plus $\int_0^1 \langle a_s, db_s \rangle$, where $a_s$ is bounded and piecewise continuous and $a_s \equiv 0$ if $s \geq s_{n-1}$. Hence it follows by replacing $a_s$ by $a_s + 1_{(s_{n-1}, s_n)s}\alpha_s$ that

$$F(b) = C + \int_0^{s_n} \langle a_s, db_s \rangle$$

for some constant $C$. Taking expectations of both sides of this equation then shows $C = \mathbb{E}[F(b)]$.                    □

**Remark 7.18.** By being more careful in the proof of the Lemma 7.17 (as is done in more generality later in Theorem 7.47) it is possible to show $a_s$ in (7.12) may be written as

$$a_s = \mathbb{E} \left[ \sum_{i=1}^n 1_{s \leq s_i} \mathrm{grad}_i f (b_{s_1}, \ldots, b_{s_n}) \Big| \mathcal{F}_s \right]. \qquad (7.14)$$

This will also be explained, by indirect means, in Theorem 7.21 below.

**Corollary 7.19.** *Let F be a smooth cylinder function on W, then there is a predictable, piecewise continuously differentiable Cameron–Martin process h such that $F = \mathbb{E}F + D^*h$.*

*Proof.* Let $h_s := \int_0^s a_r dr$ where $a$ is the process as in Lemma 7.17.                    □

**Corollary 7.20 (Martingale Representation Property).** *Let $F \in L^2(\mu)$, then there is a predictable process, $a_s$, such that $\mathbb{E} \int_0^1 |a_s|^2 ds < \infty$, and*

$$F = \mathbb{E}F + \int_0^1 \langle a, db \rangle. \qquad (7.15)$$

*Proof.* Choose a sequence of smooth cylinder functions $\{F_n\}$ such that $F_n \to F$ as $n \to \infty$. By replacing $F$ by $F - \mathbb{E}F$ and $F_n$ by $F_n - \mathbb{E}F_n$, we may assume that $\mathbb{E}F = 0$ and $\mathbb{E}F_n = 0$. Let $a^n$ be predictable processes such that $F_n = \int_0^1 \langle a^n, db \rangle$ for all $n$. Notice that

$$\mathbb{E} \int_0^1 |a_s^n - a_s^m|^2 ds = \mathbb{E}(F_n - F_m)^2 \to 0 \text{ as } m, n \to \infty.$$

Hence, if $a := L^2(ds \times d\mu) - \lim_{n \to \infty} a^n$, then

$$F_n = \int_0^1 a^n \cdot db \to \int_0^1 \langle a, db \rangle \text{ as } n \to \infty.$$

This shows that $F = \int_0^1 \langle a, db \rangle$.    $\square$

**Theorem 7.21 (Clark–Ocone Formula).** *Suppose that $F \in \mathcal{D}(\bar{D})$, then*[8]

$$F = \mathbb{E}F + \int_0^1 \left\langle \mathbb{E}\left[\frac{d}{ds}(\bar{D}F)_s(b)\Big| \mathcal{F}_s\right], db_s \right\rangle. \tag{7.16}$$

*In particular if $F = f(b_{s_1}, \ldots, b_{s_n})$ is a smooth cylinder function on $W(M)$ then*

$$F = \mathbb{E}F + \int_0^1 \left\langle \mathbb{E}\left[\sum_{i=1}^n 1_{s \le s_i} \operatorname{grad}_i f(b_{s_1}, \ldots, b_{s_n})\Big| \mathcal{F}_s\right], db_s \right\rangle. \tag{7.17}$$

*Proof.* Let $h$ be a predictable Cameron–Martin valued process such that $\mathbb{E} \int_0^1 |h_s'|^2 ds < \infty$. Then using Theorem 7.15 and the Itô isometry property,

$$\mathbb{E}\langle \bar{D}F, h \rangle_H = \mathbb{E}[FD^*h] = \mathbb{E}\left[F \int_0^1 \langle h_s', db_s \rangle\right]$$

$$= \mathbb{E}\left[\left(\mathbb{E}F + \int_0^1 \langle a, db \rangle\right) \int_0^1 \langle h_s', db_s \rangle\right] = \mathbb{E}\left[\int_0^1 \langle a_s, h_s' \rangle ds\right] \tag{7.18}$$

where $a$ is the predictable process in Corollary 7.20. Since $h$ is predictable,

$$\mathbb{E}\langle \bar{D}F, h \rangle_H = \mathbb{E}\left[\int_0^1 \left\langle \frac{d}{ds}(\bar{D}F)_s, h_s' \right\rangle ds\right]$$

$$= \mathbb{E}\left[\int_0^1 \left\langle \mathbb{E}\left[\frac{d}{ds}(\bar{D}F)_s\Big| \mathcal{F}_s\right], h_s' \right\rangle ds\right]. \tag{7.19}$$

Since $h$ is an arbitrary predictable Cameron–Martin valued process, comparing (7.18) and (7.18) shows

---

[8] Here we are abusing notation and writing $\mathbb{E}\left[\frac{d}{ds}\bar{D}F_s(b)\Big| \mathcal{F}_s\right]$ for the "predictable" projection of the process $s \to \frac{d}{ds}\bar{D}F_s(b)$. Since we will only really use (7.17) in these notes, this technicality need not concern us here.

$$a_s = \mathbb{E}\left[\frac{d}{ds}(\bar{D}F)_s \Big| \mathcal{F}_s\right]$$

which combined with (7.12) completes the proof.    □

**Remark 7.22.** As mentioned in Remark 7.18 it is possible to prove (7.17) by an inductive procedure. On the other hand if we were to know that (7.17) was valid for all $F \in \mathcal{F}C^1(W)$, then for $h \in \mathcal{X}_a$,

$$\mathbb{E}\left[F\int_0^1 \langle h_s', db_s\rangle\right] = \mathbb{E}\left[\left(\mathbb{E}F + \int_0^1 \left\langle \mathbb{E}\left[\frac{d}{ds}DF_s|\mathcal{F}_s\right], db_s\right\rangle\right)\int_0^1 \langle h_s', db_s\rangle\right]$$

$$= \mathbb{E}\left[\int_0^1 \left\langle \mathbb{E}\left[\frac{d}{ds}DF_s|\mathcal{F}_s\right], h_s'\right\rangle ds\right]$$

$$= \mathbb{E}\left[\int_0^1 \left\langle \frac{d}{ds}DF_s, h_s'\right\rangle ds\right] = \langle DF, h\rangle_{\mathcal{X}}.$$

This identity shows $h \in \mathcal{D}(D^*)$ and that $D^*h = \int_0^1 \langle h_s', db_s\rangle$, i.e., we have recovered Theorem 7.15. In this way we see that the Clark–Ocone formula may be used to recover integration by parts on a Wiener space.

Let $\mathcal{L}$ be the infinite dimensional Ornstein–Uhlenbeck operator defined as the self-adjoint operator on $L^2(\mu)$ given by $\mathcal{L} = D^*\bar{D}$. The following spectral gap inequality for $\mathcal{L}$ has been known since the early days of quantum mechanics. This is because $\mathcal{L}$ is unitarily equivalent to a "harmonic oscillator Hamiltonian" for which the full spectrum may be found, see for example [162]. However, these explicit computations will not in general be available when we consider analogous spectral gap inequalities when $\mathbb{R}^d$ is replaced by a general compact Riemannian manifold $M$.

**Theorem 7.23 (Ornstein Uhlenbeck Spectral Gap Inequality).** *The null space of $\mathcal{L}$ consists of the constant functions on $W$ and $\mathcal{L}$ has a spectral gap of size 1, i.e.,*

$$\langle \mathcal{L}F, F\rangle_{L^2(\mu)} \geq \langle F, F\rangle_{L^2(\mu)} \tag{7.20}$$

*for all $F \in \mathcal{D}(\mathcal{L})$ such that $F \in \mathrm{Nul}(\mathcal{L})^{\perp} = \{1\}^{\perp}$.*

*Proof.* Let $F \in \mathcal{D}(\bar{D})$, then by the Clark–Ocone formula in (7.16), the isometry property of the Itô integral and the contractive properties of conditional expectation,

$$\mathbb{E}(F - \mathbb{E}F)^2 = \mathbb{E}\left[\int_0^1 \left\langle \mathbb{E}\left[\frac{d}{ds}\bar{D}F_s(b)|\mathcal{F}_s\right], db_s\right\rangle\right]^2$$

$$= \mathbb{E}\left[\int_0^1 \left|\mathbb{E}\left[\frac{d}{ds}\bar{D}F_s(b)|\mathcal{F}_s\right]\right|^2 ds\right]$$

$$\leq \mathbb{E}\left[\int_0^1 \left(\mathbb{E}\left[\left|\frac{d}{ds}\bar{D}F_s(b)\right||\mathcal{F}_s\right]\right)^2 ds\right]$$

$$\leq \mathbb{E}\left[\int_0^1 \mathbb{E}\left[\left|\frac{d}{ds}\bar{D}F_s(b)\right|^2 \Big| \mathcal{F}_s\right] ds\right]$$

$$= \mathbb{E}\left[\int_0^1 \left|\frac{d}{ds}\bar{D}F_s(b)\right|^2 ds\right] = \langle \bar{D}F, \bar{D}F\rangle_{\mathcal{X}}.$$

In particular if $F \in \mathcal{D}(\mathcal{L})$, then $\langle \bar{D}F, \bar{D}F\rangle_{\mathcal{X}} = \mathbb{E}[\mathcal{L}F \cdot F]$, and hence

$$\langle \mathcal{L}F, F\rangle_{L^2(\mu)} \geq \langle F - \mathbb{E}F, F - \mathbb{E}F\rangle_{L^2(\mu)}. \tag{7.21}$$

Therefore, if $F \in \text{Nul}(\mathcal{L})$, it follows that $F = \mathbb{E}F$, i.e., $F$ is a constant. Moreover if $F \perp 1$ (i.e., $\mathbb{E}F = 0$) then (7.20) becomes (7.21). $\qquad\square$

It turns out that using a method which is attributed to Maurey and Neveu in [29], it is possible to use the Clark–Ocone formula as the starting point for a proof of Gross' logarithmic Sobolev inequality which by general theory is known to be stronger than the spectral gap inequality in Theorem 7.23.

**Theorem 7.24 (Gross' Logarithmic Sobolev Inequality for $W(\mathbb{R}^d)$).** *For all* $F \in \mathcal{D}(\bar{D})$,

$$\mathbb{E}\left[F^2 \log F^2\right] \leq 2\mathbb{E}\left[\langle DF, DF\rangle_H\right] + \mathbb{E}F^2 \cdot \log \mathbb{E}F^2. \tag{7.22}$$

*Proof.* Let $F \in \mathcal{F}C^1(W)$, $\varepsilon > 0$, $H_\varepsilon := F^2 + \varepsilon \in \mathcal{D}(\bar{D})$ and $a_s = \mathbb{E}\left[\frac{d}{ds}(DH_\varepsilon)_s | \mathcal{F}_s\right]$. By Theorem 7.21,

$$H_\varepsilon = \mathbb{E}H_\varepsilon + \int_0^1 \langle a, db\rangle$$

and hence

$$M_s := \mathbb{E}[H_\varepsilon | \mathcal{F}_s] = \mathbb{E}\left[F^2 + \varepsilon | \mathcal{F}_s\right] \geq \varepsilon$$

is a positive martingale which may be written as

$$M_s := M_0 + \int_0^s \langle a, db\rangle$$

where $M_0 = \mathbb{E}H_\varepsilon$.

Let $\phi(x) = x \ln x$ so that $\phi'(x) = \ln x + 1$ and $\phi''(x) = x^{-1}$. Then by Itô's formula,

$$d\left[\phi(M_s)\right] = \phi(M_0) + \phi'(M_s) dM_s + \frac{1}{2}\phi''(M_s)|a_s|^2 ds$$

$$= \phi(M_0) + \phi'(M_s) dM_s + \frac{1}{2}\frac{1}{M_s}|a_s|^2 ds.$$

Integrating this equation on $s$ and then taking expectations shows

$$\mathbb{E}[\phi(M_1)] = \phi(\mathbb{E}M_1) + \frac{1}{2}\mathbb{E}\left[\int_0^1 \frac{1}{M_s}|a_s|^2 ds\right]. \tag{7.23}$$

Since $\bar{D}H_\varepsilon = 2F\bar{D}F$, (7.23) is equivalent to

$$\mathbb{E}\left[\phi\left(H_\varepsilon\right)\right] = \phi\left(\mathbb{E}H_\varepsilon\right) + \frac{1}{2}\mathbb{E}\left[\int_0^1 \frac{1}{\mathbb{E}\left[H_\varepsilon|\mathcal{F}_s\right]}\left|\mathbb{E}\left[2F\left(\bar{D}F\right)_s'|\mathcal{F}_s\right]\right|^2 ds\right].$$

Using the Cauchy–Schwarz inequality and the contractive properties of conditional expectations,

$$\left|\mathbb{E}\left[2F\frac{d}{ds}\left(\bar{D}F\right)_s|\mathcal{F}_s\right]\right|^2 \le 4\left(\mathbb{E}\left[F\left|\frac{d}{ds}\left(\bar{D}F\right)_s\right||\mathcal{F}_s\right]\right)^2$$

$$\le 4\mathbb{E}\left[F^2|\mathcal{F}_s\right]\cdot\mathbb{E}\left[\left|\frac{d}{ds}\left(\bar{D}F\right)_s\right|^2|\mathcal{F}_s\right].$$

Combining the last two equations, using

$$\frac{\mathbb{E}\left[F^2|\mathcal{F}_s\right]}{\mathbb{E}\left[H_\varepsilon|\mathcal{F}_s\right]} = \frac{\mathbb{E}\left[F^2|\mathcal{F}_s\right]}{\mathbb{E}\left[F^2|\mathcal{F}_s\right] + \varepsilon} \le 1 \tag{7.24}$$

gives,

$$\mathbb{E}\left[\phi\left(H_\varepsilon\right)\right] \le \phi\left(\mathbb{E}H_\varepsilon\right) + 2\mathbb{E}\int_0^1 \mathbb{E}\left[\left|\frac{d}{ds}\left(\bar{D}F\right)_s\right|^2|\mathcal{F}_s\right]ds$$

$$= \phi\left(\mathbb{E}H_\varepsilon\right) + 2\mathbb{E}\int_0^1 \left|\frac{d}{ds}\left(\bar{D}F\right)_s\right|^2 ds.$$

We may now let $\varepsilon \downarrow 0$ in this inequality to find (7.22) is valid for $F \in \mathcal{F}C^1\left(W\right)$. Since $\mathcal{F}C^1\left(W\right)$ is a core for $\bar{D}$, standard limiting arguments show that (7.22) is valid in general.                                                                              □

The main objective for the rest of this section is to generalize the previous theorems to the setting of general compact Riemannian manifolds. Before doing this we need to record the stochastic analogues of the differentiation formula in Theorems 4.7, 4.12, and 4.13.

## 7.2  Differentials of stochastic flows and developments

**Notation 7.25.** Let $T_s^\beta\left(m\right) = \Sigma_s$ where $\Sigma_s$ is the solution to (5.1) with $\Sigma_0 = m$ and $\beta_s$ is an $\mathbb{R}^n$-valued semi-martingale, i.e.,

$$\delta\Sigma_s = \sum_{i=1}^n X_i\left(\Sigma_s\right)\delta\beta_s^i + X_0\left(\Sigma_s\right)ds \text{ with } \Sigma_0 = m.$$

**Theorem 7.26 (Differentiating $\Sigma$ in $B$).** Let $\beta_s = B_s$ be an $\mathbb{R}^n$-valued Brownian motion and $h$ be an adapted Cameron–Martin process, $h_s \in \mathbb{R}^n$ with $\left|h_s'\right|$ bounded. Then there is a version of $T_s^{B+th}\left(m\right)$ which is continuous in $s$ and differentiable in $\left(t, m\right)$. Moreover if we define $\partial_h T_s^B\left(o\right) := \frac{d}{dt}|_0 T_s^{B+sh}\left(o\right)$, then

$$\partial_h T_s^B(o) = Z_s \int_0^s Z_r^{-1} X_{h_r'}(\Sigma_r)\, dr = //_s z_s \int_0^s z_r^{-1} //_r^{-1} X_{h_r'}(\Sigma_r)\, dr \qquad (7.25)$$

where $Z_s := \left(T_s^B\right)_{*o}$, $//_s$ is stochastic parallel translation along $\Sigma$, and $z_s := //_s^{-1} Z_s$. (See Theorem 5.41 for more on the processes $Z$ and $z$.) Recall from Notation 5.4 that

$$X_a(m) := \sum_{i=1}^n a_i X_i(m) = \mathbf{X}(m)\, a.$$

*Proof.* This is a stochastic analogue of Theorem 4.7. Formally, if $B_s$ were piecewise differentiable it would follow from Theorem 4.7 with $s = t$, that

$$X_s(m) = \mathbf{X}(m)\, B_s' + X_0(m) \text{ and } Y_s(m) = \mathbf{X}(m)\, h_s'.$$

(Notice that $\frac{d}{dt}|_0 \left[\mathbf{X}(m)\left(B_s' + t h_s'\right) + X_0(m)\right] = Y_s$.) For a rigorous proof of this theorem in the flat case, which is essentially applicable here because $M$ is an imbedded submanifold, see Bell [12] or Nualart [148] for example. For this theorem in this geometric context see Bismut [20] or Driver [47] for example. $\qquad\square$

**Notation 7.27.** Let $b$ be an $T_o M \cong \mathbb{R}^d$-valued Brownian motion. A $T_o M$-valued semi-martingale $Y$ is called an *adapted vector field* or *tangent process* to $b$ if $Y$ can be written as

$$Y_s = \int_0^s q_r\, db_r + \int_0^s \alpha_r\, dr \qquad (7.26)$$

where $q_r$ is an $so(d)$-valued adapted process and $\alpha_s$ is a $T_o M$ such that

$$\int_0^1 |\alpha_s|^2\, ds < \infty \text{ a.e.}$$

A key point of a tangent process $Y$ as above is that it gives rise to natural perturbations of the underlying Brownian motion $b$. Namely, following Bismut (also see Fang and Malliavin [78]), for $t \in \mathbb{R}$ let $b_s^t$ be the process given by

$$b_s^t := \int_0^s e^{tq_r} b_r + t \int_0^s \alpha_r\, dr. \qquad (7.27)$$

Then (under some integrability restrictions on $\alpha$) by Lévy's criteria and Girsanov's theorem, the law of $b^t$ is absolutely continuous relative to the law of $b$. Moreover $b^0 = b$ and, with some additional integrability assumptions on $q_r$, $\frac{d}{dt}|_0 b^t = Y$.

Let $b$ be an $T_o M \cong \mathbb{R}^d$-valued Brownian motion, $\Sigma := \phi(b)$ be the stochastic development map as in Notation 5.30 and suppose that $X^h = //h$ is a Cameron–Martin vector field on $W(M)$. Using Theorem 4.12 as motivation (see (4.16)), the pull back of $X$ under the stochastic development map should be the process $Y$ defined by

$$Y_s = h_s + \int_0^s \left(\int_0^r R_{//_\rho}(h_\rho, \delta b_\rho)\right) \delta b_r \qquad (7.28)$$

where

$$R_{//_s}(h_s, \delta b_s) = //_s^{-1} R(//_s h_s, //_s \delta b_s) //_s \qquad (7.29)$$

like in (4.15). Since

$$\left( \int_0^r R_{//_\rho}(h_\rho, \delta b_\rho) \right) \delta b_r = \left( \int_0^r R_{//_\rho}(h_\rho, \delta b_\rho) \right) db_r + \frac{1}{2} R_{//_\rho}(h_\rho, db_\rho) db_\rho$$

$$= \left( \int_0^r R_{//_\rho}(h_\rho, \delta b_\rho) \right) db_r + \frac{1}{2} \sum_{i=1}^d R_{//_\rho}(h_\rho, e_i) e_i d\rho$$

where $\{e_i\}_{i=1}^d$ is an orthonormal basis for $T_o M$, (7.28) may be written in Itô's form as

$$Y_\cdot = \int_0^\cdot C_s db_s + \int_0^\cdot r_s ds, \qquad (7.30)$$

where

$$C_s := \int_0^s R_{//_\sigma}(h_\sigma, \delta b_\sigma), \quad r_s = h_s' + \frac{1}{2} \mathrm{Ric}_{//_s} h_s \text{ and} \qquad (7.31)$$

$$\mathrm{Ric}_{//_s} a := //_s^{-1} \mathrm{Ric} //_s a \; \forall \, a \in T_o M. \qquad (7.32)$$

By the symmetry property in item 4b of Proposition 3.36, the matrix $C_s$ is skew symmetric and therefore $Y$ is a tangent process. Here is a theorem which relates $Y$ in (7.30) to $X^h = //h$.

**Theorem 7.28 (Differential of the development map).** *Assume M is a compact manifold, $o \in M$ is fixed, b is $T_o M \cong \mathbb{R}^d$-valued Brownian motion, $\Sigma := \phi(b)$, h is a Cameron–Martin process with $|h_s'| \le K < \infty$ (K is a non-random constant) and Y is as in (7.30). As in (7.27) let*

$$b_s^t := \int_0^s e^{tC_r} db_r + t \int_0^s r_u du. \qquad (7.33)$$

*Then there exists a version of $\phi_s(b^t)$ which is continuous in $(s, t)$, differentiable in $t$ and $\frac{d}{dt}|_0 \phi(b^t) = X^h$.*

*Proof.* For the proof of this theorem and its generalization to more general $h$, the reader is referred to Section 3.1 of [45] and to [47]. Let me just point out here that formally the proof is very analogous to the deterministic version in Theorems 4.12 and 4.13. □

### 7.3 Quasi-invariance flow theorem for $W(M)$

In this section, we will discuss the $W(M)$ analogues of Theorems 7.13 and 7.14.

**Theorem 7.29 (Cameron–Martin Theorem for $M$).** *Let $h \in H(T_oM)$ and $X^h$ be the $\mu_{W(M)}$ — a.e. well defined vector field on $W(M)$ given by*

$$X_s^h(\sigma) = //_s(\sigma)h_s \text{ for } s \in [0, 1], \tag{7.34}$$

*where $//_s(\sigma)$ is stochastic parallel translation along $\sigma \in W(M)$. Then $X^h$ admits a flow $e^{tX^h}$ on $W(M)$ (see Figure 14) and this flow leaves the Wiener measure, $\mu_{W(M)}$, quasi-invariant.*

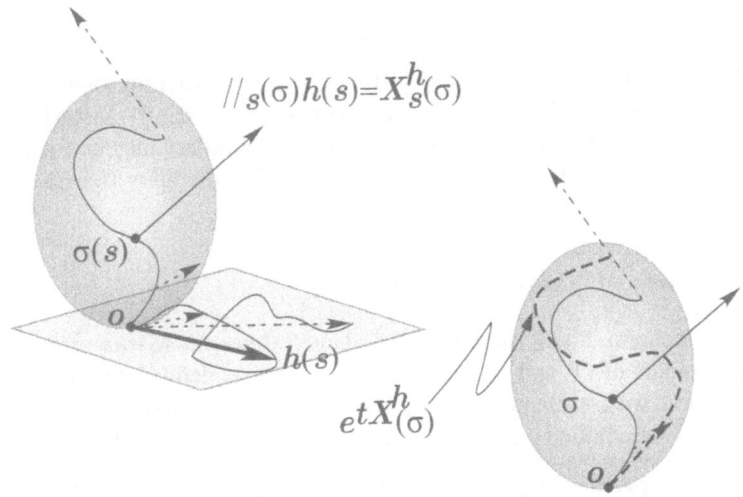

**Figure 14.** Constructing a vector field, $X^h$, on $W(M)$ from a vector field $h$ on $W(T_oM)$. The dotted path indicates the flow of $\sigma$ under this vector field.

This theorem first appeared in Driver [47] for $h \in H(T_oM) \cap C^1([0, 1], T_oM)$ and was soon extended to all $h \in H(T_oM)$ by E. Hsu [96, 97]. Other proofs may also be found in [76, 127, 146]. The proof of this theorem is rather involved and will not be given here. A sketch of the argument and more information on the technicalities involved may be found in [49].

**Example 7.30.** When $M = \mathbb{R}^d$, $//_s(\sigma)v_o = v_{\sigma_s}$ for all $v \in \mathbb{R}^d$ and $\sigma \in W(\mathbb{R}^d)$. Thus $X_s^h(\sigma) = (h_s)_{\sigma_s}$ and $e^{tX^h}(\sigma) = \sigma + th$ and so Theorem 7.29 becomes the classical Cameron–Martin Theorem 7.13.

**Corollary 7.31 (Integration by Parts for $\mu_{W(M)}$).** *For $h \in H(T_oM)$ and $F \in \mathcal{F}C^1(W(M))$ as in equation (7.2), let*

$$(X^h F)(\sigma) = \frac{d}{dt}|_0 F(e^{tX^h}(\sigma)) = G\left(DF, X^h\right)$$

*as in Notation 7.11. Then*

$$\int_{W(M)} X^h F \, d\mu_{W(M)} = \int_{W(M)} F \, z^h \, d\mu_{W(M)}$$

*where*

$$z^h := \int_0^1 \langle h_s' + \frac{1}{2} \mathrm{Ric}_{//s} h_s', db_s \rangle,$$

$$b_s(\sigma) := \Psi_s(\sigma) = \int_0^s //_r^{-1} \delta\sigma_r$$

*and* $\mathrm{Ric}_{//s} \in \mathrm{End}(T_oM)$ *is as in (5.60).*

*Proof.* A special case of this Corollary 7.31 with $F(\sigma) = f(\sigma_s)$ for some $f \in C^\infty(M)$ first appeared in Bismut [21]. The result stated here was proved in [47] as an infinitesimal form of the flow Theorem 7.29. Other proofs of this corollary may be found in [2, 5, 50, 72, 73, 70, 76, 78, 96, 97, 122, 123, 127, 146]. This corollary is a special case of Theorem 7.32 below.                                                                                □

### 7.4 Divergence and integration by parts

In the next theorem, it will be shown that adapted Cameron–Martin vector fields, $X$, are in the domain of $D^*$ and consequently $D^*$ is densely defined. For the purposes of this subsection, we assume that $b$ is a $T_oM$-valued Brownian motion, $\Sigma = \phi(b)$ is the evolved Brownian motion on $M$ and $//_s$ is stochastic parallel translation along $\Sigma$.

**Theorem 7.32.** *Let* $X \in \mathcal{X}_a$ *be an adapted Cameron–Martin vector field on* $W(M)$ *and* $h := //^{-1}X$. *Then* $X \in \mathcal{D}(D^*)$ *and*

$$X^*1 = D^*X = \int_0^1 \langle B(h), db \rangle = \int_0^1 \langle h_s' + \frac{1}{2} \mathrm{Ric}_{//s} h_s, db_s \rangle, \tag{7.35}$$

*where $B$ is the random linear operator mapping $H$ to $L^2(ds, T_oM)$ given by*

$$[B(h)]_s := h_s' + \frac{1}{2} \mathrm{Ric}_{//s} h_s. \tag{7.36}$$

**Remark 7.33.** There is a non-random constant $C < \infty$ depending only on the bound on the Ricci tensor such that $\|B\|_{H \to L^2(ds, T_oM)} \leq C$.

*Proof.* I will give a sketch of the proof here, the interested reader may find complete details of this proof in [45]. Moreover, we will give two more proofs of this theorem, see Theorem 7.40 and Corollary 7.50 below.

We start by proving the theorem under the additional assumption that $h := //^{-1}X$ satisfies $\sup_{s \in [0,1]} |h_s'| \leq K$, where $K$ is a non-random constant.

Let $b_s^t$ be defined as in (7.33). (Notice that $b^t$ is *not* the flow of the vector field $Y$ in (7.30) but does have the property that $\frac{d}{dt}|_0 b_s^t = Y_s$.) Since $C_s$ is skew-symmetric,

$e^{tC_s}$ is orthogonal and so by Levy's criteria, $s \to \int_0^s e^{tC_r} db_r$ is a Brownian motion. Combining this with Girsanov's theorem, $s \to b_s^t$ (for fixed $t$) is a Brownian motion relative to the measure $Z_t \cdot \mu$, where

$$Z_t := \exp\left(-\int_0^1 t\langle r, e^{tC} db \rangle - \frac{1}{2} t^2 \int_0^1 \langle r, r \rangle ds\right). \tag{7.37}$$

For $t \in \mathbb{R}$, let $\Sigma(t, \cdot) := \phi(b^t)$ where $\phi$ is the stochastic development map as in Theorem 5.29. Then by Theorem 7.28, $X^h = \frac{d}{dt}|_0 \Sigma(t, \cdot)$ and in particular if $F$ is a smooth cylinder function then $X^h F = \frac{d}{dt}|_0 F(\Sigma(t, \cdot))$. So differentiating the identity,

$$\mathbb{E}[F(\Sigma(t, \cdot)) Z_t] = \mathbb{E}[F(\Sigma)],$$

at $t = 0$ gives:

$$\mathbb{E}[XF] - \mathbb{E}\left[F \int_0^1 \langle r, db \rangle\right] = 0.$$

This last equation may be written alternatively as

$$\langle DF, X \rangle_{\mathcal{X}} = \mathbb{E}[G(DF, X)] = \mathbb{E}\left[F \cdot \int_0^1 \langle B(h), db \rangle\right].$$

Hence it follows that $X \in \mathcal{D}(D^*)$ and

$$D^* X = \int_0^1 \langle B(h), db \rangle.$$

This proves the theorem in the special case that $h'$ is uniformly bounded.

Let $X$ be a general adapted Cameron–Martin vector field and $h := //^{-1} X$. For each $n \in \mathbb{N}$, let $h_n(s) := \int_0^s h'(r) \cdot 1_{|h'(r)| \le n} dr$ be as in (7.11). Set $X^n := // h_n$, then by the special case above we know that $X^n \in \mathcal{D}(D^*)$ and $D^* X^n = \int_0^1 \langle B(h_n), db \rangle$. It is easy to check that

$$\langle X - X^n, X - X^n \rangle_{\mathcal{X}} = \mathbb{E}\langle h - h_n, h - h_n \rangle_H \to 0 \text{ as } n \to \infty.$$

Furthermore,

$$\mathbb{E}|D^*(X^m - X^n)|^2 = \mathbb{E} \int_0^1 |B(h_m - h_n)|^2 ds \le C\mathbb{E}\langle h_m - h_n, h_m - h_n \rangle_H,$$

from which it follows that $D^* X^m$ is convergent. Because $D^*$ is a closed operator, it follows that $X \in \mathcal{D}(D^*)$ and

$$D^* X = \lim_{n \to \infty} D^* X^n = \lim_{n \to \infty} \int_0^1 \langle B(h_n), db \rangle = \int_0^1 \langle B(h), db \rangle. \qquad \square$$

**Corollary 7.34.** *The operator* $D^* : \mathcal{X} \to L^2\left(W(M), \mu_{W(M)}\right)$ *is densely defined. In particular $D$ is closable. (Let $\bar{D}$ denote the closure of $D$.)*

*Proof.* Let $h \in H$, $X^h := //h$, and $F$ and $K$ be smooth cylinder functions. Then, by the product rule,

$$\langle DF, KX^h \rangle_{\mathcal{X}} = \mathbb{E}[G\left(KDF, X^h\right)] = \mathbb{E}[G\left(D\left(KF\right) - FDK, X^h\right)]$$

$$= \mathbb{E}[F \cdot KD^*X^h - FG\left(DK, X^h\right)].$$

Therefore $KX^h \in \mathcal{D}(D^*)$ ($\mathcal{D}(D^*)$ is the domain of $D^*$) and

$$D^*(KX^h) = KD^*X^h - G(DK, X^h).$$

Since

$$\mathrm{span}\{KX^h | h \in H \text{ and } K \in \mathcal{F}C^\infty\} \subset \mathcal{D}(D^*)$$

is a dense subspace of $\mathcal{X}$, $D^*$ is densely defined.     □

**Corollary 7.35.** *Let $h$ be an adapted Cameron–Martin valued process and $Q_s$ be defined as in (6.1). Then*

$$\left(X^{Q^{\mathrm{tr}}h}\right)^* 1 = \int_0^1 \langle Q^{\mathrm{tr}}h', db \rangle. \tag{7.38}$$

*Proof.* Taking the transpose of (6.1) shows $Q^{\mathrm{tr}}$ solves,

$$\frac{d}{ds}Q^{\mathrm{tr}} + \frac{1}{2}\mathrm{Ric}_{//}Q^{\mathrm{tr}} = 0 \text{ with } Q_0^{\mathrm{tr}} = Id. \tag{7.39}$$

Therefore, from (7.35),

$$\left(X^{Q^{\mathrm{tr}}h}\right)^* 1 = \int_0^1 \langle (Q^{\mathrm{tr}}h)' + \frac{1}{2}\mathrm{Ric}_{//}Q^{\mathrm{tr}}h, db \rangle$$

$$= \int_0^1 \langle \left[\frac{d}{ds} + \frac{1}{2}\mathrm{Ric}_{//}\right](Q^{\mathrm{tr}}h), db \rangle$$

$$= \int_0^1 \langle Q^{\mathrm{tr}}h', db \rangle. \quad □$$

Theorem 7.32 may be extended to allow for vector fields on the paths of $M$ which are not based. This theorem and its Corollary 7.37 will not be used in the sequel and may safely be skipped.

**Theorem 7.36.** *Let $h$ be an adapted $T_oM$-valued process such that $h(0)$ is non-random and $h - h(0)$ is a Cameron–Martin process, $X := X^h := //h$, $\mathbb{E}_x$ denote the path space expectation for a Brownian motion starting at $x \in M$, $F : C([0, 1] \to M) \to \mathbb{R}$ be a cylinder function as in Definition 7.4 and $X^h F$ be defined as in (7.7). Then (writing $\langle df, v \rangle$ for $df(v)$)*

$$\mathbb{E}_o[X^h F] = \mathbb{E}_o[FD^*X^h] + \langle d(\mathbb{E}_{(\cdot)}F), h(0)_o \rangle, \tag{7.40}$$

*where*

$$D^*X^h := \int_0^1 \langle h_s' + \frac{1}{2}\mathrm{Ric}_{//_s}h_s, db_s \rangle := \int_0^1 \langle B(h), db \rangle,$$

*as in (7.35) and $B(h)$ is defined in (7.36).*

*Proof.* Start by choosing a smooth path $\alpha$ in $M$ such that $\dot{\alpha}(0) = h(0)_o$. Let

$$C := \int R_{//}(h, \delta b),$$

$$r = h' + \frac{1}{2} \text{Ric}_{//}(h),$$

$$b_s^t = \int_0^s e^{tC} db + t \int_0^s r d\lambda \text{ and}$$

$$Z_t = \exp - \left\{ \int_0^1 t \langle r, e^{tC} db \rangle + \frac{1}{2} t^2 \int_0^1 \langle r, r \rangle ds \right\}$$

be defined by the same formulas as in the proof of Theorem 7.32. Let $u_0(t)$ denote parallel translation along $\alpha$, that is

$$du_0(t)/dt + \Gamma(\dot{\alpha}(t)) u_0(t) = 0 \quad \text{with} \quad u_0(0) = id.$$

For $t \in \mathbb{R}$, define $\Sigma(t, \cdot)$ by

$$\Sigma(t, \delta s) = u(t, s) \delta b_s^t \quad \text{with} \quad \Sigma(t, 0) = \alpha(t)$$

and

$$u(t, \delta s) + \Gamma(u(t, s) \delta_s b_s^t) u(t, s) = 0 \quad \text{with} \quad u(t, 0) = u_o(t).$$

Appealing to a stochastic version of Theorem 4.14 (after choosing a good version of $\Sigma$) it is possible to show that $\dot{\Sigma}(0, \cdot) = X$, so the $XF = \frac{d}{dt}|_0 F[\Sigma(t, \cdot)]$. As in the proof of Theorem 7.32, $b^t$ is a Brownian motion relative to the expectation $\mathbb{E}_t$ defined by $\mathbb{E}_t(F) := \mathbb{E}[Z_t F]$. From this it is easy to see that $\Sigma(t, \cdot)$ is a Brownian motion on $M$ starting at $\alpha(t)$ relative to the expectation $\mathbb{E}_t$. Therefore, for all $t$,

$$\mathbb{E}[F(\Sigma(t, \cdot)) Z_t] = \mathbb{E}_{\alpha(t)} F$$

and differentiating this last expression at $t = 0$ gives:

$$\mathbb{E}[XF(\Sigma)] - \mathbb{E}\left[F \int_0^1 \langle r, db \rangle\right] = \langle d\mathbb{E}_{(\cdot)} F, h(0)_o \rangle.$$

The rest of the proof is identical to the previous proof.     □

As a corollary to Theorem 7.36 we get Elton Hsu's derivative formula which played a key role in the original proof of his logarithmic Sobolev inequality on $W(M)$, see Theorem 7.52 below and [98].

**Corollary 7.37 (Hsu's Derivative Formula).** *Let $v_o \in T_oM$. Define $h$ to be the adapted $T_oM$ – valued process solving the differential equation:*

$$h_s' + \frac{1}{2} \text{Ric}_{//_s} h_s = 0 \quad \text{with} \quad h_0 = v_o. \tag{7.41}$$

*Then*

$$\langle d(\mathbb{E}_{(\cdot)} F), v_o \rangle = \mathbb{E}_o[X^h F]. \tag{7.42}$$

*Proof.* Apply Theorem 7.36 to $X^h$ with $h$ defined by (7.41). Notice that $h$ has been constructed so that $B(h) \equiv 0$, i.e., $D^* X^h = 0$.                    $\square$

The idea for the proof used here is similar to Hsu's proof, the only question is how one describes the perturbed process $\Sigma(t, \cdot)$ in the proof of Theorem 7.36 above. It is also possible to give a much more elementary proof of (7.42) based on the ideas in Section 6, see for example [58].

## 7.5 Elworthy–Li integration by parts formula

In this subsection, let $\{X_i\}_{i=0}^n \subset \Gamma(TM)$, $B$ be an $\mathbb{R}^n$-valued Brownian motion and $T_s^B(m)$ denote the solution to (5.1) with $\beta = B$ as in Notation 7.25. We will further assume that $\mathbf{X}(m) : \mathbb{R}^n \to T_m M$ (as in Notation 5.4) is surjective for all $m \in M$ and let $\mathbf{X}(m)^\# = \left[ \mathbf{X}(m) \,|_{\mathrm{Nul}(\mathbf{X}(m))^\perp} \right]^{-1}$, as in (6.5). The following Lemma is an elementary exercise in linear algebra.

**Lemma 7.38.** *For $m \in M$ and $v, w \in T_m M$ let*

$$\langle v, w \rangle_m := \langle \mathbf{X}(m)^\# v, \mathbf{X}(m)^\# w \rangle_{\mathbb{R}^n}.$$

*Then*

1. *$m \to \langle \cdot, \cdot \rangle_m$ is a smooth Riemannian metric on $M$.*
2. *$\mathbf{X}(m)^{\mathrm{tr}} = \mathbf{X}(m)^\#$ and in particular $\mathbf{X}(m)\mathbf{X}(m)^{\mathrm{tr}} = id_{T_m M}$ for all $m \in M$.*
3. *Every $v \in T_m M$ may be expanded as*

$$v = \sum_{j=1}^n \langle v, X_j(m) \rangle X_j(m) = \sum_{j=1}^n \langle v, \mathbf{X}(m) e_j \rangle \mathbf{X}(m) e_j \qquad (7.43)$$

*where $\{e_j\}_{j=1}^n$ is the standard basis for $\mathbb{R}^n$.*

The proof of this lemma is left to the reader with the comment that (7.43) is proved in the same manner as item (1) in Proposition 3.48.

**Theorem 7.39 (Elworthy–Li).** *Suppose $k_s$ is a $T_o M$ valued Cameron–Martin process such that $\mathbb{E} \int_0^1 |k_s'|^2 \, ds < \infty$ and $F : W(M) \to \mathbb{R}$ is a bounded $C^1$-function with bounded derivative on $W$, for example $F$ could be a cylinder function. Then*

$$\mathbb{E}\left[ (d_{W(M)} F)(Z.k.) \right] = \mathbb{E}\left[ F(\Sigma) \int_0^T \langle Z_s k_s', \mathbf{X}(\Sigma_s) \, dB_s \rangle \right]$$

$$= \mathbb{E}\left[ F(\Sigma) \int_0^T \langle \mathbf{X}(\Sigma_s)^{\mathrm{tr}} Z_s k_s', dB_s \rangle \right] \qquad (7.44)$$

*where $Z_s = \left( T_s^B \right)_{*o}$ is the differential of $m \to T_s^B(m)$ at $o$.*

*Proof.* Notice that $Z_s k_s \in T_{\Sigma_s} M$ for all $s$ as it should be. By the reduction argument used in the proof of Theorem 7.32, it suffices to consider the case when $|k_s'| \le K$ where $K$ is a non-random constant. Let $h_s$ be the $T_o M$-valued Cameron–Martin process defined by

$$h_s := \int_0^s \mathbf{X}(\Sigma_r)^{\mathrm{tr}} Z_r k_r' dr.$$

Then by Lemma 7.38 and Theorem 7.26,

$$\begin{aligned}
\partial_h T_s^B(o) &= Z_s \int_0^s Z_r^{-1} \mathbf{X}(\Sigma_r) h_r' dr \\
&= Z_s \int_0^s Z_r^{-1} \mathbf{X}(\Sigma_r) \mathbf{X}(\Sigma_r)^{\mathrm{tr}} Z_r k_r' dr = Z_s k_s.
\end{aligned}$$

In particular this implies

$$\partial_h F\left(T_{(\cdot)}^B(o)\right) = \langle dF(\Sigma), \partial_h T_s^B(o)\rangle = \langle d_{W(M)} F(\Sigma), Zk\rangle$$

and therefore by integration by parts on the flat Wiener space (Theorem 7.32 with $M = \mathbb{R}^n$) implies

$$\begin{aligned}
\mathbb{E}\left[(d_{W(M)} F)(\Sigma)(Z.k.)\right] &= \mathbb{E}\left[\partial_h [F(\Sigma)]\right] = \mathbb{E}\left[F(\Sigma) \int_0^T \langle h_s', dB_s\rangle\right] \\
&= \mathbb{E}\left[F(\Sigma) \int_0^T \langle \mathbf{X}(\Sigma_s)^{\mathrm{tr}} Z_s k_s', dB_s\rangle\right]. \qquad \square
\end{aligned}$$

By factoring out the redundant noise in Theorem 7.39, we get yet another proof of Corollary 7.35 which also easily gives another proof of Theorem 7.32.

**Theorem 7.40 (Factoring out the redundant noise).** *Assume* $\mathbf{X}(m) = P(m)$ *and* $X_0 = 0$, $k_s$ *is a Cameron–Martin valued process adapted to the filtration,* $\mathcal{F}_s^\Sigma := \sigma(\Sigma_r : r \le s)$, *then*

$$\mathbb{E}\left[(d_{W(M)} F)(//Q_t^{\mathrm{tr}} k)\right] = \mathbb{E}\left[F(\Sigma) \int_0^T \langle Q_s^{\mathrm{tr}} k_s', db_s\rangle\right]$$

*where* $Q_s$ *solves (6.1).*

*Proof.* By Theorems 7.39 and 5.40, we have

$$\begin{aligned}
\mathbb{E}\left[(d_{W(M)} F)(//zk)\right] &= \mathbb{E}\left[F(\Sigma) \int_0^T \langle //_s z_s k_s', P(\Sigma_s) dB_s\rangle\right] \\
&= \mathbb{E}\left[F(\Sigma) \int_0^T \langle z_s k_s', db_s\rangle\right].
\end{aligned}$$

Combining this with Theorem 5.44 implies

$$\mathbb{E}\left[\left(d_{W(M)}F\right)(//\bar{z}k)\right] = \mathbb{E}\left[F\left(\Sigma\right)\int_0^T \langle \bar{z}_s k_s', db_s\rangle\right].$$

As observed in the proof of Corollary 6.4, $\bar{z}_t = Q_t^{\text{tr}}$ which completes the proof.    □

The reader interested in seeing more of these types of arguments is referred to Elworthy, Le Jan and Li [71] where these ideas are covered in much greater detail and in full generality.

### 7.6 Fang's spectral gap theorem and proof

As in the flat case we let $\mathcal{L} = D^*\bar{D}$, an unbounded operator on $L^2\left(W\left(M\right), \mu_{W(M)}\right)$ which is a "curved" analogue of the Ornstein–Uhlenbeck operator used in Theorem 7.23. It has been shown in Driver and Röckner [56] that this operator generates a diffusion on $W(M)$. This last result also holds for pinned paths on $M$ and free loops on $\mathbb{R}^N$, see [6].

In this section, we will give a proof of S. Fang's [79] spectral gap inequality for $\mathcal{L}$. Hsu's stronger logarithmic Sobolev inequality will be covered later in Theorem 7.52 below.

**Theorem 7.41 (Fang).** *Let $\bar{D}$ be the closure of $D$ and $\mathcal{L}$ be the self-adjoint operator on $L^2\left(\mu_{W(M)}\right)$ defined by $\mathcal{L} = D^*\bar{D}$. (Note, if $M = \mathbb{R}^d$ then $\mathcal{L}$ would be an infinite dimensional Ornstein–Uhlenbeck operator.) Then the null space of $\mathcal{L}$ consists of the constant functions on $W(M)$ and $\mathcal{L}$ has a spectral gap, i.e., there is a constant $c > 0$ such that $\langle \mathcal{L}F, F\rangle_{L^2(\mu_{W(M)})} \geq c\langle F, F\rangle_{L^2(\mu_{W(M)})}$ for all $F \in \mathcal{D}(\mathcal{L})$ which are perpendicular to the constant functions.*

This theorem is the $W\left(M\right)$ analogue of Theorem 7.23. The proof of this theorem will be given at the end of this subsection. We first will need to represent $F$ in terms of $DF$. (Also see Section 7.7 below.)

**Lemma 7.42.** *For each $F \in L^2\left(W\left(M\right), \mu_{W(M)}\right)$, there is a unique adapted Cameron–Martin vector field $X$ on $W(M)$ such that*

$$F = \mathbb{E}F + D^*X.$$

*Proof.* By the martingale representation theorem (see Corollary 7.20), there is a predictable $T_oM$-valued process, $a$, (which is not in general continuous) such that

$$\mathbb{E}\int_0^1 |a_s|^2 ds < \infty,$$

and

$$F = \mathbb{E}F + \int_0^1 \langle a_s, db_s\rangle. \tag{7.45}$$

Define $h := B^{-1}(a)$, where $B$ is as in Eq. (7.36); that is to say let $h$ be the solution to the differential equation:

$$h'_s + \frac{1}{2} \mathrm{Ric}_{//_s} h_s = a_s \text{ with } h_0 = 0. \tag{7.46}$$

*Claim:* $B_\sigma^{-1}$ is a bounded linear map from $L^2(ds, T_o M) \to H$ for each $\sigma \in W(M)$, and furthermore the norm of $B_\sigma^{-1}$ is bounded independent of $\sigma \in W(M)$.

To prove the claim, use Duhamel's principle to write the solution to (7.46) as

$$h_s = \int_0^s Q_s^{\mathrm{tr}} \left(Q_\tau^{\mathrm{tr}}\right)^{-1} a_\tau d\tau, \tag{7.47}$$

where $Q_s$ is as in Eq. (6.1). Since, $W_s := Q_s^{\mathrm{tr}} \left(Q_\tau^{\mathrm{tr}}\right)^{-1}$ solves the differential equation

$$W'_s + \frac{1}{2} \mathrm{Ric}_{//_s} W_s = 0 \text{ with } W_\tau = I$$

it is easy to show from the boundedness of $\mathrm{Ric}_{//_s}$ and an application of Gronwall's inequality that

$$\left| Q_s^{\mathrm{tr}} \left(Q_\tau^{\mathrm{tr}}\right)^{-1} \right| = |W_s| \le C,$$

where $C$ is a non-random constant independent of $s$ and $\tau$. Therefore,

$$\begin{aligned}
\langle h, h \rangle_H &= \int_0^1 |a_s - \frac{1}{2} \mathrm{Ric}_{//_s} h_s|^2 ds \\
&\le 2 \int_0^1 |a_s|^2 ds + 2 \int_0^1 |\frac{1}{2} \mathrm{Ric}_{//_s} h_s|^2 ds \\
&\le 2(1 + C^2 K^2) \int_0^1 |a_s|^2 ds,
\end{aligned}$$

where $K$ is a bound on the process $\frac{1}{2} \mathrm{Ric}_{//_s}$. This proves the claim.

Because of the claim, $h := B^{-1}(a)$ satisfies $\mathbb{E}\left[\langle h, h \rangle_H\right] < \infty$ and because of (7.47), $h$ is adapted. Hence, $X := //h$ is an adapted Cameron–Martin vector field and

$$D^* X = \int_0^1 \langle B(h), db \rangle = \int_0^1 \langle a, db \rangle.$$

The existence part of the theorem now follows from this identity and (7.45).

The uniqueness assertion follows from the energy identity:

$$\mathbb{E}\left[D^* X\right]^2 = \mathbb{E} \int_0^1 |B(h)_s|^2 ds \ge C \mathbb{E}\left[\langle h, h \rangle_H\right].$$

Indeed if $D^* X = 0$, then $h = 0$ and hence $X = //h = 0$.  □

The next goal is to find an expression for the vector field $X$ in the above lemma in terms of the function $F$ itself. This will be the content of Theorem 7.45 below.

**Notation 7.43.** Let $L_a^2(\mu_{W(M)} : L^2(ds, T_oM))$ denote the $T_oM$-valued predictable processes, $v_s$ on $W(M)$ such that $\mathbb{E} \int_0^1 |v_s|^2 \, ds < \infty$. Define the bounded linear operator $\bar{B} : \mathcal{X}_a \to L_a^2(\mu_{W(M)} : L^2(ds, T_oM))$ by

$$\bar{B}(X) = B(//^{-1}X) = \frac{d}{ds}\left[//_s^{-1}X_s\right] + \frac{1}{2}//_s^{-1}\,\text{Ric}\, X_s.$$

Also let $\mathcal{Q} : \mathcal{X} \to \mathcal{X}$ denote the orthogonal projection of $\mathcal{X}$ onto $\mathcal{X}_a$.

**Remark 7.44.** Notice that $D^*X = \int_0^1 \langle \bar{B}(X), db \rangle$ for all $X \in \mathcal{X}_a$. We have seen that $\bar{B}$ has a bounded inverse, in fact $\bar{B}^{-1}(a) = //B^{-1}(a)$.

**Theorem 7.45.** *As above let $\bar{D}$ denote the closure of $D$. Also let $T : \mathcal{X} \to \mathcal{X}_a$ be the bounded linear operator defined by*

$$T(X) = (\bar{B}^*\bar{B})^{-1}\mathcal{Q}X$$

*for all $X \in \mathcal{X}$. Then for all $F \in \mathcal{D}(\bar{D})$,*

$$F = \mathbb{E}F + D^*T\bar{D}F. \tag{7.48}$$

It is worth pointing out that $\bar{B}^*$ is not $//B^*$ but is instead given by $\mathcal{Q}//B^*$. This is because $//B^*$ does not take adapted processes to adapted processes. This is the reason why it is necessary to introduce the orthogonal projection, $\mathcal{Q}$.

*Proof.* Let $Y \in \mathcal{X}_a$ be given and $X \in \mathcal{X}_a$ be chosen so that $F = \mathbb{E}F + D^*X$. Then

$$\langle Y, \mathcal{Q}\bar{D}F \rangle_{\mathcal{X}} = \langle Y, \bar{D}F \rangle_{\mathcal{X}} = \mathbb{E}\left[D^*Y \cdot F\right]$$
$$= \mathbb{E}\left[D^*Y \cdot D^*X\right] = \mathbb{E}\left[\langle \bar{B}(Y), \bar{B}(X) \rangle_{L^2(ds)}\right]$$
$$= \langle Y, \bar{B}^*\bar{B}(X) \rangle_{\mathcal{X}},$$

where in going from the first to the second line we have used $\mathbb{E}\left[D^*Y\right] = 0$. From the above displayed equation it follows that $\mathcal{Q}\bar{D}F = \bar{B}^*\bar{B}(X)$ and hence $X = (\bar{B}^*\bar{B})^{-1}\mathcal{Q}\bar{D}F = T(\bar{D}F)$. □

### 7.6.1 Proof of theorem 7.41

Let $F \in \mathcal{D}(\bar{D})$. By Theorem 7.45,

$$\mathbb{E}\left[F - \mathbb{E}F\right]^2 = \mathbb{E}\left[D^*T\bar{D}F\right]^2 = \mathbb{E}|\bar{B}(T\bar{D}F)|^2_{L^2(ds, T_oM)} \leq C\langle \bar{D}F, \bar{D}F \rangle_{\mathcal{X}}$$

where $C$ is the operator norm of $\bar{B}T$. In particular if $F \in \mathcal{D}(\mathcal{L})$, then $\langle \bar{D}F, \bar{D}F \rangle_{\mathcal{X}} = \mathbb{E}[\mathcal{L}F \cdot F]$, and hence

$$\langle \mathcal{L}F, F \rangle_{L^2(\mu_{W(M)})} \geq C^{-1}\langle F - \mathbb{E}F, F - \mathbb{E}F \rangle_{L^2(\mu_{W(M)})}.$$

Therefore, if $F \in \text{Nul}(\mathcal{L})$, it follows that $F = \mathbb{E}F$, i.e., $F$ is a constant. Moreover if $F \perp 1$ (i.e., $\mathbb{E}F = 0$) then

$$\langle \mathcal{L}F, F \rangle_{L^2(\mu_{W(M)})} \geq C^{-1}\langle F, F \rangle_{L^2(\mu_{W(M)})},$$

proving Theorem 7.41 with $c = C^{-1}$.

### 7.7  W (M)–Martingale representation theorem

In this subsection, $\Sigma$ is a Brownian motion on $M$ starting at $o \in M$, $//_s$ is stochastic parallel translation along $\Sigma$ and

$$b_s = [\Psi(\Sigma)]_s = \int_0^s //_r^{-1} \delta \Sigma_r$$

is the undeveloped $T_o M$ – valued Brownian motion associated to $\Sigma$ as described before Theorem 5.29.

**Lemma 7.46.** *If* $f \in C^\infty(M^{n+1})$ *and* $i \le n$, *then*

$$\mathbb{E}\left[ //_{s_i}^{-1} \mathrm{grad}_i f\left(\Sigma_{s_1}, \ldots, \Sigma_{s_n}, \Sigma_{s_{n+1}}\right) \bigg| \mathcal{F}_{s_n} \right]$$

$$= //_{s_i}^{-1} \mathrm{grad}_i (e^{(s_{n+1}-s_n)\bar{\Delta}_{n+1}/2} f)\left(\Sigma_{s_1}, \ldots, \Sigma_{s_n}, \Sigma_{s_n}\right). \tag{7.49}$$

*Proof.* Let us begin with the special case where $f = g \otimes h$ for some $g \in C^\infty(M^n)$ and $h \in C^\infty(M)$ where $g \otimes h (x_1, \ldots, x_{n+1}) := g(x_1, \ldots, x_n) h(x_{n+1})$. In this case

$$//_{s_i}^{-1} \mathrm{grad}_i f\left(\Sigma_{s_1}, \ldots, \Sigma_{s_n}, \Sigma_{s_{n+1}}\right) = //_{s_i}^{-1} \mathrm{grad}_i g\left(\Sigma_{s_1}, \ldots, \Sigma_{s_n}\right) \cdot h\left(\Sigma_{s_{n+1}}\right)$$

where $//_{s_i}^{-1} \mathrm{grad}_i g\left(\Sigma_{s_1}, \ldots, \Sigma_{s_n}\right)$ is $\mathcal{F}_{s_n}$-measurable. Hence by the Markov property we have

$$\mathbb{E}\left[ //_{s_i}^{-1} \mathrm{grad}_i f\left(\Sigma_{s_1}, \ldots, \Sigma_{s_n}, \Sigma_{s_{n+1}}\right) \bigg| \mathcal{F}_{s_n} \right]$$

$$= //_{s_i}^{-1} \mathrm{grad}_i g\left(\Sigma_{s_1}, \ldots, \Sigma_{s_n}\right) \mathbb{E}\left[ h\left(\Sigma_{s_{n+1}}\right) \big| \mathcal{F}_{s_n} \right]$$

$$= //_{s_i}^{-1} \mathrm{grad}_i g\left(\Sigma_{s_1}, \ldots, \Sigma_{s_n}\right) (e^{(s_{n+1}-s_n)\bar{\Delta}/2} h)\left(\Sigma_{s_n}\right)$$

$$= //_{s_i}^{-1} \mathrm{grad}_i (e^{(s_{n+1}-s_n)\bar{\Delta}_{n+1}/2} f)\left(\Sigma_{s_1}, \ldots, \Sigma_{s_n}, \Sigma_{s_n}\right).$$

*Alternatively,* as we have already seen, $M_s := (e^{(s_{n+1}-s)\bar{\Delta}/2} h)(\Sigma_s)$ is a martingale for $s \le s_{n+1}$, and therefore,

$$\mathbb{E}\left[ h\left(\Sigma_{s_{n+1}}\right) \big| \mathcal{F}_{s_n} \right] = \mathbb{E}\left[ M_{s_{n+1}} \big| \mathcal{F}_{s_n} \right] = M_{s_n} = (e^{(s_{n+1}-s_n)\bar{\Delta}/2} h)\left(\Sigma_{s_n}\right).$$

Since (7.49) is linear in $f$, this proves (7.49) when $f$ is a linear combination of functions of the form $g \otimes h$ as above.

Using a partition of unity argument along with the standard convolution approximation methods, to any $f \in C^\infty(M^{n+1})$ there exists a sequence $f_k \in C^\infty(M^{n+1})$ with each $f_k$ being a linear combination of functions of the form $g \otimes h$ such that $f_k$ along with all of its derivatives converges uniformly to $f$. Passing to the limit in (7.49) with $f$ being replaced by $f_k$, shows that (7.49) holds for all $f \in C^\infty(M^{n+1})$. □

Recall that $Q_s$ is the End $(T_o M)$-valued process determined in (6.1) and since

$$\frac{d}{ds} Q_s^{-1} = -Q_s^{-1} \left[ \frac{d}{ds} Q_s \right] Q_s^{-1},$$

$Q_s^{-1}$ solves the equation,

$$\frac{d}{ds}Q_s^{-1} = \frac{1}{2}\mathrm{Ric}_{//_s}\, Q_s^{-1} \text{ with } Q_0^{-1} = I. \tag{7.50}$$

**Theorem 7.47 (Representation Formula).** *Suppose that F is a smooth cylinder function of the form* $F(\sigma) = f\left(\sigma_{s_1}, \ldots, \sigma_{s_n}\right)$, *then*

$$F(\Sigma) = \mathbb{E}F + \int_0^1 \langle a_s, db_s \rangle \tag{7.51}$$

*where* $a_s$ *is a bounded predictable process,* $a_s$ *is zero if* $s \geq s_n$ *and* $s \to a_s$ *is continuous off the partition set,* $\{s_1, \ldots, s_n\}$. *Moreover* $a_s$ *may be expressed as*

$$a_s := Q_s^{-1}\mathbb{E}\left[\left.\sum_{i=1}^n 1_{s \leq s_i} Q_{s_i}//_{s_i}^{-1}\mathrm{grad}_i\, f\left(\Sigma_{s_1}, \ldots, \Sigma_{s_n}\right)\right| \mathcal{F}_s\right]. \tag{7.52}$$

*Proof.* The proof will be by induction on $n$. For $n = 1$ suppose $F(\Sigma) = f(\Sigma_t)$ for some $t \in (0, 1]$. Integrating (5.38) from $[0, t]$ with $g = f$ implies

$$F(\Sigma) = f(\Sigma_t) = e^{t\tilde{\Delta}/2} f(o) + \int_0^t \langle //_s^{-1}\,\mathrm{grad}\, e^{(t-s)\tilde{\Delta}/2} f(\Sigma_s), db_s \rangle. \tag{7.53}$$

Since $e^{t\tilde{\Delta}/2} f(o) = \mathbb{E}F$, (7.53) shows (7.51) holds with

$$a_s = 1_{0 \leq s \leq t}//_s^{-1}\,\mathrm{grad}\, e^{(t-s)\tilde{\Delta}/2} f(\Sigma_s).$$

By Lemma 6.1, $Q_s//_s^{-1}\,\mathrm{grad}\, e^{(t-s)\tilde{\Delta}/2} f(\Sigma_s)$ is a martingale, and hence

$$Q_s//_s^{-1}\,\mathrm{grad}\, e^{(t-s)\tilde{\Delta}/2} f(\Sigma_s) = \mathbb{E}\left[\left.Q_t//_t^{-1}\,\mathrm{grad}\, f(\Sigma_t)\right| \mathcal{F}_s\right]$$

from which it follows that

$$a_s = 1_{0 \leq s \leq t}//_s^{-1}\,\mathrm{grad}\, e^{(t-s)\tilde{\Delta}/2} f(\Sigma_s) = 1_{0 \leq s \leq t} Q_s^{-1}\mathbb{E}\left[\left.Q_t//_t^{-1}\,\mathrm{grad}\, f(\Sigma_t)\right| \mathcal{F}_s\right].$$

This shows that (7.52) is valid for $n = 1$.

To carry out the inductive step, suppose the result holds for level $n$ and now suppose that

$$F(\Sigma) = f\left(\Sigma_{s_1}, \ldots, \Sigma_{s_{n+1}}\right)$$

with $0 < s_1 < s_2 \cdots < s_{n+1} \leq 1$. Let

$$(\Delta_{n+1}f)(x_1, x_2, \ldots, x_{n+1}) = (\Delta g)(x_{n+1})$$

where $g(x) := f(x_1, x_2, \ldots, x_n, x)$. Similarly, let $\mathrm{grad}_{n+1}$ denote the gradient acting on the $(n+1)^{\text{th}}$-variable of a function $f \in C^\infty(M^{n+1})$. Set

$$H(s, \Sigma) := (e^{(s_{n+1}-s)\bar{\Delta}_{n+1}/2} f)(\Sigma_{s_1}, \dots, \Sigma_{s_n}, \Sigma_s)$$

for $s_n \leq s \leq s_{n+1}$. By Itô's Lemma, (see Corollary 5.18 and equation (5.38)),

$$d\,[H(s, \Sigma_s)] = \langle \mathrm{grad}_{n+1} e^{(s_{n+1}-s)\bar{\Delta}_{n+1}/2} f)(\Sigma_{s_1}, \dots, \Sigma_{s_n}, \Sigma_s, //_s db_s \rangle$$

for $s_n \leq s \leq s_{n+1}$. Integrating this last expression from $s_n$ to $s_{n+1}$ yields:

$$F(\Sigma) = (e^{(s_{n+1}-s_n)\bar{\Delta}_{n+1}/2} f)(\Sigma_{s_1}, \dots, \Sigma_{s_n}, \Sigma_{s_n})$$
$$+ \int_{s_n}^{s_{n+1}} \langle //_s^{-1} \mathrm{grad}_{n+1} e^{(s_{n+1}-s)\bar{\Delta}_{n+1}/2} f) \left( \Sigma_{s_1}, \dots, \Sigma_{s_n}, \Sigma_s \right), db_s \rangle$$

$$(7.54)$$

$$= (e^{(s_{n+1}-s_n)\bar{\Delta}_{n+1}/2} f)(\Sigma_{s_1}, \dots, \Sigma_{s_n}, \Sigma_{s_n}) + \int_{s_n}^{s_{n+1}} \langle \alpha_s, db_s \rangle, \qquad (7.55)$$

where $\alpha_s := //_s^{-1} (\mathrm{grad}_{n+1} e^{(s_{n+1}-s)\bar{\Delta}_{n+1}/2} f)(\Sigma_{s_1}, \dots, \Sigma_{s_n}, \Sigma_s)$. By the induction hypothesis, the smooth cylinder function,

$$(e^{(s_{n+1}-s_n)\bar{\Delta}_{n+1}/2} f)(\Sigma_{s_1}, \dots, \Sigma_{s_n}, \Sigma_{s_n}),$$

may be written as a constant plus $\int_0^1 \langle \tilde{a}_s, db_s \rangle$, where $\tilde{a}_s$ is bounded and piecewise continuous and $\tilde{a}_s \equiv 0$ if $s \geq s_n$. Thus if we let $a_s := \tilde{a}_s + 1_{s_n < s \leq s_{n+1}} \alpha_s$, we have shown

$$F(\Sigma) = C + \int_0^{s_{n+1}} \langle a_s, db_s \rangle$$

for some constant $C$. Taking expectations of both sides of this equation then shows $C = \mathbb{E}[F(\Sigma)]$ and the proof of (7.51) is complete. So to finish the proof it only remains to verify (7.52).

Again by Lemma 6.1,

$$s \to M_s := Q_s //_s^{-1} (\mathrm{grad}_{n+1} e^{(s_{n+1}-s)\bar{\Delta}_{n+1}/2} f)(\Sigma_{s_1}, \dots, \Sigma_{s_n}, \Sigma_s)$$

is a martingale for $s \in [s_n, s_{n+1}]$ and therefore,

$$M_s = Q_s //_s^{-1} (\mathrm{grad}_{n+1} e^{(s_{n+1}-s)\bar{\Delta}_{n+1}/2} f)(\Sigma_{s_1}, \dots, \Sigma_{s_n}, \Sigma_s)$$
$$= \mathbb{E}[M_{s_{n+1}} | \mathcal{F}_s] = \mathbb{E}\left[ Q_{s_{n+1}} //_{s_{n+1}}^{-1} (\mathrm{grad}_{n+1} f)(\Sigma_{s_1}, \dots, \Sigma_{s_n}, \Sigma_{s_{n+1}}) \Big| \mathcal{F}_s \right],$$

$$(7.56)$$

i.e.

$$//_s^{-1} (\mathrm{grad}_{n+1} e^{(s_{n+1}-s)\bar{\Delta}_{n+1}/2} f)(\Sigma_{s_1}, \dots, \Sigma_{s_n}, \Sigma_s)$$
$$= Q_s^{-1} \mathbb{E}\left[ Q_{s_{n+1}} //_{s_{n+1}}^{-1} (\mathrm{grad}_{n+1} f)(\Sigma_{s_1}, \dots, \Sigma_{s_n}, \Sigma_{s_{n+1}}) \Big| \mathcal{F}_s \right]. \qquad (7.57)$$

Using this identity, (7.54) may be written as

$$F(\Sigma) = g(\Sigma_{s_1}, \ldots, \Sigma_{s_n})$$
$$+ \int_{s_n}^{s_{n+1}} \left\langle Q_s^{-1} \mathbb{E}\left[ Q_{s_{n+1}} //_{s_{n+1}}^{-1} (\mathrm{grad}_{n+1} f)(\Sigma_{s_1}, \ldots, \Sigma_{s_n}, \Sigma_{s_{n+1}}) \middle| \mathcal{F}_s \right], db_s \right\rangle.$$

$$(7.58)$$

where

$$g(x_1, \ldots, x_n) := (e^{(s_{n+1} - s_n)\bar{\Delta}_{n+1}/2} f)(x_1, \ldots, x_n, x_n).$$

By the induction hypothesis,

$$g(\Sigma_{s_1}, \ldots, \Sigma_{s_n})$$
$$= C + \int_0^1 \left\langle Q_s^{-1} \mathbb{E}\left[ \sum_{i=1}^n 1_{s \le s_i} Q_{s_i} //_{s_i}^{-1} \mathrm{grad}_i g (\Sigma_{s_1}, \ldots, \Sigma_{s_n}) \middle| \mathcal{F}_s \right], db_s \right\rangle \quad (7.59)$$

where $C = \mathbb{E}[F(\Sigma)]$ as we have already seen or alternatively, by the Markov property,

$$C := \mathbb{E}(e^{(s_{n+1}-s_n)\bar{\Delta}_{n+1}/2} f)(\Sigma_{s_1}, \ldots, \Sigma_{s_n}, \Sigma_{s_n})$$
$$= \mathbb{E}f(\Sigma_{s_1}, \ldots, \Sigma_{s_n}, \Sigma_{s_{n+1}}) = \mathbb{E}[F(\Sigma)]. \quad (7.60)$$

By Lemma 7.46, for $s \le s_n$ and $i < n$

$$\mathbb{E}\left[ Q_{s_i} //_{s_i}^{-1} \mathrm{grad}_i g (\Sigma_{s_1}, \ldots, \Sigma_{s_n}) \middle| \mathcal{F}_s \right]$$
$$= \mathbb{E}\left[ Q_{s_i} \mathbb{E}\left[ //_{s_i}^{-1} \mathrm{grad}_i (e^{(s_{n+1}-s_n)\bar{\Delta}_{n+1}/2} f)(\Sigma_{s_1}, \ldots, \Sigma_{s_n}, \Sigma_{s_n}) \middle| \mathcal{F}_{s_n} \right] \middle| \mathcal{F}_s \right]$$
$$= \mathbb{E}\left[ Q_{s_i} //_{s_i}^{-1} \mathrm{grad}_i f (\Sigma_{s_1}, \ldots, \Sigma_{s_n}, \Sigma_{s_{n+1}}) \middle| \mathcal{F}_s \right]. \quad (7.61)$$

While for $s \le s_n$ and $i = n$, we have:

$$\mathrm{grad}_n g (\Sigma_{s_1}, \ldots, \Sigma_{s_n}) = \mathrm{grad}_n (e^{(s_{n+1}-s_n)\bar{\Delta}_{n+1}/2} f)(\Sigma_{s_1}, \ldots, \Sigma_{s_n}, \Sigma_{s_n})$$
$$+ \mathrm{grad}_{n+1} (e^{(s_{n+1}-s_n)\bar{\Delta}_{n+1}/2} f)(\Sigma_{s_1}, \ldots, \Sigma_{s_n}, \Sigma_{s_n}),$$

$$\mathbb{E}\left[ Q_{s_n} //_{s_n}^{-1} \mathrm{grad}_n (e^{(s_{n+1}-s_n)\bar{\Delta}_{n+1}/2} f)(\Sigma_{s_1}, \ldots, \Sigma_{s_n}, \Sigma_{s_n}) \middle| \mathcal{F}_s \right]$$
$$= \mathbb{E}\left[ Q_{s_n} \mathbb{E}\left[ //_{s_n}^{-1} \mathrm{grad}_n (e^{(s_{n+1}-s_n)\bar{\Delta}_{n+1}/2} f)(\Sigma_{s_1}, \ldots, \Sigma_{s_n}, \Sigma_{s_n}) \middle| \mathcal{F}_{s_n} \right] \middle| \mathcal{F}_s \right]$$
$$= \mathbb{E}\left[ Q_{s_n} //_{s_n}^{-1} \mathrm{grad}_n f (\Sigma_{s_1}, \ldots, \Sigma_{s_n}, \Sigma_{s_{n+1}}) \middle| \mathcal{F}_s \right]$$

by Lemma 7.46 and

$$\mathbb{E}\left[ \mathbb{E}\left[ Q_{s_n} //_{s_n}^{-1} \mathrm{grad}_{n+1} (e^{(s_{n+1}-s_n)\bar{\Delta}_{n+1}/2} f)(\Sigma_{s_1}, \ldots, \Sigma_{s_n}, \Sigma_{s_n}) \middle| \mathcal{F}_{s_n} \right] \middle| \mathcal{F}_s \right]$$
$$= \mathbb{E}\left[ Q_{s_{n+1}} //_{s_{n+1}}^{-1} (\mathrm{grad}_{n+1} f)(\Sigma_{s_1}, \ldots, \Sigma_{s_n}, \Sigma_{s_{n+1}}) \middle| \mathcal{F}_s \right]$$

from (7.57) with $s = s_n$. Combining the previous three displayed equations shows,

$$\mathbb{E}\left[ Q_{s_n}//_{s_n}^{-1}\mathrm{grad}_n g\left(\Sigma_{s_1},\ldots,\Sigma_{s_n}\right)\Big|\mathcal{F}_s\right]$$

$$=\mathbb{E}\left[ Q_{s_n}//_{s_n}^{-1}\mathrm{grad}_n f\left(\Sigma_{s_1},\ldots,\Sigma_{s_n},\Sigma_{s_{n+1}}\right)\Big|\mathcal{F}_s\right]$$

$$+\mathbb{E}\left[ Q_{s_{n+1}}//_{s_{n+1}}^{-1}\left(\mathrm{grad}_{n+1}f\right)\left(\Sigma_{s_1},\ldots,\Sigma_{s_n},\Sigma_{s_{n+1}}\right)\Big|\mathcal{F}_s\right] \qquad (7.62)$$

Assembling (7.59), (7.60), (7.61) and (7.62) implies

$$g(\Sigma_{s_1},\ldots,\Sigma_{s_n})=\mathbb{E}\left[F\left(\Sigma\right)\right]$$

$$+\int_0^1\sum_{i=1}^n\left\langle Q_s^{-1}\mathbb{E}\left[1_{s\le s_i}\,Q_{s_i}//_{s_i}^{-1}\mathrm{grad}_i f\left(\Sigma_{s_1},\ldots,\Sigma_{s_n},\Sigma_{s_{n+1}}\right)\Big|\mathcal{F}_s\right],db_s\right\rangle$$

$$+\int_0^1\left\langle Q_s^{-1}\mathbb{E}\left[1_{s\le s_n}\,Q_{s_{n+1}}//_{s_{n+1}}^{-1}\left(\mathrm{grad}_{n+1}f\right)\left(\Sigma_{s_1},\ldots,\Sigma_{s_n},\Sigma_{s_{n+1}}\right)\Big|\mathcal{F}_s\right],db_s\right\rangle$$

which combined with (7.58) shows

$$F\left(\Sigma\right)=\mathbb{E}\left[F\left(\Sigma\right)\right]$$

$$+\int_0^1\left\langle Q_s^{-1}\mathbb{E}\left[\sum_{i=1}^{n+1}1_{s\le s_i}\,Q_{s_i}//_{s_i}^{-1}\mathrm{grad}_i f\left(\Sigma_{s_1},\ldots,\Sigma_{s_n},\Sigma_{s_{n+1}}\right)\Big|\mathcal{F}_s\right],db_s\right\rangle.$$

This completes the induction argument and hence the proof.    □

**Proposition 7.48.** *Equation (7.51) may also be written as*

$$F\left(\Sigma\right)=\mathbb{E}\left[F\left(\Sigma\right)\right]+\int_0^1\left\langle\mathbb{E}\left[\xi_s-\frac{1}{2}\int_s^1 Q_s^{-1}Q_r\,\mathrm{Ric}_{//_r}\,\xi_r dr\Big|\mathcal{F}_s\right],db_s\right\rangle. \qquad (7.63)$$

*where*

$$\xi_s:=//_s^{-1}\frac{d}{ds}\left(DF\right)_s.$$

*Proof.* Let $v_i:=//_{s_i}^{-1}\mathrm{grad}_i f\left(\Sigma_{s_1},\ldots,\Sigma_{s_n}\right)$, so that

$$\xi_s:=//_s^{-1}\frac{d}{ds}\left(DF\right)_s=\sum_{i=1}^n 1_{s<s_i}v_i,$$

and let

$$\alpha_s:=\sum_{i=1}^n 1_{s\le s_i}\,Q_s^{-1}Q_{s_i}//_{s_i}^{-1}\mathrm{grad}_i f\left(\Sigma_{s_1},\ldots,\Sigma_{s_n}\right)-\sum_{i=1}^n 1_{s\le s_i}\,Q_s^{-1}Q_{s_i}v_i.$$

Then the Lebesgue–Stieljtes measure associate to $\xi_s$ is

$$d\xi_s=-\sum_{i=1}^n \delta_{s_i}\left(ds\right)v_i$$

and therefore

$$\alpha_s = -Q_s^{-1} \int_s^1 Q_r d\xi_r = -\int_s^1 Q_s^{-1} Q_r d\xi_r.$$

So using integration by parts we have, for $s \notin \{0, s_1, \ldots, s_n, 1\}$,

$$\alpha_s = -\int_s^1 Q_s^{-1} Q_r d\xi_r = -\left[Q_s^{-1} Q_r \xi_r\right]|_{r=s}^{r=1} + \int_s^1 Q_s^{-1} \left[\frac{d}{dr} Q_r\right] \xi_r$$

$$= \xi_s - \frac{1}{2} \int_s^1 Q_s^{-1} Q_r \operatorname{Ric}_{//_r} \xi_r$$

where we have used $\xi_1 = 0$. This completes the proof since from (7.51) and (7.52),

$$F(\Sigma) = \mathbb{E}[F(\Sigma)] + \int_0^1 \langle E[\alpha_s | \mathcal{F}_s], db_s \rangle. \qquad \square$$

**Corollary 7.49.** *Let $F$ be a smooth cylinder function, then there is a predictable, piecewise continuously differentiable Cameron–Martin vector field $X$ such that $F = \mathbb{E}[F] + D^*X$.*

*Proof.* Just follow the proof of Lemma 7.42 using Theorem 7.47 in place of Corollary 7.20. $\qquad \square$

### 7.7.1 The equivalence of integration by parts and the representation formula

**Corollary 7.50.** *The representation formula in Theorem 7.47 may be used to prove the integration by parts Theorem 7.32 in the case $F$ is a cylinder function.*

*Proof.* Let $F$ be a cylinder function, $a_s$ be as in (7.52), $h$ be an adapted Cameron–Martin process and $k_s := (Q_s^{\mathrm{tr}})^{-1} h_s$. Then, by the product rule and (7.39),

$$h'_s + \frac{1}{2} \operatorname{Ric}_{//_s} h_s = \left(\frac{d}{ds} + \frac{1}{2} \operatorname{Ric}_{//_s}\right) Q_s^{\mathrm{tr}} k_s = Q_s^{\mathrm{tr}} k'_s.$$

Hence,

$$\mathbb{E}\left[F \int_0^1 \langle h'_s + \frac{1}{2} \operatorname{Ric}_{//_s} h_s, db_s \rangle\right]$$

$$= \mathbb{E}\left[\left(\mathbb{E}F + \int_0^1 \langle a_s, db_s \rangle\right) \int_0^1 \langle Q_s^{\mathrm{tr}} k'_s, db_s \rangle\right]$$

$$= \mathbb{E}\left[\int_0^1 \langle Q_s^{\mathrm{tr}} k'_s, a_s \rangle ds\right]$$

$$= \mathbb{E}\left[\int_0^1 \langle Q_s^{\mathrm{tr}} k'_s, \sum_{i=1}^n 1_{s \le s_i} Q_s^{-1} Q_{s_i} //_{s_i}^{-1} \operatorname{grad}_i f(\Sigma_{s_1}, \ldots, \Sigma_{s_n}) \rangle ds\right]$$

$$= \mathbb{E}\left[\int_0^1 \langle k_s', \sum_{i=1}^n 1_{s \leq s_i} Q_{s_i} //_{s_i}^{-1} \mathrm{grad}_i f\left(\Sigma_{s_1}, \ldots, \Sigma_{s_n}\right)\rangle ds\right]$$

$$= \mathbb{E}\left[\sum_{i=1}^n \langle k_{s_i}, Q_{s_i} //_{s_i}^{-1} \mathrm{grad}_i f\left(\Sigma_{s_1}, \ldots, \Sigma_{s_n}\right)\rangle\right]$$

$$= \mathbb{E}\left[\sum_{i=1}^n \langle //_{s_i} h_{s_i}, \mathrm{grad}_i f\left(\Sigma_{s_1}, \ldots, \Sigma_{s_n}\right)\rangle\right] = \mathbb{E}\left[X^h F\right]. \quad \square$$

Conversely we may give a proof of Theorem 7.47 which is based on the integration by parts Theorem 7.32.

**Theorem 7.51 (Representation Formula).** *Suppose $F$ is a cylinder function on $W(M)$ as in (7.2) and $\xi_s := //_s^{-1}\frac{d}{ds}(DF)_s$, then*

$$F = \mathbb{E}F + \int_0^1 \left\langle \mathbb{E}\left[\xi_s - \frac{1}{2}\int_s^1 Q_s^{-1} Q_r \, \mathrm{Ric}_{//_r} \xi_r dr \,\Big|\, \mathcal{F}_s\right], db_s\right\rangle. \tag{7.64}$$

*where $Q_s$ is the solution to (6.1).*

*Proof.* Let $h \in \mathcal{X}_a$ be a predictable adapted Cameron–Martin space valued process such that $\mathbb{E}\int_0^1 |h_s'|^2 \, ds < \infty$. By the martingale representation property in Corollary 7.20,

$$F = \mathbb{E}F + \int_0^1 \langle a, db\rangle \tag{7.65}$$

for some predictable process $a$ such that $\mathbb{E}\int_0^1 |a_s|^2 \, ds < \infty$. Then from Corollary 7.35 and the Itô isometry property,

$$\mathbb{E}\left[X^{Q^{\mathrm{tr}}h} F\right] = \mathbb{E}\left[F \cdot \left(X^{Q^{\mathrm{tr}}h}\right)^* 1\right] = \mathbb{E}\left[F \cdot \int_0^1 \langle Q^{\mathrm{tr}}h', db\rangle\right]$$

$$= \mathbb{E}\left[\int_0^1 \langle Q_s^{\mathrm{tr}}h_s', a_s\rangle ds\right] = \mathbb{E}\left[\int_0^1 \langle h_s', Q_s a_s\rangle ds\right]. \tag{7.66}$$

On the other hand we may compute $\mathbb{E}\left[X^{Q^{\mathrm{tr}}h} F\right]$ as

$$\mathbb{E}\left[X^{Q^{\mathrm{tr}}h} F\right] = \mathbb{E}\left[\langle DF, //Q^{\mathrm{tr}}h\rangle_H\right] = \mathbb{E}\int_0^1 \langle \xi_s, \frac{d}{ds}\left(Q^{\mathrm{tr}}h\right)_s\rangle ds$$

$$= \mathbb{E}\int_0^1 \langle \xi_s, Q_s^{\mathrm{tr}}h_s' - \frac{1}{2}\mathrm{Ric}_{//_s} Q_s^{\mathrm{tr}}h_s\rangle ds \tag{7.67}$$

where we have used (7.39) in the last equality. We will now rewrite the right side of (7.67) so that it has the same form as (7.66) To do this let $\rho_s := \frac{1}{2}\mathrm{Ric}_{//_s}$ and notice that

$$\int_0^1 \langle \xi_s, \rho_s Q_s^{\text{tr}} h_s \rangle ds = \int_0^1 \left\langle Q_s \rho_s^* \xi_s, \left( \int_0^s h'_r dr \right) \right\rangle ds$$

$$= \int drds 1_{0 \le r \le s \le 1} \langle Q_s \rho_s^* \xi_s, h'_r \rangle = \int_0^1 \left\langle \int_s^1 Q_r \rho_r^* \xi_r dr, h'_s \right\rangle ds$$

wherein the last equality we have interchanged the role of $r$ and $s$. Using this result back in (7.67) implies

$$\mathbb{E}\left[ X^{Q^{\text{tr}} h} F \right] = \mathbb{E} \int_0^1 \left\langle Q_s \xi_s - \int_s^1 Q_r \rho_r^* \xi_r dr, h'_s \right\rangle ds. \tag{7.68}$$

and comparing this with (7.66) shows

$$\mathbb{E} \int_0^1 \left\langle Q_s a_s - Q_s \xi_s + \int_s^1 Q_r \rho_r^* \xi_r dr, h'_s \right\rangle ds = 0 \tag{7.69}$$

for all $h \in \mathcal{X}_a$.

Up to now we have only used $F \in \mathcal{D}(D)$ and not the fact that $F$ is a cylinder function. We will use this hypothesis now. From the easy part of Theorem 7.47 we know that $a_s$ satisfies the additional properties of being 1) bounded, 2) zero if $s \ge s_n$ and most importantly 3) $s \to a_s$ is continuous off the partition set, $\{s_1, \dots, s_n\}$.

Fix $\tau \in (0, 1) \setminus \{s_1, \dots, s_n\}$, $v \in T_o M$ and let $G$ be a bounded $\mathcal{F}_\tau$-measurable function. For $n \in \mathbb{N}$ let

$$l_n(s) := \int_0^s n 1_{\tau \le r \le \tau + \frac{1}{n}} dr.$$

Replacing $h$ in (7.69) by $h_n(s) := G \cdot l_n(s) v$ and then passing to the limit as $n \to \infty$, implies

$$0 = \lim_{n \to \infty} \mathbb{E} \int_0^1 \left\langle Q_s a_s - Q_s \xi_s + \int_s^1 Q_r \rho_r^* \xi_r dr, h'_n(s) \right\rangle ds$$

$$= \mathbb{E}\left[ G \left\langle Q_\tau a_\tau - Q_\tau \xi_\tau + \int_\tau^1 Q_r \rho_r^* \xi_r dr, v \right\rangle \right]$$

and since $G$ and $v$ were arbitrary we conclude from this equation that

$$\mathbb{E}\left[ Q_\tau \xi_\tau - \int_\tau^1 Q_r \rho_r \xi_r dr \,\middle|\, \mathcal{F}_\tau \right] = Q_\tau a_\tau.$$

Thus for all but finitely many $s \in [0, 1]$,

$$a_s = Q_s^{-1} \mathbb{E}\left[ Q_s \xi_s - \int_s^1 Q_r \rho_r \xi_r dr \,\middle|\, \mathcal{F}_s \right]$$

$$= \mathbb{E}\left[ \xi_s - \frac{1}{2} \int_s^1 Q_s^{-1} Q_r \, \text{Ric}_{//_r} \xi_r dr \,\middle|\, \mathcal{F}_s \right].$$

Combining this with (7.65) proves (7.64). $\qquad\qquad\square$

## 7.8 Logarithmic-Sobolev inequality for $W(M)$

The next theorem is the "curved" generalization of Theorem 7.24.

**Theorem 7.52 (Hsu's Logarithmic Sobolev Inequality).** *Let $M$ be a compact Riemannian manifold, then for all $F \in \mathcal{D}(\bar{D})$*

$$\mathbb{E}\left[F^2 \log F^2\right] \leq \mathbb{E}F^2 \cdot \log \mathbb{E}F^2$$

$$+ 2\mathbb{E}\int_0^1 \left|//_s^{-1}(DF)_s' - \frac{1}{2}\int_s^1 Q_s^{-1}Q_r \operatorname{Ric}_{//_r}//_r^{-1}(DF)_r' \, dr\right|^2 ds, \tag{7.70}$$

*where $(DF)_s' := \frac{d}{ds}(DF)_s$. Moreover, there is a constant $C = C\,(\mathrm{Ric})$ such that*

$$\mathbb{E}\left[F^2 \log F^2\right] \leq C\mathbb{E}\left[\langle DF, DF \rangle_{H(T_oM)}\right] + \mathbb{E}F^2 \cdot \log \mathbb{E}F^2. \tag{7.71}$$

*Proof.* The proof we give here follows the paper of Capitaine, Hsu and Ledoux [29]. We begin in the same way as the proof of Theorem 7.24. Let $F \in \mathcal{F}C^1(W(M))$, $\varepsilon > 0$, $H_\varepsilon := F^2 + \varepsilon \in \mathcal{D}(\bar{D})$ and

$$a_s := \mathbb{E}\left[\xi_s - \frac{1}{2}\int_s^1 Q_s^{-1}Q_r \operatorname{Ric}_{//_r}\xi_r dr \,\bigg|\, \mathcal{F}_s\right]$$

where

$$\xi_s = //_s^{-1}\frac{d}{ds}(DH_\varepsilon)_s = 2F \cdot //_s^{-1}\frac{d}{ds}(DF)_s.$$

Then by Theorem 7.47,

$$H_\varepsilon = \mathbb{E}H_\varepsilon + \int_0^1 \langle a, db \rangle.$$

The same proof used to derive (7.23), with $\phi(x) = x \ln x$, shows

$$\mathbb{E}[\phi(H_\varepsilon)] = \mathbb{E}[\phi(M_1)] = \phi(\mathbb{E}M_1) + \frac{1}{2}\mathbb{E}\left[\int_0^1 \frac{1}{M_s}|a_s|^2 \, ds\right]$$

$$= \phi(\mathbb{E}H_\varepsilon) + \frac{1}{2}\mathbb{E}\left[\int_0^1 \frac{1}{\mathbb{E}[H_\varepsilon|\mathcal{F}_s]}|a_s|^2 \, ds\right].$$

By the Cauchy–Schwarz inequality and the contractive properties of conditional expectations,

$$|a_s|^2 = \left|\mathbb{E}\left[2F\left\{//_s^{-1}(DF)_s' - \frac{1}{2}\int_s^1 Q_s^{-1}Q_r \operatorname{Ric}_{//_r}//_r^{-1}(DF)_r' \, dr\right\}\,\bigg|\,\mathcal{F}_s\right]\right|^2$$

$$\leq 4\mathbb{E}\left[F^2|\mathcal{F}_s\right] \cdot \mathbb{E}\left[\left|//_s^{-1}(DF)_s' - \frac{1}{2}\int_s^1 Q_s^{-1}Q_r \operatorname{Ric}_{//_r}//_r^{-1}(DF)_r' \, dr\right|^2\,\bigg|\,\mathcal{F}_s\right]$$

Combining the last two equations along with (7.24) implies

$$\mathbb{E}\phi\left(H_\varepsilon\right) \le \phi\left(\mathbb{E}H_\varepsilon\right)$$

$$+ 2\mathbb{E}\int_0^1 \mathbb{E}\left[\left|//_s^{-1}\left(DF\right)_s' - \frac{1}{2}\int_s^1 Q_s^{-1}Q_r\,\mathrm{Ric}_{//_r}\,//_r^{-1}\left(DF\right)_r'\,dr\right|^2\bigg|\,\mathcal{F}_s\right]ds$$

$$= \phi\left(\mathbb{E}H_\varepsilon\right)$$

$$+ 2\mathbb{E}\int_0^1 \left|//_s^{-1}\left(DF\right)_s' - \frac{1}{2}\int_s^1 Q_s^{-1}Q_r\,\mathrm{Ric}_{//_r}\,//_r^{-1}\left(DF\right)_r'\,dr\right|^2 ds.$$

We may now let $\varepsilon \downarrow 0$ in this inequality to learn (7.70) holds for all $F \in \mathcal{F}C^1\left(W\right)$. By compactness of $M$, $\mathrm{Ric}_m$ is bounded on $M$ and so by simple Gronwall type estimates on $Q$ and $Q^{-1}$, there is a non-random constant $K < \infty$ such that

$$\left\|Q_s^{-1}Q_r\,\mathrm{Ric}_{//_r}\right\|_{op} \le K \text{ for all } r, s.$$

Therefore,

$$\left|//_s^{-1}\left(DF\right)_s' - \frac{1}{2}\int_s^1 Q_s^{-1}Q_r\,\mathrm{Ric}_{//_r}\,//_r^{-1}\left(DF\right)_r'\,dr\right|^2$$

$$\le \left[\left|\left(DF\right)_s'\right| + \frac{1}{2}K\int_0^1 \left|\left(DF\right)_s'\right|ds\right]^2$$

$$\le 2\left|\left(DF\right)_s'\right|^2 + \frac{1}{2}K^2\left[\int_0^1 \left|\left(DF\right)_s'\right|ds\right]^2$$

$$\le 2\left|\left(DF\right)_s'\right|^2 + \frac{1}{2}K^2\int_0^1 \left|\left(DF\right)_s'\right|^2 ds$$

and hence

$$2\mathbb{E}\int_0^1 \left|\left(DF\right)_s' - \frac{1}{2}\int_s^1 Q_s^{-1}Q_r\,\mathrm{Ric}_{//_r}\left(DF\right)_r'\,dr\right|^2 ds$$

$$\le \left(4 + K^2\right)\int_0^1 \left|\left(DF\right)_s'\right|^2 ds.$$

Combining this estimate with (7.70) implies that (7.71) holds with $C = \left(4 + K^2\right)$. Again, since $\mathcal{F}C^1\left(W\right)$ is a core for $\bar{D}$, standard limiting arguments show that (7.70) and (7.71) are valid for all $F \in \mathcal{D}\left(\bar{D}\right)$.                                          $\square$

Theorem 7.52 was first proved by Hsu [98] with an independent proof given shortly thereafter by Aida and Elworthy [4]. Hsu's original proof relied on a Markov dependence version of a standard additivity property for logarithmic Sobolev inequalities and makes key use of Corollary 7.37. On the other hand Aida and Elworthy show, using the projection construction of Brownian motion, the logarithmic Sobolev inequality on $W\left(M\right)$ is a consequence of Gross' [92] original logarithmic Sobolev inequality on the classical Wiener space $W\left(\mathbb{R}^N\right)$, see Theorem 7.24. In Aida's and Elworthy's proof, Theorem 5.43 plays an important role.

### 7.9 More references

Many people have now proved some version of integration by parts for path and loop spaces in one context or another, see for example [21, 28, 32, 26, 28, 27, 47, 48, 49, 76, 75, 78, 85, 122, 128, 146, 161, 159, 160, 163, 102]. We have followed Bismut in these notes who proved integration by parts formulas for cylinder functions depending on one time. However, as is pointed out by Leandre and Malliavin and Fang, Bismut's technique works without any essential change for arbitrary cylinder functions. In [47, 48], the flow associated to a general class of vector fields on paths and loop spaces of a manifold were constructed. The reader is also referred to the texts [71, 100, 171] and the related articles [81, 80, 35, 77, 82, 83, 84, 34, 37, 33, 38, 36, 39, 125].

Many of the results in this section extend to pinned Wiener measure on loop spaces, see [48] for example. Loop spaces are more interesting than path spaces since they have nontrivial topology. The issue of the spectral gap and logarithmic Sobolev inequalities for general loop spaces is still an open problem. In [93], Gross has proved a logarithmic Sobolev inequality on Loop groups with an added "potential term" for a special geometry on loop groups. Here Gross uses pinned Wiener measure as the reference measure. In Driver and Lohrenz [54], it is shown that a logarithmic Sobolev inequality *without* a potential term does hold on the Loop group provided one replaces pinned Wiener measure by a "heat kernel" measure. The quasi-invariance properties of the heat kernel measure on loop groups was first established in [50, 51]. For more results on heat kernel measures on the loop groups see for example, [57, 3, 30, 31, 82, 83, 106].

The question as to when or if the potential is needed in Gross's setting for logarithmic Sobolev inequalities is still an open question, but see Gong, Röckner and Wu [89] for a positive result in this direction. Eberle [59, 60, 61, 62] has provided examples of Riemannian manifolds where the spectral gap inequality fails in the loop space setting. The reader is referred to [52, 53] and the references therein for some more perspective on the stochastic analysis on loop spaces.

## 8 Malliavin's methods for hypoelliptic operators

In this section we will be concerned with determining smoothness properties of the Law $(\Sigma_t)$ where $\Sigma_t$ denotes the solution to (5.1) with $\Sigma_0 = o$ and $\beta = B$ being an $\mathbb{R}^n$-valued Brownian motion. Unlike the previous sections in these notes, the map $X(m) : \mathbb{R}^n \to T_m M$ is *not* assumed to be surjective. Equivalently put, the diffusion generator $L := \frac{1}{2} \sum_{i=1}^{n} X_i^2 + X_0$ is no longer assumed to be elliptic. However we will always be assuming that the vector fields $\{X_i\}_{i=0}^{n}$ satisfy Hörmander's restricted bracket condition at $o \in M$ as in Definition 8.1 below. Let $\mathcal{K}_1 := \{X_1, \ldots, X_n\}$ and $\mathcal{K}_l$ be defined inductively by

$$\mathcal{K}_{l+1} = \{[X_i, K] : K \in \mathcal{K}_l\} \cup \mathcal{K}_l.$$

For example

$$\mathcal{K}_2 = \{X_1, \ldots, X_n\} \cup \{[X_j, X_i] : i, j = 1, \ldots, n\} \text{ and}$$
$$\mathcal{K}_3 = \{X_1, \ldots, X_n\} \cup \{[X_j, X_i] : i, j = 1, \ldots, n\}$$
$$\cup \{[X_k, [X_j, X_i]] : i, j, k = 1, \ldots, n\} \text{ etc.}$$

**Definition 8.1.** The collection of vector fields, $\{X_i\}_{i=0}^n \subset \Gamma(TM)$, satisfies *Hörmander's restricted bracket condition at* $m \in M$ if there exist $l \in \mathbb{N}$ such that

$$\text{span}(\{K(m) : K \in \mathcal{K}_l\}) = T_m M.$$

Under this condition it follows from a classical theorem of Hörmander that solutions to the heat equation $\partial_t u = Lu$ are necessarily smooth. Since the fundamental solution to this equation at $o \in M$ is the law of the process $\Sigma_t$, it follows that the Law $(\Sigma_t)$ is absolutely continuous relative to the volume measure $\lambda$ on $M$ and its Radon–Nikodym derivative is a smooth function on $M$. Malliavin, in his 1976 pioneering paper [130], gave a probabilistic proof of this fact. Malliavin's original paper was followed by an avalanche of papers carrying out and extending Malliavin's program including the fundamental works of Stroock [169, 170, 168], Kusuoka and Stroock [121, 119, 120], and Bismut [21]. See also [13, 12, 23, 104, 132, 152, 147, 148, 157, 158, 179] (and the references therein) along with Bell's article in this volume. The purpose of this section is to briefly explain (omitting some details) Malliavin methods.

### 8.1 Malliavin's ideas in finite dimensions

To understand Malliavin's methods it is best to begin with a finite dimensional analogue.

**Theorem 8.2 (Malliavin's Ideas in Finite Dimensions).** *Let* $W = \mathbb{R}^N$, $\mu$ *be the Gaussian measure on* $W$ *defined by*

$$d\mu(x) := (2\pi)^{-N/2} e^{-\frac{1}{2}|x|^2} dm(x).$$

*Further suppose* $F : W \to \mathbb{R}^d$ *(think* $F = \Sigma_t$*) is a function satisfying:*

*1. F is smooth and all of its partial derivatives are in*

$$L^{\infty-}(\mu) := \cap_{1 \le p < \infty} L^p(W, \mu).$$

*2. F is a submersion or equivalently assume the "Malliavin" matrix*

$$C(\omega) := DF(\omega)DF(\omega)^*$$

*is invertible for all* $\omega \in W$.
*3. Let*

$$\Delta(\omega) := \det C(\omega) = \det(DF(\omega)DF(\omega)^*)$$

*and assume* $\Delta^{-1} \in L^{\infty-}(\mu)$.

*Then the law* $(\mu_F = F_*\mu = \mu \circ F^{-1})$ *of* $F$ *is absolutely continuous relative to Lebesgue measure,* $\lambda$, *on* $\mathbb{R}^d$ *and the Radon–Nikodym derivative,* $\rho := d\mu_F/d\lambda$, *is smooth.*

*Proof.* For each vector field $Y \in \Gamma\left(T\mathbb{R}^d\right)$, define

$$\mathbb{Y}(\omega) = DF(\omega)^*C(\omega)^{-1}Y\left(F\left(\omega\right)\right), \tag{8.1}$$

a smooth vector field on $W$ such that $DF(\omega)\mathbb{Y}(\omega) = Y\left(F\left(\omega\right)\right)$ or in more geometric notation,

$$F_*\mathbb{Y}(\omega) = Y\left(F\left(\omega\right)\right). \tag{8.2}$$

For the purposes of this proof, it is sufficient to restrict our attention to the case where $Y$ is a constant vector field.

Explicit computations using the chain rule and Cramer's rule for computing $C(\omega)^{-1}$ shows that $D^k\mathbb{Y}$ may be expressed as a polynomial in $\Delta^{-1}$ and $D^\ell F$ for $\ell = 0, 1, 2\ldots, k$. In particular $D^k\mathbb{Y}$ is in $L^{\infty-}\left(\mu\right)$. Suppose $f, g : W \to \mathbb{R}$ are $C^1$ functions such that $f$, $g$, and their first order derivatives are in $L^{\infty-}\left(\mu\right)$. Then by a standard truncation argument and integration by parts, one shows that

$$\int_W (\mathbb{Y}f)g\, d\mu = \int_W f(\mathbb{Y}^*g)\, d\mu,$$

where

$$\mathbb{Y}^* = -\mathbb{Y} + \delta(\mathbb{Y}) \quad \text{and} \quad \delta(\mathbb{Y})(\omega) := -\operatorname{div}(\mathbb{Y})(\omega) + \mathbb{Y}(\omega) \cdot \omega.$$

Suppose that $\phi \in C_c^\infty(\mathbb{R}^d)$ and $Y_i \in \mathbb{R}^d \subset \Gamma\left(\mathbb{R}^d\right)$, then from (8.2) and induction,

$$(Y_1Y_2\cdots Y_k\phi)(F(\omega)) = (\mathbb{Y}_1\mathbb{Y}_2\cdots\mathbb{Y}_k(\phi \circ F))(\omega)$$

and therefore,

$$\begin{aligned}
\int_{\mathbb{R}^d} (Y_1Y_2\cdots Y_k\phi)d\mu_F &= \int_W (Y_1Y_2\cdots Y_k\phi)(F(\omega))\, d\mu(\omega) \\
&= \int_W (\mathbb{Y}_1\mathbb{Y}_2\cdots\mathbb{Y}_k(\phi \circ F))(\omega)\, d\mu(\omega) \\
&= \int_W \phi(F(\omega)) \cdot (\mathbb{Y}_k^*\mathbb{Y}_{k-1}^*\cdots\mathbb{Y}_1^*1)(\omega)\, d\mu(\omega). \tag{8.3}
\end{aligned}$$

By the remarks in the previous paragraph, $(\mathbb{Y}_k^*\mathbb{Y}_{k-1}^*\cdots\mathbb{Y}_1^*1) \in L^{\infty-}\left(\mu\right)$ which along with (8.3) shows

$$\left|\int_{\mathbb{R}^d} (Y_1Y_2\cdots Y_k\phi)d\mu_F\right| \le C\, \|\phi\|_{L^\infty(\mathbb{R}^d)},$$

where $C = \left\|\mathbb{Y}_k^*\mathbb{Y}_{k-1}^*\cdots\mathbb{Y}_1^*1\right\|_{L^1(\mu)} < \infty$. It now follows from Sobolev imbedding theorems or simple Fourier analysis that $\mu_F \lambda$ and that $\rho := d\mu_F/d\lambda$ is a smooth function. $\square$

The remainder of Section 8 will be devoted to an infinite dimensional analogue of Theorem 8.2 (see Theorem 8.9) where $\mathbb{R}^d$ is replaced by a manifold $M^d$,

$$W := \left\{\omega \in C\left([0, \infty), \mathbb{R}^n\right) : \omega\left(0\right) = 0\right\},$$

$\mu$ is taken to be Wiener measure on $W$, $B_t : W \to \mathbb{R}^n$ be defined by $B_t(\omega) = \omega_t$ and $F := \Sigma_t : W(\mathbb{R}^n) \to M$ is a solution to (5.1) with $\Sigma_0 = o \in M$ and $\beta = B$. Recall that $\mu$ is the unique measure on $\mathcal{F} := \sigma(B_t : t \in [0, \infty))$ such that $\{B_t\}_{t \geq 0}$ is a Brownian motion. We will use $t$ as the dominant parameter rather than $s$ to be in better agreement with the literature on this subject.

## 8.2 Smoothness of densities for Hörmander type diffusions

For simplicity of the exposition, it will be assumed that $M^d$ is a compact Riemannian manifold of dimensions $d$. However this can and should be relaxed. For example almost everything we are going to say would work if $M$ is an imbedded submanifold in $\mathbb{R}^N$ and the vector fields $\{X_i\}_{i=0}^n$ are the restrictions of smooth vector fields on $\mathbb{R}^N$ whose partial derivatives to any order greater than 0 are all bounded.

**Remark 8.3.** The choice of Riemannian metric here is somewhat arbitrary and is an artifact of the method to be described below. It is the author's belief that this issue has still not been adequately addressed in the literature.

To abbreviate the notation, let

$$H = \left\{ h \in W : \langle h, h \rangle_H := \int_0^\infty |\dot{h}(t)|^2 \, dt < \infty \right\}$$

and $D\Sigma_t : H \to T_{\Sigma_t} M$ be defined by $(D\Sigma_t) h := \partial_h T_t^B(o)$, as defined in Theorem 7.26. Recall from Theorem 7.26 that

$$(D\Sigma_t) h := Z_t \int_0^t Z_\tau^{-1} \mathbf{X}(\Sigma_\tau) \dot{h}_\tau d\tau = //_t z_t \int_0^t z_\tau^{-1} //_\tau^{-1} \mathbf{X}(\Sigma_\tau) \dot{h}_\tau d\tau, \qquad (8.4)$$

where $\dot{h}_\tau := \frac{d}{d\tau} h_\tau$, $Z_t := (T_t^B)_{*o} : T_o M \to T_{\Sigma_t} M$, $//_t$ is stochastic parallel translation along $\Sigma$ and $z_t := //_t^{-1} Z_t$. In the following, adjoints will be denoted by either " $*$ " or " $\text{tr}$ " with the former being used if an infinite dimensional space is involved and the latter if all spaces involved are finite dimensional.

**Definition 8.4 (Reduced Malliavin Covariance).** The End $(T_o M)$-valued random variable,

$$\bar{C}_t := \int_0^t Z_\tau^{-1} \mathbf{X}(\Sigma_\tau) \mathbf{X}(\Sigma_\tau)^{\text{tr}} \left( Z_\tau^{-1} \right)^{\text{tr}} d\tau \qquad (8.5)$$

$$= \int_0^t z_\tau^{-1} //_\tau^{-1} \mathbf{X}(\Sigma_\tau) \mathbf{X}(\Sigma_\tau)^{\text{tr}} //_\tau \left( z_\tau^{-1} \right)^{\text{tr}} d\tau, \qquad (8.6)$$

will be called the *reduced Malliavin covariance matrix*.

**Theorem 8.5.** *The adjoint,* $(D\Sigma_t)^* : T_{\Sigma_t} M \to H$, *of the map* $D\Sigma_t$ *is determined by*

$$\frac{d}{d\tau} \left[ (D\Sigma_t)^* //_t v \right]_\tau = 1_{\tau \leq t} \mathbf{X}(\Sigma_\tau)^{\text{tr}} //_\tau \left( z_t z_\tau^{-1} \right)^{\text{tr}} v \qquad (8.7)$$

*for all $v \in T_oM$. The Malliavin covariance matrix $C_t := D\Sigma_t (D\Sigma_t)^* : T_{\Sigma_t} M \to T_{\Sigma_t} M$ is given by $C_t = Z_t \bar{C}_t Z_t^{\mathrm{tr}}$ or equivalently*

$$C_t = D\Sigma_t (D\Sigma_t)^* = //_t z_t \bar{C}_t z_t^{\mathrm{tr}} //_t^{-1}. \tag{8.8}$$

*Proof.* Using (8.4),

$$
\begin{aligned}
\langle D\Sigma_t h, //_t v \rangle_{T_{\Sigma_t} M} &= \left\langle Z_t \int_0^t Z_\tau^{-1} \mathbf{X}(\Sigma_\tau) \dot{h}_\tau d\tau, //_t v \right\rangle_{T_{\Sigma_t} M} \\
&= \left\langle //_t z_t \int_0^t z_\tau^{-1} //_\tau^{-1} \mathbf{X}(\Sigma_\tau) \dot{h}_\tau d\tau, //_t v \right\rangle_{T_{\Sigma_t} M} \\
&= \int_0^t \left\langle z_t z_\tau^{-1} //_\tau^{-1} \mathbf{X}(\Sigma_\tau) \dot{h}_\tau, v \right\rangle_{T_oM} d\tau \\
&= \int_0^t \left\langle \dot{h}_\tau, \mathbf{X}(\Sigma_\tau)^{\mathrm{tr}} //_\tau \left( z_t z_\tau^{-1} \right)^{\mathrm{tr}} v \right\rangle_{\mathbb{R}^n} d\tau \tag{8.9}
\end{aligned}
$$

which implies (8.7). Combining (8.4) and (8.7), using

$$Z_\tau^{\mathrm{tr}} = (//_\tau z_\tau)^{\mathrm{tr}} = z_\tau^{\mathrm{tr}} //_\tau^{\mathrm{tr}} = z_\tau^{\mathrm{tr}} //_\tau^{-1},$$

shows

$$
\begin{aligned}
D\Sigma_t (D\Sigma_t)^* //_t v &= Z_t \int_0^t Z_\tau^{-1} \mathbf{X}(\Sigma_\tau) \mathbf{X}(\Sigma_\tau)^{\mathrm{tr}} //_\tau \left( z_t z_\tau^{-1} \right)^{\mathrm{tr}} v d\tau \\
&= Z_t \int_0^t Z_\tau^{-1} \mathbf{X}(\Sigma_\tau) \mathbf{X}(\Sigma_\tau)^{\mathrm{tr}} \left( Z_\tau^{-1} \right)^{\mathrm{tr}} Z_t^{\mathrm{tr}} //_t v d\tau.
\end{aligned}
$$

Therefore,

$$C_t = Z_t \bar{C}_t Z_t^{\mathrm{tr}} = //_t z_t \bar{C}_t z_t^{\mathrm{tr}} //_t^{-1}$$

from which (8.8) follows.    □

The next crucial theorem is at the heart of Malliavin's method and constitutes the deepest part of the theory. The proof of this theorem will be postponed until Section 8.4 below.

**Theorem 8.6 (Non-degeneracy of $\bar{C}_t$).** *Let $\bar{\Delta}_t := \det \left( \bar{C}_t \right)$. If Hörmander's restricted bracket condition at $o \in M$ holds then $\bar{\Delta}_t > 0$ a.e. (i.e., $\bar{C}_t$ is invertible a.e.) and moreover $\bar{\Delta}_t^{-1} \in L^{\infty-} (\mu)$.*

Following the general strategy outlined in Theorem 8.2, given a vector field $Y \in \Gamma(TM)$ we wish to lift it via the map $\Sigma_t : W \to M$ to a vector field $\mathbb{Y}^t$ on $W := W(\mathbb{R}^n)$. According to the prescription used in (8.1) in Theorem 8.2,

$$\mathbb{Y}^t := (D\Sigma_t)^* \left( D\Sigma_t (D\Sigma_t)^* \right)^{-1} Y (\Sigma_t) = (D\Sigma_t)^* C_t^{-1} Y (\Sigma_t) \in H. \tag{8.10}$$

From (8.8)

$$C_t^{-1} = //_t \left(z_t^{\mathrm{tr}}\right)^{-1} \bar{C}_t^{-1} z_t^{-1} //_t^{-1}$$

and combining this with (8.10), using (8.7), implies

$$
\begin{aligned}
\frac{d}{d\tau}\mathbb{Y}_\tau^t &= 1_{\tau \le t}\frac{d}{d\tau}\left[(D\Sigma_t) //_t \left(z_t^{\mathrm{tr}}\right)^{-1} \bar{C}_t^{-1} z_t^{-1} //_t^{-1} Y\left(\Sigma_t\right)\right]_\tau \\
&= 1_{\tau \le t}\mathbf{X}\left(\Sigma_\tau\right)^{\mathrm{tr}} //_\tau \left(z_\tau z_\tau^{-1}\right)^{\mathrm{tr}} \left(z_t^{\mathrm{tr}}\right)^{-1} \bar{C}_t^{-1} z_t^{-1} //_t^{-1} Y\left(\Sigma_t\right) \\
&= 1_{\tau \le t}\mathbf{X}\left(\Sigma_\tau\right)^{\mathrm{tr}} //_\tau \left(z_\tau^{-1}\right)^{\mathrm{tr}} \bar{C}_t^{-1} Z_t^{-1} Y\left(\Sigma_t\right) \\
&= 1_{\tau \le t}\mathbf{X}\left(\Sigma_\tau\right)^{\mathrm{tr}} \left(Z_\tau^{-1}\right)^{\mathrm{tr}} \bar{C}_t^{-1} Z_t^{-1} Y\left(\Sigma_t\right).
\end{aligned}
$$

Hence, the formula for $\mathbb{Y}^t$ in (8.10) may be explicitly written as

$$\mathbb{Y}_s^t = \left[\int_0^{s \wedge t} \left(Z_\tau^{-1}\mathbf{X}\left(\Sigma_\tau\right)\right)^{\mathrm{tr}} d\tau\right]\bar{C}_t^{-1} Z_t^{-1} Y\left(\Sigma_t\right). \tag{8.11}$$

The reader should observe that the process $s \to \mathbb{Y}_s^t$ is non-adapted since $\bar{C}_t^{-1} Z_t^{-1} Y\left(\Sigma_t\right)$ depends on the entire path of $\Sigma$ up to time $t$.

**Theorem 8.7.** *Let $Y \in \Gamma\left(TM\right)$ and $\mathbb{Y}^t$ be the non-adapted Cameron–Martin process defined in (8.11). Then $\mathbb{Y}^t$ is "Malliavin smooth," i.e., $\mathbb{Y}^t$ is H-differentiable (in the sense of Theorem 7.14) to all orders with all differentials being in $L^{\infty-}\left(\mu\right)$, (see Nualart [148] for more precise definitions). Moreover if $f \in C^\infty\left(M\right)$, then $f\left(\Sigma_t\right)$ is Malliavin smooth and*

$$\langle \bar{D}\left[f\left(\Sigma_t\right)\right], \mathbb{Y}^t\rangle_H = Yf\left(\Sigma_t\right) \tag{8.12}$$

*where $\bar{D}$ is the closure of the gradient operator defined in Corollary 7.16.*

*Proof.* We only sketch the proof here and refer the reader to [147, 12, 148] with regard to some of the technical details which are omitted below. Let $\{e_i\}_{i=1}^d$ be an orthonormal basis for $T_oM$, then

$$\mathbb{Y}_s^t = \sum_{i=1}^d \langle e_i, \bar{C}_t^{-1} Z_t^{-1} Y\left(\Sigma_t\right)\rangle \int_0^s \left(Z_\tau^{-1}\mathbf{X}\left(\Sigma_\tau\right)\right)^{\mathrm{tr}} e_i d\tau = \sum_{i=1}^d a_i h_s^i \tag{8.13}$$

where

$$a_i := \langle e_i, \bar{C}_t^{-1} Z_t^{-1} Y\left(\Sigma_t\right)\rangle \text{ and } h_s^i := \int_0^{s \wedge t} \left(Z_\tau^{-1}\mathbf{X}\left(\Sigma_\tau\right)\right)^{\mathrm{tr}} e_i d\tau.$$

It is well known that solutions to SDEs with smooth coefficients are Malliavin smooth from which it follows that $h^i$, $Z_t^{-1}Y\left(\Sigma_t\right)$, and $\bar{C}_t$ are Malliavin smooth. It also follows from the general theory, under the conclusion of Theorem 8.6, that $\bar{C}_t^{-1}$ is Malliavin smooth and hence so are each of the functions $a_i$ for $i = 1, \ldots d$. Therefore, $\mathbb{Y}^t = \sum_{i=1}^d a_i h^i$ is Malliavin smooth as well and in particular $\mathbb{Y}^t \in \mathcal{D}\left(D^*\right)$. It now only remains to verify (8.12).

Let $h$ be a non-random element of $H$. Then from Theorems 7.14, 7.15, 7.26 and the chain rule for Wiener calculus,

$$\mathbb{E}\left[f\left(\Sigma_t\right)\cdot D^*h\right] - \mathbb{E}\left[\partial_h\left[f\left(\Sigma_t\right)\right]\right] = \mathbb{E}\left[df\left(D\Sigma_t h\right)\right]$$

$$= \mathbb{E}\left[df\left(Z_t\int_0^t Z_\tau^{-1}\mathbf{X}\left(\Sigma_\tau\right)\dot{h}_\tau d\tau\right)\right]$$

$$= \mathbb{E}\left[\left\langle\nabla f\left(\Sigma_t\right), Z_t\int_0^t Z_\tau^{-1}\mathbf{X}\left(\Sigma_\tau\right)\dot{h}_\tau d\tau\right\rangle_{T_{\Sigma_t}M}\right]$$

$$= \mathbb{E}\left[\int_0^t\left\langle\mathbf{X}\left(\Sigma_\tau\right)^{\mathrm{tr}}\left(Z_\tau^{-1}\right)^{\mathrm{tr}}Z_t^{\mathrm{tr}}\nabla f\left(\Sigma_t\right)\nabla f\left(\Sigma_t\right), \dot{h}_\tau\right\rangle_{\mathbb{R}^n} d\tau\right]$$

from which we conclude that $f\left(\Sigma_t\right)\in\mathcal{D}\left(D^{**}\right)=\mathcal{D}\left(\bar{D}\right)$ and

$$\left(\bar{D}\left[f\left(\Sigma_t\right)\right]\right)_s = \int_0^{s\wedge t}\mathbf{X}\left(\Sigma_\tau\right)^{\mathrm{tr}}\left(Z_\tau^{-1}\right)^{\mathrm{tr}}Z_t^{\mathrm{tr}}\nabla f\left(\Sigma_t\right) d\tau.$$

From this formula and the definition of $\mathbb{Y}^t$ it follows that

$$\langle\bar{D}\left[f\left(\Sigma_t\right)\right], \mathbb{Y}^t\rangle_H$$

$$= \int_0^t\left\langle\mathbf{X}\left(\Sigma_\tau\right)^{\mathrm{tr}}\left(Z_\tau^{-1}\right)^{\mathrm{tr}}Z_t^{\mathrm{tr}}\nabla f\left(\Sigma_t\right), \mathbf{X}\left(\Sigma_\tau\right)^{\mathrm{tr}}\left(Z_\tau^{-1}\right)^{\mathrm{tr}}\bar{C}_t^{-1}Z_t^{-1}Y\left(\Sigma_t\right)\right\rangle d\tau$$

$$= \left\langle\nabla f\left(\Sigma_t\right), Z_t\left(\int_0^t Z_\tau^{-1}\mathbf{X}\left(\Sigma_\tau\right)\left(Z_\tau^{-1}\mathbf{X}\left(\Sigma_\tau\right)\right)^{\mathrm{tr}} d\tau\right)\bar{C}_t^{-1}Z_t^{-1}Y\left(\Sigma_t\right)\right\rangle$$

$$= \left\langle\nabla f\left(\Sigma_t\right), Z_t\bar{C}_t\bar{C}_t^{-1}Z_t^{-1}Y\left(\Sigma_t\right)\right\rangle = \langle\nabla f\left(\Sigma_t\right), Y\left(\Sigma_t\right)\rangle$$

$$= \left(Yf\right)\left(\Sigma_t\right). \qquad \square$$

**Notation 8.8.** Let $\mathbb{Y}^t$ act on Malliavin smooth functions by the formula, $\mathbb{Y}^t F := \langle\bar{D}F, \mathbb{Y}^t\rangle_H$ and let $\left(\mathbb{Y}^t\right)^*$ denote the $L^2\left(\mu\right)$-adjoint of $\mathbb{Y}^t$.

With this notation, Theorem 8.7 asserts that

$$\mathbb{Y}^t\left[f\left(\Sigma_t\right)\right] = \left(Yf\right)\left(\Sigma_t\right). \tag{8.14}$$

Now suppose $F, G : W\to\mathbb{R}$ are Malliavin smooth functions, then

$$\mathbb{E}\left[\mathbb{Y}^t F\cdot G + F\cdot\mathbb{Y}^t G\right] = \mathbb{E}\left[\mathbb{Y}^t\left[FG\right]\right] = \mathbb{E}\left[\langle\bar{D}\left[FG\right], \mathbb{Y}^t\rangle_H\right]$$

$$= \mathbb{E}\left[F\cdot GD^*\mathbb{Y}^t\right]$$

from which it follows that $G\in\mathcal{D}\left(\left(\mathbb{Y}^t\right)^*\right)$ and

$$\left(\mathbb{Y}^t\right)^* G = -\mathbb{Y}^t G + GD^*\mathbb{Y}^t. \tag{8.15}$$

From the general theory (see [148] for example), $D^*U$ is Malliavin smooth if $U$ is Malliavin smooth. In particular $\left(\mathbb{Y}^t\right)^* G$ is Malliavin smooth if $G$ is Malliavin smooth.

**Theorem 8.9 (Smoothness of Densities).** *Assume the restricted Hörmander condition holds at $o \in M$ (see Definition 8.1) and suppose $f \in C^\infty(M)$ and $\{Y_i\}_{i=1}^k \subset \Gamma(TM)$. Then*

$$\mathbb{E}\left[(Y_1 \ldots Y_k f)(\Sigma_t)\right] = \mathbb{E}\left[\mathbb{Y}_1^t \ldots \mathbb{Y}_k^t \left[f(\Sigma_t)\right]\right]$$
$$= \mathbb{E}\left[\left[f(\Sigma_t)\right]\left(\mathbb{Y}_k^t\right)^* \ldots \left(\mathbb{Y}_1^t\right)^* 1\right]. \qquad (8.16)$$

*Moreover, the law of $\Sigma_t$ is smooth.*

*Proof.* By an induction argument using (8.14),

$$\mathbb{Y}_1^t \ldots \mathbb{Y}_k^t \left[f(\Sigma_t)\right] = (Y_1 \ldots Y_k f)(\Sigma_t)$$

from which (8.16) is a simple consequence. As has already been observed, $\left(\mathbb{Y}_k^t\right)^* \ldots \left(\mathbb{Y}_1^t\right)^* 1$ is Malliavin smooth and in particular $\left(\mathbb{Y}_k^t\right)^* \ldots \left(\mathbb{Y}_1^t\right)^* 1 \in L^1(\mu)$. Therefore it follows from (8.16) that

$$\left|\mathbb{E}\left[(Y_1 \ldots Y_k f)(\Sigma_t)\right]\right| \leq \left\|\left(\mathbb{Y}_k^t\right)^* \ldots \left(\mathbb{Y}_1^t\right)^* 1\right\|_{L^1(\mu)} \|f\|_\infty. \qquad (8.17)$$

Since the argument used in the proof of Theorem 8.2 after (8.16) is local in nature, it follows from (8.17) that the Law($\Sigma_t$) has a smooth density relative to any smooth measure on $M$ and in particular the Riemannian volume measure. $\qquad \square$

## 8.3 The invertability of $\bar{C}_t$ in the elliptic case

As a warm-up to the proof of the full version of Theorem 8.6 let us first consider the special case where $\mathbf{X}(m) : \mathbb{R}^n \to T_m M$ is surjective for all $m \in M$. Since $M$ is compact this will imply there exists an $\varepsilon > 0$ such that

$$\mathbf{X}(m)\mathbf{X}^{\mathrm{tr}}(m) \geq \varepsilon I_{T_m M} \text{ for all } m \in M.$$

**Notation 8.10.** We will write $f(\varepsilon) = O\left(\varepsilon^{\infty-}\right)$ if, for all $p < \infty$,

$$\lim_{\varepsilon \downarrow 0} \frac{|f(\varepsilon)|}{\varepsilon^p} = 0.$$

**Proposition 8.11 (Elliptic Case).** *Suppose there is an $\varepsilon > 0$ such that*

$$\mathbf{X}(m)\mathbf{X}^{\mathrm{tr}}(m) \geq \varepsilon I_{T_m M}$$

*for all $m \in M$, then $\left[\det\left(\bar{C}_t\right)\right]^{-1} \in L^{\infty-}(\mu)$.*

*Proof.* Let $\delta \in (0, 1)$ and

$$T_\delta := \inf\left\{t > 0 : \left|z_t - I_{T_o M}\right| > \delta\right\} \qquad (8.18)$$

where, as usual,

$$z_t := //_t^{-1} Z_t = //_t^{-1} \left( T_t^B \right)_{*o}.$$

Since for all $a \in T_o M$,

$$\langle Z_\tau^{-1} \mathbf{X}(\Sigma_\tau) \mathbf{X}^{tr}(\Sigma_\tau) \left( Z_\tau^{tr} \right)^{-1} a, a \rangle$$

$$= \left\langle \mathbf{X}(\Sigma_\tau) \mathbf{X}^{tr}(\Sigma_\tau) \left( Z_\tau^{tr} \right)^{-1} a, \left( Z_\tau^{tr} \right)^{-1} a \right\rangle$$

$$\geq \varepsilon \left\langle \left( Z_\tau^{tr} \right)^{-1} a, \left( Z_\tau^{tr} \right)^{-1} a \right\rangle = \varepsilon \left\langle a, Z_\tau^{tr} \left( Z_\tau^{tr} \right)^{-1} a \right\rangle,$$

we have

$$Z_\tau^{-1} \mathbf{X}(\Sigma_\tau) \mathbf{X}^{tr}(\Sigma_\tau) \left( Z_\tau^{tr} \right)^{-1}$$

$$\geq \varepsilon Z_\tau^{tr} \left( Z_\tau^{tr} \right)^{-1} = \varepsilon z_t^{tr} //_t^{tr} \left( //_t^{tr} \right)^{-1} \left( z_t^{tr} \right)^{-1} = \varepsilon z_t^{tr} \left( z_t^{tr} \right)^{-1}.$$

Hence

$$\bar{C}_t = \int_0^t Z_\tau^{-1} \mathbf{X}(\Sigma_\tau) \mathbf{X}^{tr}(\Sigma_\tau) \left( Z_\tau^{tr} \right)^{-1} d\tau$$

$$\geq \varepsilon \int_0^t Z_\tau^{-1} \left( Z_\tau^{tr} \right)^{-1} d\tau \geq \varepsilon \int_0^{t \wedge T_\delta} z_\tau^{tr} \left( z_\tau^{tr} \right)^{-1} d\tau$$

and therefore,

$$\bar{\Delta}_t = \det \left( \bar{C}_t \right) \geq \varepsilon^d \det \left( \int_0^{t \wedge T_\delta} z_\tau^{tr} \left( z_\tau^{tr} \right)^{-1} d\tau \right).$$

By choosing $\delta > 0$ sufficiently small we may arrange that

$$\left\| z_\tau^{tr} \left( z_\tau^{tr} \right)^{-1} - I \right\| \leq 1/2$$

for all $\tau \leq t \wedge T_\delta$ in which case

$$\int_0^{t \wedge T_\delta} z_\tau^{tr} \left( z_\tau^{tr} \right)^{-1} d\tau \geq \frac{1}{2} t \wedge T_\delta \cdot Id$$

and hence $\bar{\Delta}_t = \det \left( \bar{C}_t \right) \geq \varepsilon^d \left( \frac{1}{2} t \wedge T_\delta \right)^d$. From this it follows, with $q = p \cdot d$, that

$$\mathbb{E} \left[ \bar{\Delta}_t^{-q} \right] \leq 2^q \varepsilon^{-nq} \mathbb{E} \left( \left( \frac{1}{t \wedge T_\delta} \right)^q \right).$$

Now

$$\mathbb{E} \left( \left( \frac{1}{t \wedge T_\delta} \right)^q \right) - \mathbb{E} \left( - \int_{t \wedge T_\delta}^\infty \frac{d}{d\tau} \tau^{-q} d\tau \right)$$

$$= \mathbb{E} \left( q \int_0^\infty 1_{t \wedge T_\delta \leq \tau} \cdot \tau^{-q-1} d\tau \right)$$

$$= q \int_0^\infty \tau^{-q-1} \mu \left( t \wedge T_\delta \leq \tau \right) d\tau$$

which will be finite for all $q > 1$ iff $\mu(t \wedge T_\delta \leq \tau) = \mu(T_\delta \leq \tau) = O(\tau^k)$ as $\tau \downarrow 0$ for all $k > 0$.

By Chebyschev's inequalities and (9.10) of Proposition 9.5 below,

$$\mu(T_\delta \leq \tau) = \mu\left(\sup_{s \leq \tau} |z_s - I| > \delta\right) \leq \delta^{-q} \mathbb{E}\left[\sup_{s \leq \tau} |z_s - I|^q\right] = O(\tau^{q/2}). \quad (8.19)$$

Since $q \geq 2$ was arbitrary it follows that $\mu(T_\delta \leq \tau) = O(\tau^{\infty-})$ which completes the proof. $\qquad\qquad\square$

### 8.4 Proof of theorem 8.6

**Notation 8.12.** Let $S := \{v \in T_oM : \langle v, v \rangle = 1\}$, i.e., $S$ is the unit sphere in $T_oM$.

*Proof of Theorem 8.6.* To show $\bar{C}_t^{-1} \in L^{\infty-}(\mu)$ it suffices to show

$$\mu(\inf_{v \in S} \langle \bar{C}_t v, v \rangle < \varepsilon) = O(\varepsilon^{\infty-}).$$

To verify this claim, notice that $\lambda_0 := \inf_{v \in S} \langle \bar{C}_t v, v \rangle$ is the smallest eigenvalue of $\bar{C}_t$. Since $\det \bar{C}_t$ is the product of the eigenvalues of $\bar{C}_t$ it follows that $\bar{\Delta}_t := \det \bar{C}_t \geq \lambda_0^d$ and so $\{\det \bar{C}_t < \varepsilon^d\} \subset \{\lambda_0 < \varepsilon\}$ and hence

$$\mu\left(\det \bar{C}_t < \varepsilon^d\right) \leq \mu(\lambda_0 < \varepsilon) = O(\varepsilon^{\infty-}).$$

By replacing $\varepsilon$ by $\varepsilon^{1/d}$ above this implies $\mu\left(\bar{\Delta}_t < \varepsilon\right) = O(\varepsilon^{\infty-})$. From this estimate it then follows that

$$\mathbb{E}\left[\bar{\Delta}_t^{-q}\right] = \mathbb{E}\int_{\bar{\Delta}_t}^{\infty} q\tau^{-q-1}d\tau = q\mathbb{E}\int_0^{\infty} 1_{\bar{\Delta}_t \leq \tau}\, \tau^{-q-1}d\tau$$

$$= q\int_0^{\infty} \mu(\bar{\Delta}_t \leq \tau)\, \tau^{-q-1}d\tau = q\int_0^{\infty} O(\tau^p)\, \tau^{-q-1}d\tau$$

which is seen to be finite by taking $p \geq q + 1$.

More generally if $T$ is any stopping time with $T \leq t$, since $\langle \bar{C}_T v, v \rangle \leq \langle \bar{C}_t v, v \rangle$ for all $v \in S$ it suffices to prove

$$\mu\left(\inf_{v \in S} \langle \bar{C}_T v, v \rangle < \varepsilon\right) = O(\varepsilon^{\infty-}). \quad (8.20)$$

According to Lemma 8.13 and Proposition 8.15 below, (8.20) holds with

$$T = T_\delta := \inf\left\{t > 0 : \max\left\{\left|z_t - I_{T_oM}\right|, \operatorname{dist}(\Sigma_t, \Sigma_0)\right\} > \delta\right\} \quad (8.21)$$

provided $\delta > 0$ is chosen sufficiently small. $\qquad\qquad\square$

The rest of this section is now devoted to the proof of Lemma 8.13 and Proposition 8.15 below. In what follows we will make repeated use of the identity,

$$\langle \bar{C}_T v, v \rangle = \sum_{i=1}^{n} \int_0^T \left\langle Z_\tau^{-1} X_i(\Sigma_\tau), v \right\rangle^2 d\tau. \tag{8.22}$$

To prove this, let $\{e_i\}_{i=1}^n$ be the standard basis for $\mathbb{R}^n$. Then

$$Z_\tau^{-1} \mathbf{X}(\Sigma_\tau) \mathbf{X}^{\mathrm{tr}}(\Sigma_\tau) \left(Z_\tau^{\mathrm{tr}}\right)^{-1} v = \sum_{i=1}^{n} Z_\tau^{-1} \mathbf{X}(\Sigma_\tau) e_i \left\langle e_i, \mathbf{X}^{\mathrm{tr}}(\Sigma_\tau) \left(Z_\tau^{\mathrm{tr}}\right)^{-1} v \right\rangle$$

$$= \sum_{i=1}^{n} \langle Z_\tau^{-1} X_i(\Sigma_\tau), v \rangle \, Z_\tau^{-1} X_i(\Sigma_\tau)$$

so that

$$\left\langle Z_\tau^{-1} \mathbf{X}(\Sigma_\tau) \mathbf{X}^{\mathrm{tr}}(\Sigma_\tau) \left(Z_\tau^{\mathrm{tr}}\right)^{-1} v, v \right\rangle = \sum_{i=1}^{n} \left\langle Z_\tau^{-1} X_i(\Sigma_\tau), v \right\rangle^2$$

which upon integrating on $\tau$ gives (8.22).

In the proofs below, there will always be an implied sum on repeated indices.

**Lemma 8.13 (Compactness Argument).** *Let $T_\delta$ be as in (8.21) and suppose for all $v \in S$ there exists $i \in \{1, \ldots, n\}$ and an open neighborhood $N \subset_o S$ of $v$ such that*

$$\sup_{u \in N} \mu \left( \int_0^{T_\delta} \left\langle Z_\tau^{-1} X_i(\Sigma_\tau), u \right\rangle^2 d\tau < \varepsilon \right) = O\left(\varepsilon^{\infty-}\right), \tag{8.23}$$

*then (8.20) holds provided $\delta > 0$ is sufficiently small.*

*Proof.* By compactness of $S$, it follows from (8.23) that

$$\sup_{u \in S} \mu \left( \int_0^{T_\delta} \left\langle Z_\tau^{-1} X_i(\Sigma_\tau), u \right\rangle^2 d\tau < \varepsilon \right) = O\left(\varepsilon^{\infty-}\right). \tag{8.24}$$

For $w \in T_o M$, let $\partial_w$ denote the directional derivative acting on functions $f(v)$ with $v \in T_o M$. Because for all $v, w \in \mathbb{R}^n$ with $|v| \le 1$ and $|w| \le 1$ (using (8.22)),

$$\left| \partial_w \left\langle \bar{C}_{T_\delta} v, v \right\rangle \right| \le 2 \sum_{i=1}^{n} \int_0^{T_\delta} \left| \left\langle Z_\tau^{-1} X_i(\Sigma_\tau), v \right\rangle \left\langle Z_\tau^{-1} X_i(\Sigma_\tau), w \right\rangle \right| d\tau$$

$$\le 2 \sum_{i=1}^{n} \int_0^{T_\delta} \left| Z_\tau^{-1} X_i(\Sigma_\tau) \right|^2_{\mathrm{Hom}(\mathbb{R}^n, T_o M)} d\tau$$

$$= 2 \sum_{i=1}^{n} \int_0^{T_\delta} \left| z_\tau^{-1} // _\tau^{-1} X_i(\Sigma_\tau) \right|^2_{\mathrm{Hom}(\mathbb{R}^n, T_o M)} d\tau,$$

by choosing $\delta > 0$ in (8.21) sufficiently small we may assume there is a non-random constant $\theta < \infty$ such that

$$\sup_{|v|,|w|\leq 1} \left| \partial_w \langle \bar{C}_{T_\delta} v, v \rangle \right| \leq \theta < \infty.$$

With this choice of $\delta$, if $v, w \in S$ satisfy $|v - w| < \theta/\varepsilon$ then

$$\left| \langle \bar{C}_{T_\delta} v, v \rangle - \langle \bar{C}_{T_\delta} w, w \rangle \right| < \varepsilon. \tag{8.25}$$

There exists $D < \infty$ satisfying: for any $\varepsilon > 0$, there is an open cover of $S$ with at most $D \cdot (\theta/\varepsilon)^d$ balls of the form $B(v_j, \varepsilon/\theta)$. From (8.25), for any $v \in S$ there exists $j$ such that $v \in B(v_j, \varepsilon/\theta) \cap S$ and

$$\left| \langle \bar{C}_{T_\delta} v, v \rangle - \langle \bar{C}_{T_\delta} v_j, v_j \rangle \right| < \varepsilon.$$

So if $\inf_{v \in S} \langle \bar{C}_{T_\delta} v, v \rangle < \varepsilon$ then $\min_j \langle \bar{C}_{T_\delta} v_j, v_j \rangle < 2\varepsilon$, i.e.,

$$\left\{ \inf_{v \in S} \langle \bar{C}_{T_\delta} v, v \rangle < \varepsilon \right\} \subset \left\{ \min_j \langle \bar{C}_{T_\delta} v_j, v_j \rangle < 2\varepsilon \right\} \subset \bigcup_j \left\{ \langle \bar{C}_{T_\delta} v_j, v_j \rangle < 2\varepsilon \right\}.$$

Therefore,

$$\mu \left( \inf_{v \in S} \langle \bar{C}_{T_\delta} v, v \rangle < \varepsilon \right) \leq \sum_j \mu \left( \langle \bar{C}_{T_\delta} v_j, v_j \rangle < 2\varepsilon \right)$$

$$\leq D \cdot (\theta/\varepsilon)^d \cdot \sup_{v \in S} \mu \left( \langle \bar{C}_{T_\delta} v, v \rangle < 2\varepsilon \right)$$

$$\leq D \cdot (\theta/\varepsilon)^d O(\varepsilon^{\infty-}) = O(\varepsilon^{\infty-}). \qquad \square$$

The following important proposition is the stochastic version of Theorem 4.9. It gives the first hint that Hörmander's condition in Definition 8.1 is relevant to showing $\bar{\Delta}_t^{-1} \in L^{\infty-}(\mu)$ or equivalently that $\bar{C}_t^{-1} \in L^{\infty-}(\mu)$.

**Proposition 8.14 (The appearance of commutators).** *Let $W \in \Gamma(TM)$, then*

$$\delta \left[ Z_s^{-1} W(\Sigma_s) \right] = Z_s^{-1} [X_0, W](\Sigma_s) ds + Z_s^{-1} \sum_{i=1}^n [X_i, W](\Sigma_s) \delta B_s^i. \tag{8.26}$$

*This may also be written in Itô's form as*

$$d \left[ Z_s^{-1} W(\Sigma_s) \right] = Z_s^{-1} [X_i, W](\Sigma_s) dB_s^i$$

$$+ \left\{ Z_s^{-1} [X_0, W](\Sigma_s) + \frac{1}{2} \sum_{i=1}^n Z_s^{-1} \left( L_{X_i}^2 W \right) (\Sigma_s) \right\} ds, \tag{8.27}$$

*where $L_X W := [X, W]$ as in Theorem 4.9.*

*Proof.* Write $W(\Sigma_s) = Z_s w_s$, i.e., let $w_s := Z_s^{-1} W(\Sigma_s)$. By Proposition 5.36 and Theorem 5.41,

$$\nabla_{\delta \Sigma_s} W = \delta^{\nabla} [W(\Sigma_s)] = \delta^{\nabla} [Z_s w_s] = \left(\delta^{\nabla} Z_s\right) w_s + Z_s \delta w_s$$

$$= \left(\nabla_{Z_s w_s} \mathbf{X}\right) \delta B_s + \left(\nabla_{Z_s w_s} X_0\right) ds + Z_s \delta w_s.$$

Therefore, using the fact that $\nabla$ has zero torsion (see Proposition 3.36),

$$\delta w_s = Z_s^{-1} \left[\nabla_{\delta \Sigma_s} W - \left(\nabla_{Z_s w_s} \mathbf{X}\right) \delta B_s + \left(\nabla_{Z_s w_s} X_0\right) ds\right]$$

$$= Z_s^{-1} \left[\nabla_{\mathbf{X}(\Sigma_s) \delta B_s + X_0(\Sigma_s) ds} W - \left(\nabla_{W(\Sigma_s)} \mathbf{X}\right) \delta B_s + \left(\nabla_{W(\Sigma_s)} X_0\right) ds\right]$$

$$= Z_s^{-1} \left[\left(\nabla_{X_i(\Sigma_s)} W - \nabla_{W(\Sigma_s)} X_i\right) \delta B_s^i + \left(\nabla_{X_0(\Sigma_s)} W - \nabla_{W(\Sigma_s)} X_0\right) ds\right]$$

$$= Z_s^{-1} \left([X_i, W](\Sigma_s) \delta B_s^i + [X_0, W](\Sigma_s) ds\right)$$

which proves (8.26).

Applying (8.26) with $W$ replaced by $[X_i, W]$ implies

$$d\left[Z_s^{-1} [X_i, W](\Sigma_s)\right] = Z_s^{-1} [X_j, [X_i, W]](\Sigma_s) dB_s^j + d[BV],$$

where $BV$ denotes process of bounded variation. Hence

$$Z_s^{-1} [X_i, W](\Sigma_s) \delta B_s^i = Z_s^{-1} [X_i, W](\Sigma_s) dB_s^i + \frac{1}{2} d\left\{Z_s^{-1} [X_i, W](\Sigma_s)\right\} dB_s^i$$

$$= Z_s^{-1} [X_i, W](\Sigma_s) dB_s^i + \frac{1}{2} Z_s^{-1} [X_j, [X_i, W]](\Sigma_s) dB_s^j dB_s^i$$

$$= Z_s^{-1} [X_i, W](\Sigma_s) dB_s^i + \frac{1}{2} Z_s^{-1} [X_i, [X_i, W]](\Sigma_s) ds$$

which combined with (8.26) proves (8.27).  $\square$

**Proposition 8.15.** *Let $T_\delta$ be as in (8.21). If Hörmander's restricted bracket condition holds at $o \in M$ and $v \in S$ is given, there exists $i \in \{1, 2, \ldots, n\}$ and an open neighborhood $U \subset_o S$ of $v$ such that*

$$\sup_{u \in U} \mu \left(\int_0^{T_\delta} \left\langle Z_\tau^{-1} X_i(\Sigma_\tau), u\right\rangle^2 d\tau \le \varepsilon\right) = O\left(\varepsilon^{\infty-}\right).$$

*Proof.* The proof given here will follow Norris [147]. Hörmander's condition implies there exist $l \in \mathbb{N}$ and $\beta > 0$ such that

$$\frac{1}{|\mathcal{K}_l|} \sum_{K \in \mathcal{K}_l} K(o) K(o)^{\mathrm{tr}} \ge 3\beta I$$

or equivalently put for all $v \in S$,

$$3\beta \le \frac{1}{|\mathcal{K}_l|} \sum_{K \in \mathcal{K}_l} \langle K(o), v\rangle^2 \le \max_{K \in \mathcal{K}_l} \langle K(o), v\rangle^2.$$

By choosing $\delta > 0$ in (8.21) sufficiently small we may assume that

$$\max_{K \in \mathcal{K}_l} \inf_{\tau \leq T_\delta} \left\langle Z_\tau^{-1} K(\Sigma_\tau), v \right\rangle^2 \geq 2\beta \text{ for all } v \in S.$$

Fix a $v \in S$ and $K \in \mathcal{K}_l$ such that

$$\inf_{\tau \leq T_\delta} \left\langle Z_\tau^{-1} K(\Sigma_\tau), v \right\rangle^2 \geq 2\beta$$

and choose an open neighborhood $U \subset S$ of $v$ such that

$$\inf_{\tau \leq T_\delta} \left\langle Z_\tau^{-1} K(\Sigma_\tau), u \right\rangle^2 \geq \beta \text{ for all } u \in U.$$

Then, using (8.19),

$$\sup_{u \in U} \mu \left( \int_0^{T_\delta} \left\langle Z_\tau^{-1} K(\Sigma_\tau), u \right\rangle^2 d\tau \leq \varepsilon \right)$$
$$\leq \mu \left( \int_0^{T_\delta} \beta dt \leq \varepsilon \right) = \mu \left( T_\delta \leq \varepsilon/\beta \right) = O \left( \varepsilon^{\infty-} \right). \tag{8.28}$$

Write $K = L_{X_{i_r}} \ldots L_{X_{i_2}} X_{i_1}$ with $r \leq l$. If it happens that $r = 1$ then (8.28) becomes

$$\sup_{u \in U} \mu \left( \langle \bar{C}_{T_\delta} u, u \rangle \leq \varepsilon \right) \leq \sup_{u \in U} \mu \left( \int_0^{T_\delta} \left\langle Z_\tau^{-1} X_{i_1}(\Sigma_\tau), u \right\rangle^2 dt \leq \varepsilon \right) = O \left( \varepsilon^{\infty-} \right)$$

and we are done. So now suppose $r > 1$ and set

$$K_j = L_{X_{i_j}} \ldots L_{X_{i_2}} X_{i_1} \text{ for } j = 1, 2, \ldots, r$$

so that $K_r = K$. We will now show by (decreasing) induction on $j$ that

$$\sup_{u \in U} \mu \left( \int_0^{T_\delta} \left\langle Z_\tau^{-1} K_j(\Sigma_\tau), u \right\rangle^2 dt \leq \varepsilon \right) = O \left( \varepsilon^{\infty-} \right). \tag{8.29}$$

From Proposition 8.14 we have

$$d \left[ Z_t^{-1} K_{j-1}(\Sigma_t) \right] = Z_t^{-1} [X_i, K_{j-1}](\Sigma_t) dB^i(t)$$
$$+ \left\{ Z_t^{-1} [X_0, K_{j-1}](\Sigma_t) + \frac{1}{2} Z_t^{-1} \left( L_{X_i}^2 K_{j-1} \right)(\Sigma_t) \right\} dt$$

which upon integrating on $t$ gives

$$\left\langle Z_t^{-1} K_{j-1}(\Sigma_t), u \right\rangle = \left\langle K_{j-1}(\Sigma_0), u \right\rangle + \int_0^t \left\langle Z_\tau^{-1} [X_i, K_{j-1}](\Sigma_\tau), u \right\rangle dB_\tau^i$$
$$+ \int_0^t \left\langle Z_\tau^{-1} [X_0, K_{j-1}](\Sigma_\tau) + \frac{1}{2} Z_\tau^{-1} \left( L_{X_i}^2 K_{j-1} \right)(\Sigma_\tau), u \right\rangle d\tau.$$

Applying Proposition 9.13 of the appendix with $T = T_\delta$,

$$Y_t := \left\langle Z_t^{-1} K_{j-1}(\Sigma_t), u \right\rangle, \quad y = \left\langle K_{j-1}(\Sigma_0), u \right\rangle,$$

$$M_t = \int_0^t \left\langle Z_\tau^{-1}[X_i, K_{j-1}](\Sigma_\tau), u \right\rangle dB_\tau^i \text{ and}$$

$$A_t := \int_0^t \left\langle Z_\tau^{-1}[X_0, K_{j-1}](\Sigma_\tau) + \frac{1}{2} Z_\tau^{-1} \left( L_{X_i}^2 K_{j-1} \right)(\Sigma_\tau), u \right\rangle dt$$

implies

$$\sup_{u \in U} \mu\left( \Omega_1(u) \cap \Omega_2(u) \right) = O\left( \varepsilon^{\infty-} \right), \tag{8.30}$$

where

$$\Omega_1(u) := \left\{ \int_0^{T_\delta} \left\langle Z_t^{-1} K_{j-1}(\Sigma_t), u \right\rangle^2 dt < \varepsilon^q \right\},$$

$$\Omega_2(u) := \left\{ \int_0^{T_\delta} \sum_{i=1}^n \left\langle Z_\tau^{-1}[X_i, K_{j-1}](\Sigma_\tau), u \right\rangle^2 d\tau \geq \varepsilon \right\}$$

and $q > 4$. Since

$$\sup_{u \in U} \mu\left( [\Omega_2(u)]^c \right) = \sup_{u \in U} \mu\left( \int_0^{T_\delta} \sum_{i=1}^n \left\langle Z_\tau^{-1}[X_i, K_{j-1}](\Sigma_\tau), u \right\rangle^2 d\tau < \varepsilon \right)$$

$$\leq \sup_{u \in U} \mu\left( \int_0^{T_\delta} \left\langle Z_\tau^{-1} K_j(\Sigma_\tau), u \right\rangle^2 d\tau < \varepsilon \right)$$

we may apply the induction hypothesis to learn,

$$\sup_{u \in U} \mu\left( [\Omega_2(u)]^c \right) = O\left( \varepsilon^{\infty-} \right). \tag{8.31}$$

It now follows from (8.30) and (8.31) that

$$\sup_{u \in U} \mu(\Omega_1(u)) \leq \sup_{u \in U} \mu(\Omega_1(u) \cap \Omega_2(u)) + \sup_{u \in U} \mu(\Omega_1(u) \cap [\Omega_2(u)]^c)$$

$$\leq \sup_{u \in U} \mu(\Omega_1(u) \cap \Omega_2(u)) + \sup_{u \in U} \mu([\Omega_2(u)]^c)$$

$$= O\left( \varepsilon^{\infty-} \right) + O\left( \varepsilon^{\infty-} \right) = O\left( \varepsilon^{\infty-} \right),$$

which is to say

$$\sup_{u \in U} \mu\left( \int_0^{T_\delta} \left\langle Z_t^{-1} K_{j-1}(\Sigma_t), u \right\rangle^2 dt < \varepsilon^q \right) = O\left( \varepsilon^{\infty-} \right).$$

Replacing $\varepsilon$ by $\varepsilon^{1/q}$ in the previous equation, using $O\left( (\varepsilon^{1/q})^{\infty-} \right) = O\left( \varepsilon^{\infty-} \right)$, completes the induction argument and hence the proof.  □

## 8.5 More references

The literature on the "Malliavin calculus" is very extensive and I will not make any attempt at summarizing it here. Let me just add to references already mentioned the articles in [176, 105, 153] which carry out Malliavin's method in the geometric context of these notes. Also see [150] for another method which works if Hörmander's bracket condition holds at level 2, namely when

$$\mathrm{span}(\{K(m) : K \in \mathcal{K}_2\}) = T_m M \text{ for all } m \in M$$

(see Definition 8.1). The reader should also be aware of the deep results of Ben Arous and Leandre in [17, 18, 16, 15, 124].

# 9 Appendix: Martingale and SDE estimates

In this appendix $\{B_t : t \geq 0\}$ will denote and $\mathbb{R}^n$-valued Brownian motion, $\{\beta_t : t \geq 0\}$ will be a one dimensional Brownian motion and, unlike in the text, we will use the more standard letter $P$ rather than $\mu$ to denote the underlying probability measure.

**Notation 9.1.** When $M_t$ is a martingale and $A_t$ is a process of bounded variation let $\langle M \rangle_t$ be the quadratic variation of $M$ and $|A|_t$ be the total variation of $A$ up to time $t$.

## 9.1 Estimates of Wiener functionals associated to SDE's

**Proposition 9.2.** Suppose $p \in [2, \infty)$, $\alpha_\tau$ and $A_\tau$ are predictable $\mathbb{R}^d$ and Hom $(\mathbb{R}^n, \mathbb{R}^d)$-valued processes respectively and

$$Y_t := \int_0^t A_\tau dB_\tau + \int_0^t \alpha_\tau d\tau. \tag{9.1}$$

Then, letting $Y_t^* := \sup_{\tau \leq t} |Y_\tau|$, there exists $C_p < \infty$ such that

$$\mathbb{E}\left(Y_t^*\right)^p \leq C_p \left\{ \mathbb{E}\left(\int_0^t |A_\tau|^2 \, d\tau\right)^{p/2} + \mathbb{E}\left(\int_0^t |\alpha_\tau| \, d\tau\right)^p \right\} \tag{9.2}$$

where

$$|A|^2 = \mathrm{tr}\left(AA^*\right) = \sum_{i=1}^n (AA^*)_{ii} = \sum_{i,j} A_{ij} A_{ij} = \mathrm{tr}\left(A^*A\right).$$

*Proof.* We may assume the right side of (9.2) is finite for otherwise there is nothing to prove. For the moment also assume $\alpha \equiv 0$. By a standard limiting argument involving stopping times we may further assume there is a non-random constant $C < \infty$ such that

$$Y_T^* + \int_0^T |A_\tau|^2 \, d\tau \leq C.$$

Let $f(y) = |y|^p$ and $\hat{y} := y/|y|$ for $y \in \mathbb{R}^d$. Then, for $a, b \in \mathbb{R}^d$,

$$\partial_a f(y) = p \, |y|^{p-1} \, \hat{y} \cdot a = p \, |y|^{p-2} \, y \cdot a$$

and

$$\partial_b \partial_a f(y) = p \, (p-2) \, |y|^{p-4} \, (y \cdot a) \, (y \cdot b) + p \, |y|^{p-2} \, b \cdot a$$
$$= p \, |y|^{p-2} \left[ (p-2) \, (\hat{y} \cdot a) \, (\hat{y} \cdot b) + b \cdot a \right].$$

So by Itô's formula

$$d \, |Y_t|^p = d \, [f(Y_t)]$$
$$= p \, |Y_t|^{p-1} \, \hat{Y}_t \cdot dY_t + \frac{p}{2} \, |Y_t|^{p-2} \left[ (p-2) \left( \hat{Y}_t \cdot dY_t \right) \left( \hat{Y}_t \cdot dY_t \right) + dY_t \cdot dY_t \right].$$

Taking expectations of this formula ($Y$ being a martingale) then gives

$$\mathbb{E} \, |Y_t|^p = \frac{p}{2} \int_0^t \mathbb{E} \left( |Y|^{p-2} \left[ (p-2) \left( \hat{Y} \cdot dY \right) \left( \hat{Y} \cdot dY \right) + dY \cdot dY \right] \right). \tag{9.3}$$

Using $dY = AdB$, we have

$$dY \cdot dY = Ae_i \cdot Ae_j dB^i dB^j = e_i \cdot A^* Ae_i dt = \mathrm{tr}(A^*A)dt = |A|^2 \, dt$$

and

$$\left( \hat{Y} \cdot dY \right)^2 = \left( \hat{Y} \cdot Ae_i \right) \left( \hat{Y} \cdot Ae_j \right) dB^i dB^j = \left( A^* \hat{Y} \cdot e_i \right) \left( A^* \hat{Y} \cdot e_i \right) dt$$
$$= \left( A^* \hat{Y} \cdot A^* \hat{Y} \right) dt = \left( AA^* \hat{Y} \cdot \hat{Y} \right) dt \leq |A|^2 \, dt.$$

Putting these results back into (9.3) implies

$$\mathbb{E} \, |Y_t|^p \leq \frac{p}{2} (p-1) \int_0^t \mathbb{E} \left( |Y_\tau|^{p-2} \, |A_\tau|^2 \right) d\tau.$$

By Doob's inequality there is a constant $C_p$ (for example $C_p = \left[ \frac{p}{p-1} \right]^p$ will work) such that

$$\mathbb{E} \left| Y_t^* \right|^p \leq C_p \mathbb{E} \, |Y_t|^p.$$

Combining the last two displayed equations implies

$$\mathbb{E} \left| Y_t^* \right|^p \leq C \int_0^t \mathbb{E} \left( |Y_\tau|^{p-2} \, |A_\tau|^2 \right) d\tau \leq C \mathbb{E} \left( |Y_t^*|^{p-2} \int_0^t |A_\tau|^2 \, d\tau \right). \tag{9.4}$$

Now applying Hölder's inequality to the result, with exponents $q = p \, (p-2)^{-1}$ and conjugate exponent $q' = p/2$ gives

$$\mathbb{E} \left| Y_t^* \right|^p \leq C \left[ \mathbb{E} \left| Y_t^* \right|^p \right]^{\frac{p-2}{p}} \left[ \mathbb{E} \left( \int_0^t |A_\tau|^2 \, d\tau \right)^{p/2} \right]^{2/p}$$

or equivalently, using $1 - (p - 2)/p = 2/p$,

$$\left(\mathbb{E}\,|Y_t^*|^p\right)^{2/p} \le C\left[\mathbb{E}\left(\int_0^t |A_\tau|^2\, d\tau\right)^{p/2}\right]^{2/p}.$$

Taking the $2/p$ roots of this equation then shows

$$\mathbb{E}\,|Y_t^*|^p \le C\mathbb{E}\left(\int_0^t |A_\tau|^2\, d\tau\right)^{p/2}. \tag{9.5}$$

The general case now follows, since when $Y$ is given as in (9.1) we have

$$Y_t^* \le \left(\int_0^\cdot A_\tau dB_\tau\right)_t^* + \int_0^t |\alpha_\tau|\, d\tau$$

so that

$$\begin{aligned}
\|Y_t^*\|_p &\le \left\|\left(\int_0^\cdot A_\tau dB_\tau\right)_t^*\right\|_p + \left\|\int_0^t |\alpha_\tau|\, d\tau\right\|_p \\
&\le C\left[\mathbb{E}\left(\int_0^t |A_\tau|^2\, d\tau\right)^{p/2}\right]^{1/p} + \left[\mathbb{E}\left(\int_0^t |\alpha_\tau|\, d\tau\right)^p\right]^{1/p}
\end{aligned}$$

and taking the $p^{\text{th}}$-power of this equation proves (9.2).     $\square$

**Remark 9.3.** A slightly different application of Hölder's inequality to the right side of (9.4) gives

$$\begin{aligned}
\mathbb{E}\,|Y_t^*|^p &\le C\left(\int_0^t \mathbb{E}\left[|Y_t^*|^{p-2}|A_\tau|^2\right] d\tau\right) \le C\left(\int_0^t \left[\mathbb{E}\,|Y_t^*|^p\right]^{\frac{p-2}{p}}\left[\mathbb{E}\,|A_\tau|^p\right]^{2/p} d\tau\right) \\
&= \left[\mathbb{E}\,|Y_t^*|^p\right]^{\frac{p-2}{p}} C\int_0^t \left[\mathbb{E}\,|A_\tau|^p\right]^{2/p} d\tau
\end{aligned}$$

which leads to the estimate

$$\mathbb{E}\,|Y_t^*|^p \le C\left(\int_0^t \left[\mathbb{E}\,|A_\tau|^p\right]^{2/p} d\tau\right)^{p/2}.$$

Here are some applications of Proposition 9.2.

**Proposition 9.4.** Let $\{X_i\}_{i=0}^n$ be a collection of smooth vector fields on $\mathbb{R}^N$ for which $D^k X_i$ is bounded for all $k \ge 1$ and suppose $\Sigma_t$ denotes the solution to (5.1) with $\Sigma_0 = x \in M := \mathbb{R}^N$ and $\beta = B$. Then for all $T < \infty$ and $p \in [2, \infty)$,

$$\mathbb{E}\left(\Sigma_T^*\right)^p := \mathbb{E}\left[\sup_{t \le T} |\Sigma_t|^p\right] < \infty. \tag{9.6}$$

*Proof.* Since

$$X_i(\Sigma_t)\delta B^i(t) = X_i(\Sigma_t)dB^i(t) + \frac{1}{2}d\,[X_i(\Sigma_t)] \cdot dB^i(t)$$

$$= X_i(\Sigma_t)dB^i(t) + \frac{1}{2}\left(\partial_{X_i(\Sigma_t)}X_i\right)(\Sigma_t)dt,$$

the Itô form of (5.1) is

$$\delta\Sigma_t = \left[X_0(\Sigma_t) + \frac{1}{2}\left(\partial_{X_i(\Sigma_t)}X_i\right)(\Sigma_t)\right]dt + X_i(\Sigma_t)dB^i(t) \text{ with } \Sigma_0 = x,$$

or equivalently,

$$\Sigma_t = x + \int_0^t X_i(\Sigma_\tau)dB^i_\tau + \int_0^t\left[X_0(\Sigma_\tau) + \frac{1}{2}\left(\partial_{X_i(\Sigma_\tau)}X_i\right)(\Sigma_\tau)\right]d\tau.$$

By Proposition 9.2,

$$\mathbb{E}\,|\Sigma_t|^p \le \mathbb{E}\left(\Sigma_t^*\right)^p \le C_p\,|x|^p + C_p\mathbb{E}\left(\int_0^t |\mathbf{X}(\Sigma_\tau)|^2\,d\tau\right)^{p/2}$$

$$+ C_p\mathbb{E}\left(\int_0^t\left|X_0(\Sigma_\tau) + \frac{1}{2}\left(\partial_{X_i(\Sigma_\tau)}X_i\right)(\Sigma_\tau)\right|d\tau\right)^p. \tag{9.7}$$

Using the bounds on the derivatives of $X$ we learn

$$|\mathbf{X}(\Sigma_\tau)|^2 \le C\left(1 + |\Sigma_\tau|^2\right) \text{ and }$$

$$\left|X_0(\Sigma_\tau) + \frac{1}{2}\left(\partial_{X_i(\Sigma_\tau)}X_i\right)(\Sigma_\tau)\right| \le C\left(1 + |\Sigma_\tau|\right)$$

which combined with (9.7) gives the estimate

$$\mathbb{E}\,|\Sigma_t|^p \le \mathbb{E}\left(\Sigma_t^*\right)^p$$

$$\le C_p\,|x|^p + C_p\mathbb{E}\left(\int_0^t C\left(1 + |\Sigma_\tau|^2\right)d\tau\right)^{p/2} + C_p\mathbb{E}\left(\int_0^t C\left(1 + |\Sigma_\tau|\right)d\tau\right)^p.$$

Now assuming $t \le T < \infty$, we have by Jensen's (or Hölder's) inequality that

$$\mathbb{E}\,|\Sigma_t|^p \le \mathbb{E}\left(\Sigma_t^*\right)^p$$

$$\le C\,|x|^p + Ct^{p/2}\mathbb{E}\int_0^t\left(1 + |\Sigma_\tau|^2\right)^{p/2}\frac{d\tau}{t}$$

$$+ Ct^p\mathbb{E}\int_0^t\left(1 + |\Sigma_\tau|\right)^p\frac{d\tau}{t}$$

$$\le C\,|x|^p + CT^{(p/2-1)}\mathbb{E}\int_0^t\left(1 + |\Sigma_\tau|^2\right)^{p/2}d\tau$$

$$+ CT^{(p-1)}\mathbb{E}\int_0^t\left(1 + |\Sigma_\tau|\right)^p\,d\tau$$

from which it follows that

$$\mathbb{E}\,|\Sigma_t|^p \le \mathbb{E}\left(\Sigma_t^*\right)^p \le C\,|x|^p + C(T)\int_0^t \left(1 + \mathbb{E}\,|\Sigma_\tau|^p\right) d\tau. \tag{9.8}$$

An application of Gronwall's inequality now shows $\sup_{t \le T} \mathbb{E}\,|\Sigma_t|^p < \infty$ for all $p < \infty$ and feeding this back into (9.8) with $t = T$ proves (9.6).     $\square$

**Proposition 9.5.** *Suppose* $\{X_i\}_{i=0}^n$ *is a collection of smooth vector fields on* $M$, $\Sigma_t$ *solves (5.1) with* $\Sigma_0 = o \in M$ *and* $\beta = B$, $z_t$ *is the solution to (5.59) (i.e.,* $z_t := //_t^{-1} T_{t*o}^B$*) and further assume*[9] *there is a constant* $K < \infty$ *such that* $\|A(m)\|_{op} \le K < \infty$ *for all* $m \in M$, *where* $A(m) \in \mathrm{End}\,(T_m M)$ *is defined by*

$$A(m)\,v := \frac{1}{2}\left[\nabla_v\left(\sum_{i=1}^n \nabla_{X_i} X_i + X_0\right) - \sum_{i=1}^n R^\nabla\,(v, X_i\,(m))\,X_i\,(m)\right]$$

*and*

$$\sum_{i=1}^n |\nabla_v X_i| \le K\,|v| \text{ for all } v \in TM.$$

*Then for all* $p < \infty$ *and* $T < \infty$,

$$\mathbb{E}\left[\sup_{t \le T} |z_t|^p\right] < \infty \tag{9.9}$$

*and*

$$\mathbb{E}\left[(z. - I)_t^{*p}\right] = O\left(t^{p/2}\right) \text{ as } t \downarrow 0. \tag{9.10}$$

*Proof.* In what follows $C$ will denote a constant depending on $K$, $T$ and $p$. From Theorem 5.43, we know that the integrated Itô form of (5.59) is

$$z_t = I_{T_o M} + \int_0^t //_\tau^{-1}\left(\nabla_{//_\tau z_\tau(\cdot)}\mathbf{X}\right) dB_\tau + \frac{1}{2} A_{//_t} z_t v\,d\tau \tag{9.11}$$

where $A_{//_t} := //_t^{-1} A(\Sigma_t) //_t$. By Proposition 9.2 and the assumed bounds on $A$ and $\nabla.\mathbf{X}$,

$$\mathbb{E}\left(z_t^*\right)^p \le C\,|I|^p + C\mathbb{E}\left(\int_0^t \sum_{i=1}^n \left|//_\tau^{-1}\left(\nabla_{//_\tau z_\tau(\cdot)}X_i\right)\right|^2 d\tau\right)^{p/2}$$

$$+ C\mathbb{E}\left(\int_0^t |A_{//_\tau} z_\tau| d\tau\right)^p$$

$$\le C + C\mathbb{E}\left(\int_0^t |z_\tau|^2 d\tau\right)^{p/2} + C\mathbb{E}\left(\int_0^t |z_\tau| d\tau\right)^p$$

$$\le C + C\int_0^t \mathbb{E}\,|z_\tau|^p\,d\tau$$

---

[9] This will always be true when $M$ is compact.

and

$$\mathbb{E}\left[(z. - I)_t^{*p}\right] \le C\mathbb{E}\left(\int_0^t |z_\tau|^2 \, d\tau\right)^{p/2} + C\mathbb{E}\left(\int_0^t |z_\tau| \, d\tau\right)^p$$

$$\le C \cdot \mathbb{E}\left|z_t^*\right|^p \cdot \left(t^{p/2} + t^p\right) \tag{9.12}$$

where we have made use of Hölder's (or Jensen's) inequality. Since

$$\mathbb{E}|z_t|^p \le \mathbb{E}\left(z_t^*\right)^p \le C + C\int_0^t \mathbb{E}|z_\tau|^p \, d\tau, \tag{9.13}$$

Gronwall's inequality implies

$$\sup_{t \le T} \mathbb{E}\left[|z_t|^p\right] \le Ce^{CT} < \infty.$$

Feeding the last inequality back into (9.13) shows (9.9). (9.10) now follows from (9.9), and (9.12).                    □

**Exercise 9.6.** Show under the same hypothesis of Proposition 9.5 that

$$\mathbb{E}\left[\sup_{t \le T}\left|z_t^{-1}\right|^p\right] < \infty$$

for all $p$, $T < \infty$. *Hint:* Show $z_t^{-1}$ satisfies an equation similar to (9.11) with coefficients satisfying the same type of bounds.

### 9.2 Martingale etimates

This section follows the presentation in Norris [147].

**Lemma 9.7 (Reflection Principle).** *Let $\beta_t$ be a 1-dimensional Brownian motion starting at 0, $a > 0$ and $T_a = \inf\{t > 0 : \beta_t = a\}$ be the first time $\beta_t$ hits height $a$ (see Figure 15). Then*

$$P(T_a < t) = 2P(\beta_t > a) = \frac{2}{\sqrt{2\pi t}}\int_a^\infty e^{-x^2/2t} \, dx$$

*Proof.* Since $P(\beta_t = a) = 0$,

$$P(T_a < t) = P(T_a < t \ \& \ \beta_t > a) + P(T_a < t \ \& \ \beta_t < a)$$
$$= P(\beta_t > a) + P(T_a < t \ \& \ \beta_t < a),$$

it suffices to prove

$$P(T_a < t \ \& \ \beta_t < a) = P(\beta_t > a).$$

To do this define a new process $\tilde{\beta}_t$ by

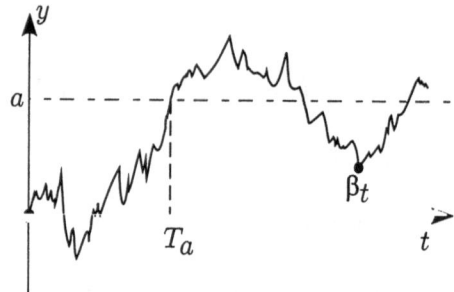

**Figure 15.** The first hitting time $T_a$ of level $a$ by $\beta_t$.

$$\tilde{\beta}_t = \begin{cases} \beta_t & \text{for } t < T_a \\ 2a - \beta_t & \text{for } t \geq T_a \end{cases}$$

(see Figure 16) and notice that $\tilde{\beta}_t$ may also be expressed as

$$\tilde{\beta}_t = \beta_{t \wedge T_a} - 1_{t \geq T_a}(\beta_t - \beta_{t \wedge T_a}) = \int_0^t \left(1_{\tau < T_a} - 1_{\tau \geq T_a}\right) d\beta_\tau. \qquad (9.14)$$

So $\tilde{\beta}_t = \beta_t$ for $t \leq T_a$ and $\tilde{\beta}_t$ is $\beta_t$ reflected across the line $y = a$ for $t \geq T_a$.

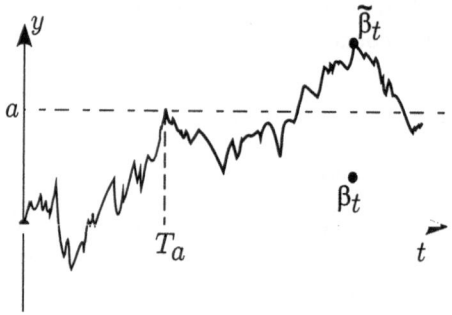

**Figure 16.** The Brownian motion $\beta_t$ and its reflection $\tilde{\beta}_t$ about the line $y = a$. Note that after time $T_a$, the labellings of the $\beta_t$ and the $\tilde{\beta}_t$ could be interchanged and the picture would still be possible. This should help alleviate the readers fears that Brownian motion has some funny asymmetry after the first hitting of level $a$.

From (9.14) it follows that $\tilde{\beta}_t$ is a martingale and

$$\left(d\tilde{\beta}_t\right)^2 = \left(1_{\tau < T_a} - 1_{\tau \geq T_a}\right)^2 dt = dt$$

and hence that $\tilde{\beta}_t$ is another Brownian motion. Since $\tilde{\beta}_t$ hits level $a$ for the first time exactly when $\beta_t$ hits level $a$,

$$T_a = \tilde{T}_a := \inf\left\{t > 0 : \tilde{\beta}_t = a\right\}$$

and $\left\{\tilde{T}_a < t\right\} = \{T_a < t\}$. Furthermore (see Figure 16),

$$\{T_a < t \,\&\, \beta_t < a\} = \left\{\tilde{T}_a < t \,\&\, \tilde{\beta}_t > a\right\} = \left\{\tilde{\beta}_t > a\right\}.$$

Therefore,

$$P(T_a < t \,\&\, \beta_t < a) = P(\tilde{\beta}_t > a) = P(\beta_t > a)$$

which completes the proof.  □

**Remark 9.8.** An alternate way to get a handle on the stopping time $T_a$ is to compute its Laplace transform. This can be done by considering the martingale

$$M_t := e^{\lambda\beta_t - \frac{1}{2}\lambda^2 t}.$$

Since $M_t$ is bounded by $e^{\lambda a}$ for $t \in [0, T_a]$, the optional sampling theorem may be applied to show

$$e^{\lambda a} E\left[e^{-\frac{1}{2}\lambda^2 T_a}\right] = E\left[e^{\lambda a - \frac{1}{2}\lambda^2 T_a}\right] = E M_{T_a} = E M_0 = 1,$$

i.e., this implies that $E\left[e^{-\frac{1}{2}\lambda^2 T_a}\right] = e^{-\lambda a}$. This is equivalent to

$$E\left[e^{-\lambda T_a}\right] = e^{-a\sqrt{2\lambda}}.$$

From this point of view one would now have to invert the Laplace transform to get the density of the law of $T_a$.

**Corollary 9.9.** *Suppose now that* $T = \inf\{t > 0 : |\beta_t| = a\}$, *i.e., the first time* $\beta_t$ *leaves the strip* $(-a, a)$. *Then*

$$P(T < t) \le 4P(\beta_t > a) = \frac{4}{\sqrt{2\pi t}} \int_a^\infty e^{-x^2/2t} dx$$

$$\le \min\left(\sqrt{\frac{8t}{\pi a^2}} e^{-a^2/2t}, 1\right). \qquad (9.15)$$

*Notice that* $P(T < t) = P(\beta_t^* \ge a)$ *where* $\beta_t^* = \max\{|\beta_\tau| : \tau \le t\}$. *So* (9.15) *may be rewritten as*

$$P(\beta_t^* \ge a) \le 4P(\beta_t > a) \le \min\left(\sqrt{\frac{8t}{\pi a^2}} e^{-a^2/2t}, 1\right) \le 2e^{-a^2/2t}. \qquad (9.16)$$

*Proof.* By definition $T = T_a \wedge T_{-a}$ so that $\{T < t\} = \{T_a < t\} \cup \{T_{-a} < t\}$ and therefore

$$P(T < t) \leq P(T_a < t) + P(T_{-a} < t)$$

$$= 2P(T_a < t) = 4P(\beta_t > a) = \frac{4}{\sqrt{2\pi t}} \int_a^\infty e^{-x^2/2t} dx$$

$$\leq \frac{4}{\sqrt{2\pi t}} \int_a^\infty \frac{x}{a} e^{-x^2/2t} dx = \frac{4}{\sqrt{2\pi t}} \left( -\frac{t}{a} e^{-x^2/2t} \right) \Big|_a^\infty = \sqrt{\frac{8t}{\pi a^2}} e^{-a^2/2t}.$$

This proves everything but the very last inequality in (9.16). To prove this inequality first observe the elementary calculus inequality:

$$\min \left( \frac{4}{\sqrt{2\pi} y} e^{-y^2/2}, 1 \right) \leq 2e^{-y^2/2}. \tag{9.17}$$

Indeed (9.17) holds $\frac{4}{\sqrt{2\pi} y} \leq 2$, i.e., if $y \geq y_0 := 2/\sqrt{2\pi}$. The fact that (9.17) holds for $y \leq y_0$ follows from the following trivial inequality

$$1 \leq 1.4552 \cong 2e^{-\frac{1}{\pi}} = e^{-y_0^2/2}.$$

Finally letting $y = a/\sqrt{t}$ in (9.17) gives the last inequality in (9.16). $\qquad\square$

**Theorem 9.10.** *Let $N$ be a continuous martingale such that $N_0 = 0$ and $T$ be a stopping time. Then for all $\varepsilon, \delta > 0$,*

$$P\left( \langle N \rangle_T < \varepsilon \ \& \ N_T^* \geq \delta \right) \leq P(\beta_\varepsilon^* \geq \delta) \leq 2e^{-\delta^2/2\varepsilon}.$$

*Proof.* By the Dambis, Dubins & Schwarz's theorem (see p.174 of [109]) we may write $N_t = \beta_{\langle N \rangle_t}$ where $\beta$ is a Brownian motion (on a possibly "augmented" probability space). Therefore

$$\left\{ \langle N \rangle_T < \varepsilon \ \& \ N_T^* \geq \delta \right\} \subset \left\{ \beta_\varepsilon^* \geq \delta \right\}$$

and hence from (9.16),

$$P\left( \langle N \rangle_T < \varepsilon \ \& \ N_T^* \geq \delta \right) \leq P(\beta_\varepsilon^* \geq \delta) \leq 2e^{-\delta^2/2\varepsilon}. \qquad\square$$

**Theorem 9.11.** *Suppose that $Y_t = M_t + A_t$ where $M_t$ is a martingale and $A_t$ is a process of bounded variation which satisfy: $M_0 = A_0 = 0$, $|A|_t \leq ct$ and $\langle M \rangle_t \leq ct$ for some constant $c < \infty$. If $T_a := \inf \{t > 0 : |Y_t| = a\}$ and $t < a/2c$, then*

$$P(Y_t^* \geq a) = P(T_a \leq t) \leq \frac{4}{\sqrt{\pi a}} \exp \left( -\frac{a^2}{8ct} \right)$$

*Proof.* Since

$$Y_t^* \leq M_t^* + A_t^* \leq M_t^* + |A|_t \leq M_t^* + ct$$

it follows that

$$\left\{ Y_t^* \geq a \right\} \subset \left\{ M_t^* \geq a/2 \right\} \cup \{ct \geq a/2\} = \left\{ M_t^* \geq a/2 \right\}$$

when $t < a/2c$. Again by the Dambis, Dubins and Schwarz's theorem (see p.174 of [109]), we may write $M_t = \beta_{\langle M \rangle_t}$ where $\beta$ is a Brownian motion on a possibly augmented probability space. Since

$$M_t^* = \max_{\tau \le \langle M \rangle_t} |\beta_\tau| \le \max_{\tau \le ct} |\beta_\tau| = \beta_{ct}^*$$

we learn

$$P(Y_t^* \ge a) \le P\left(M_t^* \ge a/2\right) \le P\left(\beta_{ct}^* \ge a/2\right)$$

$$\le \sqrt{\frac{8ct}{\pi (a/2)^2}} e^{-(a/2)^2/2ct} = \sqrt{\frac{8ct}{\pi (a/2)^2}} e^{-(a/2)^2/2ct}$$

$$\le \sqrt{\frac{8c (a/2c)}{\pi (a/2)^2}} e^{-(a/2)^2/2ct} = \frac{4}{\sqrt{\pi a}} \exp\left(-\frac{a^2}{8ct}\right)$$

wherein the last inequality we have used the restriction $t < a/2c$. $\qquad\square$

**Lemma 9.12.** *If $f : [0, \infty) \to \mathbb{R}$ is a locally absolutely continuous function such that $f(0) = 0$, then*

$$|f(t)| \le \sqrt{2 \|\dot{f}\|_{L^\infty([0,t])} \|f\|_{L^1([0,t])}} \ \forall \, t \ge 0.$$

*Proof.* By the fundamental theorem of calculus,

$$f^2(t) = 2 \int_0^t f(\tau) \dot{f}(\tau) d\tau \le 2 \|\dot{f}\|_{L^\infty([0,t])} \|f\|_{L^1([0,t])} . \qquad\square$$

We are now ready for a key result needed in the probabilistic proof of Hörmander's theorem. Loosely speaking it states that if $Y$ is a Brownian semi-martingale, then it can happen *only* with small probability that the $L^2$-norm of $Y$ is small while the quadratic variation of $Y$ is relatively large.

**Proposition 9.13 (A key martingale inequality).** *Let $T$ be a stopping time bounded by $t_0 < \infty$, $Y = y + M + A$ where $M$ is a continuous martingale and $A$ is a process of bounded variation such that $M_0 = A_0 = 0$. Further assume, on the set $\{t \le T\}$, that $\langle M \rangle_t$ and $|A|_t$ are absolutely continuous functions and there exists finite positive constants, $c_1$ and $c_2$, such that*

$$\frac{d\langle M \rangle_t}{dt} \le c_1 \ and \ \frac{d|A|_t}{dt} \le c_2.$$

*Then for all $v > 0$ and $q > v + 4$ there exists constants $c = c(t_0, q, v, c_1, c_2) > 0$ and $\varepsilon_0 = \varepsilon_0(t_0, q, v, c_1, c_2) > 0$ such that*

$$P\left(\int_0^T Y_t^2 dt < \varepsilon^q, \ \langle Y \rangle_T = \langle M \rangle_T \ge \varepsilon\right) \le 2 \exp\left(-\frac{1}{2c_1 \varepsilon^v}\right) = O\left(\varepsilon^{-\infty}\right) \quad (9.18)$$

*for all $\varepsilon \in (0, \varepsilon_0]$.*

*Proof.* Let $q_0 = \frac{q-v}{2}$ (so that $q_0 \in (2, q/2)$), $N := \int_0^\cdot Y \, dM$ and

$$C_\varepsilon := \left\{ \langle N \rangle_T \leq c_1 \varepsilon^q, \ N_T^* \geq \varepsilon^{q_0} \right\}. \tag{9.19}$$

We will show shortly that for $\varepsilon$ sufficiently small,

$$B_\varepsilon := \left\{ \int_0^T Y_t^2 \, dt < \varepsilon^q, \ \langle Y \rangle_T \geq \varepsilon \right\} \subset C_\varepsilon. \tag{9.20}$$

By an application of Theorem 9.10,

$$P(C_\varepsilon) \leq 2 \exp\left( -\frac{\varepsilon^{2q_0}}{2c_1 \varepsilon^q} \right) = 2 \exp\left( -\frac{1}{2c_1 \varepsilon^v} \right)$$

and so assuming the validity of (9.20),

$$P\left( \int_0^T Y_t^2 \, dt < \varepsilon^q, \ \langle Y \rangle_T \geq \varepsilon \right) \leq P(C_\varepsilon) \leq 2 \exp\left( -\frac{1}{2c_1 \varepsilon^v} \right) \tag{9.21}$$

which proves (9.18). So to finish the proof it only remains to verify (9.20) which will be done by showing $B_\varepsilon \cap C_\varepsilon^c = \emptyset$.

For the rest of the proof, it will be assumed that we are on the set $B_\varepsilon \cap C_\varepsilon^c$. Since $\langle N \rangle_T = \int_0^T |Y_t|^2 \, d\langle M \rangle_t$, we have

$$B_\varepsilon \cap C_\varepsilon^c = \left\{ \int_0^T Y_t^2 \, dt < \varepsilon^q, \ \langle Y \rangle_T \geq \varepsilon, \ \int_0^T |Y_t|^2 \, d\langle M \rangle_t > c_1 \varepsilon^q, \ N_T^* < \varepsilon^{q_0} \right\}. \tag{9.22}$$

From Lemma 9.12 with $f(t) = \langle Y \rangle_t$ and the assumption that $d\langle Y \rangle_t / dt \leq c_1$,

$$\langle Y \rangle_T \leq \sqrt{2 \|\dot{f}\|_{L^\infty([0,T])} \|f\|_{L^1([0,T])}} \leq \sqrt{2c_1 \int_0^T \langle Y \rangle_t \, dt}. \tag{9.23}$$

By Itô's formula, the quadratic variation, $\langle Y \rangle_t$, of $Y$ satisfies

$$\langle Y \rangle_t = Y_t^2 - y^2 - 2 \int_0^t Y \, dY \leq Y_t^2 + 2 \left| \int_0^t Y \, dY \right| \tag{9.24}$$

and on the set $\{t \leq T\} \cap B_\varepsilon \cap C_\varepsilon^c$,

$$\left| \int_0^t Y \, dY \right| = \left| \int_0^t Y \, dM + \int_0^t Y \, dA \right| \leq |N_t| + \int_0^t |Y| \, dA$$

$$\leq N_T^* + c_2 \int_0^T |Y_\tau| \, d\tau \leq \varepsilon^{q_0} + c_2 T^{1/2} \sqrt{\int_0^T Y_\tau^2 \, d\tau}$$

$$\leq \varepsilon^{q_0} + c_2 t_0^{1/2} \varepsilon^q. \tag{9.25}$$

Combining Eqs. (9.24) and (9.25) shows, on the set $\{t \leq T\} \cap B_\varepsilon \cap C_\varepsilon^c$ that

$$\langle Y \rangle_t \le Y_t^2 + 2 \left[ \varepsilon^{q_0} + c_2 t_0^{1/2} \varepsilon^q \right]$$

and using this in (9.23) implies

$$\langle Y \rangle_T \le \sqrt{2c_1 \int_0^T \left( Y_t^2 + 2 \left[ \varepsilon^{q_0} + c_2 t_0^{1/2} \varepsilon^q \right] \right) dt}$$

$$\le \sqrt{2c_1 \left[ \varepsilon^q + 2 \left[ \varepsilon^{q_0} + c_2 t_0^{1/2} \varepsilon^q \right] t_0 \right]} = O\left( \varepsilon^{q_0/2} \right) = o\left( \varepsilon \right). \qquad (9.26)$$

Hence we may choose $\varepsilon_0 = \varepsilon_0 \left( c_1, c_2, t_o, q, \nu \right) > 0$ such that if $\varepsilon \le \varepsilon_0$ then

$$\sqrt{2c_1 \left( \varepsilon^q + 2\varepsilon^{q_0} t_0 + 2c_2 t_0^{3/2} \varepsilon^{q/2} \right)} < \varepsilon$$

and hence on $B_\varepsilon \cap C_\varepsilon^c$ we learn $\varepsilon \le \langle Y \rangle_T < \varepsilon$ which is absurd. So we must conclude that $B_\varepsilon \cap C_\varepsilon^c = \emptyset$. □

*Acknowledgment.* It is a pleasure to thank Professor A. Sznitman and the ETH for their hospitality, support, and the opportunity to give the talks that started these notes. I also would like to thank Professor E. Bolthausen for his hospitality and his role in arranging the first lecture held at the University of Zürich. The author also gratefully acknowledges support from the NSF which, through grants DMS 9223177, DMS 96-12651, DMS 99-71036 and DMS 0202939, has partially supported the research by the author that has been described in this paper.

# References

[1] R. Abraham and J. E. Marsden, *Foundations of mechanics*, Benjamin/Cummings Publishing Co. Inc. Advanced Book Program, Reading, Mass., 1978, Second edition, revised and enlarged, With the assistance of Tudor Raţiu and Richard Cushman. MR 81e:58025

[2] S. Aida, *On the irreducibility of certain dirichlet forms on loop spaces over compact homogeneous spaces*, New Trends in Stochastic Analysis (New Jersey) (K. D. Elworthy, S. Kusuoka, and I. Shigekawa, eds.), Proceedings of the 1994 Taniguchi Symposium, World Scientific, 1997, pp. 3–42.

[3] S. Aida and B. K. Driver, Equivalence of heat kernel measure and pinned Wiener measure on loop groups, *C. R. Acad. Sci. Paris Sér. I Math.* **331** (2000), no. 9, 709–712. MR 1 797 756

[4] S. Aida and D. Elworthy, Differential calculus on path and loop spaces. I. Logarithmic Sobolev inequalities on path spaces, *C. R. Acad. Sci. Paris Sér. I Math.* **321** (1995), no. 1, 97 102.

[5] H. Airault and P. Malliavin,Integration by parts formulas and dilatation vector fields on elliptic probability spaces, *Probab. Theory Related Fields* **106** (1996), no. 4, 447–494.

[6] S. Albeverio, R. Léandre, and M. Röckner, Construction of a rotational invariant diffusion on the free loop space, *C. R. Acad. Sci. Paris Sér. I Math.* **316** (1993), no. 3, 287–292.

[7] Y. Amit, A multiflow approximation to diffusions, *Stochastic Process. Appl.* **37** (1991), no. 2, 213–237.

[8] L. Andersson and B. K. Driver, Finite-dimensional approximations to Wiener measure and path integral formulas on manifolds, *J. Funct. Anal.* **165** (1999), no. 2, 430–498. MR 2000j:58059

[9] L. Auslander and R. E. MacKenzie, *Introduction to Differentiable Manifolds*, Dover Publications Inc., New York, 1977, Corrected reprinting. MR 57 #10717

[10] V. Bally, Approximation for the solutions of stochastic differential equations. I. $L^p$-convergence, *Stochastics Stochastics Rep.* **28** (1989), no. 3, 209–246.

[11] ———, Approximation for the solutions of stochastic differential equations. II. Strong convergence, *Stochastics Stochastics Rep.* **28** (1989), no. 4, 357–385.

[12] D.R. Bell, Degenerate stochastic differential equations and hypoellipticity, in: *Pitman Monographs and Surveys in Pure and Applied Mathematics*, vol. 79, Longman, Harlow, 1995. MR 99e:60124

[13] D.R. Bell, S. Eldin, and A. Mohammed, The Malliavin calculus and stochastic delay equations, *J. Funct. Anal.* **99** (1991), no. 1, 75–99. MR 92k:60124

[14] Ya. I. Belopolśkaya and Yu. L. Dalecky, Stochastic equations and differential geometry, *Mathematics and its Applications (Soviet Series)*, Vol. 30, Kluwer Academic Publishers Group, Dordrecht, 1990, Translated from the Russian. MR 91b:58271

[15] G. Ben Arous and R. Léandre, Décroissance exponentielle du noyau de la chaleur sur la diagonale. I, *Probab. Theory Related Fields* **90** (1991), no. 2, 175–202. MR 93b:60136a

[16] ———, Décroissance exponentielle du noyau de la chaleur sur la diagonale. II, *Probab. Theory Related Fields* **90** (1991), no. 3, 377–402. MR 93b:60136b

[17] G. Ben Arous, Développement asymptotique du noyau de la chaleur hypoelliptique sur la diagonale, *Ann. Inst. Fourier (Grenoble)* **39** (1989), no. 1, 73–99. MR 91b:58272

[18] ———, Flots et séries de Taylor stochastiques, *Probab. Theory Related Fields* **81** (1989), no. 1, 29–77. MR 90a:60106

[19] R.L. Bishop and R.J. Crittenden, *Geometry of Manifolds*, AMS Chelsea Publishing, Providence, RI, 2001, Reprint of the 1964 original. MR 2002d:53001

[20] J-M Bismut, *Mecanique aleatoire. (french) [random mechanics]*, Lecture Notes in Mathematics, Vol. 866. Springer-Verlag, Berlin, New York, 1981,

[21] ———, *Large deviations and the Malliavin calculus*, Progress in Mathematics, vol. 45, Birkhäuser Boston Inc., Cambridge, MA, 1984.

[22] G. Blum, A note on the central limit theorem for geodesic random walks, *Bull. Austral. Math. Soc.* **30** (1984), no. 2, 169–173.

[23] N. Bouleau and F. Hirsch, *Dirichlet forms and analysis on Wiener space*, de Gruyter Studies in Mathematics, vol. 14, Walter de Gruyter & Co., Berlin, 1991. MR 93e:60107

[24] T. Bröcker and T. tom Dieck, *Representations of compact Lie groups*, Graduate Texts in Mathematics, vol. 98, Springer-Verlag, NY, 1995, Translated from the German manuscript, Corrected reprint of the 1985 translation. MR 97i:22005

[25] R. H. Cameron and W. T. Martin, Transformations of Wiener integrals under translations, *Ann. of Math.* (2) **45** (1944), 386–396. MR 6,5f

[26] ———, The transformation of Wiener integrals by nonlinear transformations, *Trans. Amer. Math. Soc.* **66** (1949), 253–283. MR 11,116b

[27] ———, Non-linear integral equations, *Ann. of Math.* (2) **51** (1950), 629–642. MR 11,728d

[28] R.H. Cameron, The first variation of an indefinite Wiener integral, *Proc. Amer. Math. Soc.* **2** (1951), 914–924. MR 13,659b

[29] M. Capitaine, E. P. Hsu, and M. Ledoux, Martingale representation and a simple proof of logarithmic Sobolev inequalities on path spaces, *Electron. Comm. Probab.* **2** (1997), 71–81 (electronic).

[30] T. R. Carson, *Logarithmic sobolev inequalities for free loop groups*, University of California at San Diego Ph.D. thesis. This may be retrieved at http://math.ucsd.edu/ driver/driver/thesis.htm, 1997.

[31] _____, A logarithmic Sobolev inequality for the free loop group, *C.R. Acad. Sci. Paris Sér. I Math.* **326** (1998), no. 2, 223–228. MR 99g:60108

[32] C. Cross, *Differentials of measure-preserving flows on path space*, University of California at San Diego Ph.D. thesis. This may be retrieved at http://math.ucsd.edu/ driver/driver/thesis.htm, 1996.

[33] A. Cruzeiro, S. Fang, and P. Malliavin, A probabilistic Weitzenböck formula on Riemannian path space, *J. Anal. Math.* **80** (2000), 87–100. MR 2002d:58045

[34] A. Cruzeiro and P. Malliavin, *Riesz transforms, commutators, and stochastic integrals*, Harmonic analysis and partial differential equations (Chicago, IL, 1996), Chicago Lectures in Math., Univ. Chicago Press, Chicago, IL, 1999, 151–162. MR 2001d:60062

[35] A. Cruzeiro and S. Fang, An $L^2$ estimate for Riemannian anticipative stochastic integrals, *J. Funct. Anal.* **143** (1997), no. 2, 400–414. MR 98d:60106

[36] A. Cruzeiro and S. Fang, A Weitzenböck formula for the damped Ornstein-Uhlenbeck operator in adapted differential geometry, *C.R. Acad. Sci. Paris Sér. I Math.* **332** (2001), no. 5, 447–452. MR 2002a:60090

[37] A. Cruzeiro and P. Malliavin, Frame bundle of Riemannian path space and Ricci tensor in adapted differential geometry, *J. Funct. Anal.* **177** (2000), no. 1, 219–253. MR 2001h:60103

[38] _____, A class of anticipative tangent processes on the Wiener space, *C.R. Acad. Sci. Paris Sér. I Math.* **333** (2001), no. 4, 353–358. MR 2002g:60084

[39] _____, Stochastic calculus of variations and Harnack inequality on Riemannian path spaces, *C.R. Math. Acad. Sci. Paris* **335** (2002), no. 10, 817–820. MR 2003k:58058

[40] R. W. R. Darling, *Martingales in manifolds—definition, examples, and behaviour under maps*, Seminar on Probability, XVI, Supplement, Lecture Notes in Math., vol. 921, Springer-Verlag, Berlin, 217–236, 1982. MR 84j:58133

[41] E. B. Davies, *Heat kernels and spectral theory*, Cambridge Tracts in Mathematics, vol. 92, Cambridge University Press, Cambridge, UK, 1990. MR 92a:35035

[42] M.P. do Carmo, *Riemannian Geometry*,in: Mathematics: Theory & Applications, Birkhäuser Boston Inc., Cambridge, MA, 1992; translated from the second Portuguese edition by F. Flaherty. MR 92i:53001

[43] J. Dodziuk, Maximum principle for parabolic inequalities and the heat flow on open manifolds, *Indiana Univ. Math. J.* **32** (1983), no. 5, 703–716. MR 85e:58140

[44] H. Doss, *Connections between stochastic and ordinary integral equations*, Biological Growth and Spread (Proc. Conf., Heidelberg, 1979), Lecture Notes in Biomath., 38, pp. 443–448, Springer-Verlag, New York, Berlin, 1979.

[45] B. K. Driver, The Lie bracket of adapted vector fields on Wiener spaces, *Appl. Math. Optim.* **39** (1999), no. 2, 179–210. MR 2000b:58063

[46] B.K. Driver, Classifications of bundle connection pairs by parallel translation and lassos, *J. Funct. Anal.* **83** (1989), no. 1, 185–231.

[47] _____,A Cameron-Martin type quasi-invariance theorem for Brownian motion on a compact Riemannian manifold, *J. Funct. Anal.* **110** (1992), no. 2, 272–376.

[48] _____, A Cameron-Martin type quasi-invariance theorem for pinned Brownian motion on a compact Riemannian manifold, *Trans. Amer. Math. Soc.* **342** (1994), no. 1, 375–395.

[49] _____, *Towards calculus and geometry on path spaces*, Stochastic analysis (Ithaca, NY, 1993), Proc. Sympos. Pure Math., vol. 57, Amer. Math. Soc., Providence, RI, 1995, 405–422.

[50] _____, Integration by parts and quasi-invariance for heat kernel measures on loop groups, *J. Funct. Anal.* **149** (1997), no. 2, 470–547.

[51] _____, A correction to the paper: "Integration by parts and quasi-invariance for heat kernel measures on loop groups", *J. Funct. Anal.* **155** (1998), no. 1, 297–301. MR 99a:60054b

[52] _____, *Analysis of Wiener measure on path and loop groups*, Finite and infinite dimensional analysis in honor of Leonard Gross (New Orleans, LA, 2001), Contemp. Math., vol. 317, Amer. Math. Soc., Providence, RI, 2003, pp. 57–85. MR 2003m:58055

[53] _____, Heat kernels measures and infinite dimensional analysis, to appear in *Heat Kernels and Analysis on Manifolds, Graphs, and Metric Spaces*, Contemp. Math., vol. 338, Amer. Math. Soc., Providence, RI, 2004, 41 pages. MR 2003m:58055

[54] B.K. Driver and T. Lohrenz, Logarithmic Sobolev inequalities for pinned loop groups, *J. Funct. Anal.* **140** (1996), no. 2, 381–448.

[55] B.K. Driver and T. Melcher, Hypoelliptic heat kernel inequalities on the Heisenberg group, to appear *J. Funct. Anal.* (2004).

[56] B.K. Driver and M. Röckner, Construction of diffusions on path and loop spaces of compact Riemannian manifolds, *C.R. Acad. Sci. Paris Sér. I Math.* **315** (1992), no. 5, 603–608.

[57] B. K. Driver and V. K. Srimurthy, Absolute continuity of heat kernel measure with pinned wiener measure on loop groups, *Ann. Probab.* **29** (2001), no. 2, 691–723.

[58] B.K. Driver and A. Thalmaier, Heat equation derivative formulas for vector bundles, *J. Funct. Anal.* **183** (2001), no. 1, 42–108. MR 1 837 533

[59] A. Eberle, Local Poincaré inequalities on loop spaces, *C.R. Acad. Sci. Paris Sér. I Math.* **333** (2001), no. 11, 1023–1028. MR 2003c:58025

[60] _____, Absence of spectral gaps on a class of loop spaces, *J. Math. Pures Appl.* (9) **81** (2002), no. 10, 915–955. MR 2003k:58059

[61] _____, Local spectral gaps on loop spaces, *J. Math. Pures Appl.* (9) **82** (2003), no. 3, 313–365. MR 1 993 285

[62] _____, Spectral gaps on discretized loop spaces, *Infin. Dimens. Anal. Quantum Probab. Relat. Top.* **6** (2003), no. 2, 265–300. MR 1 991 495

[63] J. Eells and K. D. Elworthy, *Wiener integration on certain manifolds*, Problems in nonlinear analysis (C.I.M.E., IV Ciclo, Varenna, 1970), Edizioni Cremonese, 1971, 67–94.

[64] J. Eells, *Integration on Banach manifolds*, Proceedings of the Thirteenth Biennial Seminar of the Canadian Mathematical Congress (Dalhousie Univ., Halifax, N.S., 1971), Vol. 1, Canad. Math. Congr., Montreal, Que., 1972, 41–49. MR 51 #9112

[65] D. Elworthy, *Geometric aspects of diffusions on manifolds*, École d'Été de Probabilités de Saint-Flour XV–XVII, 1985–87, Lecture Notes in Math., vol. 1362, Springer-Verlag, Berlin, 1988, 277–425. MR 90c:58187

[66] K. D. Elworthy, *Gaussian measures on Banach spaces and manifolds*, Global analysis and its applications (Lectures, Internat. Sem. Course, Internat. Centre Theoret. Phys., Trieste, 1972), Vol. II, Internat. Atomic Energy Agency, Vienna, 1974, 151–166.

[67] _____, *Measures on infinite-dimensional manifolds*, Functional integration and its applications (Proc. Internat. Conf., London, 1974), Clarendon Press, Oxford, 1975, pp. 60–68.

[68] _____, *Stochastic dynamical systems and their flows*, Stochastic Analysis (Proc. Internat. Conf., Northwestern Univ., Evanston, Ill., 1978), Academic Press, London, New York, 1978, pp. 79–95.

[69] _____, *Stochastic differential equations on manifolds*, London Mathematical Society Lecture Note Series, vol. 70, Cambridge University Press, Cambridge, 1982. MR 84d:58080

[70] K. D. Elworthy, Y. Le Jan, and X.-M. Li, Integration by parts formulae for degenerate diffusion measures on path spaces and diffeomorphism groups, *C.R. Acad. Sci. Paris Sér. I Math.* **323** (1996), no. 8, 921–926.

[71] K. D. Elworthy, Y. Le Jan, and X.-M. Li, *On the geometry of diffusion operators and stochastic flows*, Lecture Notes in Mathematics, vol. 1720, Springer-Verlag, Berlin, 1999. MR 2001f:58072

[72] K. D. Elworthy and X.-M. Li, Formulae for the derivatives of heat semigroups, *J. Funct. Anal.* **125** (1994), no. 1, 252–286.

[73] K. D. Elworthy and X.-M. Li, *A class of integration by parts formulae in stochastic analysis. I*, Itô's stochastic calculus and probability theory, Springer, Tokyo, 1996, pp. 15–30.

[74] M. Emery, *Stochastic Calculus in Manifolds*, (with an appendix by P.-A. Meyer), Universitext, Springer-Verlag, Berlin, 1989.

[75] O. Enchev and D. Stroock, Integration by parts for pinned Brownian motion, *Math. Res. Lett.* **2** (1995), no. 2, 161–169.

[76] O. Enchev and D. W. Stroock, Towards a Riemannian geometry on the path space over a Riemannian manifold, *J. Funct. Anal.* **134** (1995), no. 2, 392–416.

[77] S. Fang, Stochastic anticipative calculus on the path space over a compact Riemannian manifold, *J. Math. Pures Appl.* (9) **77** (1998), no. 3, 249–282. MR 99i:60110

[78] S. Fang and P. Malliavin, Stochastic analysis on the path space of a Riemannian manifold. I. Markovian stochastic calculus, *J. Funct. Anal.* **118** (1993), no. 1, 249–274.

[79] S. Fang, Inégalité du type de Poincaré sur l'espace des chemins riemanniens, *C.R. Acad. Sci. Paris Sér. I Math.* **318** (1994), no. 3, 257–260.

[80] S. Fang, Rotations et quasi-invariance sur l'espace des chemins, *Potential Anal.* **4** (1995), no. 1, 67–77. MR 96d:60080

[81] _____, Stochastic anticipative integrals on a Riemannian manifold, *J. Funct. Anal.* **131** (1995), no. 1, 228–253. MR 96i:58178

[82] _____, Integration by parts for heat measures over loop groups, *J. Math. Pures Appl.* (9) **78** (1999), no. 9, 877–894. MR 1 725 745

[83] _____, Integration by parts formula and logarithmic Sobolev inequality on the path space over loop groups, *Ann. Probab.* **27** (1999), no. 2, 664–683. MR 1 698 951

[84] _____, Ricci tensors on some infinite-dimensional Lie algebras, *J. Funct. Anal.* **161** (1999), no. 1, 132–151. MR 2000f:58013

[85] I. B. Frenkel, Orbital theory for affine Lie algebras, *Invent. Math.* **77** (1984), no. 2, 301–352. MR 86d:17014

[86] S. Gallot, D. Hulin, and J. Lafontaine, *Riemannian Geometry*, Second ed., Universitext, Springer-Verlag, Berlin, 1990. MR 91j:53001

[87] R. Gangolli, On the construction of certain diffusions on a differenitiable manifold, *Z. Wahrscheinlichkeitstheorie und Verw. Gebiete* **2** (1964), 406–419.

[88] I. V. Girsanov, On transforming a class of stochastic processes by absolutely continuous substitution of measures, *Teor. Verojatnost. i Primenen.* **5** (1960), 314–330. MR 24 #A2986

[89] F. Gong, M. Röckner, and L. Wu, Poincaré inequality for weighted first order Sobolev spaces on loop spaces, *J. Funct. Anal.* **185** (2001), no. 2, 527–563. MR 2002j:47074

[90] L. Gross, *Abstract Wiener spaces*, Proc. Fifth Berkeley Sympos. Math. Statist. and Probability (Berkeley, Calif., 1965/66), Vol. II: Contributions to Probability Theory, Part 1, Univ. California Press, Berkeley, CA., 1967, 31–42. MR 35 #3027

[91] _____, Potential theory on Hilbert space, *J. Functional Analysis* **1** (1967), 123–181. MR 37 #3331

[92] _____, Logarithmic Sobolev inequalities, *Amer. J. Math.* **97** (1975), no. 4, 1061–1083. MR 54 #8263

[93] _____, Logarithmic Sobolev inequalities on loop groups, *J. Funct. Anal.* **102** (1991), no. 2, 268–313. MR 93b:22037

[94] S. J. Guo, On the mollifier approximation for solutions of stochastic differential equations, *J. Math. Kyoto Univ.* **22**, no. 2 (1982), 243–254.

[95] N. J. Hicks, *Notes on differential geometry*, Van Nostrand Mathematical Studies, No. 3, D. Van Nostrand Co., Inc., Princeton, NJ, Toronto, London, 1965. MR 31 #3936

[96] E. P. Hsu, *Flows and quasi-invariance of the Wiener measure on path spaces*, Stochastic analysis (Ithaca, NY, 1993), Proc. Sympos. Pure Math., vol. 57, Amer. Math. Soc., Providence, RI, 1995, pp. 265–279.

[97] _____, Quasi-invariance of the Wiener measure on the path space over a compact Riemannian manifold, *J. Funct. Anal.* **134** (1995), no. 2, 417–450.

[98] E. P. Hsu, Inégalités de Sobolev logarithmiques sur un espace de chemins, *C.R. Acad. Sci. Paris Sér. I Math.* **320** (1995), no. 8, 1009–1012.

[99] _____, Estimates of derivatives of the heat kernel on a compact Riemannian manifold, *Proc. Amer. Math. Soc.* **127** (1999), no. 12, 3739–3744. MR 2000c:58047

[100] _____, Quasi-invariance of the Wiener measure on path spaces: noncompact case, *J. Funct. Anal.* **193** (2002), no. 2, 278–290. MR 2003i:58069

[101] _____, *Stochastic analysis on manifolds*, Graduate Studies in Mathematics, vol. 38, American Mathematical Society, Providence, RI, 2002. MR 2003c:58026

[102] Y. Hu, A. S. Üstünel, and M. Zakai, Tangent processes on Wiener space, *J. Funct. Anal.* **192** (2002), no. 1, 234–270. MR 2003e:60117

[103] N. Ikeda and S. Watanabe, *Stochastic Differential Equations and Diffusion Processes*, North Holland, Amsterdam, 1981.

[104] N. Ikeda and S. Watanabe, *Stochastic Differential Equations and Diffusion Processes*, Second ed., North-Holland Mathematical Library, vol. 24, North-Holland Publishing Co., Amsterdam, 1989. MR 90m:60069

[105] P. Imkeller, *Enlargement of the Wiener filtration by a manifold valued random element via Malliavin's calculus*, Statistics and control of stochastic processes (Moscow, 1995/1996), World Sci. Publishing, River Edge, NJ, 1997, 157–171. MR 99h:60112

[106] Y. Inahama, Logarithmic Sobolev inequality on free loop groups for heat kernel measures associated with the general Sobolev spaces, *J. Funct. Anal.* **179** (2001), no. 1, 170–213. MR 2001k:60077

[107] E. Jørgensen, The central limit problem for geodesic random walks, *Z. Wahrscheinlichkeitstheorie und Verw. Gebiete* **32** (1975), 1–64.

[108] H. Kaneko and S. Nakao, *A note on approximation for stochastic differential equations*, Séminaire de Probabilités, XXII, Lecture Notes in Math., Vol. 1321, Springer-Verlag, Berlin, 1988, 155–162.

[109] I. Karatzas and S. E. Shreve, *Brownian Motion and Stochastic Calculus*, 2nd Ed., Graduate Texts in Mathematics, No. 113, Springer-Verlag, Berlin, 1991.

[110] W. S. Kendall, Stochastic differential geometry: an introduction, *Acta Appl. Math.* **9** (1987), no. 1-2, 29–60. MR 88m:58203

[111] W. Klingenberg, *Lectures on Closed Geodesics*, Grundlehren der Mathematischen Wissenschaften, Vol. 230, Springer-Verlag, Berlin, 1978.

[112] W. P. A. Klingenberg, *Riemannian Geometry*, de Gruyter Studies in Mathematics, Walter de Gruyter & Co., Berlin, New York, 1982.

[113] _____, *Riemannian Geometry*, Second Ed., de Gruyter Studies in Mathematics, vol. 1, Walter de Gruyter & Co., Berlin, 1995.

[114] S. Kobayashi and K. Nomizu, *Foundations of Differential Geometry. Vol. I*, John Wiley & Sons Inc., New York, 1996, Reprint of the 1963 original, A Wiley-Interscience Publication. MR 97c:53001a

[115] _____, *Foundations of Differential Geometry. Vol. II*, John Wiley & Sons Inc., New York, 1996, Reprint of the 1969 original, A Wiley-Interscience Publication. MR 97c:53001b

[116] H. Kunita, *Stochastic Flows and Stochastic Differential Equations*, Cambridge Studies in Advanced Mathematics, vol. 24, Cambridge University Press, Cambridge, UK, 1990.

[117] T. G. Kurtz and P. Protter, Weak limit theorems for stochastic integrals and stochastic differential equations, *Ann. Probab.* **19** (1991), no. 3, 1035–1070.

[118] _____, *Wong-Zakai Corrections, Random Evolutions, and Simulation Schemes for SDEs*, Stochastic Analysis, Academic Press, Boston, MA, 1991, 331–346.

[119] S. Kusuoka and D. Stroock, Applications of the Malliavin calculus. II, *J. Fac. Sci. Univ. Tokyo Sect. IA Math.* **32** (1985), no. 1, 1–76. MR 86k:60100b

[120] _____, Applications of the Malliavin calculus. III, *J. Fac. Sci. Univ. Tokyo Sect. IA Math.* **34** (1987), no. 2, 391–442. MR 89c:60093

[121] S. Kusuoka and D. Stroock, *Applications of the Malliavin calculus. I*, Stochastic analysis (Katata/Kyoto, 1982), North-Holland Math. Library, vol. 32, North-Holland, Amsterdam, 1984, 271–306. MR 86k:60100a

[122] R. Léandre, Integration by parts formulas and rotationally invariant Sobolev calculus on free loop spaces, *J. Geom. Phys.* **11** (1993), no. 1-4, 517–528, Infinite-dimensional geometry in physics (Karpacz, 1992).

[123] R. Léandre and J. R. Norris, *Integration by parts Cameron–Martin formulas for the free path space of a compact Riemannian manifold*, 1995 Warwick Univ. Preprint, 1995.

[124] R. Léandre, Développement asymptotique de la densité d'une diffusion dégénérée, *Forum Math.* **4** (1992), no. 1, 45–75. MR 93d:60100

[125] X. D. Li, Existence and uniqueness of geodesics on path spaces, *J. Funct. Anal.* **173** (2000), no. 1, 182–202. MR 2001f:58074

[126] T. J. Lyons and Z. M. Qian, *Calculus for Multiplicative Functionals, Itô's Formula and Differential Equations*, Itô's stochastic calculus and probability theory, Springer, Tokyo, 1996, 233–250.

[127] _____, *Stochastic Jacobi fields and vector fields induced by varying area on path spaces*, Imperial College of Science, 1996.

[128] M.-P. Malliavin and P. Malliavin, *An infinitesimally quasi-invariant measure on the group of diffeomorphisms of the circle*, Special functions (Okayama, 1990), ICM-90 Satell. Conf. Proc., Springer, Tokyo, 1991, 234–244. MR 93h:58027

[129] P. Malliavin, *Geometrie differentielle stochastique*, Séminaire de Mathématiques Supérieures, Presses de l'Université de Montréal, Montreal, Que, 1978, Notes prepared by Danièle Dehen and Dominique Michel.

[130] _____, *Stochastic calculus of variation and hypoelliptic operators*, Proceedings of the International Symposium on Stochastic Differential Equations (Res. Inst. Math. Sci., Kyoto Univ., Kyoto, 1976), Wiley, New York, Chichester, Brisbane, 1978, 195–263.

[131] _____, *Stochastic Jacobi Fields*, Partial Differential Equations and Geometry (Proc. Conf., Park City, Utah, 1977 (New York), Dekker, 1979, Lecture Notes in Pure and Appl. Math., 48, pp. 203–235.

[132] _____, *Stochastic Analysis*, Grundlehren der Mathematischen Wissenschaften [Fundamental Principles of Mathematical Sciences], Vol. 313, Springer-Verlag, Berlin, 1997.

[133] G. Maruyama, Notes on Wiener integrals, *Kōdai Math. Sem. Rep.* **1950** (1950), 41–44. MR 12,343d

[134] E. J. McShane, *Stochastic differential equations and models of random processes*, Proceedings of the Sixth Berkeley Symposium on Mathematical Statistics and Probability (Univ. California, Berkeley, Calif., 1970/1971), Vol. III: Probability theory, Univ. California Press, Berkeley, CA, 1972, 263–294.

[135] _____, *Stochastic Calculus and Stochastic Models*, Academic Press, NY, Probability and Mathematical Statistics, Vol. 25, 1974.

[136] T. Melcher, Hypoelliptic heat kernel inequalities on Lie groups, Ph.D. Thesis, 2004. See www.math.ucsd.edu/˜driver/driver/thesis.html.

[137] P.-A. Meyer, *A differential geometric formalism for the Itô calculus*, Stochastic integrals (Proc. Sympos., Univ. Durham, Durham, 1980), Lecture Notes in Math., Vol. 851, Springer-Verlag, Berlin, 1981, 256–270. MR 84e:60084

[138] J. Moser, A new technique for the construction of solutions of nonlinear differential equations, *Proc. Nat. Acad. Sci. U.S.A.* **47** (1961), 1824–1831. MR 24 #A2695

[139] _____, A rapidly convergent iteration method and non-linear differential equations. II, *Ann. Scuola Norm. Sup. Pisa* (3) **20** (1966), 499–535. MR 34 #6280

[140] _____, A rapidly convergent iteration method and non-linear partial differential equations. I, *Ann. Scuola Norm. Sup. Pisa* (3) **20** (1966), 265–315. MR 33 #7667

[141] J.-M. Moulinier, Théorème limite pour les équations différentielles stochastiques, *Bull. Sci. Math.* (2) **112** (1988), no. 2, 185–209.

[142] S. Nakao and Y. Yamato, *Approximation theorem on stochastic differential equations*, Proceedings of the International Symposium on Stochastic Differential Equations (Res. Inst. Math. Sci., Kyoto Univ., Kyoto, 1976), Wiley, New York, Chichester, Brisbane, 1978, 283–296.

[143] J. Nash, The imbedding problem for Riemannian manifolds, *Ann. of Math.* (2) **63** (1956), 20–63. MR 17,782b

[144] J. R. Norris, *A complete differential formalism for stochastic calculus in manifolds*, Séminaire de Probabilités, XXVI, Lecture Notes in Math., Vol. 1526, Springer-Verlag, Berlin, 1992, 189–209. MR 94g:58254

[145] _____, Path integral formulae for heat kernels and their derivatives, *Probab. Theory Related Fields* **94** (1993), no. 4, 525–541.

[146] _____, Twisted sheets, *J. Funct. Anal.* **132** (1995), no. 2, 273–334. MR 96f:60094

[147] J. Norris, *Simplified Malliavin calculus*, Séminaire de Probabilités, XX, 1984/85, Lecture Notes in Math., Vol. 1204, Springer-Verlag, Berlin, 1986, 101–130.

[148] D. Nualart, *The Malliavin Calculus and Related Topics*, Probability and its Applications, Springer-Verlag, New York, 1995. MR 96k:60130

[149] O'Neill, *Semi-Riemannian geometry*, Pure and Applied Mathematics, Vol. 103, Academic Press Inc. [Harcourt Brace Jovanovich Publishers], New York, 1983, With applications to relativity. MR 85f:53002

[150] J. Picard, Gradient estimates for some diffusion semigroups, *Probab. Theory Related Fields* **122** (2002), no. 4, 593–612. MR 2003d:58056

[151] M.A. Pinsky, *Stochastic Riemannian geometry*, Probabilistic analysis and related topics, Vol. 1 (New York), Academic Press, 1978, 199–236.

[152] M. M. Rao, *Stochastic processes: general theory*, Mathematics and its Applications, Vol. 342, Kluwer Academic Publishers, Dordrecht, 1995. MR 97c:60092

[153] J. Schiltz, Time dependent Malliavin calculus on manifolds and application to nonlinear filtering, *Probab. Math. Statist.* **18** (1998), no. 2, Acta Univ. Wratislav. No. 2111, 319–334. MR 2000b:60144

[154] L. Schwartz, *Semi-martingales sur des variétés, et martingales conformes sur des variétés analytiques complexes*, Lecture Notes in Mathematics, Vol. 780, Springer-Verlag, Berlin, 1980. MR 82m:60051

[155] _____, *Géométrie différentielle du 2ème ordre, semi-martingales et équations différentielles stochastiques sur une variété différentielle*, Seminar on Probability, XVI, Supplement, Lecture Notes in Math., Vol. 921, Springer-Verlag, Berlin, 1982, 1–148. MR 83k:60064

[156] _____, *Semimartingales and their stochastic calculus on manifolds*, Collection de la Chaire Aisenstadt. [Aisenstadt Chair Collection], Presses de l'Université de Montréal, Montreal, QC, 1984, Edited and with a preface by Ian Iscoe. MR 86b:60085

[157] I. Shigekawa, Absolute continuity of probability laws of Wiener functionals, *Proc. Japan Acad. Ser. A Math. Sci.* **54** (1978), no. 8, 230–233. MR 81m:60097

[158] _____, Derivatives of Wiener functionals and absolute continuity of induced measures, *J. Math. Kyoto Univ.* **20** (1980), no. 2, 263–289. MR 83g:60051

[159] _____, On stochastic horizontal lifts, *Z. Wahrsch. Verw. Gebiete* **59** (1982), no. 2, 211–221. MR 83i:58102

[160] _____, Transformations of the Brownian motion on a Riemannian symmetric space, *Z. Wahrsch. Verw. Gebiete* **65** (1984), no. 4, 493–522.

[161] _____, *Transformations of the Brownian Motion on the Lie Group*, Stochastic analysis (Katata/Kyoto, 1982), North-Holland Math. Library, vol. 32, North-Holland, Amsterdam, 1984, 409–422.

[162] I. Shigekawa, de Rham-Hodge-Kodaira's decomposition on an abstract Wiener space, *J. Math. Kyoto Univ.* **26** (1986), no. 2, 191–202. MR 88h:58009

[163] I. Shigekawa, *Differential Calculus on a Based Loop Group*, New Trends in Stochastic Analysis, (Charingworth, 1994), World Sci. Publishing, River Edge, NJ, 1997, 375–398. MR 99k:60146

[164] M. Spivak, *A Comprehensive Introduction to Differential Geometry. Vol. I*, Second Ed., Publish or Perish Inc., Wilmington, DE, 1979. MR 82g:53003a

[165] R. S. Strichartz, Analysis of the Laplacian on the complete Riemannian manifold, *J. Funct. Anal.* **52** (1983), no. 1, 48–79. MR 84m:58138

[166] D. Stroock and S. Taniguchi, *Diffusions as integral curves, or Stratonovich without Itô*, The Dynkin Festschrift, Progr. Probab., Vol. 34, Birkhäuser Boston, Cambridge, MA, 1994, 333–369.

[167] D. W. Stroock and S. R. S. Varadhan, *On the support of diffusion processes with applications to the strong maximum principle*, Proceedings of the Sixth Berkeley Symposium on Mathematical Statistics and Probability (Univ. California, Berkeley, Calif., 1970/1971), Vol. III: Probability theory, Univ. California Press, Berkeley, Calif., 1972, 333–359.

[168] D. W. Stroock, The Malliavin calculus, a functional analytic approach, *J. Funct. Anal.* **44** (1981), no. 2, 212–257. MR 83h:60076

[169] _____, The Malliavin calculus and its application to second order parabolic differential equations. I, *Math. Systems Theory* **14** (1981), no. 1, 25–65. MR 84d:60092a

[170] _____, The Malliavin calculus and its application to second order parabolic differential equations. II, *Math. Systems Theory* **14** (1981), no. 2, 141–171. MR 84d:60092b

[171] _____, *An Introduction to the Analysis of Paths on a Riemannian Manifold*, Mathematical Surveys and Monographs, Vol. 74, American Mathematical Society, Providence, RI, 2000. MR 2001m:60187

[172] D. W. Stroock and J. Turetsky, Short time behavior of logarithmic derivatives of the heat kernel, *Asian J. Math.* **1** (1997), no. 1, 17–33. MR 99b:58225

[173] _____, Upper bounds on derivatives of the logarithm of the heat kernel, *Comm. Anal. Geom.* **6** (1998), no. 4, 669–685. MR 99k:58174

[174] D. W. Stroock and S. R. S. Varadhan, Diffusion processes with continuous coefficients. II, *Comm. Pure Appl. Math.* **22** (1969), 479–530.

[175] H. J. Sussmann, *Limits of the Wong-Zakai Type with a Modified Drift Term*, Stochastic Analysis, Academic Press, Boston, MA, 1991, 475–493.

[176] S. Taniguchi, Malliavin's stochastic calculus of variations for manifold-valued Wiener functionals and its applications, *Z. Wahrsch. Verw. Gebiete* **65** (1983), no. 2, 269–290. MR 85d:58088

[177] K. Twardowska, Approximation theorems of Wong-Zakai type for stochastic differential equations in infinite dimensions, *Dissertationes Math. (Rozprawy Mat.)* **325** (1993), 54. MR 94d:60092

[178] N. R. Wallach, *Harmonic Analysis on Homogeneous Spaces*, Pure and Applied Mathematics, No. 19., Marcel Dekker Inc., New York, 1973. MR 58 #16978

[179] S. Watanabe, *Lectures on stochastic differential equations and Malliavin calculus*, Tata Institute of Fundamental Research Lectures on Mathematics and Physics, Vol. 73, Published for the Tata Institute of Fundamental Research, Bombay, 1984, Notes by M. Gopalan Nair and B. Rajeev. MR 86b:60113

[180] E. Wong and M. Zakai, On the relation between ordinary and stochastic differential equations, *Internat. J. Engrg. Sci.* **3** (1965), 213–229.

[181] _____ , *On the relation between ordinary and stochastic differential equations and applications to stochastic problems in control theory*, Automatic and remote control III (Proc. Third Congr. Internat. Fed. Automat. Control (IFAC), London, 1966), Vol. 1, p. 5, Paper 3B, Inst. Mech. Engrs., London, 1967, p. 8.

# Noncommutative Probability and Applications

Stanley Gudder

Department of Mathematics University of Denver Denver, Colorado 80208
sgudder@math.du.edu

**Abstract.** Various versions of noncommutative probability theory are surveyed. It is stressed that the main motivation and applications of these noncommutative theories is quantum mechanics. A review of traditional probability theory and its unsharp version are presented. Sharp and unsharp Hilbert space probability theories are considered next. We then present a general discussion of observables and statistical maps. Finally, we consider sequential effect algebras and show that they unify and generalize all of these types of probability theories.

## 1 Introduction

This article presents a survey of various versions of noncommutative probability theory. Other names have been used for this subject such as operational probability theory and quantum probability theory. The latter designation stems from the fact that quantum mechanics provides the main motivation and applications of this formalism. In order to set the stage and make comparisons, we begin with a review of traditional probability theory and its unsharp (or fuzzy) version. We then study sharp and unsharp Hilbert space probability theories with some applications to quantum computation. We next present a general discussion of observables and statistical maps. Finally, a new framework called a sequential effect algebra, that generalizes all of these probability theories, is considered.

The article begins with the more concrete and familiar and then gradually progresses to the more general and abstract. To save space and spare the reader from technical details we have decided to omit most of the proofs, especially at the beginning well known stages. However, we provide references for the proofs of all stated results and encourage the reader to further pursue topics of interest. A few proofs are included for some of the latest results in Sections 8–12 to give the reader a flavor of the techniques that are employed. We also include some open problems for further research.

The noncommutative probability that we present generalizes traditional probability theory in two ways. First, a traditional random variable $f$ describes a measurement that is absolutely precise. After all, the values $f(\omega)$ are sharply defined with no fuzziness involved. The uncertainty comes from our lack of knowledge of the situation. However, real measurements are never completely precise. There are always errors involved and these should be taken into account in a realistic theory. Second, in a quantum mechanical world, two different measurements $A$ and $B$ frequently interfere with each other; one way this is manifested is that they cannot be performed simultaneously. We must specify which measurement is performed first and which is performed next. This results in a sequential or temporal product $A \circ B$ of the measurements. Due to interference, in general we have $A \circ B \neq B \circ A$ so that the theory becomes noncommutative. Traditional probability theory contains no mechanism for describing such situations. Until recently the main framework for such descriptions has been operators on a Hilbert space, which provide an obvious noncommutativity. In this case, fuzzy quantum events are represented by operators satisfying $0 \leq A \leq I$ and observables are described by normalized positive operator-valued measures. These are discussed in Sections 4–10. In the last few years, researchers in the foundations of quantum mechanics have developed a more general formalism called a sequential effect algebra. In this case, fuzzy quantum events are represented by abstract objects, with certain natural properties, called effects and observables are described by $\sigma$-morphisms. These are discussed in Sections 11 and 12.

Most of the material presented in this survey is not new and we have relied heavily on the work in [1, 15, 16, 17, 18, 19]. However, we have included a few new results and we hope we have organized the material in a way that gives a new perspective. The author would like to thank his recent collaborators Alvaro Arias, Aurelian Gheondea, Richard Greechie and Gabriel Nagy for many useful discussions on these subjects.

## 2 Traditional probability theory

The basic structure in traditional (or classical) probability theory is a measurable space $(\Omega, \mathcal{A})$. The set $\Omega$ corresponds to a set of possible outcomes for a classical experiment or collection of such experiments. The $\sigma$-algebra $\mathcal{A}$ of subsets of $\Omega$ corresponds to the set of events that may occur when an experiment is performed. There are three algebraic operations that are used to form various combinations of events. These are set theoretic complementation $A^c$, union $A \cup B$, and intersection $A \cap B$.

The set of events $\mathcal{A}$ with set theoretic inclusion $A \subseteq B$ forms a distributive lattice (in fact, $\mathcal{A}$ is a Boolean algebra) where the distributive law is

$$A \cap (B \cup C) = (A \cap B) \cup (A \cap C).$$

The order $A \subseteq B$ has the natural interpretation that $B$ occurs whenever $A$ occurs. The complement $A^c$ has the natural interpretation that $A^c$ occurs if and only if $A$ does not occur ($A^c$ is sometimes called the negation of $A$). Complementation satisfies: $A^{cc} = A$, $A \subseteq B$ implies $B^c \subseteq A^c$ and $A \cap A^c = \emptyset$. Moreover, De Morgan's laws $(A \cap B)^c = A^c \cup B^c$ and $(A \cup B)^c = A^c \cap B^c$ hold. Of course, $A \cup B$ occurs if and only if $A$ or $B$ occurs and $A \cap B$ occurs if and only if $A$ and $B$ occur. If $A \subseteq B^c$ we say that $A$ and $B$ are *orthogonal* and write $A \perp B$. Notice that $A \perp B$ if and only if $A$ and $B$ are disjoint ($A \cap B = \emptyset$). We interpret $A \perp B$ as meaning that if $A$ occurs then

$B$ does not occur (or vice versa). If $A \perp B$ we denote the *orthogonal* (or *disjoint*) *sum* by $A \oplus B = A \cup B$.

Probabilities are determined by a probability measure $\mu$ on $\mathcal{A}$. Algebraically, $\mu$ can be thought of as a $\sigma$-morphism $\mu \colon \mathcal{A} \to [0, 1]$. That is, $\mu(\Omega) = 1$ and $\mu(\oplus A_n) = \sum \mu(A_n)$. If $A \in \mathcal{A}$ is an event, the probability that $A$ occurs when an experiment is performed is given by $\mu(A)$. We also have the important concept of conditional probability. The *conditional probability* that $A \in \mathcal{A}$ occurs given that $B \in \mathcal{A}$ has occurred is $\mu(A \mid B) = \mu(A \cap B)/\mu(B)$. (Whenever we consider the conditional probability $\mu(A \mid B)$ we always assume that $\mu(B) \neq 0$.) Two easily proved but useful results are Bayes' law

$$\mu(B \mid A) = \frac{\mu(B)\mu(A \mid B)}{\mu(A)}$$

and the law of total probability

$$P(A) = \sum P(B_n)P(A \mid B_n)$$

whenever $B_i \perp B_j, i \neq j$ and $\cup B_i = \Omega$.

A measurement in classical probability theory is described by a random variable $f \colon \Omega \to \mathbb{R}$. The measurement is considered to be sharp (precise) even though it is described by a random variable. The randomness is caused by our lack of knowledge of the system and not by any inaccuracies of the measurement. For example, after we flip a coin, we have complete certainty about whether the result is heads or tails. Our measuring apparatus, in this case our eyes, can be assumed to be perfectly accurate. (If they are not, we have an unsharp theory considered later.) The probability (or randomness) is present because we have a lack of knowledge and cannot predict beforehand the result of a coin flip. For a given random variable $f$, probability measure $\mu$ and Borel set $B \in \mathcal{B}(\mathbb{R})$, $\mu_f(B) = \mu\left(f^{-1}(B)\right)$ gives the probability of the event $f^{-1}(B)$. Thus, $\mu_f(B)$ is the probability that $f$ has a value in $B$ and $\mu_f \colon \mathcal{B}(\mathbb{R}) \to [0, 1]$ is a probability measure on $\mathcal{B}(\mathbb{R})$ called the *distribution* of $f$. As is well known, $E(f) = \int f \, d\mu$ can be interpreted as the expectation of $f$ (if the integral exists) and this coincides with the mean $\int \lambda \mu_f(d\lambda)$ of the distribution $\mu_f$. Of course, $E(f)$ exists if and only if $\int |f| d\mu < \infty$. Other important statistical quantities such as the variance Var $(f)$ and standard deviation $\sigma(f) = $ Var $(f)^{1/2}$ can be defined. If it exists,

$$\text{Var} (f) = E\left[f - E(f)\right]^2 = E(f^2) - E(f)^2$$

A $\sigma$-morphism from $\mathcal{B}(\mathbb{R})$ to $\mathcal{A}$ is a map $h \colon \mathcal{B}(\mathbb{R}) \to \mathcal{A}$ that satisfies $h(\mathbb{R}) = \Omega$ and $h(\oplus B_n) = \oplus h(B_n)$. It is important to note that there is a one-to-one correspondence between $\sigma$-morphisms from $\mathcal{B}(\mathbb{R})$ to $\mathcal{A}$ and random variables (classical measurements). If $f \colon \Omega \to \mathbb{R}$ is a random variable, then $h = f^{-1} \colon \mathcal{B}(\mathbb{R}) \to \mathcal{A}$ is a $\sigma$-morphism and conversely, any $\sigma$-morphism $h \colon \mathcal{B}(\mathbb{R}) \to \mathcal{A}$ has this form for a unique random variable $f$. We can write statistical formulas concerning $f$ in terms of the corresponding $\sigma$-morphism $h$. For example, the distribution of $f$ becomes

$$\mu_f(B) = \mu\left(f^{-1}(B)\right) = \mu\left(h(B)\right)$$

and we also call $\mu \circ h$ the *distribution* of $h$. The expectation of $f$ becomes

$$\int f \, d\mu = \int \lambda \mu_f(d\lambda) = \int \lambda \mu \left( f^{-1}(d\lambda) \right) = \int \lambda \mu \left( h(d\lambda) \right)$$

which we also call the *expectation* of $h$. In fact we can reformulate the entire framework of traditional probability theory in terms of $\sigma$-morphisms. This is not important for the study of traditional probability theory, but it is crucial for noncommutative probability theory.

## 3 Unsharp traditional probability theory

Suppose we perform a position measurement for a particle $p$ using a particle detector $d$ such as a geiger counter or photon counter. If $d$ is perfectly accurate with sensitivity domain $A \subseteq \mathbb{R}^3$, then $d$ clicks if and only if $p$ is in $A$ so $d$ corresponds to the characteristic function $\chi_A$. Any real detector is not perfectly accurate (it is unsharp or fuzzy) so $d$ may not click when $p$ is in $A$ or $d$ may give a false click when $p$ is not in $A$. We define the confidence function $f_d : \mathbb{R}^3 \to [0, 1]$, where $f_d(\lambda)$ gives the confidence that $d$ will click when $p$ is at the point $\lambda$ on a scale between zero and one. The function $f_d$ can be determined using calibration experiments with $d$ and is characteristic of any particular counter. In general, if $X$ is a nonempty set, we call an element of $[0, 1]^X$ a *fuzzy subset* of $X$ [15, 31]. In fuzzy set theory, a function $f \in [0, 1]^X$ corresponds to a degree of membership for a fuzzy set in $X$. In this way, a counter corresponds to a fuzzy subset of $\mathbb{R}^3$. The *sharp* (or *crisp*) elements of $[0, 1]^X$ are the characteristic functions or equivalently, the subsets of $X$.

Motivated by the previous paragraph, we now introduce an unsharp classical probability theory. Let $(\Omega, \mathcal{A})$ be a measurable space and let $\mathcal{E}(\Omega, \mathcal{A})$ be the set of all random variables in $[0, 1]^\Omega$ . The elements of $\mathcal{E}(\Omega, \mathcal{A})$ correspond to measurable confidence functions and are called *classical effects* on $(\Omega, \mathcal{A})$. We define a natural partial order $\leq$ on $\mathcal{E}(\Omega, \mathcal{A})$ by $f \leq g$ if $f(\omega) \leq g(\omega)$ for all $\omega \in \Omega$. Then $(\mathcal{E}(\Omega, \mathcal{A}), \leq)$ becomes a distributive lattice in which $f \wedge g = \min(f, g)$ and $f \vee g = \max(f, g)$. We define a natural negation on $\mathcal{E}(\Omega, \mathcal{A})$ given by the complementation $f' = 1 - f$. We then have $f'' = f$, $f \leq g$ implies $g' \leq f'$, and De Morgan's laws hold. The main difference between this algebraic structure on $\mathcal{E}(\Omega, \mathcal{A})$ and the algebraic structure of the sharp theory on $\mathcal{A}$ is that $f \wedge f'$ need not be 0 (equivalently, $f \vee f'$ need not be 1). For example, let $f$ be the constant function $f = 1/2$ in which case $f' = 1/2$. As before, we write $f \perp g$ if $f \leq g'$ or equivalently if $f + g \leq 1$. As in the sharp classical theory we have $f \wedge g = 0$ implies $f \perp g$, but unlike the sharp theory $f \perp g$ does not imply $f \wedge g = 0$. As usual, if $f \perp g$, we define the *orthogonal sum* $f \oplus g = f + g$. The following lemma characterizes the sharp classical effects in $\mathcal{E}(\Omega, \mathcal{A})$.

**Lemma 3.1.** *The following statements for $f \in \mathcal{E}(\Omega, \mathcal{A})$ are equivalent.* (a) $f$ *is sharp.* (b) $f^2 = f$. (c) $f \wedge f' = 0$. (d) $f \vee f' = 1$.

If $f_d$ is the confidence function for a detector $d$, it is not hard to justify that the probability that $d$ clicks should be the average value of $f_d$. In general, if $\mu$ is a probability measure on $\mathcal{A}$, we define the probability that the effect $f \in \mathcal{E}(\Omega, \mathcal{A})$ is observed to be $\mu(f) = E(f) = \int f \, d\mu$. Analogous to the sharp theories, we define a $\sigma$-*morphism* $h \colon \mathcal{B} \to \mathcal{E}(\Omega, \mathcal{A})$ as follows.

(M1) $h(\mathbb{R}) = 1$.
(M2) If $A_i \in \mathcal{B}$ are mutually orthogonal, then $h(\bigcup A_i) = \sum h(A_i)$ where the convergence of the summation is pointwise.

Notice that the condition $A \perp B$ implies $h(A) \perp h(B)$ follows automatically from (M2). As expected, an unsharp classical measurement corresponds to a $\sigma$-morphism from $\mathcal{B}$ to $\mathcal{E}(\Omega, \mathcal{A})$. For example, suppose such a measurement $M$ has only finitely many values $\lambda_1, \ldots, \lambda_n$. Denote by $M(\{\lambda_i\})$ the effect or fuzzy event that $h$ has the value $\lambda_i$. Then $\sum M(\{\lambda_i\})$ corresponds to the effect that $M$ has some value so we should have $\sum M(\{\lambda_i\}) = 1$. Then $M$ has a unique extension to a $\sigma$-morphism $h \colon \mathcal{B} \to \mathcal{E}(\Omega, \mathcal{A})$ given by

$$h(A) = \sum \{M(\{\lambda_i\}) : \lambda_i \in A\}$$

Now let $h \colon \mathcal{B} \to \mathcal{E}(\Omega, \mathcal{A})$ be an unsharp measurement ($\sigma$-morphism) and let $\mu$ be a probability measure on $\mathcal{A}$. Then the probability that $h$ has a value in $B \in \mathcal{B}$ is $\mu_h(B) = E(h(B))$. It follows from the monotone convergence theorem, that $\mu_h \colon \mathcal{B} \to [0, 1]$ is a probability measure on $\mathcal{B}$ which we call the *distribution* of $h$. As usual, we can now define various statistical quantities for $h$. For example, the *expectation* of $h$ becomes $E(h) = \int \lambda \mu_h(d\lambda)$.

We now briefly describe how other important probabilistic concepts can be formulated in this context. If $f, g \in \mathcal{E}(\Omega, \mathcal{A})$ and $\mu$ is a probability measure on $\mathcal{A}$, we define the *conditional probability* $\mu(f \mid g) = \mu(fg)/\mu(g)$ if $\mu(g) \neq 0$. This reduces to the usual definition for sharp effects. Moreover, if $\mu(g) \neq 0$, then $\mu(\cdot \mid g)$ is a probability measure on $\mathcal{E}(\Omega, \mathcal{A})$ in the sense that $\mu(1 \mid g) = 1$ and if $f_i \in \mathcal{E}(\Omega, \mathcal{A})$ with $\sum f_i = 1$, then $\mu(\sum f_i \mid g) = \sum \mu(f_i \mid g)$. It is easy to show that Bayes' fuzzy rules hold in the following sense. If $\mu(f), \mu(g) \neq 0$, then

$$\mu(f \mid g) = \frac{\mu(f)\mu(g \mid f)}{\mu(g)}$$

and if $\mu(g_i) \neq 0$ and $\sum g_i = 1$, then

$$\mu(f) = \sum \mu(g_i)\mu(f \mid g_i)$$

There are natural definitions of independence, functions of several measurements, and joint distributions. Moreover, one can prove versions of the law of large numbers and the central limit theorem [15], but these carry us too far from the main goals of this survey.

We now give an example of an unsharp measurement that has practical applications. Let $M$ be a pattern recognition apparatus that recognizes the symbols $\omega_1, \ldots, \omega_n$. Suppose that in a specific situation, the symbols occur with probabilities $\mu(\omega_i)$. Letting $\Omega = \{\omega_1, \ldots, \omega_n\}$, $\mathcal{A} = 2^\Omega$ and extending $\mu$ to $\mathcal{A}$ in the usual way, $(\Omega, \mathcal{A}, \mu)$ becomes a probability space. When $\omega_i$ occurs, a perfectly accurate apparatus registers a number

$\lambda_i \in \mathbb{R}$ where $\lambda_i, \ldots, \lambda_n$ are distinct. However, no apparatus is perfectly accurate so we consider an unsharp apparatus. In this case, when $\omega_i$ occurs, $M$ will usually register $\lambda_i$, but on occasion might register $\lambda_j \neq \lambda_i$. A calibration experiment gives the probabilities $\mu_M(\lambda_j \mid \omega_i) \geq 0$ where

$$\sum_{j=1}^{n} \mu_M(\lambda_j \mid \omega_i) = 1, \qquad i = 1, \ldots, n.$$

Of course, $\mu_M(\lambda_j \mid \omega_i)$ is the probability that $M$ registers $\lambda_j$ when $\omega_i$ occurs. If $M$ were sharp, then $\mu_M(\lambda_j \mid \omega_i) = \delta_{ji}$ but in general, this is not the case. We call $f_{M,j}(\omega_i) = \mu_M(\lambda_j \mid \omega_i)$, $i, j = 1, \ldots, n$, the *confidence functions* for $M$. Given the probabilities $\mu_M(\omega_i)$, $\mu_M(\lambda_j \mid \omega_i)$, $i, j = 1 \ldots, n$, it is important to find the probabilities $\mu_M(\lambda_i)$ that $M$ registers $\lambda_i$ and the probabilities $\mu_M(\omega_i \mid \lambda_j)$ that the symbol $\omega_i$ appeared given that $M$ registers $\lambda_j$, $i, j = 1, \ldots, n$. These probabilities can be found by Bayes' rules to be

$$\mu_M(\lambda_i) = \sum_{j=1}^{n} \mu(\omega_j) \mu_M(\lambda_i \mid \omega_j)$$

and

$$\mu_M(\omega_i \mid \lambda_j) = \frac{\mu(\omega_i) \mu_M(\lambda_j \mid \omega_i)}{\mu_M(\lambda_j)}.$$

It can be shown that, in general, no single random variable on $(\Omega, \mathcal{A})$ can give these probabilities. This is because we are describing an unsharp measurement and random variables only describe sharp measurements on $(\Omega, \mathcal{A})$. However, there is a larger probability space $(\widehat{\Omega}, \widehat{\mathcal{A}}, \widehat{\mu})$ with a different probability measure $\widehat{\mu}$ and a random variable on this space that does give these probabilities. This larger space is required in order to incorporate the fuzziness of $M$. But this is an awkward and inefficient method. The set $\widehat{\Omega}$ has twice as many points as $\Omega$ and $\widehat{\mu}$ depends on $M$. This method would be impractical if one wants to consider several unsharp apparatus or a sequence of unsharp apparatus. However, we can construct a single unsharp measurement ($\sigma$-morphism) on $(\Omega, \mathcal{A}, \mu)$ that does the job. Defining $h \colon \mathcal{B} \to \mathcal{E}(\Omega, \mathcal{A})$ by

$$h(A) = \sum \{f_{M,j} \colon \lambda_j \in A\}$$

it is clear that $h$ is a $\sigma$-morphism. By our previous discussion, we have

$$\mu_M(\lambda_j \mid \omega_i) = \mu\left[h\left(\{\lambda_j\}\right) \mid \chi_{\{\omega_i\}}\right] = \frac{\mu(f_{M,j}\chi_{\omega_i})}{\mu(\chi_{\{\omega_i\}})} = f_{M,j}(\omega_i)$$

which is consistent with our definition of $f_{M,j}$. We also obtain

$$\mu_M(\lambda_i) = \mu[h(\{\lambda_i\})] = \mu(f_{M,j}) = \sum_{j=1}^{n} f_{M,i}(\omega_j)\mu(\omega_j)$$

$$= \sum_{j=1}^{n} \mu(\omega_j)\mu_M(\lambda_i \mid \omega_j)$$

which agrees with our derived formula. Finally,

$$\mu_M(\omega_i \mid \lambda_j) = \mu\left[\chi_{\{\omega_i\}} \mid h(\{\lambda_j\})\right] = \frac{\mu(\chi_{\{\omega_i\}} f_{M,j})}{\mu(f_{M,j})}$$

$$= \frac{\mu(\omega_i) f_{M,j}(\omega_i)}{\mu_M(\lambda_j)} = \frac{\mu(\omega_i)\mu_M(\lambda_j \mid \omega_i)}{\mu_M(\lambda_j)}$$

which again agrees with our derived formula. These formulas could also be obtained using Bayes' fuzzy rules.

## 4 Sharp quantum probability

The basic structure for sharp quantum probability (or sharp Hilbert space probability) is usually taken to be a complex separable Hilbert space $\mathcal{H}$. The motivation for $\mathcal{H}$ is not as clear cut as the basic structure for traditional probability theory. Although a considerable amount of literature is devoted to a deeper physical motivation for $\mathcal{H}$ [11, 12, 24], this Hilbert space is frequently just thought of as the arena in which quantum events and probabilities may be displayed.

A sharp quantum event is described by an orthogonal projection operator $P$ on $\mathcal{H}$. We denote the set of projection operators on $\mathcal{H}$ by $\mathcal{P}(\mathcal{H})$ and we let $I \in \mathcal{P}(\mathcal{H})$ be the identity operator. A *quantum state* is represented by a probability measure $m: \mathcal{P}(\mathcal{H}) \to [0, 1]$ in the sense that $m(I) = 1$ and if $P_i \in \mathcal{P}(\mathcal{H})$ are mutually orthogonal, then

$$m\left(\sum P_i\right) = \sum m(P_i).$$

The summation $\sum P_i$ converges in the strong operator topology. It follows from a deep and beautiful theorem due to A. Gleason [14] that every state $m$ has the form $m(P) = \mathrm{tr}(\rho P)$ where $\rho$ is a unique positive trace class operator with $\mathrm{tr}(\rho) = 1$. Such an operator $\rho$ is called a *density operator* and we denote the set of density operators on $\mathcal{H}$ by $\mathcal{D}(\mathcal{H})$. The set $\mathcal{D}(\mathcal{H})$ is convex and its extreme points are called *pure states*. It is not hard to show that $m$ is a pure state if and only if there exists a unit vector $\psi \in \mathcal{H}$ such that $m(P) = \langle P\psi, \psi \rangle = \|P\psi\|^2$. In this case the density operator corresponding to $m$ is the one-dimensional projection $P_\psi$ onto $\psi$. We denote the pure state corresponding to $\psi$ by $m_\psi$ and the mixed state corresponding to $\rho \in \mathcal{D}(\mathcal{H})$ by $m_\rho$.

If $M$ is a quantum measurement (or observable) and $B \in \mathcal{B}(\mathbb{R})$, we denote the quantum event that $M$ has a value in $B$ when this measurement is performed by $M(B) \in \mathcal{P}(\mathcal{H})$. It is easy to justify on physical grounds that $M: \mathcal{B}(\mathbb{R}) \to \mathcal{P}(\mathcal{H})$ should satisfy the following conditions.

(O1) $M(\mathbb{R}) = I$.

(O2) If $A \cap B = \emptyset$, then $M(A) \perp M(B)$ (that is, $M(A)M(B) = 0$).

(O3) If $A_i \in \mathcal{B}(\mathbb{R})$ are mutually disjoint, then $M(\cup A_i) = \sum M(A_i)$ where the convergence in the summation is in the strong operator topology.

We conclude that a quantum measurement corresponds to a spectral measure (or projection-valued (PV) measure). According to the spectral theorem [28, 29] there is a one-to-one correspondence between spectral measures and self-adjoint operators on $\mathcal{H}$. If $T$ is a self-adjoint operator, we denote the corresponding spectral measure by $P^T : \mathcal{B}(\mathbb{R}) \to \mathcal{P}(\mathcal{H})$. Then $T$ can be written in terms of the spectral integral $T = \int \lambda P^T(d\lambda)$. We conclude that a quantum measurement $M$ can be described by a self-adjoint operator $T$ where $P^T = M$. If a system is in the quantum state $m_\rho$, then the probability that $M$ has a value in $B \in \mathcal{B}(\mathbb{R})$ is given by

$$m_\rho(M(B)) = m_\rho\left(P^T(B)\right) = \mathrm{tr}\left(\rho P^T(B)\right). \tag{4.1}$$

Using the notation $\rho_T(B) = \mathrm{tr}\left(\rho P^T(B)\right)$ it follows from (O1), (O2) and (O3) that $\rho_T : \mathcal{B}(\mathbb{R}) \to [0, 1]$ is a probability measure on $\mathcal{B}(\mathbb{R})$ which we call the *distribution* of $T$ (or $M$). If the integral exists, then just as in the sharp traditional theory, we define the *expectation* of $T$ (or $M$) by

$$E(T) = \int \lambda \rho_T(d\lambda) = \int \lambda \mathrm{tr}\left(\rho P^T(d\lambda)\right) = \int \lambda m_\rho\left(P^T(d\lambda)\right). \tag{4.2}$$

In the case of a pure state $m_\psi$, (4.1) becomes

$$m_\psi(M(B)) = \left\langle P^T(B)\psi, \psi \right\rangle \tag{4.3}$$

and using the notation $\psi_T(B) = \left\langle P^T(B)\psi, \psi \right\rangle$, (4.2) becomes

$$E(T) = \int \lambda \left\langle P^T(d\lambda)\psi, \psi \right\rangle \tag{4.4}$$

By the spectral theorem we can write (4.4) as

$$E(T) = \left\langle \int \lambda P^T(d\lambda)\psi, \psi \right\rangle = \langle T\psi, \psi \rangle$$

assuming that $\psi$ is in the domain of $T$. If $\psi$ is also in the domain of $T^2$, then Var$(T)$ is given by

$$\mathrm{Var}(T) = E(T^2) - E(T)^2 = \left\langle T^2\psi, \psi \right\rangle - \langle T\psi, \psi \rangle^2$$

and $\sigma(T) = \mathrm{Var}(T)^{1/2}$. In a similar way, for a mixed state $\rho \in \mathcal{D}(\mathcal{H})$ we have $E(T) = \mathrm{tr}(\rho T)$.

As an illustration of these concepts, let $\mathcal{H} = L^2(\mathbb{R}, d\lambda)$. Two important observables are the position $S$ and the momentum $T$ observables. On the appropriate domains, $S$ is the self-adjoint operator given by $Sf(\lambda) = \lambda f(\lambda)$ and $Tf = -i\hbar \, df/d\lambda$ where $\hbar$ is

Planck's constant. The *commutator* of $S$ and $T$ is defined as $[S, T] = ST - TS$ and if $\psi$ is in the domain of $[S, T]$, then a simple calculation give $[S, T]\psi = i\hbar\psi$. If $\|\psi\| = 1$, applying Schwarz's inequality, we have

$$\hbar = |\langle [S, T]\psi, \psi \rangle| \le |\langle ST\psi, \psi \rangle| + |\langle TS\psi, \psi \rangle| = 2|\langle S\psi, T\psi \rangle|$$

$$\le 2\|S\psi\| \|T\psi\| = 2\langle S^2\psi, \psi \rangle^{1/2} \langle T^2\psi, \psi \rangle^{1/2}.$$

If $E(S) = E(T) = 0$, we conclude that $\sigma(S)\sigma(T) \ge \hbar/2$ which is the famous Heisenberg uncertainty principle. If $E(S), E(T) \ne 0$, just replace $S$ and $T$ by $S - E(S)I, T - E(T)I$, respectively, to again obtain the uncertainty principle. Notice that the uncertainty principle does not say that $S$ and $T$ themselves are not sharp but that $S$ and $T$ are not simultaneously sharp. That is, if a very accurate position measurement is performed, then a simultaneous momentum measurement must be inaccurate to the extent of the uncertainty principle.

The following table summarizes a comparison of the sharp traditional and sharp quantum probability theories.

**Table 1**
**Comparison of sharp classical and quantum probability**

| Concept | Sharp Classical Probability | Sharp Quantum Probability |
|---|---|---|
| Basic Structure | Measurable Space $(\Omega, \mathcal{A})$ | Hilbert Space $\mathcal{H}$ |
| Event | Measurable Set $A \in \mathcal{A}$ | Projection Operator $P$ |
| Probability | Probability Measure $\mu$ | Pure State $\psi \in \mathcal{H}, \|\psi\| = 1$<br>Mixed State $\rho \in \mathcal{D}(\mathcal{H})$ |
| Probability of Event | $\mu(A)$ | $\langle P\psi, \psi \rangle = \|P\psi\|^2$<br>or $\mathrm{tr}(\rho P)$ |
| Measurement | Random Variable $f$ | Self-Adjoint Operator $T$ |
| Distribution | $\mu_f(B) = \mu[f^{-1}(B)]$ | $\psi_T(B) = \langle P^T(B)\psi, \psi \rangle$<br>or $\mathrm{tr}\left(\rho P^T(B)\right)$ |
| Expectation | $\int f \, d\mu = \int \lambda\mu_f(d\lambda)$ | $\langle T\psi, \psi \rangle$ or $\mathrm{tr}(\rho T)$ |

The set of sharp quantum events $\mathcal{P}(\mathcal{H})$ shares some of its properties with the set of sharp classical events $\mathcal{A}$, but also has important differences. We can again define a natural partial order $\le$ on $\mathcal{P}(\mathcal{H})$ by defining $P \le Q$ if $PQ = P$. This is equivalent to $\langle P\psi, \psi \rangle \le \langle Q\psi, \psi \rangle$ for every $\psi \in \mathcal{H}$ which is equivalent to $m_\psi(P) \le m_\psi(Q)$ for every $\psi \in \mathcal{H}$ with $\|\psi\| = 1$. Thus, $P \le Q$ has the natural interpretation that the probability of the occurrence of $P$ is not greater than the probability of the occurrence of $Q$ in every state of the system. Another condition that is equivalent to $P \le Q$ is that $R(P) \subseteq R(Q)$ where $R(P)$ denotes the range of $P$. Under the partial order $\le$, $(\mathcal{P}(\mathcal{H}), \le)$ is a lattice where $P \wedge Q$ is the projection onto $R(P) \cap R(Q)$ and $P \vee Q$ is the projection onto the closed span $\overline{\mathrm{sp}}\,[R(P) \cup R(Q)]$. However, unlike the classical theory, simple examples show that $(\mathcal{P}(\mathcal{H}), \le)$ is not distributive.

We can still define a natural negation $P \mapsto P'$ on $\mathcal{P}(\mathcal{H})$ given by the orthocomplementation $P' = I - P$. Then for any state $m_\psi$, we have $m_\psi(P') = 1 - m_\psi(P)$. As in the classical theory, we have $P'' = P$, $P \leq Q$ implies $Q' \leq P'$ and $P \wedge P' = 0$. Also, De Morgan's laws $(A \wedge B)' = A' \vee B'$ and $(A \vee B)' = A' \wedge B'$ still hold. We again define $P \perp Q$ if $P \leq Q'$ or equivalently if $P + Q \leq I$. Although $P \perp Q$ implies $P \wedge Q = 0$, unlike the classical theory, $P \wedge Q = 0$ does not imply $P \perp Q$ so disjointness is not equivalent to orthogonality. In fact, it can be shown that $P \wedge Q = 0$ implies $P \perp Q$ if and only if the distributive law holds. If $P \perp Q$ we again define the orthogonal sum $P \oplus Q = P + Q$. A $\sigma$-morphism $h \colon \mathcal{B} \to \mathcal{P}(\mathcal{H})$ is just a PV-measure, so as previously discussed, there is a one-to-one correspondence between $\sigma$-morphisms from $\mathcal{B}$ to $\mathcal{P}(\mathcal{H})$ and self-adjoint operators (quantum measurements or observables). In terms of the orthogonal sum, we can write Condition (O3) as $h(\bigoplus A_i) = \bigoplus h(A_i)$.

## 5 Unsharp quantum probability

We now consider unsharp quantum (or unsharp Hilbert space) probability theory. A sharp quantum event is represented by a projection operator $P \in \mathcal{P}(\mathcal{H})$. The spectrum of $P$ is contained in the set $\{0, 1\}$ which is analogous to the values of a characteristic function. A classical effect is represented by a function with values in $[0, 1]$ so it is reasonable to represent a quantum effect by a self-adjoint operator with spectrum in $[0, 1]$, $[8, 11, 12, 21, 23, 24]$.

Let $\mathcal{S}(\mathcal{H})$ be the set of bounded self-adjoint operators on $\mathcal{H}$. Define the partial order $\leq$ on $\mathcal{S}(\mathcal{H})$ by $S \leq T$ if $\langle S\psi, \psi \rangle \leq \langle T\psi, \psi \rangle$ for all $\psi \in \mathcal{H}$. An element $A \in \mathcal{S}(\mathcal{H})$ is a *quantum effect* if $0 \leq A \leq I$ and we denote the set of quantum effects by $\mathcal{E}(\mathcal{H})$. It can be shown that $A \in \mathcal{S}(\mathcal{H})$ is in $\mathcal{E}(\mathcal{H})$ if and only if the spectrum of $A$ is contained in $[0, 1]$. We also have $\mathcal{P}(\mathcal{H}) \subseteq \mathcal{E}(\mathcal{H})$ and the elements of $\mathcal{P}(\mathcal{H})$ are called *sharp effects*.

Unlike our three previous theories, $(\mathcal{E}(\mathcal{H}), \leq)$ is a partially ordered set that is not a lattice (unless $\dim \mathcal{H} = 1$). However, we can still define a natural negation $\mathcal{E}(\mathcal{H})$ by the complementation $A' = I - A$. As before, we have $A'' = A$ and $A \leq B$ implies $B' \leq A'$. Then as in the traditional unsharp theory but unlike the sharp theories $A \wedge A'$ need not be 0 and $A \vee A'$ need not be $I$ ($A \wedge A'$ and $A \vee A'$ may not even exist). For $A, B \in \mathcal{E}(\mathcal{H})$ we write $A \perp B$ if $A \leq B'$ or equivalently if $A \perp B \leq I$. The next lemma characterizes sharp elements of $\mathcal{E}(\mathcal{H})$ and is directly analogous to Lemma 3.1 [16].

**Lemma 5.1.** *The following statements for* $A \in \mathcal{E}(\mathcal{H})$ *are equivalent.* (a) $A$ *is sharp.* (b) $A^2 = A$. (c) $A \wedge A' = 0$. (d) $A \vee A' = I$.

As in the three previous theories, we define a $\sigma$-*morphism* $h \colon \mathcal{B}(\mathbb{R}) \to \mathcal{E}(\mathcal{H})$ by $h(\mathbb{R}) = I$ and if $A_n \in \mathcal{B}(\mathbb{R})$ are mutually disjoint, then $h(\cup A_n) = \sum h(A_n)$ where the summation converges in the strong operator topology. A $\sigma$-morphism from $\mathcal{B}(\mathbb{R})$ to $\mathcal{E}(\mathcal{H})$ corresponds to an unsharp quantum measurement and is called a *normalized positive operator-valued* (POV) *measure*. Probabilities, distributions and expectations are defined analogously to our previous theories and are summarized in Table 2.

**Table 2**

**Comparison of unsharp classical and quantum probability**

| Concept | Unsharp Classical Probability | Unsharp Quantum Probability |
|---|---|---|
| Basic Structure | Measurable Space $(\Omega, \mathcal{A})$ | Hilbert Space $\mathcal{H}$ |
| Event | Classical Effect $f \in \mathcal{E}(\Omega, \mathcal{A})$ | Quantum Effect $A \in \mathcal{E}(\mathcal{H})$ |
| Sharp Event | Measurable Set $A \in \mathcal{A}$ | Projection Operator $P \in \mathcal{P}(\mathcal{H})$ |
| Probability | Probability Measure $\mu$ | Pure State $\psi \in \mathcal{H}$, $\|\psi\| = 1$ or Mixed State $\rho \in \mathcal{D}(\mathcal{H})$ |
| Probability of Event | $E(f) = \int f \, d\mu$ | $\langle A\psi, \psi \rangle$ or $\mathrm{tr}(\rho A)$ |
| Measurement | $\sigma$-morphism $h: \mathcal{B}(\mathbb{R}) \to \mathcal{E}(\Omega, \mathcal{A})$ | POV-measure $h: \mathcal{B}(\mathbb{R}) \to \mathcal{E}(\mathcal{H})$ |
| Distribution | $\mu_h(B) = \int h(B) \, d\mu$ | $\mu_h(B) = \langle h(B)\psi, \psi \rangle$ or $\mathrm{tr}(\rho h(B))$ |
| Expectation | $\int \lambda \mu_h(d\lambda)$ | $\int \lambda \langle h(d\lambda)\psi, \psi \rangle$ or $\int \lambda \mathrm{tr}(\rho h(d\lambda))$ |

We now discuss the lattice properties of $\mathcal{E}(\mathcal{H})$. It was already mentioned that $\mathcal{E}(\mathcal{H})$ is not a lattice unless $\dim \mathcal{H} = 1$. For a simple example, let $\mathcal{H} = \mathbb{C}^2$ and let $A$ and $B$ be the diagonal matrices $A = \mathrm{diag}(1/2, 1/2)$, $B = \mathrm{diag}(3/4, 1/4)$. It is clear that $A, B \in \mathcal{E}(\mathcal{H})$ and it is not hard to show that $A \wedge B$ does not exist. This means that there is no largest effect whose occurrence implies the occurrences of both $A$ and $B$. We now consider characterizations of the pairs $A, B \in \mathcal{E}(\mathcal{H})$ such that $A \wedge B$ exists. These also characterize the existence of $A \vee B$ because by De Morgan's law $A \vee B = (A' \wedge B')'$ in the sense that if one side exists, then the other side exists and they are equal. One reason that such a characterization is mathematically interesting is the following surprising result due to R. Kadison [22].

**Theorem 5.2.** *For $A, B \in \mathcal{S}(\mathcal{H})$, $A \wedge_S B$ exists if and only if $A$ and $B$ are comparable (that is, either $A \le B$ or $B \le A$).*

We used the notation $A \wedge_S B$ because the greatest lower bound is computed relative to $\mathcal{S}(\mathcal{H})$ and we have used and will continue to use $\wedge$ for the greatest lower bound relative to $\mathcal{E}(\mathcal{H})$. Kadison's theorem states that $A \wedge_S B$ exists only when it must exist and for this reason $(\mathcal{S}(\mathcal{H}), \le)$ is called an *antilattice*. Thus, $(\mathcal{S}(\mathcal{H}), \le)$ is as far from a lattice as it can get. Now let $A, B \in \mathcal{E}(\mathcal{H})$ be incomparable. Just because $A \wedge_S B$ does not exist, it need not follow that $A \wedge B$ does not exist. After all, $\mathcal{E}(\mathcal{H})$ is a smaller set than $\mathcal{S}(\mathcal{H})$ so it might be easier for a greatest lower bound to exist in $\mathcal{E}(\mathcal{H})$ than in $\mathcal{S}(\mathcal{H})$. In fact, it is well known that if $A, B \in \mathcal{P}(\mathcal{H})$, then $A \wedge B$ exists even if $A$ and $B$ are incomparable and $A \wedge B$ is the projection onto $R(A) \cap R(B)$. We conclude that Kadison's theorem does not hold on $\mathcal{E}(\mathcal{H})$ and what we present is a counterpart to Kadison's theorem on $\mathcal{E}(\mathcal{H})$. For $A, B \in \mathcal{E}(\mathcal{H})$ we denote the projection onto the intersection of the closure of their ranges by $P_{A,B}$.

**Theorem 5.3.** [25] *If $A \in \mathcal{E}(\mathcal{H})$, $P \in \mathcal{P}(\mathcal{H})$, then $A \wedge P$ exists.*

**Theorem 5.4.** [25] *If* $\dim \mathcal{H} < \infty$ *and* $A, B \in \mathcal{E}(\mathcal{H})$, *then* $A \wedge B$ *exists if and only if* $A \wedge P_{A,B}$ *and* $B \wedge P_{A,B}$ *are comparable. In this case,* $A \wedge B$ *is the smaller of* $A \wedge P_{A,B}$ *and* $B \wedge P_{A,B}$.

A generalization of Theorem 5.4 to the case $\dim \mathcal{H} = \infty$ is considerably more complicated and may be found in [12].

# 6 Effects and observables

This section discusses a probabilistic formalism that unifies the work of the previous four sections. In particular, we show that traditional unsharp probability theory generalizes all the structures that we have considered so far.

Let $\mathcal{E} = \mathcal{E}(\Omega, \mathcal{A})$ be a set of effects (or fuzzy events) considered in Section 3. For $f_i \in \mathcal{E}(\Omega, \mathcal{A}), i = 1, 2, \ldots$, we write $\oplus f_i$ for $\sum f_i$ whenever $\sum f_i \leq 1$. Denote the set of probability measures on $(\Omega, \mathcal{A})$ by $M_1^+(\Omega, \mathcal{A})$. If $\mu \in M_1^+(\Omega, \mathcal{A})$ and $f \in \mathcal{E}(\Omega, \mathcal{A})$, as in Section 3, we define the probability of $f$ by $\mu(f) = \int f \, d\mu$. We call a map $\phi: \mathcal{E}(\Omega, \mathcal{A}) \to [0, 1]$ a *state* if $\phi(1) = 1$ and $\phi(\oplus f_n) = \sum \phi(f_n)$. It is shown in [17] that every state $\phi$ on $\mathcal{E}(\Omega, \mathcal{A})$ has the form $\phi(f) = \mu(f)$ for a unique $\mu \in M_1^+(\Omega, \mathcal{A})$. If $(\Lambda, \mathcal{B})$ is another measurable space, a $\sigma$-*morphism* $\phi: \mathcal{E}(\Lambda, \mathcal{B}) \to \mathcal{E}(\Omega, \mathcal{A})$ satisfies $\phi(1) = 1$ and $\phi(\oplus f_n) = \sum \phi(f_n)$. A map $X: \mathcal{B} \to \mathcal{E}(\Omega, \mathcal{A})$ is an *observable on* $\mathcal{E}(\Omega, \mathcal{A})$ *with value space* $(\Lambda, \mathcal{B})$ if $X(\Lambda) = 1$ and $X(\oplus B_n) = \sum X(B_n)$ where the convergence in the summation is pointwise. We interpret $X(B) \in \mathcal{E}(\Omega, \mathcal{A})$ as the effect or fuzzy event that occurs when $X$ has a value in $B \in \mathcal{B}$. We sometimes use the notation $X(\omega, B) = X(B)(\omega)$.

A *probability kernel on* $(\Omega, \mathcal{A})$ *with value space* $(\Lambda, \mathcal{B})$ is a map $K: \Omega \times \mathcal{B} \to [0, 1]$ such that $K(\cdot, B)$ is measurable for every $B \in \mathcal{B}$ and $K(\omega, \cdot) \in M_1^+(\Lambda, \mathcal{B})$ for every $\omega \in \Omega$. Observables and probability kernels are equivalent concepts. Indeed, if $X: \mathcal{B} \to \mathcal{E}(\Omega, \mathcal{A})$ is an observable, then $X(\omega, B)$ is a probability kernel and conversely, if $K: \Omega \times \mathcal{B} \to [0, 1]$ is a probability kernel, then $X(B)(\omega) = K(\omega, B)$ is an observable. An observable $X: \mathcal{B} \to \mathcal{E}(\Omega, \mathcal{A})$ is *sharp* (or *crisp*) if $X(B)$ is sharp for every $B \in \mathcal{B}$. If $(\Lambda, \mathcal{B})$ is a Polish measurable space, then it can be shown that $X: \mathcal{B} \to \mathcal{E}(\Omega, \mathcal{A})$ is sharp if and only if there exists a measurable function $f: \Omega \to \Lambda$ such that $X(B) = \chi_{f^{-1}(B)}$ for every $B \in \mathcal{B}$, see [2, 3]. We use the notation $X_f$ for the sharp observable corresponding to $f$.

If $\mu \in M_1^+(\Omega, \mathcal{A})$ and $X: \mathcal{B} \to \mathcal{E}(\Omega, \mathcal{A})$ is an observable, then

$$D_X(\mu) = \mu \circ X \in M_1^+(\Lambda, \mathcal{B})$$

is called the *distribution of* $X$ *in the state* $\mu$. We interpret $D_X(\mu)(B) = \mu(X(B))$ as the probability that $X$ has a value in $B$ when the system is in the state $\mu$. When $X = X_f$ is crisp, we have $D_X(\mu)(B) = \mu(f^{-1}(B))$ which is the usual distribution of the random variable $f$. The next result, which is proved in [17], shows that there is a natural one-to-one correspondence between observables and $\sigma$-morphisms.

**Theorem 6.1.** *If* $X: \mathcal{B} \to \mathcal{E}(\Omega, \mathcal{A})$ *is an observable, then* $X$ *has a unique extension to a* $\sigma$-*morphism* $\bar{X}: \mathcal{E}(\Lambda, \mathcal{B}) \to \mathcal{E}(\Omega, \mathcal{A})$. *If* $Y: \mathcal{E}(\Lambda, \mathcal{B}) \to \mathcal{E}(\Omega, \mathcal{A})$ *is a* $\sigma$-*morphism, then* $Y \mid \mathcal{B}$ *is an observable.*

It is shown in [17] that the unique extension $\widetilde{X}$ is given by

$$(\widetilde{X}g)(\omega) = \int g(\lambda)X(\omega, d\lambda) \tag{6.1}$$

and if $X_f$ is crisp, then $\widetilde{X}_f g = g \circ f$. In the sequel, we shall omit the $\sim$ on $\widetilde{X}$ and shall frequently identify an observable with its corresponding unique $\sigma$-morphism.

Let $(\Omega, \mathcal{A})$, $(\Lambda_1, \mathcal{B}_1)$, $(\Lambda_2, \mathcal{B}_2)$ be measurable spaces. If $X: \mathcal{B}_1 \to \mathcal{E}(\Omega, \mathcal{A})$ is an observable and $u: \Lambda_1 \to \Lambda_2$ is a measurable function, we define the observable $u(X): \mathcal{B}_1 \to \mathcal{E}(\Omega, \mathcal{A})$ by $u(X)(B) = X(u^{-1}(B))$. We shall see that $u(X)$ can be viewed as a composition of the observables $X$ and $X_u$.

Let $(\Omega, \mathcal{A})$, $(\Lambda_1, \mathcal{B}_1)$, $(\Lambda_2, \mathcal{B}_2)$ be measurable spaces and let $Y: \mathcal{B}_2 \to \mathcal{E}(\Lambda_1, \mathcal{B}_1)$ and $X: \mathcal{B}_1 \to \mathcal{E}(\Omega, \mathcal{A})$ be observables. Although we cannot directly compose $X$ and $Y$, we can compose them if they are thought of as $\sigma$-morphisms. Doing this, we have the $\sigma$-morphism $X \circ Y: \mathcal{E}(\Lambda_2, \mathcal{B}_2) \to \mathcal{E}(\Omega, \mathcal{A})$, which we identify with the observable $X \circ Y: \mathcal{B}_2 \to \mathcal{E}(\Omega, \mathcal{A})$. We call $X \circ Y$ the *composition* of $X$ and $Y$. We thus have

$$(X \circ Y)(\omega, B) = [X(Y(B))](\omega) = \int Y(B)(\lambda_1)X(\omega, d\lambda_1)$$
$$= \int Y(\lambda_1, B)X(\omega, d\lambda_1) \tag{6.2}$$

which is the usual way of composing probability kernels. We now consider the special cases in which $X$ or $Y$ is crisp. Suppose that $Y$ is crisp and $Y = X_u$ where $u: \Lambda_1 \to \Lambda_2$ is a measurable function. We then have

$$(X \circ X_u)(B) = X(u^{-1}(B)) = u(X)(B).$$

Hence, $X \circ X_u = u(X)$ and $(X \circ X_u)(\omega, B) = X(\omega, u^{-1}(B))$. Next, suppose that $X$ is crisp and $X = X_f$ where $f: \Omega \to \Lambda_1$ is a random variable. We then have

$$(X_f \circ Y)(B) = X_f(Y(B)) = Y(B) \circ f$$

and $(X_f \circ Y)(\omega, B) = Y(f(\omega), B)$. Finally, if both $X$ and $Y$ are crisp, we have

$$(X_f \circ X_u)(B) = u(X_f)(B) = f^{-1}(u^{-1}(B)) = (u \circ f)^{-1}(B) = X_{u \circ f}.$$

Hence, $X_f \circ X_u = X_{u \circ f}$.

An observable $\varepsilon: \mathcal{B} \to \mathcal{E}(\Lambda, \mathcal{B})$ is called a *fuzzification*. The terminology stems from the fact that in general $\varepsilon$ takes crisp events on $(\Lambda, \mathcal{B})$ to fuzzy events on $(\Lambda, \mathcal{B})$. If $X: \mathcal{B} \to \mathcal{E}(\Omega, \mathcal{A})$ is an observable, then $X \circ \varepsilon: \mathcal{B} \to \mathcal{E}(\Omega, \mathcal{A})$ is the observable $X$ *fuzzified by* $\varepsilon$. In particular, if $f: \Omega \to \Lambda$ is a random variable, then $X_f \circ \varepsilon: \mathcal{B} \to \mathcal{E}(\Omega, \mathcal{A})$ is the random variable $f$ *fuzzified by* $\varepsilon$, see [2, 7]. We then have

$$(X_f \circ \varepsilon)(\omega, B) = \varepsilon(f(\omega), B)$$

An observable $\phi: \mathcal{A} \to \mathcal{E}(\Omega, \mathcal{A})$ is called a *noise* and if $X: \mathcal{B} \to \mathcal{E}(\Omega, \mathcal{A})$ is an observable, then $\phi \circ X: \mathcal{B} \to \mathcal{E}(\Omega, \mathcal{A})$ is called the observable $X$ *with noise* $\phi$. In particular, if $f: \Omega \rightarrow \Lambda$ is a random variable, then $\phi \circ X_f: \mathcal{B} \rightarrow \mathcal{E}(\Omega, \mathcal{A})$ is the random variable $f$ with noise $\phi$, see [2, 7]. We then have $\phi \circ X_f = f(\phi)$. Corresponding to a noise $\phi: \mathcal{A} \rightarrow \mathcal{E}(\Omega, \mathcal{A})$ we have a sequence of observables $\phi, \phi \circ \phi, \phi \circ \phi \circ \phi, \ldots$ which we denote by $\phi, \phi^{(2)}, \phi^{(3)}, \ldots$. Such a sequence is called a *discrete fuzzy dynamical system*. In particular, if $f: \Omega \rightarrow \Omega$ is a random variable, we use the notation $f^{(n)} = f \circ f \circ \cdots \circ f$ ($n$ factors). Then $X_f^{(n)} = X_{f^{(n)}}$ corresponds to the discrete crisp dynamical system $f, f^{(2)}, f^{(3)}, \ldots$.

Let $(\Omega_i, \mathcal{A}_i), (\Lambda_i, \mathcal{B}_i), i = 1, 2$, be measurable spaces and let $X_i: \mathcal{B}_i \rightarrow \mathcal{E}(\Omega_i, \mathcal{A}_i)$, $i = 1, 2$, be observables. Denote the corresponding product spaces by $(\Omega_1 \times \Omega_2, \mathcal{A}_1 \times \mathcal{A}_2), (\Lambda_1 \times \Lambda_2, \mathcal{B}_1 \times \mathcal{B}_2)$. Using standard results on product measures, it can be shown that there exists a unique observable

$$X_1 \times X_2: \mathcal{B}_1 \times \mathcal{B}_2 \rightarrow (\Omega_1 \times \Omega_2, \mathcal{A}_1 \times \mathcal{A}_2)$$

such that

$$[(X_1 \times X_2)(B_1 \times B_2)](\omega_1, \omega_2) = X_1(B_1)(\omega_1) X_2(B_2)(\omega_2)$$

for all $B_i \in \mathcal{B}_i, i = 1, 2$, see [2, 17]. We call $X_1 \times X_2$ the *product* of $X_1$ and $X_2$. If $Y: \mathcal{B}_1 \times \mathcal{B}_2 \rightarrow \mathcal{E}(\Omega_1 \times \Omega_2, \mathcal{A}_1 \times \mathcal{A}_2)$ is an arbitrary observable, then the *marginal observables* $Y_i: \mathcal{B}_i \rightarrow \mathcal{E}(\Omega_1 \times \Omega_2, \mathcal{A}_1 \times \mathcal{A}_2), i = 1, 2$, for $Y$ are given by $Y_1(B_1) = Y(B_1 \times \Lambda_2)$ and $Y_2(B_2) = Y(\Lambda_1 \times B_2)$. In general, $Y \neq Y_1 \times Y_2$. However, if $Y = X_1 \times X_2$, then the marginal observables for $Y$ are $X_1, X_2$. This construction can easily be extended to a product $X_1 \times X_2 \times \cdots \times X_n$ of a finite number of observables. More generally, if $(\Omega_t, \mathcal{A}_t), (\Lambda_t, \mathcal{B}_t), t \in T$, are indexed families of measurable spaces, we can form the product spaces $(\times \Omega_t, \times \mathcal{A}_t), (\times \Lambda_t, \times \mathcal{B}_t)$ where the $\sigma$-algebras $\times \mathcal{A}_t$, $\times \mathcal{B}_t$ are generated by the cylinder sets. We then extend the product construction to form the product $\times X_t$ of observables $X_t: \mathcal{B}_t \rightarrow \mathcal{E}(\Omega_t, \mathcal{A}_t), t \in T$.

A similar construction applies for observables $X_i: \mathcal{B}_i \rightarrow \mathcal{E}(\Omega, \mathcal{A}), i = 1, 2$, on the same measurable space. In this case, we have the *direct product observable* $X_1 \otimes X_2: \mathcal{B}_1 \times \mathcal{B}_2 \rightarrow \mathcal{E}(\Omega, \mathcal{A})$ which is the unique observable that satisfies

$$(X_1 \otimes X_2)(B_1 \times B_2) = X_1(B_1) X_2(B_2)$$

for every $B_i \in \mathcal{B}_i, i = 1, 2$. Again, we can extend this to direct products $X_1 \otimes X_2 \otimes \cdots \otimes X_n$ of a finite number of observables $X_1, \ldots, X_n$ and to direct products $\otimes X_t$ of an indexed family of observables $X_t, t \in T$.

Let $(\Omega, \mathcal{A}), (\Lambda, \mathcal{B})$ be measurable spaces and let $T = \mathbb{R}^+$ or $\mathbb{Z}^+$. Letting $\Lambda_t = \Lambda$ and $\mathcal{B}_t = \mathcal{B}$ for all $t \in T$, we use the notation $\Lambda^T = \times \Lambda_t$ and $\mathcal{B}^T = \times \mathcal{B}_t$ where $\mathcal{B}^T$ is the $\sigma$-algebra on $\Lambda^T$ generated by the cylinder sets. We then form the product space $(\Lambda^T, \mathcal{B}^T)$. The elements $\tilde{\lambda} \in \Lambda^T$ are functions $\tilde{\lambda}: T \rightarrow \Lambda$, which we call *paths* in $\Lambda$. Recall that $\mathcal{B}^T$ is the smallest $\sigma$-algebra on $\Lambda^T$ such that the projections $\pi_t: \Lambda^T \rightarrow \Lambda$ given by $\pi_t(\tilde{\lambda}) = \tilde{\lambda}(t)$ are measurable. We call $\tilde{\lambda}(t)$ the *coordinate of* $\tilde{\lambda}$ *at time* $t \in T$. A *fuzzy stochastic process* is an observable $X: \mathcal{B}^T \rightarrow \mathcal{E}(\Omega, \mathcal{A})$. For $B \in \mathcal{B}, t \in T$, we define $B_t \in \mathcal{B}^T$ by

$$B_t = \left\{ \tilde{\lambda} \in \Lambda^T : \tilde{\lambda}(t) \in B \right\}$$

Then $\{B_t : B \in \mathcal{B}\}$ is a $\sigma$-subalgebra of $\mathcal{B}^T$ that is isomorphic to $\mathcal{B}$. The observable $X_t : \mathcal{B} \to \mathcal{E}(\Omega, \mathcal{A})$ defined by $X_t(B) = X(B_t)$ is the *marginal observable for $X$ at time $t$*. In general, the marginal observables $X_t$, $t \in T$, do not determine the process $X$. Conversely, let $X_t : \mathcal{B} \to \mathcal{E}(\Omega, \mathcal{A})$ be a family of observables, $t \in T$. Then this family generates a fuzzy stochastic process $Y = \otimes X_t$, $Y : \mathcal{B}^T \to \mathcal{E}(\Omega, \mathcal{A})$ such that $Y_t = X_t$, $t \in T$. However, in general there are other processes with marginals $X_t$, $t \in T$. If $X : \mathcal{B}^T \to \mathcal{E}(\Omega, \mathcal{A})$ is crisp (which corresponds to a standard stochastic process), then this ambiguity disappears and $X = \otimes X_t$, see [2, 3]. If $X = \otimes X_t$, we call $X$ a *factorizable* fuzzy stochastic process. If $X$ is factorizable, $(\Lambda, \mathcal{B}) = (\Omega, \mathcal{A})$ and $X_{s+t} = X_s \circ X_t$ for all $s, t \in T$, then $X$ is a *Markov process*. In this case, by (6.2) we have

$$X_{s+t}(\omega, A) = (X_s \circ X_t)(\omega, A) = \int X_t(\omega', A) X_s(\omega, d\omega')$$

which is the Chapman-Kolmogorov equation. A Markov process $X$ for which $T = \mathbb{Z}^+$ is called a *Markov chain*. In this case $X_2 = X_1 \circ X_1 = X_1^{(2)}$, $X_3 = X_1 \circ X_2 = X_1^{(3)}, \ldots, X_n = X^{(n)}$.

We can compose a fuzzy stochastic process with an observable to form a new stochastic process. For example, let $X : \mathcal{B}^T \to \mathcal{E}(\Omega, \mathcal{A})$ be a fuzzy stochastic process and let $Y : \mathcal{A} \to \mathcal{E}(\Lambda', \mathcal{B}')$ be an observable. Then $Y \circ X : \mathcal{B}^T \to \mathcal{E}(\Lambda', \mathcal{B}')$ is the fuzzy stochastic process $X$ *transferred* by $Y$. As another example, let $Y : \mathcal{B}' \to \mathcal{E}(\Lambda, \mathcal{B})$ be an observable and let $Y_{(t)} = Y$ for every $t \in T$. Then $X \circ (\times Y_{(t)}) : \mathcal{B}'^T \to \mathcal{E}(\Omega, \mathcal{A})$ is the process $X$ *pretransferred* by $Y$. In particular, if $X : \mathcal{A}^T \to \mathcal{E}(\Omega, \mathcal{A})$ gives the evolution of a system and $Y : \mathcal{B} \to \mathcal{E}(\Omega, \mathcal{A})$ is an observable, then $X \circ (\times Y_{(t)}) : \mathcal{B}^T \to \mathcal{E}(\Omega, \mathcal{A})$ gives the evolution of $Y$.

Let $\mathcal{H}$ be a complex Hilbert space and let $\Omega(\mathcal{H}) = \{\omega \in \mathcal{H} : \|\omega\| = 1\}$. Endow $\Omega(\mathcal{H})$ with the norm topology $\tau$ and let $\mathcal{A}(\mathcal{H})$ be the $\sigma$-algebra generated by the open sets in $\tau$. We now examine the relationship between $\mathcal{E}(\mathcal{H})$ and $\mathcal{E}(\Omega(\mathcal{H}), \mathcal{A}(\mathcal{H}))$. For $F \in \mathcal{E}(\mathcal{H})$, define $\tilde{F} : \Omega(\mathcal{H}) \to [0, 1]$ by $\tilde{F}(\omega) = \langle F\omega, \omega \rangle$. If a sequence $\omega_i \in \Omega(\mathcal{H})$ converges to $\omega \in \Omega(\mathcal{H})$ in the topology $\tau$, then $F^{1/2}\omega_i$ converges to $F^{1/2}\omega$ and hence $\lim \|F^{1/2}\omega_i\|^2 = \|F^{1/2}\omega\|^2$. But

$$\|F^{1/2}\omega\|^2 = \left\langle F^{1/2}\omega, F^{1/2}\omega \right\rangle = \langle F\omega, \omega \rangle = \tilde{F}(\omega)$$

and similarly, $\|F^{1/2}\omega_i\|^2 = \tilde{F}(\omega_i)$. Hence, $\lim \tilde{F}(\omega_i) = \tilde{F}(\omega)$ so $\tilde{F}$ is continuous in the $\tau$ topology. It follows that $\tilde{F}$ is measurable so $\tilde{F} \in \mathcal{E}(\Omega(\mathcal{H}), \mathcal{A}(\mathcal{H}))$. It is easy to show that $\tilde{\ } : \mathcal{E}(\mathcal{H}) \to \mathcal{E}(\Omega(\mathcal{H}), \mathcal{A}(\mathcal{H}))$ is a $\sigma$-morphism. Moreover, if $\tilde{F} \perp \tilde{G}$, then $F \perp G$. It follows that $\tilde{\ } : \mathcal{E}(\mathcal{H}) \to \mathcal{E}(\Omega(\mathcal{H}), \mathcal{A}(\mathcal{H}))$ is a $\sigma$-isomorphism from $\mathcal{E}(\mathcal{H})$ onto its range $\mathcal{E}(\mathcal{H})^{\tilde{\ }}$ in $\mathcal{E}(\Omega(\mathcal{H}), \mathcal{A}(\mathcal{H}))$. Let $(\Lambda, \mathcal{B})$ be a measurable space and let $X : \mathcal{B} \to \mathcal{E}(\mathcal{H})$ be a normalized, positive operator-valued measure. Then $\tilde{X} : \mathcal{B} \to \mathcal{E}(\Omega(\mathcal{H}), \mathcal{A}(\mathcal{H}))$ defined by $\tilde{X}(B) = X(B)^{\tilde{\ }}$ is an observable on $\mathcal{E}(\Omega(\mathcal{H}), \mathcal{A}(\mathcal{H}))$, which we call a *quantum observable*. The distribution $D_{\tilde{X}}(\mu)$ of $\tilde{X}$ for $\mu \in M_1^+(\Omega(\mathcal{H}), \mathcal{A}(\mathcal{H}))$ becomes

$$D_{\tilde{X}}(\mu)(B) = \mu\left(\tilde{X}(B)\right) = \int \tilde{X}(B)(\omega)\mu(d\omega) = \int \langle X(B)\omega, \omega \rangle \mu(d\omega)$$

Since $\mathcal{E}(\mathcal{H})$ and $\mathcal{E}(\mathcal{H})\tilde{}$ are isomorphic, we also call $X$ a *quantum observable*.

As we have seen, any state $s$ on $\mathcal{E}(\mathcal{H})$ has the form $s(F) = \mathrm{tr}(\rho F)$ for a unique positive trace class operator $\rho$. By the spectral theorem, $\rho$ has the unique representation $\rho = \sum \lambda_i P_i$ where $\lambda_i > 0$, $\sum \lambda_i = 1$ and the $P_i$ are mutually orthogonal one-dimensional projections. Let $\omega_i$ be unit vectors in the range of $P_i$, $i = 1, 2, \ldots$, and define the probability measure $\tilde{s}$ on $(\Omega(\mathcal{H}), \mathcal{A}(\mathcal{H}))$ by $\tilde{s} = \sum \lambda_i \delta_{\omega_i}$, where $\delta_\omega$ denotes the Dirac measure concentrated at $\omega$. Then for $F \in \mathcal{E}(\mathcal{H})$ we have

$$\tilde{s}(\tilde{F}) = \sum \lambda_i \tilde{F}(\omega_i) = \sum \lambda_i \langle F\omega_i, \omega_i \rangle = s(F)$$

It follows that if $X: \mathcal{B} \to \mathcal{E}(\mathcal{H})$ is a quantum observable and $s$ is a state on $\mathcal{E}(\mathcal{H})$, then the distribution $B \mapsto s(X(B))$ of $X$ coincides with the distribution $D_{\tilde{X}}(\tilde{s})$ of $\tilde{X}$ relative to $\tilde{s} \in M_1^+(\Omega(\mathcal{H}), \mathcal{A}(\mathcal{H}))$. In this case, we have

$$D_{\tilde{X}}(\tilde{s})(B) = \sum \lambda_i \langle X(B)\omega_i, \omega_i \rangle = \mathrm{tr}\,(X(B)\rho)$$

In particular, $B \mapsto \tilde{X}(B)(\omega) = \langle X(B)\omega, \omega \rangle$ is the distribution of $\tilde{X}$ (and $X$) in the pure state $\omega$.

We now consider the important question of characterizing the elements of $\mathcal{E}(\mathcal{H})\tilde{}$ in $\mathcal{E}(\Omega(\mathcal{H}), \mathcal{A}(\mathcal{H}))$. That is, we would like to characterize the effects $f \in \mathcal{E}(\Omega(\mathcal{H}), \mathcal{A}(\mathcal{H}))$ that are quantum effects. For $f \in \mathcal{E}(\Omega(\mathcal{H}), \mathcal{A}(\mathcal{H}))$, define $\bar{f}: \mathcal{H} \to \mathbb{R}$ by $\bar{f}(0) = 0$ and if $\psi \neq 0$, then $\bar{f}(\psi) = \|\psi\| f(\psi/\|\psi\|)^{1/2}$.

**Theorem 6.2.** [17] *For $f \in \mathcal{E}(\Omega(\mathcal{H}), \mathcal{A}(\mathcal{H}))$, we have $f \in \mathcal{E}(\mathcal{H})\tilde{}$ if and only if $\bar{f}$ is a seminorm that satisfies the parallelogram law.*

We now consider quantum dynamics. If $U: \mathcal{H} \to \mathcal{H}$ is a unitary operator, then $U: \Omega(\mathcal{H}) \to \Omega(\mathcal{H})$ is continuous and hence measurable. Thus, $X_U: \mathcal{A}(\mathcal{H}) \to \mathcal{E}(\Omega(\mathcal{H}), \mathcal{A}(\mathcal{H}))$ is a crisp observable where, by definition, $X_U(A) = I_{U^{-1}(A)}$. Now suppose that $U(t), t \in \mathbb{R}$, is a dynamical group. That is, $U(t)$ is a unitary operator and $U(s + t) = U(s)U(t)$ for all $s, t \in \mathbb{R}$. For example, $U(t) = e^{-itK}$ the group of unitary transformations generated by the Schrödinger equation, where $K$ is the energy operator. Then $\otimes X_{U(t)}: \mathcal{A}(\mathcal{H})^{\mathbb{R}} \to \mathcal{E}(\Omega(\mathcal{H}), \mathcal{A}(\mathcal{H}))$ is a fuzzy stochastic process. If $Y: \mathcal{B} \to \mathcal{E}(\Omega(\mathcal{H}), \mathcal{A}(\mathcal{H}))$ is an observable, then

$$\otimes X_{U(t)} \circ \times Y_{(t)}: \mathcal{B}^{\mathbb{R}} \to \mathcal{E}(\Omega(\mathcal{H}), \mathcal{A}(\mathcal{H}))$$

describes the evolution of $Y$. To verify this statement for quantum observables, let $X: \mathcal{B} \to \mathcal{E}(\mathcal{H})$ be a quantum observable. Then in conventional quantum mechanics, $U(t)^* X U(t), t \in \mathbb{R}$, describes the evolution of $X$. Then for every $t \in \mathbb{R}$, we have

$$\left[ U(t)^* X U(t) \right]\tilde{} : \mathcal{B} \to \mathcal{E}(\Omega(\mathcal{H}), \mathcal{A}(\mathcal{H}))$$

and

$$\otimes \left[ U(t)^* X U(t) \right]\tilde{} : \mathcal{B}^{\mathbb{R}} \to \mathcal{E}(\Omega(\mathcal{H}), \mathcal{A}(\mathcal{H}))$$

is a fuzzy stochastic process. The next result shows that this process is given by our previous description.

**Theorem 6.3.** *In terms of our previous notation, we have*

$$\otimes \tilde{X}_{U(t)} \circ \times \tilde{X}_{(t)} = \otimes \left[U(t)^* X U(t)\right]^{\sim}.$$

# 7 Statistical maps

We now discuss the relationship between observables and statistical maps. If $(\Omega, \mathcal{A})$, $(\Lambda, \mathcal{B})$ are measurable spaces, a function $f \colon \Omega \to M_1^+(\Lambda, \mathcal{B})$ is called a *fuzzy random variable* (or a *statistical function*) (see [2, 3, 4, 6, 15]) if $w \mapsto [f(\omega)](B)$ is measurable for every $B \in \mathcal{B}$. There is a one-to-one correspondence between observables and fuzzy random variables. Indeed, if $X \colon \mathcal{B} \to \mathcal{E}(\Omega, \mathcal{A})$ is an observable, then $\widehat{X} \colon \Omega \to M_1^+(\Lambda, \mathcal{B})$ defined by $\widehat{X}(\omega)(B) = X(\omega, B)$ is a fuzzy random variable. Conversely, if $f \colon \Omega \to M_1^+(\Lambda, \mathcal{B})$ is a fuzzy random variable, then $f^\vee \colon \mathcal{B} \to \mathcal{E}(\Omega, \mathcal{A})$ defined by $f^\vee(B)(\omega) = f(\omega)(B)$ is an observable.

Let $M(\Omega, \mathcal{A})$ be the set of bounded signed measures on $(\Omega, \mathcal{A})$ and let $M^+(\Omega, \mathcal{A})$ be the set of bounded measures on $(\Omega, \mathcal{A})$. By the Jordan decomposition theorem, every $\mu \in M(\Omega, \mathcal{A})$ has the form $\mu = \mu^+ - \mu^-$ where $\mu^+, \mu^- \in M^+(\Omega, \mathcal{A})$. Now $M(\Omega, \mathcal{A})$ is a real linear space and it follows that $M^+(\Omega, \mathcal{A})$ is a generating positive cone in $M(\Omega, \mathcal{A})$. Moreover, $M_1^+(\Omega, \mathcal{A})$ is a convex subset of $M^+(\Omega, \mathcal{A})$. If $\mu \in M^+(\Omega, \mathcal{A})$, $\mu \neq 0$, then there exists a unique $\nu \in M_1^+(\Omega, \mathcal{A})$ and a unique $\lambda \in (0, 1]$ such that $\mu = \lambda\nu$. Hence, $M_1^+(\Omega, \mathcal{A})$ is a convex base for $M(\Omega, \mathcal{A})$ and $M(\Omega, \mathcal{A})$ becomes a base normed space. Let $B^+(\Omega, \mathcal{A})$ be the set of nonnegative bounded measurable functions on $(\Omega, \mathcal{A})$ and let $B(\Omega, \mathcal{A})$ be the real linear space of bounded real-valued measurable functions on $(\Omega, \mathcal{A})$. Then $B^+(\Omega, \mathcal{A})$ is a generating positive cone for $B(\Omega, \mathcal{A})$. Moreover, $\mathcal{E}(\Omega, \mathcal{A})$ is a generating convex subset of $B(\Omega, \mathcal{A})$ and $1 \in \mathcal{E}(\Omega, \mathcal{A})$ is an order unit of $B(\Omega, \mathcal{A})$. Hence, $B(\Omega, \mathcal{A})$ becomes an order unit space.

The pair of spaces $(B(\Omega, \mathcal{A}), M(\Omega, \mathcal{A}))$ possesses a natural duality $F \colon B(\Omega, \mathcal{A}) \times M(\Omega, \mathcal{A}) \to \mathbb{R}$ given by $F(f, \mu) = \int f \, d\mu$. We shall use the notation $\mu(f) = f(\mu) = F(f, \mu)$. It is easy to show that $\mu(f_1) = \mu(f_2)$ for every $\mu \in M(\Omega, \mathcal{A})$ implies $f_1 = f_2$ and that $f(\mu_1) = f(\mu_2)$ for every $f \in B(\Omega, \mathcal{A})$ implies $\mu_1 = \mu_2$. Hence, $F$ is indeed a duality. This duality induces the *weak topologies* on $B(\Omega, \mathcal{A})$ and $M(\Omega, \mathcal{A})$. A net $f_\alpha \in B(\Omega, \mathcal{A})$ converges weakly to $f \in B(\Omega, \mathcal{A})$ if $\mu(f_\alpha) \to \mu(f)$ for every $\mu \in M(\Omega, \mathcal{A})$ and a net $\mu_\alpha \in M(\Omega, \mathcal{A})$ converges weakly to $\mu \in M(\Omega, \mathcal{A})$ if $\mu_\alpha(f) \to \mu(f)$ for every $f \in B(\Omega, \mathcal{A})$. It follows that every weakly continuous linear functional $G$ on $B(\Omega, \mathcal{A})$ has the form $G(f) = \mu(f)$ for a unique $\mu \in M(\Omega, \mathcal{A})$ and every weakly continuous linear functional $H$ on $M(\Omega, \mathcal{A})$ has the form $H(\mu) = f(\mu)$ for a unique $f \in B(\Omega, \mathcal{A})$, see [28]. If $T \colon M(\Omega, \mathcal{A}) \to M(\Lambda, \mathcal{B})$ is a weakly continuous linear operator, then its *adjoint* $T' \colon B(\Lambda, \mathcal{B}) \to B(\Omega, \mathcal{A})$ is the weakly continuous linear operator defined by $(T'g)(\mu) = (T\mu)(g)$ for every $g \in B(\Lambda, \mathcal{B}), \mu \in M(\Omega, \mathcal{A})$. Conversely, if $S \colon B(\Lambda, \mathcal{B}) \to B(\Omega, \mathcal{A})$ is a weakly continuous operator, then its *adjoint* $S' \colon M(\Omega, \mathcal{A}) \to M(\Lambda, \mathcal{B})$ is the weakly continuous linear operator defined by $(S'\mu)(g) = (Sg)(\mu)$ for every $\mu \in M(\Omega, \mathcal{A}), g \in B(\Lambda, \mathcal{B})$. It follows that $T'' = T$ and $S'' = S$.

A map $T: M_1^+(\Omega, \mathcal{A}) \to M_1^+(\Lambda, \mathcal{B})$ is *measurable* if $\omega \mapsto (T\delta_\omega)(B)$ is measurable for every $B \in \mathcal{B}$. We call $T: M_1^+(\Omega, \mathcal{A}) \to M_1^+(\Lambda, \mathcal{B})$ a *statistical map* if $T$ is affine, measurable and weakly continuous. If $X: \mathcal{B} \to \mathcal{E}(\Omega, \mathcal{A})$ is an observable, it is clear that its distribution map $D_X: M_1^+(\Omega, \mathcal{A}) \to M_1^+(\Lambda, \mathcal{B})$ is affine. The next result shows that $D_X$ is a statistical map and that every statistical map has this form.

**Theorem 7.1.** [17] *A map* $T: M_1^+(\Omega, \mathcal{A}) \to M_1^+(\Lambda, \mathcal{B})$ *is a statistical map if and only if there exists an observable* $X: \mathcal{B} \to \mathcal{E}(\Omega, \mathcal{A})$ *such that* $T = D_X$. *Moreover,* $X$ *is unique.*

**Corollary 7.2.** *If* $X, Y: \mathcal{B} \to \mathcal{E}(\Omega, \mathcal{A})$ *are observables for which* $D_X = D_Y$, *then* $X = Y$.

**Corollary 7.3.** *If* $T: M_1^+(\Omega, \mathcal{A}) \to M_1^+(\Lambda, \mathcal{B})$ *is a statistical map, then for every* $\mu \in M_1^+(\Omega, \mathcal{A})$ *and* $B \in \mathcal{B}$ *we have*

$$(T\mu)(B) = \int (T\delta_\omega)(B)\mu(d\omega) \tag{7.1}$$

*Conversely, if* $T: M_1^+(\Omega, \mathcal{A}) \to M_1^+(\Lambda, \mathcal{B})$ *is measurable and satisfies (7.1), then* $T$ *is a statistical map.*

In [2, 3, 4, 6, 7] a statistical map is defined to be a measurable $T: M_1^+(\Omega, \mathcal{A}) \to M_1^+(\Lambda, \mathcal{B})$ that satisfies (7.1). (It is also assumed that $T$ is affine, but this condition is redundant.) Corollary 7.3 shows that this definition is equivalent to the one we have given. However, we believe that our definition is more basic and easier to verify.

A linear operator $L: B(\Lambda, \mathcal{B}) \to B(\Omega, \mathcal{A})$ is *positive* if $L\left[B^+(\Lambda, \mathcal{B})\right] \subseteq B^+(\Omega, \mathcal{A})$ and *unital* if $L(1) = 1$. The next result shows that there is a one-to-one correspondence between observables and positive, unital, weakly continuous operators.

**Theorem 7.4.** [17] *If* $X: \mathcal{B} \to \mathcal{E}(\Omega, \mathcal{A})$ *is an observable, then* $X$ *has a unique extension to a positive, unital, weakly continuous operator* $\overline{X}: B(\Lambda, \mathcal{B}) \to B(\Omega, \mathcal{A})$. *If* $L: B(\Lambda, \mathcal{B}) \to B(\Omega, \mathcal{A})$ *is a positive, unital, weakly continuous operator, then* $L \mid \mathcal{B}$ *is an observable.*

If $X: \mathcal{B} \to \mathcal{E}(\Omega, \mathcal{A})$ is an observable and $\mu \in M(\Omega, \mathcal{A})$, we define $\mu_X \in M(\Lambda, \mathcal{B})$ by $\mu_X(B) = \int X(\omega, B)\mu(d\omega)$. We then define the operator $\overline{D}_X: M(\Omega, \mathcal{A}) \to M(\Lambda, \mathcal{B})$ by $\overline{D}_X\mu = \mu_X$. An operator $S: M(\Omega, \mathcal{A}) \to M(\Lambda, \mathcal{B})$ is a *statistical operator* if

$$S\left[M_1^+(\Omega, \mathcal{A})\right] \subseteq M_1^+(\Lambda, \mathcal{B})$$

and $S$ is weakly continuous. An operator $S: M(\Omega, \mathcal{A}) \to M(\Lambda, \mathcal{B})$ *preserves total mass* if $(S\mu)(\Lambda) = \mu(\Omega)$ for every $\mu \in M(\Omega, \mathcal{A})$. We thus see that an operator $S: M(\Omega, \mathcal{A}) \to M(\Lambda, \mathcal{B})$ is a statistical operator if and only if $S$ is positive, weakly continuous and preserves total mass.

**Theorem 7.5.** [17] *An operator* $S: M(\Omega, \mathcal{A}) \to M(\Lambda, \mathcal{B})$ *is a statistical operator if and only if there exists an observable* $X: \mathcal{B} \to \mathcal{E}(\Omega, \mathcal{A})$ *such that* $S = \overline{D}_X$. *Moreover,* $X$ *is unique.*

**Corollary 7.6.** *If* $T: M_1^+(\Omega, \mathcal{A}) \to M_1^+(\Lambda, \mathcal{B})$ *is a statistical map, then* $T$ *has a unique extension to a statistical operator* $\overline{T}: M(\Omega, \mathcal{A}) \to M(\Lambda, \mathcal{B})$. *If* $S: M(\Omega, \mathcal{A}) \to M(\Lambda, \mathcal{B})$ *is a statistical operator, then* $S \mid M_1^+(\Omega, \mathcal{A})$ *is a statistical map.*

If $S: M(\Omega, \mathcal{A}) \to M(\Lambda, \mathcal{B})$ is a statistical operator, then by Theorem 7.5, there exists a unique observable $X: \mathcal{B} \to \mathcal{E}(\Omega, \mathcal{A})$ such that $S = \overline{D}_X$. Also, by Theorem 7.4, $X$ has a unique extension to a positive, unital, weakly continuous operator $\overline{X}: B(\Lambda, \mathcal{B}) \to B(\Omega, \mathcal{A})$. Since

$$(\overline{D}_X \mu)(g) = \mu_X(g) = (\overline{X}g)(\mu)$$

for every $\mu \in M(\Omega, \mathcal{A})$, $g \in B(\Lambda, \mathcal{B})$ we conclude that the adjoint $\overline{D}'_X = \overline{X}$ and the adjoint $\overline{X}' = \overline{D}_X$. Hence, $S' = \overline{X}$ and $\overline{X}' = S$.

We have seen that if $T: M_1^+(\Omega, \mathcal{A}) \to M_1^+(\Lambda, \mathcal{B})$ is a statistical map, then there exists a unique observable $X: \mathcal{B} \to \mathcal{E}(\Omega, \mathcal{A})$ such that $T = D_X$. We say that $T$ is *crisp* if $X$ is crisp. Thus, $T$ is crisp if and only if there exists a random variable $f: \Omega \to \Lambda$ such that $T = D_{X_f}$. In this case $T\mu = \mu_f$, the distribution of $f$ relative to $\mu$. Denote the set of Dirac $\delta$ measures on $(\Omega, \mathcal{A})$ by $\partial M_1^+(\Omega, \mathcal{A})$. The proof of the following result is contained in [7].

**Theorem 7.7.** *A statistical map* $T: M_1^+(\Omega, \mathcal{A}) \to M_1^+(\Lambda, \mathcal{B})$ *is crisp if and only if* $T\left[\partial M_1^+(\Omega, \mathcal{A})\right] \subseteq \partial M_1^+(\Lambda, \mathcal{B})$.

# 8 Sequential products on Hilbert space

Two measurements $A$ and $B$ cannot be performed simultaneously in general, so they are frequently executed sequentially. We denote by $A \circ B$ a sequential measurement in which $A$ is performed first and $B$ second. We restrict our attention to yes-no measurements that have only two possible results usually taken to be 0 and 1. A paradigm situation is an optical bench in which a beam of particles prepared in a certain state is injected at the left and then subjected to a sequence of filters $F_1, \ldots, F_n$. An individual particle either passes through a filter $F_i$ or does not, so the filters can be thought of as yes-no measurements. Particles that pass through all the filters enter a detection device at the right of $F_n$ and are counted. Because of quantum interference, the order of placement of the filters usually makes a difference. The resulting sequential measurement is $F_1 \circ F_2 \circ \cdots \circ F_n$ and the probability that a particle is detected is denoted by $P(F_1 \circ F_2 \circ \cdots \circ F_n)$. In practice, this probability is usually approximated by a relative long-run frequency. Thus, if a large number $N_{\text{in}}$ of particles is injected and $N_{\text{out}}$ particles are detected then

$$P(F_1 \circ F_2 \circ \cdots \circ F_n) \approx \frac{N_{\text{out}}}{N_{\text{in}}}$$

Following our previous terminology, we call yes-no measurements *effects*. For effects $A$ and $B$, it is reasonable to assume that

$$P(A \circ B) = P(A)P(B \mid A) \tag{8.1}$$

For a classical system, $A$ and $B$ are represented by sets and $P$ is represented by a probability measure. In this case we have

$$P(A \circ B) = \frac{P(A)P(A \cap B)}{P(A)} = P(A \cap B)$$

when $P(A) \neq 0$. Hence, $A \circ B$ is represented by $A \cap B$ and we write $A \circ B = A \cap B$. But then

$$A \circ B = A \cap B = B \cap A = B \circ A$$

and this does not describe quantum interference. For this reason, we must abandon classical probability theory and we are forced to employ quantum probability theory. For example, let $A$ and $B$ be polarizing filters in planes perpendicular to the particle beam, where $A$ polarizes vertically and $B$ at a $45°$ angle. If the incoming beam is prepared in a state of horizontal polarization, then $A \circ B$ will transmit no particles, while $B \circ A$ will transmit particles. In this case, $A \circ B \neq B \circ A$.

For $A, B \in \mathcal{E}(\mathcal{H})$, $\rho \in \mathcal{D}(\mathcal{H})$, we define the *conditional probability of $B$ given $A$* by

$$P_\rho(B \mid A) = \frac{\mathrm{tr}\left(BA^{1/2}\rho A^{1/2}\right)}{\mathrm{tr}(A\rho)} = \frac{\mathrm{tr}\left(A^{1/2}BA^{1/2}\rho\right)}{\mathrm{tr}(A\rho)} \tag{8.2}$$

when $\mathrm{tr}(A\rho) \neq 0$. As usual, $A^{1/2}$ is the unique positive square-root of $A$. Equation (8.2) generalizes the well-known von Neumann–Lüders formula, see [19, 24].

$$P_\rho(B \mid A) = \frac{\mathrm{tr}(BA\rho A)}{\mathrm{tr}(A\rho)}$$

for $A, B \in \mathcal{P}(\mathcal{H})$. Applying (8.1) and (8.2) we have

$$P_\rho(A \circ B) = P_\rho(A)P_\rho(B \mid A) = \mathrm{tr}\left(A^{1/2}BA^{1/2}\rho\right) = P_\rho\left(A^{1/2}BA^{1/2}\right) \tag{8.3}$$

Notice that $A^{1/2}BA^{1/2} \in \mathcal{E}(\mathcal{H})$ because

$$0 \leq \left\langle A^{1/2}BA^{1/2}x, x \right\rangle = \left\langle BA^{1/2}x, A^{1/2}x \right\rangle \leq \left\langle A^{1/2}x, A^{1/2}x \right\rangle = \langle Ax, x \rangle \leq \langle x, x \rangle$$

Since (8.3) holds for every $\rho \in \mathcal{D}(\mathcal{H})$, we define $A \circ B = A^{1/2}BA^{1/2}$ and we have just shown that $A \circ B \leq A$. We call $A \circ B$ the *sequential product* of $A$ and $B$.

We say that $A, B \in \mathcal{E}(\mathcal{H})$ are *compatible* if $AB = BA$. The sequential product illustrates why it is important to consider unsharp effects. Even if $A, B \in \mathcal{P}(\mathcal{H})$ are sharp, $A \circ B = ABA \notin \mathcal{P}(\mathcal{H})$ unless $A$ and $B$ are compatible. We now study various properties of the sequential product.

It is clear that the sequential product satisfies $0 \circ A = 0$, $I \circ A = A$, $A \circ (B + C) = A \circ B + A \circ C$ whenever $B + C \leq I$, and $(\lambda A) \circ B = A \circ (\lambda B) = \lambda(A \circ B)$ for every $0 \leq \lambda \leq 1$. We shall show that $A \circ B$ has practically no other algebraic properties unless compatibility conditions are imposed. To illustrate the fact that $A \circ B$ does not have properties that one might expect, we now show that $A \circ B = A \circ C$ does not imply

$B \circ A = C \circ A$ even for $A$, $B$, $C \in \mathcal{P}(\mathcal{H})$. In $\mathcal{H} = \mathbb{C}^2$, consider $A$, $B$, $C \in \mathcal{P}(\mathcal{H})$ given by the following matrices

$$A = \frac{1}{2}\begin{bmatrix} 1 & 1 \\ 1 & 1 \end{bmatrix} \quad B = \begin{bmatrix} 1 & 0 \\ 0 & 0 \end{bmatrix} \quad C = \begin{bmatrix} 0 & 0 \\ 0 & 1 \end{bmatrix}$$

We then have

$$A \circ B = ABA = \frac{1}{2}A = ACA = A \circ C$$

However,

$$B \circ A = BAB = \frac{1}{2}B \neq \frac{1}{2}C = CAC = C \circ A$$

This example also shows that $A \circ B \not\leq B$ in general, even though we always have $A \circ B \leq A$.

We say that $A$, $B \in \mathcal{E}(\mathcal{H})$ are *sequentially independent* if $A \circ B = B \circ A$. It is clear that if $A$ and $B$ are compatible then they are sequentially independent. To prove the converse, we shall need the following result due to Fuglede–Putnam–Rosenblum (for a detailed proof, cf., e.g., [29]).

**Theorem 8.1.** *If $M$, $N$, $T$ are bounded linear operators on $\mathcal{H}$ with $M$ and $N$ normal, then $MT = TN$ implies $M^*T = TN^*$.*

**Corollary 8.2.** *For $A$, $B \in \mathcal{E}(\mathcal{H})$, $A \circ B = B \circ A$ implies $AB = BA$.*

*Proof.* Since $A \circ B = B \circ A$, we have

$$A^{1/2}B^{1/2}B^{1/2}A^{1/2} = B^{1/2}A^{1/2}A^{1/2}B^{1/2}$$

Hence, $M = A^{1/2}B^{1/2}$ and $N = B^{1/2}A^{1/2}$ are normal. Letting $T = A^{1/2}$, we have $MT = TN$. Applying Theorem 8.1 we conclude that $B^{1/2}A = AB^{1/2}$. It immediately follows that $BA = AB$.

Sequential independence for three or more effects was considered in [19] and a more general result was proved. Our next result shows that if $A \circ B$ is sharp then $A$ and $B$ are compatible (and hence, sequentially independent).

**Theorem 8.3.** [19] *For $A$, $B \in \mathcal{E}(\mathcal{H})$, if $A \circ B \in \mathcal{P}(\mathcal{H})$ then $AB = BA$.*

*Proof.* Assume that $A^{1/2}BA^{1/2} = A \circ B \in \mathcal{P}(\mathcal{H})$. Suppose that $A \circ Bx = x$ where $\|x\| = 1$. We then have $\langle BA^{1/2}x, A^{1/2}x \rangle = 1$. By Schwarz's inequality we have $BA^{1/2}x = A^{1/2}x$ and hence, $Ax = A \circ Bx = x$. Since $x$ is an eigenvector of $A$ with eigenvalue 1, the same holds for $A^{1/2}$. Thus, $A^{1/2}x = x$ so that $BA^{1/2}x = A \circ Bx$. We conclude that $BA^{1/2}x = A \circ Bx$ for every $x$ in the range $R(A \circ B)$. Now suppose that $A \circ Bx = 0$. We then have

$$\|B^{1/2}A^{1/2}x\|^2 = \langle B^{1/2}A^{1/2}x, B^{1/2}A^{1/2}x \rangle = \langle A \circ Bx, x \rangle = 0$$

so that $B^{1/2}A^{1/2}x = 0$. Hence, $BA^{1/2}x = 0$ and it follows that $BA^{1/2}x = A \circ Bx$ for every $x$ in the null space $N(A \circ B)$. We conclude that $BA^{1/2} = A \circ B$. Hence,

$$BA^{1/2} = A \circ B = (A \circ B)^* = A^{1/2}B$$

so that $AB = BA$.

Simple examples show that the converse of Theorem 8.3 does not hold. However, the converse does hold for sharp effects.

**Corollary 8.4.** *For $A, B \in \mathcal{P}(\mathcal{H})$, $A \circ B \in \mathcal{P}(\mathcal{H})$ if and only if $AB = BA$.*

This last result shows that $\mathcal{P}(\mathcal{H})$ is not closed under sequential products. If $AB = BA$ we write $A \mid B$.

**Corollary 8.5.** (i) *If $A \circ B = 0$, then $B \circ A = 0$.* (ii) *If $A \mid B$ then $A \mid B'$ and $A \circ (B \circ C) = (A \circ B) \circ C$ for every $C \in \mathcal{E}(\mathcal{H})$* (iii) *If $C \mid A$ and $C \mid B$ then $C \mid A \circ B$ and $C \mid (A \oplus B)$.*

The simplest version of the law of total probability would say that

$$P_\rho(B) = P_\rho(A)P_\rho(B \mid A) + P_\rho(I - A)P_\rho(B \mid I - A)$$

In terms of the sequential product this can be stated as

$$P_\rho(B) = P_\rho(A \circ B) + P_\rho((I - A) \circ B) = P_\rho(A \circ B + (I - A) \circ B)$$

Does this equation hold for every $\rho \in \mathcal{D}(\mathcal{H})$? Equivalently does the following equation hold?

$$B = A \circ B + (I - A) \circ B \tag{8.4}$$

**Theorem 8.6.** ([9, 19]) *For $A, B \in \mathcal{E}(\mathcal{H})$, (8.4) holds if and only if $AB = BA$.*

*Proof.* It is clear that (8.4) holds if $AB = BA$. Conversely, assume that (8.4) holds and write it as

$$B = A^{1/2}BA^{1/2} + (I - A)^{1/2}B(I - A)^{1/2}$$

Multiplying by $A^{1/2}$ on the left and right, we obtain

$$
\begin{aligned}
A^{1/2}BA^{1/2} &= ABA + A^{1/2}(I - A)^{1/2}B(I - A)^{1/2}A^{1/2} \\
&= ABA + (I - A)^{1/2}A^{1/2}BA^{1/2}(I - A)^{1/2} \\
&= ABA + (I - A)^{1/2}\left[B - (I - A)^{1/2}B(I - A)^{1/2}\right](I - A)^{1/2} \\
&= ABA - (I - A)B(I - A) + (I - A)^{1/2}B(I - A)^{1/2} \\
&= ABA - (I - A)B(I - A) + B - A^{1/2}BA^{1/2}
\end{aligned}
$$

Hence,

$$2A^{1/2}BA^{1/2} = ABA - (I - A)B(I - A) + B = AB + BA \tag{8.5}$$

Using the commutator notation $[X, Y] = XY - YX$ (8.5) gives

$$
\begin{aligned}
\left[A^{1/2}, [A^{1/2}, B]\right] &= A^{1/2}(A^{1/2}B - BA^{1/2}) - (A^{1/2}B - BA^{1/2})A^{1/2} \\
&= AB - 2A^{1/2}BA^{1/2} + BA = 0
\end{aligned}
$$

It follows that for every spectral projection $E$ of $A$ we have

$$\left[E, [A^{1/2}, B]\right] = 0$$

By the Jacobi identity

$$\left[E, [A^{1/2}, B]\right] + \left[B, [E, A^{1/2}]\right] + \left[A^{1/2}, [B, E]\right] = 0$$

we have that $\left[A^{1/2}, [E, B]\right] = 0$. As before we obtain $[E, [E, B]] = 0$. Hence,

$$0 = E(EB - BE) - (EB - BE)E = EB + BE - 2EBE$$

which we can write as

$$EB = 2EBE - BE$$

Multiplying on the left by $E$ gives $EB = EBE$. Hence,

$$EB = (EBE)^* = BE$$

It follows that $AB = BA$.

Although the sequential product is always distributive on the right, Theorem 8.6 shows that it is not always distributive on the left. That is, $(A + B) \circ C \neq A \circ C + B \circ C$ in general, when $A + B \leq I$. Indeed, if $AC \neq CA$ then by Theorem 8.6 we have

$$A \circ C + (I - A) \circ C \neq C = (A + (I - A)) \circ C$$

*Example.* One might conjecture that the following generalization of Theorem 8.6 holds. If $A + B \leq I$ and $(A + B) \circ C = A \circ C + B \circ C$ then $CA = AC$ or $CB = BC$. However, this conjecture is false. Suppose that $CB \neq BC$. Nevertheless, we have

$$\left(\tfrac{1}{2}B + \tfrac{1}{2}B\right) \circ C = B \circ C = \tfrac{1}{2}B \circ C + \tfrac{1}{2}B \circ C = \left(\tfrac{1}{2}B\right) \circ C + \left(\tfrac{1}{2}B\right) \circ C$$

## 9 Quantum operations

We have seen that a general quantum measurement (or observable) is represented by a normalized POV measure (or $\sigma$-morphism) $h \colon \mathcal{B}(\mathbb{R}) \to \mathcal{E}(\mathcal{H})$. In this section we shall only consider discrete quantum measurements. These are described by a sequence $E_i \in \mathcal{E}(\mathcal{H})$, $i = 1, 2, \ldots$, satisfying $\sum E_i = I$ where the summation converges in the strong operator topology. Motivated by our discussion in Section 8, the probability that outcome $i$ occurs in the state $\rho$ is $P_\rho(E_i) = \mathrm{tr}(\rho E_i)$ and the post-measurement state given that $i$ occurs is $E_i^{1/2} \rho E_i^{1/2} / \mathrm{tr}(\rho E_i)$. Moreover, the resulting state after the measurement is executed but no observation is performed is given by

$$\phi(\rho) = \sum_i E_i^{1/2} \rho E_i^{1/2}$$

An important physical question is whether the measurement disturbs the state $\rho$. The fact that the measurement does not disturb $\rho$ is given mathematically by the equation $\phi(\rho) = \rho$. We shall show in Theorem 9.1 that $\phi(\rho) = \rho$ if and only if $\rho$ commutes with every $E_i$, $i = 1, 2, \ldots$, and this result is called a generalized Lüders theorem, see [9].

In the dual picture, the probability that an effect $A$ occurs in the state $\rho$ given that the measurement was performed is

$$P_{\phi(\rho)}(A) = \operatorname{tr}\left(A \sum_i E_i^{1/2} \rho E_i^{1/2}\right) = \operatorname{tr}\left(\sum E_i^{1/2} A E_i^{1/2} \rho\right).$$

If $A$ is not disturbed by the measurement in any state we have

$$A = \sum E_i^{1/2} A E_i^{1/2}. \tag{9.1}$$

Again, defining $\phi(A) = \sum E_i^{1/2} A E_i^{1/2}$, (9.1) reduces to $\phi(A) = A$. But now $A$ need not be a trace class operator and the proof of the generalized Lüders theorem does not go through. In fact, we shall show in Section 10 that $\phi(A) = A$ does not necessarily imply that $AE_i = E_i A$, $i = 1, 2, \ldots$. Another way that (9.1) comes about is from the law of total probability which is given by

$$\operatorname{tr}(\rho A) = P_\rho(A) = \sum P_\rho(E_i) P_\rho(A \mid E_i) = \sum \operatorname{tr}(\rho E_i) \frac{\operatorname{tr}(E_i^{1/2} \rho E_i^{1/2} \rho}{\operatorname{tr}(\rho E_i)}$$

$$= \operatorname{tr}\left(\rho \sum E_i^{1/2} A E_i^{1/2}\right).$$

If this law holds for every $\rho \in \mathcal{D}(\mathcal{H})$, we again obtain $\phi(A) = A$. Notice that when we write (9.1) in the form $A = \sum E_i \circ A$ then we have a generalization of the simple law of total probability (8.4).

An application of our previously mentioned result can be found in axiomatic quantum field theory. Suppose a measurement given by the discrete POV measure $\{E_i : i = 1, 2, \ldots\}$ is performed in a bounded spacetime region $X$ and $A \in \mathcal{E}(\mathcal{H})$ is a measurement performed in another bounded spacetime region $Y$ that is spacelike separated from $X$. According to Einstein causality, the measurement in $X$ should not disturb $A$ so that $\phi(A) = A$. But applying our result, $A$ may not be compatible with $E_i$, $i = 1, 2, \ldots$. Thus, the axiom of local commutativity does not follow from Einstein causality. We conclude that this axiom may be too strong and it should be replaced by a weaker axiom.

More general measurements are frequently considered in quantum dynamics, quantum computation and quantum information theory, see [11, 20, 23, 26]. Let $\mathcal{A}$ be a set of operators $\mathcal{A} = \{A_i, A_i^* : i = 1, 2, \ldots\}$ where $A_i \in \mathcal{B}(\mathcal{H})$ satisfy $\sum A_i A_i^* \le I$. A map $\phi \colon \mathcal{B}(\mathcal{H}) \to \mathcal{B}(\mathcal{H})$ of the form $\phi_{\mathcal{A}}(B) = \sum A_i B A_i^*$ is called a *quantum operation* and we call $\mathcal{A}$ the set of *operation elements* for $\phi_{\mathcal{A}}$. If $\phi_{\mathcal{A}}(I) = I$ or equivalently if $\sum A_i A_i^* = I$, then $\phi_{\mathcal{A}}$ is *unital*. If $\sum A_i^* A_i = I$, then $\phi_{\mathcal{A}}$ is *trace preserving*. If the $A_i$ are self-adjoint, then $\phi_{\mathcal{A}}$ is *self-adjoint*. An important example of a unital self-adjoint quantum operation is a Lüders operation

$$L_{\mathcal{A}}(B) = \sum A_i^{1/2} B A_i^{1/2}$$

where $A_i \geq 0$ and $\sum A_i = I$. A quantum operation $\phi_{\mathcal{A}}$ is *faithful* if $\phi_{\mathcal{A}}(B^*B) = 0$ implies that $B = 0$. We say that $B \subset \mathcal{B}(\mathcal{H})$ is a $\phi_{\mathcal{A}}$ *fixed point* if $\phi_{\mathcal{A}}(B) = B$ and denote the set of $\phi_{\mathcal{A}}$ fixed points by $\mathcal{B}(\mathcal{H})^{\phi_{\mathcal{A}}}$. It is clear that the commutant $\mathcal{A}' \subseteq \mathcal{B}(\mathcal{H})^{\phi_{\mathcal{A}}}$. One of our purposes is to decide whether $\mathcal{A}' = \mathcal{B}(\mathcal{H})^{\phi_{\mathcal{A}}}$. The next theorem gives a partial result.

**Theorem 9.1.** ([1, 9]) *Let $\phi_{\mathcal{A}}$ be a self-adjoint quantum operation. If $B \in \mathcal{B}(\mathcal{H})^{\phi_{\mathcal{A}}}$ is positive and has pure point spectrum which can be totally ordered in decreasing order then $B \in \mathcal{A}'$.*

*Proof.* Let $h$ be a unit eigenvector of $B$ corresponding to the largest eigenvalue $\lambda_1 = \|B\|$. Then $\phi_{\mathcal{A}}(B) = B$ implies that

$$\lambda_1 = \sum \langle B A_i h, A_i h \rangle \leq \|B\| \sum \|A_i h\|^2 = \lambda_1 \sum \langle A_i^2 h, h \rangle \leq \lambda_1.$$

Since $\langle B A_i h, A_i h \rangle \leq \lambda_1 \langle A_i^2 h, h \rangle$ it follows that

$$\langle (\lambda_1 I - B) A_i h, A_i h \rangle = 0.$$

Hence, $(\lambda_1 I - B) A_i h = 0$ for every eigenvector $h$ corresponding to $\lambda_1$. Thus, $A_i$ leaves the $\lambda_1$-eigenspace invariant. Letting $P_1$ be the corresponding spectral projection of $B$ we have $P_1 A_i P_1 = A_i P$, which implies that $A_i P_1 = P_1 A_i$, $i = 1, 2, \ldots$. Now $B = \lambda_1 P_1 + B_1$, where $B_1$ is a positive operator with a largest eigenvalue. Since

$$\lambda_1 P_1 + B_1 = B = \phi_{\mathcal{A}}(B) = \lambda_1 \phi_{\mathcal{A}}(P_1) + \phi_{\mathcal{A}}(B_1) = \lambda_1 P_1 + \phi_{\mathcal{A}}(B_1)$$

we have that $\phi_{\mathcal{A}}(B_1) = B_1$. Proceeding by induction, $B \in \mathcal{A}'$.

We shall show in the next section that Theorem 9.1 cannot be extended to an arbitrary positive $B \in \mathcal{B}(\mathcal{H})^{\phi_{\mathcal{A}}}$. Moreover, it cannot be extended to a non-self-adjoint $\phi_{\mathcal{A}}$ even in the case where $B$ is positive with finite spectrum and $\mathcal{A}$ contains only two operation elements. The next example shows that self-adjointness cannot be deleted even when $\dim \mathcal{H} < \infty$ [5].

Let $\phi_{\mathcal{A}}(B) = \sum_{i=1}^{4} A_i B A_i^*$ be the quantum operation with

$$A_1 = \begin{bmatrix} 1 & 0 & 0 \\ 0 & 0 & 0 \\ 0 & 0 & 0 \end{bmatrix}, \qquad A_2 = \begin{bmatrix} 0 & 0 & 0 \\ 0 & 1 & 0 \\ 0 & 0 & 0 \end{bmatrix},$$

$$A_3 = \frac{1}{\sqrt{2}} \begin{bmatrix} 0 & 0 & 0 \\ 0 & 0 & 0 \\ 1 & 0 & 0 \end{bmatrix}, \qquad A_4 = \frac{1}{\sqrt{2}} \begin{bmatrix} 0 & 0 & 0 \\ 0 & 0 & 0 \\ 0 & 1 & 0 \end{bmatrix}.$$

It is easy to check that $\phi_{\mathcal{A}}$ is unital but not self-adjoint nor tracial. Let $\rho \in \mathcal{D}(\mathbb{C}^3)$ be the state

$$\rho = \frac{1}{3} \begin{bmatrix} 2 & 0 & 0 \\ 0 & 0 & 0 \\ 0 & 0 & 1 \end{bmatrix}.$$

A straightforward computation shows that $\rho \in \mathcal{B}(\mathbb{C}^3)^{\phi_A}$ but $\rho A_3 \neq A_3 \rho$ so that $\rho \notin \mathcal{A}'$. We now give examples of quantum operation descriptions of some simple noisy quantum channels [26]. As is customary in quantum information theory we shall use Dirac notation. In this notation $|\psi\rangle\langle\psi|$ is the projection onto the one-dimensional subspace spanned by the unit vector $|\psi\rangle$.

A two-dimensional quantum system is called a *qubit*. This is the most basic quantum system studied in quantum computation and quantum information theory. A qubit has a two-dimensional state space $\mathbb{C}^2$ with (computational) basis elements $|0\rangle = (1, 0)$ and $|1\rangle = (0, 1)$. The *bit flip channel* flips the state of a qubit from $|0\rangle$ to $|1\rangle$ (and vice versa) with probability $1 - p, 0 < p < 1$. Letting $X$ be the Pauli matrix

$$X = \begin{bmatrix} 0 & 1 \\ 1 & 0 \end{bmatrix}$$

we can represent the bit flip channel by the quantum operation

$$\phi_{bf}(\rho) = p\rho + (1 - p)X\rho X$$

Notice that $\phi_{bf}$ has operation elements $\{p^{1/2}I, (1 - p)^{1/2}X\}$ and that $\phi_{bf}$ is self-adjoint and tracial. It is also unital because for any self-adjoint quantum operation, tracial and unital are equivalent. Of course, $\phi_{bf}$ gives a bit flip because $X|0\rangle = |1\rangle$ and $X|1\rangle = |0\rangle$. Hence,

$$\phi_{bf}(|0\rangle\langle0|) = p|0\rangle\langle0| + (1 - p)|1\rangle\langle1|$$

so the pure state $|0\rangle\langle0|$ is left undisturbed with probability $p$ and is flipped with probability $1 - p$. Similarly,

$$\phi_{bj}(|1\rangle\langle1|) = p|1\rangle\langle1| + (1 - p)|0\rangle\langle0|.$$

The *phase flip channel* is represented by the quantum operation

$$\phi_{pf}(\rho) = p\rho + (1 - p)Z\rho Z$$

where $0 < p < 1$ and $Z$ is the Pauli matrix

$$Z = \begin{bmatrix} 1 & 0 \\ 0 & -1 \end{bmatrix}.$$

The operation elements for $\phi_{pf}$ are $\{p^{1/2}I, (1 - p)^{1/2}Z\}$ so again $\phi_{pf}$ is self-adjoint and tracial. Since $Z|0\rangle = |0\rangle$ and $Z|1\rangle = -|1\rangle$ we see that $\phi_{pf}$ changes the relative phase of the qubit states with probability $1 - p$.

The *bit-phase flip* channel is represented by the quantum operation

$$\phi_{bpf}(\rho) = p\rho + (1 - p)Y\rho Y$$

where $0 < p < 1$ and $Y$ is the Pauli matrix

$$Y = \begin{bmatrix} 0 & -i \\ i & 0 \end{bmatrix}$$

This gives a combination of a bit flip and a phase flip because $Y = iXZ$. The operation elements for $\phi_{bpf}$ are $\left\{ p^{1/2}I, (1-p)^{1/2}Y \right\}$ so $\phi_{bpf}$ is self-adjoint and tracial. We obtain an interesting quantum operation by forming the composition $\phi_{bf} \circ \phi_{pf}$. Since $XZ = -iY$ we have

$$\phi_{bf} \circ \phi_{pf}(\rho) = p^2\rho + p(1-p)Z\rho Z + p(1-p)X\rho X + (1-p)^2 Y\rho Y.$$

The operation elements become

$$\left\{ pI, \sqrt{p(1-p)}\,Z, \sqrt{p(1-p)}\,X, (1-p)Y \right\}$$

so again, $\phi_{hf} \circ \phi_{pf}$ is self-adjoint and tracial. It is also easy to check that $\phi_{pf} \circ \psi_{bf} = \phi_{bf} \circ \phi_{pf}$.

Another important type of quantum noise is the *depolarizing channel* given by the quantum operation

$$\phi_{dp}(\rho) = \frac{pI}{2} + (1-p)\rho$$

where $0 < p < 1$. This channel depolarizes a qubit state with probability $p$. That is, the state $\rho$ is replaced by the completely mixed state $I/2$ with probability $p$. By applying the identity

$$\frac{I}{2} = \frac{\rho + X\rho X + Y\rho Y + Z\rho Z}{4}$$

that holds for every $\rho \in \mathcal{D}(\mathbb{C}^2)$ we can write

$$\phi_{dp}(\rho) = \left( 1 - \frac{3}{4}p \right)\rho + \frac{p}{4}(X\rho X + Y\rho Y + Z\rho Z).$$

Thus, the operation elements for $\phi_{dp}$ become

$$\left\{ \sqrt{1 - 3p/4}\,I, \sqrt{p}\,X/2, \sqrt{p}\,Y/2, \sqrt{p}\,Z/2 \right\}.$$

As before, $\phi_{dp}$ is self-adjoint and tracial.

There are practical quantum operations that are not self-adjoint nor unital. For example, consider the *amplitude damping channel* given by the quantum operation

$$\phi_{ad}(\rho) = A_1 \rho A_1^* + A_2 \rho A_2^*$$

where

$$A_1 = \begin{bmatrix} 1 & 0 \\ 0 & \sqrt{1-\gamma} \end{bmatrix}, \qquad A_2 = \begin{bmatrix} 0 & \sqrt{\gamma} \\ 0 & 0 \end{bmatrix}$$

and $0 < \gamma < 1$. It is easy to check that $\phi_{ad}$ is tracial but not self-adjoint nor unital. Although the quantum channels (quantum operations) that we have considered appear to be quite specialized, general quantum channels and quantum operations can be constructed in terms of these simple ones and this is important for the theory of quantum error correction.

## 10 Completely positive maps

This section studies completely positive maps on von Neumann algebras. Such maps give a unifying generalization of quantum operations and many of our results will follow from these general considerations.

An *operator system* is a linear subspace of $\mathcal{B}(\mathcal{H})$ that is closed under the involution * and contains the identity operator. Let $\mathcal{M}_k$ be the $C^*$-algebra of $k \times k$ complex matrices which we identify with $\mathcal{B}(\mathbb{C}^k)$. For an operator system $S \subseteq \mathcal{B}(\mathcal{H})$ we consider $S \otimes \mathcal{M}_k$ embedded in the $C^*$-algebra $\mathcal{B}(\mathcal{H}) \otimes \mathcal{M}_k$. Then $S \otimes \mathcal{M}_k$ carries the natural operator norm and the natural operator order. Given operator systems $\mathcal{V}$ and $\mathcal{W}$ and a linear map $\phi \colon \mathcal{V} \to \mathcal{W}$, for any integer $k \geq 1$, there is defined a linear map $\phi_k \colon \mathcal{V} \otimes \mathcal{M}_k \to \mathcal{W} \otimes \mathcal{M}_k$ given by

$$\phi_k(v) = \left[\phi(v_{ij})\right] \text{ where } v = \left[v_{ij}\right] \in \mathcal{V} \otimes \mathcal{M}_k, \quad i, j = 1, \dots, k.$$

We then have a nondecreasing sequence of operator norms

$$\|\phi\| = \|\phi_1\| \leq \|\phi_2\| \leq \|\phi_3\| \leq \cdots.$$

The map $\phi$ is called *completely bounded* if

$$\|\phi\|_{\mathrm{cb}} = \sup_{k \geq 1} \|\phi_k\| < \infty.$$

It follows that $\| \cdot \|_{\mathrm{cb}}$ is a norm on the linear space $CB(\mathcal{V}, \mathcal{W})$ of completely bounded maps from $\mathcal{V}$ into $\mathcal{W}$. If $\|\phi\|_{\mathrm{cb}} \leq 1$, then $\phi$ is called *completely contractive*. If $\phi_k$ is positive for all $k$, then $\phi$ is called *completely positive*. Any completely positive map $\phi$ is completely bounded and

$$\|\phi(I)\| = \|\phi\| = \|\phi\|_{\mathrm{cb}}.$$

In particular, if $\phi(I) = I$ then $\phi$ is completely contractive, see [27, 30].

Now let $\mathcal{M} \subseteq \mathcal{B}(\mathcal{H})$ be a von Neumann algebra. A map $\psi \colon \mathcal{M} \to \mathcal{M}$ is *idempotent* if $\psi \circ \psi = \psi$. A completely contractive idempotent map $\psi$ from $\mathcal{B}(\mathcal{H})$ onto $\mathcal{M}$ is called a *projection* onto $\mathcal{M}$. If there exists a projection onto $\mathcal{M}$ then $\mathcal{M}$ is *injective*, see [27]. A linear map $\psi$ from $\mathcal{M}$ onto a $C^*$-subalgebra $\mathcal{N}$ of $\mathcal{M}$ is a *conditional expectation* if $\psi$ is positive idempotent and $\psi(CB) = \psi(C)B$ and $\psi(BC) = B\psi(C)$ for every $B \in \mathcal{N}$ and $C \in \mathcal{M}$. A *state* on $\mathcal{M}$ is a positive linear functional $\omega \colon \mathcal{M} \to \mathbb{C}$ such that $\omega(I) = 1$. We say that $\omega$ is *faithful* if $\omega(A^*A) = 0$ implies $A = 0$.

If $\phi \colon \mathcal{M} \to \mathcal{M}$ is a unital completely positive map and $\omega$ is a state on $\mathcal{M}$, then $\omega \circ \phi$ is again a state on $\mathcal{M}$. We say that $\omega$ is *$\phi$-invariant* of $\omega \circ \phi = \omega$. We say that $A \in \mathcal{M}$ is a *fixed point* of $\phi$ if $\phi(A) = A$ and denote the set of fixed points of $\phi$ by $\mathcal{M}^\phi$, see [5]. Notice that $\mathcal{M}^\phi$ is an operator system. In general, $\mathcal{M}^\phi$ is not an algebra [5]. It is easy to check that

$$\mathcal{I}(\phi) = \{A \in \mathcal{M} \colon \phi(AB) = A\phi(B), \phi(BA) = \phi(B)A \text{ for every } B \in \mathcal{M}\}$$

is a $C^*$-algebra in $\mathcal{M}$ and $\mathcal{I}(\phi) \subseteq \mathcal{M}^\phi$. Moreover, if $\phi$ is weakly continuous, then $\mathcal{I}(\phi)$ is a von Neumann subalgebra of $\mathcal{M}$.

**Lemma 10.1.** *The following statements are equivalent.* (a) $\mathcal{M}^\phi = \mathcal{I}(\phi)$. (b) $\mathcal{M}^\phi$ *is a $C^*$-algebra.* (c) *If $A \in \mathcal{M}^\phi$, then $A^*A \in \mathcal{M}^\phi$.*

*Proof.* (a)$\Rightarrow$(b)$\Rightarrow$(c) is clear. To prove that (c) implies (a) suppose (c) holds and $A \in \mathcal{M}^\phi$. Then $A^*A \in \mathcal{M}^\phi$ so that

$$\phi(A)^*\phi(A) = A^*A = \phi(A^*A).$$

It follows from [10] that for every $B \in \mathcal{M}$ we have

$$\phi(AB) = \phi(A)\phi(B) = A\phi(B)$$

and

$$\phi(BA) = \phi(B)\phi(A) = \phi(B)A$$

Hence, $A \in \mathcal{I}(\phi)$.

**Theorem 10.2.** *If $\phi$ admits a faithful invariant state $\omega$, then $\mathcal{M}^\phi = \mathcal{I}(\phi)$.*

*Proof.* Suppose that $A \in \mathcal{M}^\phi$. By [10] we have

$$A^*A = \phi(A)^*\phi(A) \leq \phi(A^*A)$$

so that $\phi(A^*A) - A^*A \in \mathcal{M}^+$. Since

$$\omega\big[\phi(A^*A) - A^*A\big] = \omega \circ \phi(A^*A) - \omega(A^*A) = 0$$

we conclude that $\phi(A^*A) = A^*A$. Hence, $A^*A \in \mathcal{M}^\phi$. Applying Lemma 10.1 we have that $\mathcal{M}^\phi = \mathcal{I}(\phi)$.

For $\phi: \mathcal{M} \to \mathcal{M}$ we denote the map obtained by composing $\phi$ with itself $n$ times by $\phi^n$. If $\psi: \mathcal{M} \to \mathcal{M}$ we denote the composition $\psi \circ \phi$ by $\psi\phi$. The proof of the next theorem is a little long and may be found in [1].

**Theorem 10.3.** *Let $\phi: \mathcal{M} \to \mathcal{M}$ be a weakly continuous, unital, completely positive map.* (a) *There exists an idempotent, unital, completely positive map $\psi: \mathcal{M} \to \mathcal{M}$ with range $\mathrm{ran}(\psi) = \mathcal{M}^\phi$.* (b) $\mathcal{M}^\phi = \mathcal{I}(\phi)$ *if and only if $\psi$ is a conditional expectation.* (c) *If $\mathcal{M} = \mathcal{B}(\mathcal{H})$ and $\mathcal{M}^\phi = \mathcal{I}(\phi)$, then $\mathcal{I}(\phi)$ is an injective von Neumann algebra.*

**Corollary 10.4.** *Let $\phi: \mathcal{B}(\mathcal{H}) \to \mathcal{B}(\mathcal{H})$ be a weakly continuous, unital, completely positive map. If $\mathcal{I}(\phi)$ is not injective, then $\mathcal{B}(\mathcal{H})^\phi \neq \mathcal{I}(\phi)$.*

**Theorem 10.5.** *([1]) Let $\phi_A$ be a quantum operation.* (a) *$\phi_A$ is a weakly continuous completely positive map.* (b) *If $\phi_A$ is trace preserving, then $\phi_A$ is faithful and $\mathrm{tr}(\phi_A(B)) = \mathrm{tr}(B)$ for every $B \in \mathcal{T}(\mathcal{H})$.*

Let $\phi_A$ be a unital quantum operation and define the fixed point set $\mathcal{B}(\mathcal{H})^{\phi_A}$ as before. We then have the von Neumann algebras $\mathcal{I}(\phi_A)$ and $\mathcal{A}'$ and it is clear that

$$\mathcal{A}' \subseteq \mathcal{I}(\phi_A) \subseteq \mathcal{B}(\mathcal{H})^{\phi_A}.$$

We are now interested in when these sets coincide; that is, when $\mathcal{B}(\mathcal{H})^{\phi_A} \subseteq \mathcal{A}'$.

**Lemma 10.6.** *If $\phi_A$ is a unital quantum operation, then the following statements are equivalent.* (a) $\mathcal{B}(\mathcal{H})^{\phi_A} = \mathcal{A}'$. (b) $\mathcal{B}(\mathcal{H})^{\phi_A}$ *is a von Neumann algebra.* (c) $\mathcal{B}(\mathcal{H})^{\phi_A} = \mathcal{I}(\phi_A)$. (d) *If $B \in \mathcal{B}(\mathcal{H})^{\phi_A}$, then $B^*B \in \mathcal{B}(\mathcal{H})^{\phi_A}$.*

*Proof.* (a)$\Rightarrow$(b) is clear, (b)$\Rightarrow$(c) follows from Lemma 10.1 and (c)$\Rightarrow$(d) is clear. To show that (d) implies (a) assume that (d) holds and $B \in \mathcal{B}(\mathcal{H})^{\phi_A}$. Then $B^*B \in \mathcal{B}(\mathcal{H})^{\phi_A}$. Notice that

$$0 \le [B, A_i][B, A_i]^* = (BA_i - A_iB)(A_i^*B^* - B^*A_i^*)$$
$$= BA_iA_i^*B^* + A_iBB^*A_i^* - A_iBA_i^*B^* - BA_iB^*A_i^* .$$

Summing over $i$ yields

$$0 \le \sum_i [B, A_i][B, A_i]^* = BB^* + \phi_A(BB^*) - \phi_A(B)B^* - B\phi_A(B^*)$$
$$= \phi_A(BB^*) - BB^* = 0.$$

Hence, $[B, A_i] = 0$ for all $i = 1, 2, \ldots$. In a similar way we have $[B, A_i^*] = 0$. Hence, $B \in \mathcal{A}'$ so that $\mathcal{B}(\mathcal{H})^{\phi_A} = \mathcal{A}'$

**Corollary 10.7.** *Let $\phi_A$ be a unital quantum operation. If $B \in \mathcal{B}(\mathcal{H})^{\phi_A}$ then $B \in \mathcal{A}'$ if and only if $B^*B, BB^* \in \mathcal{B}(\mathcal{H})^{\phi_A}$.*

We denote the set of trace class operators on $\mathcal{H}$ by $\mathcal{T}(\mathcal{H})$. An operator $\rho \in \mathcal{T}(\mathcal{H})$ is *faithful* if for any $A \in \mathcal{B}(\mathcal{H})$, $\mathrm{tr}(\rho^*A^*A\rho) = 0$ implies $A = 0$.

**Theorem 10.8.** *Let $\phi_A$ be a trace preserving, unital, quantum operation.* (a) *If $\dim(\mathcal{H}) < \infty$, then $\mathcal{B}(\mathcal{H})^{\phi_A} = \mathcal{A}'$.* (b) *If there exists a faithful operator $\rho \in \mathcal{T}(\mathcal{H}) \cap \mathcal{A}'$, then $\mathcal{B}(\mathcal{H})^{\phi_A} = \mathcal{A}'$.* (c) *If $B \in \mathcal{B}(\mathcal{H})^{\phi_A}$ and $B = C + D$ for $C \in \mathcal{A}'$, $D \in \mathcal{T}(\mathcal{H})$, then $B \in \mathcal{A}'$.* (d) $\mathcal{B}(\mathcal{H})^{\phi_A} \cap \mathcal{T}(\mathcal{H}) = \mathcal{A}' \cap \mathcal{T}(\mathcal{H})$.

*Proof.* (a) If $\dim(\mathcal{H}) = n < \infty$, then $\omega(B) = \mathrm{tr}(B)/n$ is a faithful $\phi$-invariant state. The result follows from Theorem 10.2 and Lemma 10.6. (b) By the proof of Lemma 10.6, if $B \in \mathcal{B}(\mathcal{H})^{\phi_A}$ then

$$\phi_A(B^*B) - B^*B \ge 0.$$

Since $\rho \in \mathcal{T}(\mathcal{H}) \cap \mathcal{A}'$, by Theorem 10.5(b) we have

$$\mathrm{tr}\left[\rho^*(\phi_A(B^*B) - B^*B)\rho\right] = \mathrm{tr}\left[\phi_A(\rho^*B^*B\rho)\right] - \mathrm{tr}(\rho^*B^*B\rho) = 0.$$

Hence, $\phi_A(B^*B) = B^*B$ so that $B^*B \in \mathcal{B}(\mathcal{H})^{\phi_A}$. Applying Lemma 10.6 gives $\mathcal{B}(\mathcal{H})^{\phi_A} = \mathcal{A}'$. (c) Since

$$C + D = \phi_A(C + D) = \phi_A(C) + \phi_A(D) = C + \phi_A(D)$$

we have $D \in \mathcal{B}(\mathcal{H})^{\phi_A}$. Now $D^*D \in \mathcal{T}(\mathcal{H})$ and by the proof of Lemma 10.6,

$$\phi_A(D^*D) - D^*D \ge 0$$

Since $\phi_A$ is trace preserving, we have

$$\text{tr}\left[\phi_A(D^*D) - D^*D\right] = 0$$

Hence, $\phi_A(D^*D) = D^*D$ so that $D^*D \in B(\mathcal{H})^{\phi_A}$. Similarly, $DD^* \in B(\mathcal{H})^{\phi_A}$ so by Corollary 10.7, $D \in \mathcal{A}'$. Hence, $B \in \mathcal{A}'$. (d) follows from (c).

The next result follows from Theorem 10.3 and Lemma 10.6.

**Theorem 10.9.** *Let $\phi_A$ be a unital quantum operation.* (a) *There exists an idempotent, unital, completely positive map* $\psi : B(\mathcal{H}) \rightarrow B(\mathcal{H})$ *with* $\text{ran}(\psi) = B(\mathcal{H})^{\phi_A}$. (b) $B(\mathcal{H})^{\phi_A} = \mathcal{A}'$ *if and only if $\psi$ is a conditional expectation.* (c) *If $B(\mathcal{H})^{\phi_A} = \mathcal{A}'$ then $\mathcal{A}'$ is an injective von Neumann algebra.*

It follows from Theorem 10.9(c) that if $\mathcal{A}'$ is not injective then $B(\mathcal{H})^{\phi_A} \neq \mathcal{A}'$. We can apply this observation to obtain counterexamples for various conjectures. For instance, the following counterexample shows: (a) If $\phi_A(B) = \sum A_i^{1/2} B A_i^{1/2}$, $A_i \geq 0$, $\sum A_i = I$, is a Lüders operation, then $B(\mathcal{H})^{\phi_A} \neq \mathcal{A}'$ in general. (This answers a question posed in [9, 19].) (b) If $\phi_A(B) = A_1 B A_1^* + A_2 B A_2^*$ is a trace preserving, unital, quantum operation, then $B(\mathcal{H})^{\phi_A} \neq \mathcal{A}'$ in general.

*Counterexample.* Let $\mathbb{F}_2$ be the free group with two generators $g_1$, $g_2$ and identity $e$. It is clear that $\mathbb{F}_2$ is countable. Let $H = \ell_2(\mathbb{F}_2)$ be the separable complex Hilbert space

$$H = \ell_2(\mathbb{F}_2) = \left\{ f : \mathbb{F}_2 \rightarrow \mathbb{C} : \sum |f(x)|^2 < \infty \right\}$$

For $x \in \mathbb{F}_2$ define $\delta_x : \mathbb{F}_2 \rightarrow \mathbb{C}$ by

$$\delta_x(y) = \begin{cases} 1 & \text{if } y = x \\ 0 & \text{if } y \neq x. \end{cases}$$

Then $\{\delta_x : x \in \mathbb{F}_2\}$ is an orthonormal basis for $H$. Define the unitary operators $U_1$, $U_2$ on $H$ by $U_1 \delta_x = \delta_{g_1 x}$, $U_2 \delta_x = \delta_{g_2 x}$. The von Neumann algebra generated by $U_1$ and $U_2$ is denoted by $N = L(\mathbb{F}_2)$. It is known that $N$ and hence $N'$ are not injective [10].

**Lemma 10.10.** *Suppose $B \in B(H)$ has the form $B\delta_x = \lambda_x \delta_x$, $0 \leq \delta_x \leq 1$. Then $B \in \mathcal{E}(H)$ and if $B \in N'$, then $B = \lambda_e I$.*

*Proof.* It is clear that $B \in \mathcal{E}(H)$. Now suppose that $B \in N'$. Then

$$\lambda_{g_1} \delta_{g_1} = B\delta_{g_1} = BU_1\delta_e = U_1 B\delta_e = \lambda_e U_1 \delta_e = \lambda_e \delta_{g_1}.$$

Hence, $\lambda_{g_1} = \lambda_e$ and in a similar way

$$\lambda_{g_1^{-1}} = \lambda_{g_2} = \lambda_{g_2^{-1}} = \lambda_e.$$

Now suppose $x \in \mathbb{F}_2$ has the form $x = g_1 y$ for some $y \in \mathbb{F}_2$. Then

$$\lambda_x \delta_x = B\delta_x = B\delta_{g_1 y} = BU_1\delta_y = U_1 B\delta_y = \lambda_y U_1 \delta_y = \lambda_y \delta_{g_1 y} = \lambda_y \delta_x.$$

Hence, $\lambda_{g_1 y} = \lambda_y$ for every $y \in \mathbb{F}_2$. Similarly,

$$\lambda_{g_1^{-1} y} = \lambda_{g_2 y} = \lambda_{g_2^{-1} y} = \lambda_y$$

for every $y \in \mathbb{F}_2$. Continuing by induction, we conclude that $\lambda_x = \lambda_e$ for every $x \in \mathbb{F}_2$. Hence, $B = \lambda_e I$.

Let $A_1 = 2^{-1/2} U_1$ and $A_2 = 2^{-1/2} U_2$ and let $\mathcal{A} = \{A_1, A_2\}$. Then $\mathcal{A}' = N'$ and

$$A_1 A_1^* + A_2 A_2^* = A_1^* A_1 + A_2^* A_2 = I.$$

Thus, $\phi_{\mathcal{A}}$ is a trace preserving, unital quantum operation. Now a $B$ of the form in Lemma 10.10 satisfies $B \in \mathcal{B}(H)^{\phi_{\mathcal{A}}}$ if and only if

$$\frac{1}{2} \lambda_{g_1^{-1} x} + \frac{1}{2} \lambda_{g_2^{-1} x} = \lambda_x \tag{10.1}$$

for all $x \in \mathbb{F}_2$. Define $B \in \mathcal{B}(H)$ by $B \delta_x = \lambda_x \delta_x$ where $\lambda_x = 0$ if $x$ ends in $g_2^{-1}$, $\lambda_x = 1$ if $x$ ends in $g_1^{-1}$ and $\lambda_x = 1/2$ otherwise. Then it is easy to check that (10.1) is satisfied. Hence, $B \in \mathcal{B}(H)^{\phi_{\mathcal{A}}}$ and by Lemma 10.10 $B \notin \mathcal{A}'$. Thus, $\mathcal{B}(H)^{\phi_{\mathcal{A}}} \neq \mathcal{A}'$.

**Theorem 10.11.** *There exists a Lüders operation $\phi_{\mathcal{A}}$ on $H = \ell_2(\mathbb{F}_2)$ such that $\mathcal{B}(H)^{\phi_{\mathcal{A}}} \neq \mathcal{A}'$. More precisely, there exists a $B \in \mathcal{E}(H)$ such that $B \in \mathcal{B}(H)^{\phi_{\mathcal{A}}}$ but $B \notin \mathcal{A}'$.*

*Proof.* By taking the real and imaginary parts of $U_1 = V_1 + i V_2$, $U_2 = V_3 + i V_4$ we see that $N$ is generated by four self-adjoint operators $V_1, V_2, V_3, V_4$. Moreover, $N$ is generated by the four positive operators $C_i = \|V_i\| I - V_i$, $i = 1, 2, 3, 4$. Let $A_i = C_i / 4 \|C_i\|$, $i = 1, 2, 3, 4$, and let $A_5 = I - \sum_{i=1}^{4} A_i$. Then $A_i \in \mathcal{E}(H)$, $i = 1, \dots, 5$, $\sum A_i = I$ and $\mathcal{A} = \{A_i : i = 1, \dots, 5\}$ generates $N$. Now $\phi_{\mathcal{A}}$ is a Lüders operation and since $N' = \mathcal{A}'$ is not injective we have $\mathcal{B}(H)^{\phi_{\mathcal{A}}} \neq \mathcal{A}'$. Hence, there exists a $B \in \mathcal{B}(H)$ such that $B \in \mathcal{B}(H)^{\phi_{\mathcal{A}}} \setminus \mathcal{A}'$. Now the real part or the imaginary part $B_1$ of $B$ also satisfies $B_1 \in \mathcal{B}(H)^{\phi_{\mathcal{A}}} \setminus \mathcal{A}'$. Letting $B_2 = \|B_1\| I - B_1$ we see that $B_2 \geq 0$ and that $B_3 = B_2 / \|B_2\| \in \mathcal{E}(H)$. Moreover $B_3 \in \mathcal{B}(H)^{\phi_{\mathcal{A}}} \setminus \mathcal{A}'$.

Although the $B$ in Theorem 10.11 exists, it appears to be quite difficult to construct a concrete example of such a $B$.

## 11 Sequential effect algebras

This section discusses a framework that unifies and generalizes all the probabilistic structures that we have considered. The framework is obtained by abstracting the properties of the operations of orthogonal sum $\oplus$ and sequential product $\circ$. These two operations played fundamental roles in all of our probabilistic structures.

If $\oplus$ is a partial binary operation, we write $a \perp b$ if $a \oplus b$ is defined. An *effect algebra* [12, 17] is a system $(E, 0, 1, \oplus)$ where $0, 1$ are distinct elements of $E$ and $\oplus$ is a partial binary operation on $E$ that satisfies the following conditions.

(E1) If $a \perp b$, then $b \perp a$ and $b \oplus a = a \oplus b$.
(E2) If $a \perp b$ and $c \perp (a \oplus b)$, then $b \perp c$, $a \perp (b \oplus c)$ and $a \oplus (b \oplus c) = (a \oplus b) \oplus c$.
(E3) For every $a \in E$ there exists a unique $a' \in E$ such that $a \perp a'$ and $a \oplus a' = 1$.
(E4) If $a \perp 1$, then $a = 0$.

We call the elements of $E$ *effects* and whenever we write $a \oplus b$ we are implicitly assuming that $a \perp b$. We define $a \leq b$ if there exists a $c \in E$ such that $a \oplus c = b$. If such a $c \in E$ exists, then it is unique and we write $c = b \ominus a$. It can be shown that $(E, \leq, ')$ is a partially ordered set with $0 \leq a \leq 1$ for all $a \in E$, $a'' = a$, and $a \leq b$ implies $b' \leq a'$. Moreover, we have $a \perp b$ if and only if $a \leq b'$. If $a \perp a$ we call $a$ an *isotropic* element and when 0 is the only isotropic element of $E$, then we call $E$ an *orthoalgebra*. An element $a \in E$ is *sharp* if $a \wedge a' = 0$ and we denote the set of sharp elements by $E_S$. It is easy to show that $E$ is an orthoalgebra if and only if $E = E_S$. Also, $0, 1 \in E_S$ and $a \in E_S$ if and only if $a' \in E_S$.

If $E$ and $F$ are effect algebras, we say that $\phi \colon E \to F$ is *additive* if $a \perp b$ implies $\psi(a) \perp \phi(b)$ and $\phi(a \oplus b) = \phi(a) \oplus \phi(b)$. If $\phi \colon E \to F$ is additive and $\phi(1) = 1$, then $\phi$ is a *morphism*. If $\phi \colon E \to F$ is a morphism and $\phi(a) \perp \phi(b)$ implies that $a \perp b$, then $\phi$ is a *monomorphism*. A surjective monomorphism is an *isomorphism*. It is easy to see that a morphism $\phi$ is an isomorphism if and only if $\phi$ is bijective and $\phi^{-1}$ is a morphism. A *state* on $E$ is a morphism $s \colon E \to [0, 1]$. We interpret $s(a)$ as the probability that the effect $a$ is observed (has answer yes) when the system is in state $s$. We denoted the set of states on $E$ by $\Omega(E)$. An effect algebra $E$ is a $\sigma$-*effect algebra* if for any increasing sequence $a_1 \leq a_2 \leq \cdots$ we have $\vee a_i$ exists in $E$. A morphism $\phi \colon E \to F$ between $\sigma$-effect algebras is a $\sigma$-*morphism* if whenever $a_1 \leq a_1 \leq \cdots$ we have $\phi(\vee a_i) = \vee \phi(a_i)$. Strictly speaking, in the sequel we should consider $\sigma$-effect algebras and $\sigma$-morphisms because our probabilistic structures satisfy the axioms of a $\sigma$-effect algebra. However, to avoid technicalities that the interested reader can easily fill in, we shall only consider effect algebras and their morphisms.

We now endow an effect algebra with additional structure by providing it with a sequential product. For a binary operation $\circ$, if $a \circ b = b \circ a$ we write $a \mid b$. A *sequential effect algebra* (SEA) is a system $(E, 0, 1, \oplus, \circ)$ where $(E, 0, 1, \oplus)$ is an effect algebra and $\circ \colon E \times E \to E$ is a binary operation that satisfies the following conditions.

(S1) $b \mapsto a \circ b$ is additive for all $a \in E$.
(S2) $1 \circ a = a$ for all $a \in E$.
(S3) If $a \circ b = 0$, then $a \mid b$.
(S4) If $a \mid b$, then $a \mid b'$ and $a \circ (b \circ c) = (a \circ b) \circ c$ for all $c \in E$.
(S5) If $c \mid a$ and $c \mid b$, then $c \mid a \circ b$ and $c \mid (a \oplus b)$.

We call an operation that satisfies (S1)–(S5) a *sequential* product on $E$. If $a \mid b$ for all $a, b \in E$, we call $E$ a *commutative* SEA. We now present some examples of SEA's including our previously studied probabilistic structures.

**Example 1.** For a Boolean algebra $\mathcal{B}$, define $a \perp b$ if $a \wedge b = 0$ and in this case $a \oplus b = a \vee b$. Then $(\mathcal{B}, 0, 1, \oplus)$ is an effect algebra that happens to be an orthoalgebra. In particular, if $X \neq \emptyset$, then $(2^X, \emptyset, X, \oplus)$ is an effect algebra. Defining $a \circ b = a \wedge b$, $(\mathcal{B}, 0, 1, \oplus, \circ)$ becomes a SEA. We shall show later that $\circ$ is unique.

**Example 2.**   For $[0, 1] \subseteq \mathbb{R}$, define $a \perp b$ if $a + b \leq 1$ and in this case $a \oplus b = a + b$. Then $([0, 1], 0, 1, \oplus)$ is an effect algebra. The only sharp elements are 0 and 1. Defining $a \circ b = ab$, $([0, 1], 0, 1, \oplus, \circ)$ becomes a SEA. We shall show later that $\circ$ is unique.

**Example 3.**   Let $X \neq \emptyset$ and let $\mathcal{F} \subseteq [0, 1]^X$. We call $\mathcal{F}$ a *fuzzy set system on* $X$ if the following conditions hold.
(i)  The functions $0, 1 \in \mathcal{F}$.
(ii)  If $f \in \mathcal{F}$, then $1 - f \in \mathcal{F}$.
(iii)  If $f, g \in \mathcal{F}$ with $f + g \leq 1$, then $f + g \in \mathcal{F}$.
(iv)  If $f, g \in \mathcal{F}$, then $fg \in \mathcal{F}$.
In this case, $(\mathcal{F}, 0, 1, \oplus)$ is an effect algebra with $f \oplus g = f + g$ whenever $f - g \leq 1$. If $\mathcal{F} = [0, 1]^X$ we call $\mathcal{F}$ a *full* fuzzy set system. The sharp elements of a fuzzy set system $\mathcal{F}$ are the characteristic functions in $\mathcal{F}$, which can be identified with the (sharp) subsets of $X$ in $\mathcal{F}$. Indeed, if $f \in \mathcal{F}$ is sharp, then $f(1 - f) \leq f, 1 - f$ implies that $f(1 - f) = 0$. Hence, if $f(x) \neq 0$, then $f(x) = 1$. A fuzzy set system $\mathcal{F}$ is a SEA under the operation $f \circ g = fg$. We shall show later that if $\mathcal{F}$ is full, then $\circ$ is unique.

**Example 4.**   Let $\mathcal{E}(\mathcal{H})$ be the set of quantum effects on the Hilbert space $\mathcal{H}$. For $A, B \in \mathcal{E}(\mathcal{H})$ we define $A \perp B$ if $A + B \in \mathcal{E}(\mathcal{H})$ and in this case $A \oplus B = A + B$. Then $(\mathcal{E}(\mathcal{H}), 0, I, \oplus)$ is an effect algebra. The sharp elements of $\mathcal{E}(\mathcal{H})$ are the set of projection operators $\mathcal{P}(\mathcal{H})$ on $\mathcal{H}$. Our work in Section 8 shows that $\mathcal{E}(\mathcal{H})$ is a SEA under the operation $A \circ B = A^{1/2} B A^{1/2}$. Unlike the previous three examples, $\mathcal{E}(\mathcal{H})$ is a noncommutative SEA. We do not know whether $\circ$ is unique. However, $\circ$ is unique if it satisfies some additional conditions.

There are many other examples of SEA's but these four are the important ones for our purposes. We now observe that a SEA gives a general framework for noncommutative probability. If $s \in \Omega(E)$ is a state of $E$ then $s$ gives a (finitely additive) probability measure on $E$. A morphism $h: \mathcal{B}(\mathbb{R}) \to E$ corresponds to an observable and the morphism $s \circ h: \mathcal{B}(\mathbb{R}) \to [0, 1]$ gives the distribution of $h$ in the state $s$. We can now define the expectation and variance of $h$ in the usual way. If $a \in E$ with $s(b) \neq 0$ we can define the *conditional probability of* $b \in E$ *given* $a$ as $s(b \mid a) = s(a \circ b)/s(a)$. Notice that $b \mapsto s(\cdot \mid a)$ is a state on $E$ so we can develop a theory of conditional probabilities on $E$. Motivated by Theorem 8.6 an interesting unsolved problem is the following. If $b = a \circ b \oplus a' \circ b$, does $a \mid b$?

We now state some basic results about a SEA $E$. The proofs are not difficult and may be found in [18].

**Lemma 11.1.**   (i) $a \circ 0 = 0 \circ a = 0$ *and* $a \circ 1 = 1 \circ a = a$ *for* $a \in E$. (ii) $a \circ b \leq a$ *for all* $a, b \in E$. (iii) *If* $a \leq b$, *then* $c \circ a \leq c \circ b$ *for all* $c \in E$. (iv) *The following statements are equivalent.* (a) $a \in E_S$. (b) $a \circ a' = 0$. (c) $a \circ a = a$.

**Lemma 11.2.**   (i) *If* $a \circ b = 0$ *then* $a \perp b$. (ii) *For* $a \in E$, $b \in E_S$, $a \circ b = 0$ *if and only if* $a \perp b$. (iii) *For* $a, b \in E_S$ *with* $a \perp b$ *we have* $a \oplus b \in E_S$.

It follows from Lemma 11.2(iii) that $E_S$ is a sub-effect algebra of $E$ that is an orthoalgebra. We have seen in Section 8 that in general, if $a, b \in E_S$ then $a \circ b \notin E_S$ so $E_S$ is not a sub-SEA of $E$.

**Theorem 11.3.** *Let $a \in E$ and $b \in E_S$. (i) $a \leq b$ if and only if $a \circ b = b \circ a = a$ and $b \leq a$ if and only if $a \circ b = b \circ a = b$. (ii) If $a \mid b$, then $a \wedge b = a \circ b$. (iii) If $a \perp b$, then $a \oplus b = a \vee b = (a' \circ b')'$.*

We say that $a, b \in E$ *coexist* if there exist $c, d, e \in E$ such that $c \oplus d \oplus e$ is defined and $a = c \oplus d$, $b = c \oplus e$, see [23, 24]. We say that $a, b \in E_S$ are *compatible* if there exist mutually orthogonal elements $c, d, e \in E_S$ such that $a = c \vee d$ and $b = c \vee e$.

**Theorem 11.4.** *(i) If $a \mid b$ then $a$ and $b$ coexist. (ii) For $a \in E$, $b \in E_S$, $a \mid b$ if and only if $a$ and $b$ coexist. (iii) For $a, b \in E_S$ the following statements are equivalent. (a) $a \mid b$. (b) $a$ and $b$ are compatible. (c) $a$ and $b$ coexist.*

It is well known that the converse of Theorem 11.4(i) does not hold, see [18].

## 12 Further SEA results

We begin by showing that for certain important examples of SEA's, the sequential product is unique.

**Theorem 12.1.** *For an effect algebra $E$ that is a Boolean algebra, there is a unique sequential product $a \circ b = a \wedge b$.*

*Proof.* Since $E$ is a Boolean algebra, we have $E = E_S$. For $a, b \in E$, define $c = a \wedge b$, $d = a \ominus a \wedge b$, $e = b \ominus a \wedge b$. It is easy to check that $c, d, e$ are mutually orthogonal and that $a = c \vee d$ and $b = c \vee e$. hence, any two elements of $E$ are compatible. By Theorem 11.4(iii) we have that $a \mid b$ for every $a, b \in E$ and any sequential product $\circ$. It follows from Theorem 11.3(ii) that $a \circ b = a \wedge b$.

**Theorem 12.2.** *There is a unique sequential product on the effect algebra $[0, 1] \subseteq \mathbb{R}$.*

*Proof.* Let $\circ$ be a sequential product on $[0, 1]$. Then for any integer $n \geq 1$ and $a \in [0, 1]$ we have

$$a = a \circ 1 = a \circ \left( \frac{1}{n} \oplus \cdots \oplus \frac{1}{n} \right) = n \left( a \circ \frac{1}{n} \right)$$

so that $a \circ (1/n) = (1/n)a$. Also, for any integer $1 \leq m \leq n$ we have

$$a \circ \frac{m}{n} = a \circ \left( \frac{1}{n} \oplus \cdots \oplus \frac{1}{n} \right) = m \left( a \circ \frac{1}{n} \right) = \frac{m}{n} a.$$

Hence, for any rational number $r \in Q \cap [0, 1]$ we have $a \circ r = ar$. Now let $b \in [0, 1]$ be irrational. If $r \in Q \cap [0, 1]$ and $b \leq r$ then by Lemma 11.1(iii) we obtain

$$a \circ b \leq a \circ r = ar.$$

Similarly, if $r \in Q \cap [0, 1]$ and $b > r$, then

$$a \circ b \geq a \circ r = ar$$

Since $Q \cap [0, 1]$ is dense in $[0, 1]$, we obtain $a \circ b = ab$.

A SEA *isomorphism* is an effect algebra isomorphism that preserves the sequential product. The proof of the next result is a straightforward verification.

**Theorem 12.3.** *Let $E, F$ be effect algebras and let $\phi: E \to F$ be an effect algebra isomorphism. If $\circ$ is a sequential product on $E$, then $a * b = \phi\left[\phi^{-1}(a) \circ \phi^{-1}(b)\right]$ is a sequential product on $F$. Moreover, $(E, \circ)$ and $(F, *)$ are SEA isomorphic.*

**Corollary 12.4.** *If $E$ and $F$ are isomorphic effect algebras and $E$ admits a unique sequential product $\circ$, then $F$ admits a unique sequential product.*

*Proof.* By Theorem 12.3, $F$ admits a sequential product $*$. Let $\phi: E \to F$ be an effect algebra isomorphism and define $\cdot: E \times E \to E$ by $a \cdot b = \phi^{-1}[\phi(a) * \phi(b)]$. By Theorem 12.3, $\cdot$ is a sequential product on $E$ so we have $a \cdot b = a \circ b$. Hence,

$$\phi(a \circ b) = \phi(a \cdot b) = \phi(a) * \phi(b).$$

Thus, for every $c, d \in F$ we have

$$c * d = \phi\left[\phi^{-1}(c)\right] * \left[\phi^{-1}(d)\right] = \phi\left[\phi^{-1}(c) \circ \phi^{-1}(d)\right]$$

It follows that $*$ is unique.

Let $(E_i, 0_i, 1_i, \oplus_i)$ be a collection of effect algebras. We obtain a new effect algebra by taking the cartesian product $\prod E_i$ and defining $\prod a_i \perp \prod b_i$ if $a_i \perp b_i$ for each $i$ in which case $\prod a_i \oplus \prod b_i = \prod(a_i \oplus b_i)$. Then $(\prod E_i, \prod 0_i, \prod 1_i, \oplus)$ is an effect algebra. If each of the $E_i$'s is a SEA, then $\prod E_i$ becomes a SEA with the componentwise sequential product.

**Theorem 12.5.** *Let $(E_i, 0_i, 1_i, \oplus_i, \circ_i)$ be SEAs $i \in I$. Then $E = \prod E_i$ admits a unique sequential product if and only if each $E_i, i \in I$, admits a unique sequential product.*

*Proof.* If $E_j$ admits two sequential products for some $j \in I$, then clearly $\prod E_i$ admits at least two sequential products. Conversely, suppose $E_i, i \in I$, admits a unique sequential product $\circ_i$ and let $*$ be a sequential product on $\prod E_i$. For $j \in I$, let $f_j \in \prod E_i$ be defined by

$$f_i(i) = \begin{cases} 1_j & \text{if } i = j \\ 0_j & \text{if } i \neq j. \end{cases}$$

Clearly, $f_j \in E_S$. Let $\circ$ be the sequential product on $E$ given by $(f \circ g)(i) = f(i) \circ_i g(i)$, $i = I$, for any $f, g \in E$. Since $f \circ f_j = f_j \circ f$, by Theorem 11.4(ii), $f$ and $f_j$ coexist. But coexistence is independent of the sequential product so again by Theorem 11.4(ii), $f * f_j = f_j * f$ for any $f \in E$. It follows from Theorem 11.3(ii) that $f * f_j = f \wedge f_j$. Hence,

$$f * f_i(i) = \begin{cases} 0_i & \text{if } i \neq j \\ f(j) & \text{if } i = j. \end{cases}$$

For any $f, g \in E$ we have

$$(f * g) * f_j = f_j * (f * g) = (f_j * f) * g = [(f * f_j) * f_j] * g$$
$$= (f * f_j) * (f_j * g).$$

Now $[0, f_j] \subseteq E$ is an effect algebra with greatest element $f_j$ and $\phi : E_j \to [0, f_j]$ given by

$$[\phi(a)](i) = \begin{cases} 0_i & \text{if } i \neq j \\ a & \text{if } i = j \end{cases}$$

is an effect algebra isomorphism. Since $E_j$ admits a unique sequential product, by Corollary 12.4, $[0, f_j]$ admits a unique sequential product. Hence,

$$(f * g)(j) = [(f * g) * f_j](j) = [(f * f_j) * (g * f_j)](j)$$
$$= [(f * f_j) \circ (g * f_j)](j) = (f * f_j)(j) \circ_j (g * f_j)(j)$$
$$= f(j) \circ_j g(j) = (f \circ g)(j).$$

Hence, $\circ$ is the unique sequential product on $E$.

**Corollary 12.6.** *A full fuzzy set system $E = [0, 1]^X$ admits a unique sequential product.*

*Proof.* This follows from Theorem 12.5, because by Theorem 12.2, $[0, 1]$ admits a unique sequential product.

Whether a general fuzzy set system admits a unique sequential product is an unsolved problem. One can give examples of effect algebras that admit no sequential product and effect algebras that admit many sequential products, see [18]. The next result generalizes Corollary 8.4 to an arbitrary SEA.

**Theorem 12.7.** *Let $E$ be a SEA. For $p, q \in E_S$ we have $p \circ q \in E_S$ if and only if $p \mid q$.*

*Proof.* If $p \mid q$ it is clear that $p \circ q \in E_S$. Conversely, suppose that $r = p \circ q \in E_S$. We first show that $r \leq q$. Since $r \circ q \leq r$ and $r \leq p$, we have

$$p \circ (r \ominus r \circ q) = p \circ r \ominus p \circ (r \circ q) = r \ominus (p \circ r) \circ q$$
$$= r \ominus (r \circ p) \circ q = r \ominus r \circ (p \circ q) = r \ominus r = 0$$

By Lemma 11.2(ii), $p \perp (r \ominus r \circ q)$. Hence,

$$r \ominus r \circ q \leq p' \leq r'.$$

Since $r \ominus r \circ q \leq r \in E_S$ we have that $r \ominus r \circ q = 0$. Thus, $r = r \circ q$ so that $r \circ q' = 0$. Again by Lemma 11.2(ii) we have that $r \leq (q')' = q$. We now show that $p \circ q' \leq q'$. Since $p \circ (q \ominus p \circ q) = 0$ we have that $q \ominus p \circ q \leq p'$ so that

$$q \leq (p \circ q) \oplus p' = (p \ominus p \circ q') \oplus p' = (p \circ q')'.$$

Hence, $p \circ q' \leq q'$. Applying Theorem 11.3(i) we have that $p \circ q \mid q$ and $p \circ q' \mid q$. Since $p = p \circ q \oplus p \circ q'$ we conclude that $p \mid q$.

For arbitrary elements $p, q \in E$, it is an unsolved problem whether $p \circ q \in E_S$ implies that $p \mid q$. This result does hold for $p, q \in \mathcal{E}(\mathcal{H})$ by Theorem 8.3.

**Lemma 12.8.** *If $a \circ b = a$, then $a \mid b$ and $a \leq b$.*

*Proof.* Since $a \circ b = a$ we have

$$a = a \circ b \oplus a \circ b' = a \oplus a \circ b'.$$

Hence, $a \circ b' = 0$. Thus, $a \mid b'$ so that $a \mid b$. Therefore, $a = a \circ b = b \circ a \leq b$.

It is unknown whether $b \circ a = a$ implies that $a \mid b$.

**Theorem 12.9.** *Let $E$ be a SEA and suppose $a, b \in E_S$. (i) $a \circ b \leq b$ if and only if $a \mid b$. (ii) $a \circ b \geq b$ if and only if $a \circ b = b \circ a = b$.*

*Proof.* (i) If $a \mid b$, then $a \circ b = b \circ a \leq b$. Conversely, assume that $a \circ b \leq b$. We now show that $a \circ b \in E_S$. Since $a \circ b \leq a, b$, applying Theorem 11.3(i) we have $a \circ b \mid a$ and $a \circ b \mid b$. Hence,

$$(a \circ b) \circ (a \circ b) = [(a \circ b) \circ a] \circ b = (a \circ b) \circ b = a \circ b.$$

The result follows from Theorem 12.7. (ii) Suppose that $a \circ b \geq b$. Then since $a \circ b \mid b$, $a \circ b \mid a$ as before we have

$$(a \circ b) \circ (a \circ b) = [(a \circ b) \circ a] \circ b = (a \circ b) \circ b = b.$$

Taking the sequential product on the left with $a$ gives

$$(a \circ b) \circ (a \circ b) = a \circ b.$$

Since $a \circ b \in E_S$ by Theorem 12.7, $a \mid b$. Hence,

$$a \circ b = b \circ a = (a \circ b)^2 = b.$$

The converse is trivial.

We close with some unsolved problems for a general SEA. It is shown in [19] that these problems have affirmative answers for $\mathcal{E}(\mathcal{H})$.

*Problems*

1. Does $a \circ (b \circ c) = (a \circ b) \circ c$ for every $c$ implying that $a \mid b$?
2. Are the following statements equivalent? (i) $a \circ (c \circ b) = (a \circ c) \circ b$ for every $c$. (ii) $c \circ (a \circ b) = (c \circ a) \circ b$ for every $c$. (iii) $a \mid c$ and $b \mid c$ for every $c$.
3. Suppose $\Omega(E)$ is separating. If $\omega(a \circ b) = \omega(a)\omega(b)$ for every $\omega \in \Omega(E)$ does $a \mid c$ and $b \mid c$ for every $c$?

# References

[1] A. Arias, A. Gheondea and S. Gudder, Fixed points of quantum operations, *J. Math. Phys.* **43** (2002), 5872–5881.

[2] E. G. Beltrametti and S. Bugajski, Quantum observables in classical frameworks, *Int. J. Theor. Phys.* **34** (1995) 1221–1229.

[3] E. G. Beltrametti and S. Bugajski, A classical extension of quantum mechanics, *J. Phys. A Math. Gen.* **28** (1995) 3329–3334.

[4] E. G. Beltrametti and S. Bugajski, Effect algebras and statistical physical theories, *J. Math. Phys.* **38** (1997) 3020–3030.

[5] O Bratteli, P. Jørgensen, A. Kishimoto and R. Werner, Pure states on $\mathcal{O}_d$, *J. Operator Theory* **43** (2000) 97–143.

[6] S. Bugajski, Fundamentals of fuzzy probability theory, *Int. J. Theor. Phys.* **35** (1996) 2229–2244.

[7] S. Bugajski, K.-E. Hellwig and W. Stulpe, On fuzzy random variables and statistical maps, *Rep. Math. Phys.* **41** (1998) 1–11.

[8] P. Busch, M. Grabowski and P. J. Lahti, *Operational Quantum Physics*, Springer-Verlag, Berlin, 1995.

[9] P. Busch and J. Singh, Lüders theorem for unsharp quantum effects, *Phys. Lett. A* **249** (1998) 10–24.

[10] M.-D. Choi, A Schwarz inequality for positive linear maps on $C^*$-algebras, *Illinois J. Math.* **18** (1974) 565–574.

[11] E. B. Davies, *Quantum Theory of Open Systems*, Academic Press, London, 1976.

[12] A. Dvurečenskij and S. Pulmannová, *New Trends in Quantum Structures*, Kluwer, Dordrecht, 2000.

[13] D. J. Foulis and M. K. Bennett, Effect algebras and unsharp quantum logics, *Found. Phys.* **24**, (1994), 1325–1346.

[14] A. Gleason, Measures on closed subspaces of a Hilbert space, *J. Rat. Mech. Anal.* **6**, (1975), 885–893.

[15] S. Gudder, Fuzzy probability theory, *Demon. Math.* **31** (1998), 235–254.

[16] S. Gudder, Sharp and unsharp quantum effects, *Adv. Appl. Math.* **20** (1998), 169–187.

[17] S. Gudder, Observables and statistical maps, *Found Phys.* **29** (1999), 877–897.

[18] S. Gudder and R. Greechie, Sequential products on effect algebras, *Rep. Math. Phys.* **49** (2002), 87–111.

[19] S. Gudder and G. Nagy, Sequential quantum measurements, *J. Math. Phys.* **42** (2001), 5212–5222.

[20] M. Hirvensalo, *Quantum Computing*, Springer-Verlag, Berlin, 2001.

[21] A. S. Holevo, *Probabilistic and Statistical Aspects of Quantum Theory*, North Holland, Amsterdam, 1982.

[22] R. Kadison, Order properties of bounded self-adjoint operators, *Proc. Amer. Math. Soc.* **34** (1951), 505–510.

[23] K. Kraus, *States, Effects, and Operations*, Springer-Verlag, Berlin, 1983.

[24] G. Ludwig, *Foundations of Quantum Mechanics*, Vols I and II, Springer-Verlag, Berlin, 1983/1985.

[25] T. Moreland and S. Gudder, Infima of Hilbert space effects, *Lin. Alg. Appl.* **286** (1999), 1–17.

[26] M. Nielsen and I. Chuang, *Quantum Computation and Quantum Information* Cambridge University Press, Cambridge, 2000.

[27] V. I. Paulsen, *Completely Bounded Maps and Dilations*, Longman Scientific and Technical, Harlow, 1986.

[28] M. Reed and B. Simon, *Functional Analysis*, Academic Press, New York, 1972.

[29] W. Rudin, *Functional Analysis* McGraw-Hill, New York, 1991.

[30] W. F. Stinespring, Positive functions on $C^*$-algebras, *Proc. Amer. Math. Soc.* **6** (1955), 211–216.

[31] L. A. Zadeh, Probability measures and fuzzy events, *J. Math. Anal. Appl.* **23** (1968), 412–427.

The following items will also be of interest to readers of this work. [Reference [33] is closely related to, and extends some points, of [1] above; see also [9].]

[32] E. G. Effros and Z. Ruan, *Operator Spaces*, Oxford University Press, Oxford, 2000.

[33] G. K. Pedersen, A note on fixed points of completely positive maps, (preprint, dated July 2003), Internet, arXiv:math.OA/0308053 v1.

# Advances and Applications of the Feynman Integral

Brian Jefferies

School of Mathematics The University of New South Wales NSW 2052 Australia
b.jefferies@unsw.edu.au

## 1 Introduction

We begin with a brief background of the basic ideas behind the Feynman integral. Our task is made simpler by the recent appearance of the excellent monograph [33] by G.W. Johnson and M. Lapidus. A discussion of the Feynman integral in greater depth than is given here is presented in Chapter 7 of [33] and is a basic reference at the appropriate points below. As much as possible, the notation of [33] is used when it is needed.

For a quantum particle in one dimension, the evolution of its state is expressed in terms of the continuous unitary group $e^{-itH/\hbar}$, $t \in \mathbb{R}$ of operators. The dynamics is specified by the Hamiltonian operator $H$. The integral kernel $K(x, y, t)$, $x, y \in \mathbb{R}$, of the operator $e^{-itH/\hbar}$ is known as the *propagator* of the quantum system, although this may only be a distributional expression. Knowledge of the propagator therefore specifies the evolution of quantum states.

The *quantization* of the dynamics of a classical mechanical system means expressing the quantum propagator in terms of classical quantities. Usually this is done by replacing classical observables such as position and momentum by their corresponding operator counterparts. By contrast, R. Feynman [15] tried to write the quantum propagator $K(x, y, t)$ directly in terms of the classical action $S(q, \dot{q}, t)$ by means of the formula

$$K(x, y, t) = N^{-1} \int_{C_{x,y}^{0,t}} e^{\frac{i}{\hbar} S(q, \dot{q}, t)} \mathcal{D}q. \tag{1.1}$$

Here $C_{x,y}^{0,t}$ is the set of all continuous paths $\omega$ such that $\omega(0) = x$ and $\omega(t) = y$, $N$ is a 'normalization factor' and $\mathcal{D}q$ is 'uniform measure' on $C_{x,y}^{0,t}$.

For a single particle of mass $m$ on the line with a real valued potential $V$, the action of a classical path $q : [0, t] \rightarrow \mathbb{R}$ is the expression

$$S(q, \dot{q}, t) = \int_0^t (\frac{1}{2} m \, \dot{q}(s)^2 - V(q(s))) \, ds. \tag{1.2}$$

An attempt can be made to interpret formula (1.1) by taking a positive integer $n$, dividing the interval $[0, t]$ into $n$ equal parts and setting $K_n(x, y, t)$ equal to

$$C_n \int_{\mathbb{R}^{n-1}} \exp\left\{ \frac{i}{\hbar} \sum_{j=1}^{n} \left[ \frac{m}{2(t/n)}(x_j - x_{j-1})^2 - \frac{t}{n}V(x_j) \right] \right\} dx_1 \cdots dx_{n-1}, \quad (1.3)$$

where $x_0 = x$ and $x_n = y$ and the normalization constant $C_n$ is given by

$$\left( \frac{-im}{2\pi\hbar(t/n)} \right)^{n/2}.$$

The exponent in the integrand of expression (1.3) is $i/\hbar$ times an approximation to the action (1.2) in the case that $q$ is a polygonal path [33, Equation (7.4.1)]. One might hope that the approximations $K_n(x, y, t)$, $n = 1, 2, \ldots$, converge in an appropriate sense to the propagator $K(x, y, t)$.

One difficulty is that the integrand of expression (1.3) has absolute value one, so the integral is not absolutely convergent. Nevertheless, $K_n(x, y, t)$ can be viewed as the kernel of an operator which converges in the strong operator topology as $n \to \infty$ for a large class of potentials $V$ [33, Theorem 7.5.1]. The limiting operator is $e^{-itH/\hbar}$ for the associated quantum Hamiltonian operator $H$.

The reinterpretation of formula (1.1) in terms of limits of operator products was first given by E. Nelson [37] and is actually part of a more general theory of the approximation of semigroups of operators [33, Chapter 11]. A mathematical physicist may simply declare that the story ended with the paper [37], but formula (1.1) has continued to inspire a great deal more mathematics, a part of which has a detailed exposition in the monograph [33]. Moreover, the heuristic Feynman integral, taking the essential ideas of formula (1.1), has produced useful formulae in low dimensional topology and knot theory, see [33, Section 20.2] and [49] for a discussion of these developments. The geometric nature of these applications means that there is no fundamental time parameter with which to provide subdivisions and form operator products (but see the discussion of the multiplicative property of the heuristic Feynman path integral on [33, p. 648]). In the viewpoint of E. Witten [49], understanding four-dimensional quantum gauge theory is intimately connected with understanding Feynman path integrals over the space of connections on a four-manifold.

The aim is more modest in these notes, where attention is restricted to the case for which there is a distinguished time parameter and expressions like (1.3) are associated with the kernel of an integral of a scalar valued function with respect to an operator valued set function defined on a space of *paths*. One might hope that the "correlation functions" mentioned on [49, p. 26] could be similarly expressed in terms of an integral of a scalar valued function with respect to an operator valued set function defined on a space of connections. However, even a *candidate* for such an additive operator valued set function has not been determined, so speculation on these matters is avoided here.

On a more philosophical note, the monograph [33] runs to 771 pages. A treatment of other approaches to Feynman's ideas at the level of mathematical precision of [33] would require several more volumes. Nor would it be possible to write an adequate survey in the length of these notes. For example, the subject of Hamiltonian Feynman path integrals has attracted increased attention recently as in [14, 44]. Here the idea is to replace the classical action (1.2) by the action $\int_0^t (p(s)\dot{q}(s) - H(p(s), q(s)))\, ds$

for a path $s \longmapsto (p(s), q(s))$ in phase space rather than in configuration space. The analysis of the approximations of Hamiltonian Feynman path integrals is closely related to quantization procedures sending a classical Hamiltonian to its quantum counterpart, for which a study of pseudodifferential operators is necessary.

Many developments of Feynman's ideas represent quantities of physical interest as the limit of finite dimensional approximations like (1.3) and proving the existence of such limits can be a tricky business. The emphasis in these notes is to look carefully at the integration theoretic background of approximations of this type, especially to the extent that the approximations are associated with the integral with respect to an additive set function. The additive set functions considered below, although well defined, can be highly singular. As will become clear from what follows, it is too much to expect that the existence of the necessary limits can be proved without a detailed understanding of the underlying operator theory, although 'measure theory' and 'operator theory' are often treated as separate disciplines. Advancement in the understanding of Feynman integrals requires a combination of techniques from these two subject areas.

The paper is organized as follows. Some less common terminology is mentioned at the end of this section. The operator valued Feynman integral as discussed in [33] is defined in Section 2. The connection with the Trotter product formula appears in equation (2.3). Section 3 highlights the connection between the operator valued Feynman integral and "random processes" measured by operator valued set functions—*evolution processes*. Abandoning probability measures in favor of operator valued measures is quite natural in the context of quantum physics. The theme is expanded upon in the monograph [28].

The *Feynman–Kac formula* described in Section 4 is interpreted more widely than is usual in the probability and mathematical physics literature, where it usually refers to the representation of a perturbation to the heat semigroup in terms of an integral with respect to Wiener measure. A similar proof works for operator valued measures. The material here is from [28, Chapter 3].

Sufficient conditions for a semigroup to be associated with operator valued measures are considered in Section 5. Example 5.1 shows why the set functions associated with Feynman integrals do not satisfy these conditions, even under analytic continuation in time. Theorem 5.6 gives a recent characterization of semigroups associated with operator valued measures acting on $L^p$-spaces, see [31].

An important example of operator valued measures generated by semigroups has been studied by E. Thomas [47]. Section 6 shows that these are examples of the processes treated in Section 5 and describes how there is an associated Markov process in the sense of probability theory. These turn out to be jump processes like the Poisson process.

On the other hand, Section 7 treats in some detail the example of the Dirac equation in two space-time dimensions. It has been known since the work of T. Ichinose [22] that there are associated matrix valued measures and these are naturally expressed in the form of operator valued measures considered in Section 3. The idea of a *multiplicative functional* familiar from the theory of Markov processes has a natural expression here. The usual Feynman–Kac functional $F_t = e^{-i \int_0^t V \circ X_s \, ds}$ makes no sense if the potential $V$ is too singular. A "renormalized" Feynman–Kac functional is used to represent the dynamics of a Dirac particle on the line with Coulomb-like potential $V(x) = 1/|x|$, $x \in \mathbb{R}$, $x \neq 0$. This is joint work with Z. Brzeźniak [9].

In order to make precise the idea of integration with respect to unbounded set functions, examples of unbounded additive set functions that arise naturally in several areas of analysis are given in Section 8. The well-known connections between harmonic analysis, quantization and the Weyl functional calculus arise here. This section is based on [29].

Section 9 deals with an integration theory for Feynman integrals. The essential idea is that integration with respect to a family $\langle M_\lambda^t \rangle_{\lambda > 0}$ of operator valued measures is used to control integration with respect to the unbounded set functions $M_\lambda^t$ with $\Im \lambda \neq 0$, associated with the analytic continuation in time of Wiener measure. The regularity of the complex valued multiplicative functionals $F_t$ we are trying to integrate is reflected in their integrability. It is the oscillatory nature of $F_t$ that is important here because $|F_t| = 1$. This section is based on [26, 27, 30] and [28, Chapter 8].

Feynman integrals figure significantly in quantum field theory at the heuristic level. Section 10 describes quantum field theoretic integration by analogy with its quantum mechanical counterpart described in Section 9. For a Boson quantum field theory with self-interaction, the task is to construct a suitable integrable multiplicative functional of the field. The Euclidean version is described in [20], but it seems that the consideration of Feynman integrals introduces new features into quantum field theory, even in two space–time dimensions. This is also apparent for the Dirac equation considered in some detail in Section 7. Although the ideas behind the functional integral approach to quantum field theory are set out in considerable detail in [20], the reformulation in the general framework considered in these notes appears here for the first time.

## 1.1 Notation and terminology

Because we shall be dealing with operator valued measures, we make a few remarks about additive set functions and vector integration. Details may be found in [35] and [12].

By a *semi-algebra* we mean a family $C$ of subsets of a nonempty set such that

(i) the whole set is a member of $C$;
(ii) $C$ is closed under finite intersections; and
(iii) if $A$, $B \in C$, then there exist a positive integer $n$ and sets $U_j \in C$, $j = 0, 1, \ldots, n$, such that $A \cap B = U_0$ and $A \backslash B = \cup_{j=1}^n U_j$, and such that $\cup_{j=0}^k U_j \in C$ for all $k = 1, \ldots, n$.

A typical example of a semi-algebra is the family of all half-open intervals $[a, b)$ with $0 \leq a \leq b \leq 1$.

Let $S$ be a semi-algebra of subsets of a nonempty set $\Omega$. Let $m : S \to \mathbb{C}$ be an additive set function, that is, $m(A \cup B) = m(A) + m(B)$ for disjoint $A$, $B \in S$ such that $A \cup B \in S$. The unique extension of $m$ to the algebra generated by $S$ is denoted by the same symbol. The *variation* $|m| : S \to [0, \infty]$ of $m$ is defined for each $A \in S$ by

$$|m|(A) = \sup\left\{\sum_{B\in\pi} |m(A\cap B)|\right\}.$$

The supremum is over all partitions $\pi$ of $\Omega$ by elements of $S$.

Now let $X$ be a Banach space, $(\Sigma, \mathcal{E})$ a measurable space and $m : \mathcal{E} \to X$ a vector measure, that is, an $X$-valued set function $\sigma$-additive for the norm topology of $X$. Here '$\sigma$-additive' has the same meaning as it does for scalar valued measures: $m\left(\cup_{j=1}^{\infty} E_j\right) = \sum_{j=1}^{\infty} m(E_j)$ for all pairwise disjoint elements $E_j$, where $j = 1, 2, \ldots,$ of $\mathcal{E}$. The vector space of all continuous linear functionals on $X$ is denoted by $X'$.

A function $f : \Sigma \to \mathbb{C}$ is said to be $m$-*integrable* if for each $\xi \in X'$, the function $f$ is integrable with respect to the scalar measure $\langle m, \xi\rangle : E \longmapsto \langle m(E), \xi\rangle$, $E \in \mathcal{E}$, and for each $E \in \mathcal{E}$, there exists a vector $(f.m)(E)$ belonging to $X$ such that the equality

$$\langle (f.m)(E), \xi\rangle = \int_E f\, d\langle m, \xi\rangle$$

holds for every $\xi \in X'$. It turns out that the mapping $E \longmapsto (f.m)(E)$, $E \in \mathcal{E}$, is necessarily $\sigma$-additive for the norm topology of $X$, that is, the indefinite integral $f.m = fm$ of an $m$-integrable function $f$ is again a vector measure. We shall also write $\int_E f\, dm$ for the vector $(f.m)(E)$, $E \in \mathcal{E}$.

Let $\mathcal{L}(X)$ denote the space of bounded linear operators acting on $X$ equipped with the strong operator topology. A similar notion of integrability applies to a function $f : \Sigma \to \mathbb{C}$ with respect to an operator valued measure $M : \mathcal{E} \to \mathcal{L}(X)$, that is, an $\mathcal{L}(X)$-valued set function $\sigma$-additive for the *strong operator topology*. For each $x \in X$ and $\xi \in X'$, the function $f$ is integrable with respect to the scalar measure $\langle Mx, \xi\rangle : E \longmapsto \langle M(E)x, \xi\rangle$, $E \in \mathcal{E}$, and for each $E \in \mathcal{E}$, there exists an operator $(f.M)(E)$ belonging to $\mathcal{L}(X)$ such that the equality

$$\langle (f.M)(E)x, \xi\rangle = \int_E f\, d\langle Mx, \xi\rangle$$

holds for every $x \in X$ and $\xi \in X'$. The indefinite integral $fM : E \longmapsto (f.M)(E)$, $E \in \mathcal{E}$, is $\sigma$-additive for the strong operator topology of $\mathcal{L}(X)$. We shall also write $\int_E f\, dM$ for the bounded linear operator $(f.M)(E)$, $E \in \mathcal{E}$.

With integration understood in the above sense, the usual monotone and dominated convergence theorems are valid. In particular, bounded measurable functions are integrable. If $\phi : \Sigma \to \Xi$ is a measurable map into the measurable space $(\Xi, \mathcal{S})$, then $M \circ \phi^{-1}$ denotes the operator valued measure $S \longmapsto M(\phi^{-1}(S))$, $S \in \mathcal{S}$. Notation applied to bounded linear operators will also be adopted for operator valued measures. If $S : X \to X$ and $T : X \to X$ are bounded linear operators, then the operator valued measure $E \longmapsto SM(E)T$, $E \in \mathcal{E}$, is denoted by $SMT$. The direct sum $M_1 \oplus M_2$ of two operator valued measures $M_1 : \mathcal{E} \to \mathcal{L}(H_1)$, $M_2 : \mathcal{E} \to \mathcal{L}(H_2)$ acting on Hilbert spaces $H_1$, $H_2$ is defined by $M_1 \oplus M_2 : E \longmapsto M_1(E) \oplus M_2(E)$, $E \in \mathcal{E}$.

## 2 The operator valued Feynman integral

M. Kac noticed the similarity between the expression (1.3) and Wiener measure [33, p. 115], which led to the representation of solutions of the heat equation as integrals with respect to Wiener measure and to the idea that sense could be made of formula (1.1) by analytic continuation from Wiener measure [37].

Fix a positive integer $d$ and a time $t > 0$. Set $C^t = C([0, t], \mathbb{R}^d)$. Let $w$ be Wiener measure on the space $C_0^t$ of continuous functions $\omega : [0, t] \to \mathbb{R}^d$ for which $\omega(0) = 0$ [33, Section 3.2]. The *analytic in mass operator valued Feynman integral* of a function $F : C^t \to \mathbb{C}$ is defined in the following fashion [33, Definition 13.5.1].

Suppose that for each $\lambda > 0$ and $\xi \in \mathbb{R}^d$, the function $\omega \longmapsto F(\lambda^{-1/2}\omega + \xi)$, $\omega \in C_0^t$, is $w$-measurable and there exists an operator $K_\lambda^t(F) \in \mathcal{L}(L^2(\mathbb{R}^d))$ such that for every $\psi \in L^2(\mathbb{R}^d)$, the function

$$\omega \longmapsto F(\lambda^{-1/2}\omega + \xi)\psi(\lambda^{-1/2}\omega(t) + \xi), \quad \omega \in C_0^t,$$

is $w$-integrable for almost every $\xi \in \mathbb{R}^d$ and the equality

$$(K_\lambda^t(F)\psi)(\xi) = \int_{C_0^t} F(\lambda^{-1/2}\omega + \xi)\psi(\lambda^{-1/2}\omega(t) + \xi)\, dw(\omega) \qquad (2.1)$$

holds for almost all $\xi \in \mathbb{R}^d$. If $\lambda \longmapsto K_\lambda^t(F), \lambda > 0$, is the restriction to the positive real axis of an $\mathcal{L}(L^2(\mathbb{R}^d))$-valued function analytic in the region $\mathbb{C}_+ = \{z \in \mathbb{C} : \Re z > 0\}$, then we call the operator $K_{-iq_0}^t(F) = \lim_{\lambda \to -iq_0} K_\lambda^t(F)$ the *analytic in mass operator valued Feynman integral of $F$ with parameter* $-iq_0 \neq 0$, if the limit exists in the strong operator topology. Other types of limits are possible. Indeed, nontangential limits play an important role in [33, Section 13.5].

At first glance, formula (2.1) does not bear much resemblance to formula (1.1). In order to examine (2.1) more closely, let $\Delta = \partial^2/\partial x_1^2 + \cdots + \partial^2/\partial x_d^2$ be the selfadjoint Laplacian acting in $L^2(\mathbb{R}^d)$. In a scaled coordinate system, the free Hamiltonian of a quantum system in $L^2(\mathbb{R}^d)$ is $H_0 = -\frac{1}{2}\Delta$. If $V$ is a sufficiently regular, real valued potential, then the operator closure $H$ of the sum $H_0 + V$ of the two densely defined operators $H_0$ and $V$ is selfadjoint in $L^2(\mathbb{R}^d)$. For a given function $\psi \in L^2(\mathbb{R}^d)$, the function $\left(e^{-itH_0/n}e^{-itV/n}\right)^n \psi$ is essentially equal to $x \longmapsto \int_{\mathbb{R}^d} K_n(x, y, t)\psi(y)\, dy$, $x \in \mathbb{R}^d$, with $K_n$ given by formula (1.3)—the product integral in this expression needs to be interpreted as an iterated integral with mean-square limits [33, Lemma 11.2.18]. On the other hand, by the Trotter product formula for unitary groups [33, Theorem 7.5.2], we have

$$e^{-itH}\psi = \lim_{n \to \infty} \left(e^{-itH_0/n}e^{-itV/n}\right)^n \psi \qquad (2.2)$$

in $L^2(\mathbb{R}^d)$, uniformly for $t$ in bounded subsets of $\mathbb{R}$.

If we set

$$F_{-iV}(\omega) = \exp\left\{-i\int_0^t V(\omega(s))\, ds\right\}, \quad \omega \in C^t, \qquad (2.3)$$

then it turns out that $F_{-iV}$ is defined almost everywhere, $K_{-i}^t(F_{-iV})$ exists and we have

$$e^{-itH}\psi = K^t_{-i}(F_{-iV})\psi = \lim_{n\to\infty}\left(e^{-itH_0/n}e^{-itV/n}\right)^n\psi. \tag{2.4}$$

On a technical note, equality (2.4) follows from [33, Corollary 13.5.18] if, for *every* $\lambda \in \mathbb{C}_+$ and for $\lambda = -i$, the operator $C_\lambda$ in [33, Equation 13.5.42a] is shown to be the infinitesimal generator of the $C_0$-contraction semigroup $T_\lambda$ defined by the right-hand side of [33, Equation 13.5.43]. A sufficient condition guaranteeing the equalities (2.4) is that the potential $V$ belongs to the class $S_d$ defined in [33, Definition 11.2.13]— there is a considerable amount of operator theory underlying this assertion, see [33, Proposition 11.2.14] and [34, Corollary IX.2.5, Theorem IX.2.16, Theorem VIII.1.5, Theorem V.4.3].

The analytic in mass operator valued Feynman integral of $F_{-iV}$ is discussed in [33, Section 13.5] under the assumption that the potential $V$ is continuous off a closed set of capacity zero. This permits arbitrarily strong singularities without regard to sign on the closed set of capacity zero. The draw back is that $K^t_{-iq_0}(F_{-iV})$ is only shown to exist for *almost all* $q_0 \in \mathbb{R} \setminus \{0\}$.

## 3 Evolution processes

The integral (1.3) can be rewritten so as to emphasize the operator products that underlie it. To see this, let $f_1, \ldots, f_n$ be a bounded complex valued Borel measurable functions on $\mathbb{R}^d$. Let $0 < t_1 < \cdots < t_n < t$ be $n$ distinct times before $t$.

Set $F(\omega) = f_1(\omega(t_1)) \cdots f_n(\omega(t_n))$ for every $\omega \in C^t$. Then formula (2.1) gives

$$(K^t_\lambda(F)\psi)(\xi) = \int_{C^t_0} F(\lambda^{-1/2}\omega + \xi)\psi(\lambda^{-1/2}\omega(t) + \xi)\,dw(\omega)$$

$$= \int_{C^t_0} f_1(\lambda^{-1/2}\omega(t_1) + \xi) \cdots f_n(\lambda^{-1/2}\omega(t_n) + \xi)\psi(\lambda^{-1/2}\omega(t) + \xi)\,dw(\omega)$$

$$= C_{n+1}\int_{\mathbb{R}^d} \cdots \int_{\mathbb{R}^d} f_1(\lambda^{-1/2}x_1 + \xi) \cdots f_n(\lambda^{-1/2}x_n + \xi)\psi(\lambda^{-1/2}x_{n+1} + \xi)$$

$$\times e^{-\frac{|x_{n+1}-x_n|^2}{2(t-t_n)}} e^{-\frac{|x_n-x_{n-1}|^2}{2(t_n-t_{n-1})}} \cdots e^{-\frac{|x_2-x_1|^2}{2(t_2-t_1)}} e^{-\frac{|x_1|^2}{2t_1}}\,dx_1\cdots dx_{n+1}$$

$$= C_{n+1}\lambda^{\frac{d(n+1)}{2}}\int_{\mathbb{R}^d} \cdots \int_{\mathbb{R}^d} f_1(x_1) \cdots f_n(x_n)\psi(x_{n+1})$$

$$\times e^{-\frac{\lambda|x_{n+1}-x_n|^2}{2(t-t_n)}} e^{-\frac{\lambda|x_n-x_{n-1}|^2}{2(t_n-t_{n-1})}} \cdots e^{-\frac{\lambda|x_2-x_1|^2}{2(t_2-t_1)}} e^{-\frac{\lambda|x_1-\xi|^2}{2t_1}}\,dx_1\cdots dx_{n+1}. \tag{3.1}$$

Here $C_{n+1} = (2\pi(t - t_n))^{-d/2} \ldots (2\pi t_1)^{-d/2}$. The integral (3.1) is absolutely convergent for all $\lambda \in \mathbb{C}$ for which $\Re\lambda > 0$.

Let $\Delta$ be the selfadjoint Laplacian in $L^2(\mathbb{R}^d)$ and set $S_\lambda(t) = e^{t\Delta/(2\lambda)}$ for all $t \geq 0$ and $\lambda > 0$. The exponential is defined by the functional calculus for selfadjoint operators. Then for each $\lambda > 0$, the $C_0$-semigroup $S_\lambda$ is explicitly given by

$$(S_\lambda(t)\phi)(x) = \left(\frac{\lambda}{2\pi t}\right)^{d/2}\int_{\mathbb{R}^d} e^{-\frac{\lambda}{2t}|x-y|^2}\phi(y)\,dy \quad a.e. \tag{3.2}$$

for every $\phi \in L^2(\mathbb{R}^d)$ and $t > 0$.

Let $Q(B) : L^2(\mathbb{R}^d) \to L^2(\mathbb{R}^d)$ be the operator of multiplication by the characteristic function of the Borel subset $B$ of $\mathbb{R}^d$. For a bounded Borel measurable function $f$, the operator of multiplication by $f$ is written as $Q(f)$. The operator $Q(f)$ is actually the integral $\int_{\mathbb{R}^d} f \, dQ$ of the bounded function $f$ with respect to the spectral measure $Q : B \longmapsto Q(B)$.

The operator $K_\lambda^t(F)$ given by equation (3.1) then has the representation

$$K_\lambda^t(F) = S_\lambda(t_1)Q(f_1) \cdots S_\lambda(t_n - t_{n-1})Q(f_n)S_\lambda(t - t_n) \tag{3.3}$$

as a product of operators.

Reversing the order of these operators, we arrive at the following definition. Let $\lambda > 0$. For the cylinder set

$$E = \{\omega \in C^t : \omega(t_1) \in B_1, \ldots, \omega(t_n) \in B_n\} \tag{3.4}$$

with $0 \leq t_1 < \cdots < t_n \leq t$, set $M_\lambda^t(E) \in \mathcal{L}(L^2(\mathbb{R}^d))$ equal to the operator

$$M_\lambda^t(E) := S_\lambda(t - t_n)Q(B_n)S_\lambda(t_n - t_{n-1}) \cdots Q(B_1)S_\lambda(t_1). \tag{3.5}$$

The *algebra* generated by all such cylinder sets $E$ as the times $0 \leq t_1 < \cdots < t_n \leq t$, the Borel subsets $B_1, \ldots, B_n$ of $\mathbb{R}^d$ and the positive integer $n$ vary, is denoted by $\mathcal{S}_t$. Then $M_\lambda^t : E \longmapsto M_\lambda^t(E)$ is well defined and has a unique additive extension to the algebra $\mathcal{S}_t$ of subsets of $C^t$.

On the other hand, if $\phi \in L^2(\mathbb{R}^d)$, then

$$\left( \left( \int_{C^t} F(\omega) \, M_\lambda^t(d\omega) \right) \phi \right)(\xi) = \left( \int_{C^t} F(\omega) \, (M_\lambda^t \phi)(d\omega) \right)(\xi)$$

$$= \left( \int_{C^t} f_1(\omega(t_1)) \cdots f_n(\omega(t_n)) \, (M_\lambda^t \phi)(d\omega) \right)(\xi)$$

$$= (S_\lambda(t - t_n)Q(f_n)S_\lambda(t_n - t_{n-1}) \cdots Q(f_1)S_\lambda(t_1)\phi)(\xi)$$

$$= C_{n+1}\lambda^{\frac{d(n+1)}{2}} \int_{\mathbb{R}^d} \cdots \int_{\mathbb{R}^d} f_1(x_1) \cdots f_n(x_n)$$

$$\times e^{-\frac{\lambda|\xi - x_n|^2}{2(t - t_n)}} e^{-\frac{\lambda|x_n - x_{n-1}|^2}{2(t_n - t_{n-1})}} \cdots e^{-\frac{\lambda|x_2 - x_1|^2}{2(t_2 - t_1)}} e^{-\frac{\lambda|x_1 - x|^2}{2t_1}} \phi(x) \, dx \, dx_1 \cdots dx_n. \tag{3.6}$$

Comparison between the expressions (3.1) and (3.6) above shows that in the inner product $\langle u, v \rangle = \int_{\mathbb{R}^d} u(x)\overline{v(x)} \, dx$ of $L^2(\mathbb{R}^d)$, we have

$$\left\langle \left( \int_{C^t} F(\omega) \, M_\lambda^t(d\omega) \right) \phi, \overline{\psi} \right\rangle = \langle K_\lambda^t(F)\psi, \overline{\phi} \rangle = \langle \phi, K_\lambda^t(\overline{F})\overline{\psi} \rangle.$$

Hence, $\int_{C^t} F(\omega) \, M_\lambda^t(d\omega) = K_\lambda^t(\overline{F})^*$. The $\sigma$-additive extension of $M_\lambda^t$ from $\mathcal{S}_t$ to $\sigma(\mathcal{S}_t)$ is equal to the operator valued measure $A \longmapsto K_\lambda^t(\chi_A)^*$ for all $A \in \sigma(\mathcal{S}_t)$. As is customary, the unique extension is denoted, again, by $M_\lambda^t$.

Now we check that the set function

$$A \longmapsto K_\lambda^t(\chi_A)^*, \quad A \in \sigma(\mathcal{S}_t)$$

is actually a measure in the strong operator topology for each $\lambda > 0$. Inspection of (2.1) shows that for all $\phi \in L^2(\mathbb{R}^d)$ and $A \in \sigma(\mathcal{S}_t)$, we have

$$\|K_\lambda^t(\chi_A)^*\phi\|_2$$

$$= \sup_{\|\psi\|_2 \le 1} \left| \int_{\mathbb{R}^d} \left( \int_{C_0^t} \chi_A(\lambda^{-1/2}\omega + \xi)\overline{\psi}(\lambda^{-1/2}\omega(t) + \xi)\, dw(\omega) \right) \phi(\xi)\, d\xi \right|$$

$$\le \sup_{\|\psi\|_2 \le 1} \int_{\mathbb{R}^d} \left( \int_{C_0^t} \chi_A(\lambda^{-1/2}\omega + \xi)|\psi(\lambda^{-1/2}\omega(t) + \xi)|\, |\phi(\xi)|\, dw(\omega) \right) d\xi$$

$$\le \sup_{\|\psi\|_2 \le 1} \int_{\mathbb{R}^d} \left( \int_{C_0^t} |\psi(\lambda^{-1/2}\omega(t) + \xi)|\, |\phi(\xi)|\, dw(\omega) \right) d\xi$$

$$= \sup_{\|\psi\|_2 \le 1} \langle S_\lambda(t)|\psi|, |\phi| \rangle \le \|\phi\|_2.$$

Therefore, the collection $\{K_\lambda^t(\chi_A)^* : A \in \mathcal{S}_t \}$ of operators is uniformly bounded in operator norm in $\mathcal{L}(L^2(\mathbb{R}^d))$. It is enough to check the norm $\sigma$-additivity of the set function $A \longmapsto K_\lambda^t(\chi_A)^*\phi$, $A \in \sigma(\mathcal{S}_t)$, for a dense set of $\phi \in L^2(\mathbb{R}^d)$.

For $\phi \in L^\infty(\mathbb{R}^d) \cap L^1(\mathbb{R}^d)$, the Cauchy–Schwarz inequality applied to the finite Borel measure $|\phi(\xi)|\, dw(\omega)d\xi$ on $C_0^t \times \mathbb{R}^d$ gives

$$\|K_\lambda^t(\chi_A)^*\phi\|_2$$

$$\le \sup_{\|\psi\|_2 \le 1} \int_{C_0^t \times \mathbb{R}^d} \chi_A(\lambda^{-1/2}\omega + \xi)|\psi(\lambda^{-1/2}\omega(t) + \xi)|\, |\phi(\xi)|\, dw(\omega)d\xi$$

$$\le \sup_{\|\psi\|_2 \le 1} \left( \int_{C_0^t \times \mathbb{R}^d} |\psi(\lambda^{-1/2}\omega(t) + \xi)|^2\, |\phi(\xi)|\, dw(\omega)d\xi \right)^{\frac{1}{2}}$$

$$\times \left( \int_{C_0^t \times \mathbb{R}^d} \chi_A(\lambda^{-1/2}\omega + \xi)\, |\phi(\xi)|\, dw(\omega)d\xi \right)^{\frac{1}{2}}$$

$$\le \sup_{\|\psi\|_2 \le 1} \langle S_\lambda(t)|\psi|^2, |\phi| \rangle^{\frac{1}{2}} \left( \int_{C_0^t \times \mathbb{R}^d} \chi_A(\lambda^{-1/2}\omega + \xi)\, |\phi(\xi)|\, dw(\omega)d\xi \right)^{\frac{1}{2}}$$

$$\le \left( \int_{C_0^t \times \mathbb{R}^d} \chi_A(\lambda^{-1/2}\omega + \xi)\, |\phi(\xi)|\, dw(\omega)d\xi \right)^{\frac{1}{2}} \|\phi\|_\infty.$$

If $A_n \downarrow \emptyset$ as $n \to \infty$, then the right-hand side converges to zero. The set function $A \longmapsto K_\lambda^t(\chi_A)^*\phi$, $A \in \sigma(\mathcal{S}_t)$, is clearly additive, so it is norm $\sigma$-additive in $L^2(\mathbb{R}^d)$. Hence, for each $\lambda > 0$, the additive set function $M_\lambda^t$ is the restriction to $\mathcal{S}_t$ of a unique $\mathcal{L}(L^2(\mathbb{R}^d))$-valued measure defined on the $\sigma$-algebra $\sigma(\mathcal{S}_t)$ generated by $\mathcal{S}_t$. We call $M_\lambda^t$ the $(S_\lambda, Q)$-*measure* on $\sigma(\mathcal{S}_t)$.

Each semigroup $S_\lambda$, $\lambda > 0$, has a unique analytic extension from the set of all positive real numbers $\lambda$ to the set $\mathbb{C}_+ = \{\lambda \in \mathbb{C} : \Re\lambda > 0\}$. Formula (3.2) is valid for all $\lambda \in \mathbb{C}_+$ provided we take the branch of $\lambda \longmapsto \lambda^{1/2}$ such that $\Re(\lambda^{1/2}) > 0$ on $\mathbb{C}_+$. Then the operator valued function $\lambda \longmapsto M_\lambda^t(E)$, $\lambda > 0$, has a unique analytic extension to $\mathbb{C}_+$ for each cylinder set (3.4). The set functions $M_\lambda^t$ are defined for all $\lambda \in \overline{\mathbb{C}}_+ \setminus \{0\}$ by continuity and are given by formula (3.5). In this fashion, we obtain a family $\langle M_\lambda^t \rangle_{\lambda \in \overline{\mathbb{C}}_+ \setminus \{0\}}$ of additive operator valued functions $M_\lambda^t : S_t \to \mathcal{L}(L^2(\mathbb{R}^d))$.

If $\Im\lambda \neq 0$, then the collection $\{M_\lambda^t(A) : A \in S_t\}$ of bounded linear operators is *unbounded* in the operator norm. Note here that $S_t$ is the *algebra* generated by cylinder sets (3.4). Each operator (3.5) is a contraction on $L^2(\mathbb{R}^d)$ for $\lambda \in \overline{\mathbb{C}}_+ \setminus \{0\}$, but the collection of all cylinder sets (3.4) is only a *semi-algebra* [33, Definition 3.2.1].

There is a good reason that the operator ordering of the operator product (3.5) is the reverse of the analytic operator valued Feynman integral (3.3). Suppose that $d = 3$ and an initial state $\psi$ is prepared for a free quantum particle moving in $\mathbb{R}^3$. In the case that $\lambda = -i$ and $E$ is the cylinder set (3.4), the vector $M_{-i}^t(E)\psi$ represents the state at time $t$ after observations of position have been made at times $0 \leq t_1 < \ldots < t_n \leq t$. Of course, this conclusion depends on the interpretation of quantum mechanics one is willing to accept. But even in classical mechanics, if instead of the semigroup $S_\lambda$ in the expression (3.5) one substitutes the flow $S$ acting on an initial distribution $\mu$ of classical states, the resulting measure $M^t(E)\mu := S(t-t_n)Q(B_n)S(t_n-t_{n-1}) \cdots Q(B_1)S(t_1)\mu$ represents the final distribution of states after observations have been made at times $0 \leq t_1 < \ldots < t_n \leq t$. In probability theory, if the semigroup $S_1$ now acts on nonnegative measures $\mu$, then $M_1^t(E)\mu$ represents the total mass of a diffusing substance with initial distribution $\mu$ after the substance has been eliminated (or the diffusion process "killed") outside the sets $B_1, \ldots, B_n$ at times $0 \leq t_1 < \ldots < t_n \leq t$ respectively. The same interpretation holds if $S$ is the semigroup acting on probability measures associated with a time-homogeneous Markov process. Back to quantum mechanics, if $S$ is the dynamical group of an interacting system, suppose that the operator valued distribution $\Phi((x_1, t_1), \ldots, (x_n, t_n))$ represents the density of the set function

$$(B_1 \times \cdots \times B_n) \longmapsto M^t(X_{t_1} \in B_1, \ldots, X_{t_n} \in B_n)$$

with respect to Lebesgue measure. Then for the ground state $\psi$, the distribution

$$((x_1, t_1), \ldots, (x_n, t_n)) \longmapsto \langle \Phi((x_1, t_1), \ldots, (x_n, t_n))\psi, \psi \rangle$$

is the nonrelativistic analogue of the correlation functions of central importance to quantum field theory [20].

The operator valued set functions formed from expressions like (3.5) are therefore common features of mathematical disciplines closely allied with quantum physics and are the central object of study of these notes. By analogy with the theory of time-homogeneous Markov processes in probability theory, we consider the following abstract setup described in [28].

Let $(\Sigma, \mathcal{E})$ be a measurable space. For each $s \geq 0$, suppose that $S_s$ is a semi-algebra of subsets of a nonempty set $\Omega$ such that $S_s \subseteq S_t$ for every $0 \leq s < t$. For every $s \geq 0$, there are given functions $X_s : \Omega \to \Sigma$ with the property that $X_s^{-1}(B) \in \mathcal{S}_t$ for all $0 \le s \le t$ and $B \in \mathcal{E}$. It follows that the cylinder sets

$$E = \{X_{t_1} \in B_1, \ldots, X_{t_n} \in B_n\}$$
$$:= \{\omega \in \Omega : X_{t_1}(\omega) \in B_1, \ldots, X_{t_n}(\omega) \in B_n\} = X_{t_1}^{-1}(B_1) \cap \cdots \cap X_{t_n}^{-1}(B_n) \quad (3.7)$$

belong to $\mathcal{S}_t$ for all $0 \le t_1 < \ldots < t_n \le t$ and $B_1, \ldots, B_n \in \mathcal{E}$.

Let $\mathcal{X}$ be a Banach space. A *semigroup* $S$ of operators acting on $\mathcal{X}$ is a map $S : [0, \infty) \to \mathcal{L}(\mathcal{X})$ such that $S(0) = Id_{\mathcal{X}}$, the identity map on $\mathcal{X}$ and $S(t+s) = S(t)S(s)$ for all $s, t \ge 0$. The semigroup $S$ represents the evolution of state vectors belonging to $\mathcal{X}$.

It is an inconvenience of probability theory that $\mathcal{X}$ should be the space of Borel measures with total variation norm or the space of uniformly bounded Borel measurable functions with the supremum norm, so that $S$ is not usually strongly continuous. In this case, it is preferable that $\mathcal{X}$ should carry a topology weaker than the natural norm topology for which well-posedness of the corresponding Cauchy problem is valid.

An $\mathcal{L}(\mathcal{X})$-valued *spectral measure* $Q$ on $\mathcal{E}$ is a map $Q : \mathcal{E} \to \mathcal{L}(\mathcal{X})$ that is $\sigma$-additive in the strong operator topology and satisfies $Q(\Sigma) = Id_{\mathcal{X}}$ and $Q(A \cap B) = Q(A)Q(B)$ for all $A, B \in \mathcal{E}$. In quantum theory, $Q$ is typically multiplication by characteristic functions associated with the position observables, but the spectral measures associated with momentum operators also appear.

Now suppose that $M^t : \mathcal{S}_t \to \mathcal{L}(\mathcal{X})$ is an additive operator valued set function. The system

$$\left( \Omega, \langle \mathcal{S}_t \rangle_{t \ge 0}, \langle M^t \rangle_{t \ge 0}; \langle X_t \rangle_{t \ge 0} \right)$$

is called a *time homogeneous Markov evolution process* if there exists a $\mathcal{L}(\mathcal{X})$-valued spectral measure $Q$ on $\mathcal{E}$ and a semigroup $S$ of operators acting on $\mathcal{X}$ such that for each $t \ge 0$, the operator $M^t(E) \in \mathcal{L}(\mathcal{X})$ is given by

$$M^t(E) = S(t - t_n)Q(B_n)S(t_n - t_{n-1}) \cdots Q(B_1)S(t_1) \quad (3.8)$$

for every cylinder set $E \in \mathcal{S}_t$ of the form (3.7) and the process is called an $(S, Q)$-*process*. The basic ingredients are the semigroup $S$ describing the evolution of states and the spectral measure $Q$ describing observation of states represented by vectors in $\mathcal{X}$. An explicit proof that formula (3.8) actually does define an additive set function has been written, for example, in [36, Proposition 7.1]. The remainder of these notes is concerned with various examples of time homogeneous Markov evolution process, or briefly, evolution processes, and, inspired by Feynman's formula (1.1), with the possibility of using them to represent perturbations to semigroups.

# 4 The Feynman–Kac formula

There is a distinguished $(S_1, Q)$-process

$$\left( \Omega, \langle \mathcal{S}_t \rangle_{t \ge 0}, \langle M_1^t \rangle_{t \ge 0}; \langle X_t \rangle_{t \ge 0} \right)$$

associated with the semigroup $S_1$ defined on $\mathcal{X} = L^2(\mathbb{R}^d)$ by formula (3.2) in the case that $\lambda = 1$. The spectral measure $Q$ is, again, multiplication by characteristic functions on $\mathcal{X}$. The space $\Omega$ of paths is the set of all continuous functions from $\mathbb{R}_+$ into $\mathbb{R}^d$. The random process $X$ is given by the evaluation maps $X_s : \omega \longmapsto \omega(s)$ for all $\omega \in \Omega$ and $s \geq 0$ and $S_t$ is now the $\sigma$-algebra generated by all cylinder sets $E$ of the form (3.7) with times $0 \leq t_1 < \ldots < t_n \leq t$ and $n = 1, 2, \ldots$. As observed in Section 3, the formula $M_1^t(A) = K_1^t(\chi_A)^*$ shows that the mapping $A \longmapsto M_1^t(A)$, $A \in S_t$, is an operator valued measure for the strong operator topology of $\mathcal{L}(\mathcal{X})$.

Suppose that $V$ is a suitable real valued potential and set

$$F_{-V}(\omega) = \exp\left\{-\int_0^t V(\omega(s))\, ds\right\}.$$

Then comparison with equation (2.1) for $\psi \in L^2(\mathbb{R}^d)$ shows that

$$(K_1^t(F_{-v})\psi)(\xi) = \int_{C_0^t} \exp\left\{-\int_0^t V(\omega(s) + \xi)\, ds\right\} \psi(\omega(t) + \xi)\, dw(\omega). \quad (4.1)$$

The Feynman–Kac formula [33, Theorem 12.1.1] asserts that if $H := H_0 \dotplus V$ is the form sum of the free Hamiltonian $H_0$ and the potential $V$, then $(e^{-tH}\psi)(\xi)$ is equal to the right-hand side of formula (4.1) for almost all $\xi \in \mathbb{R}^d$, that is, $e^{-tH} = K_1^t(F_{-v})$. The heat semigroup $e^{-tH}$, $t \geq 0$, is a semigroup of selfadjoint bounded linear operators, so

$$e^{-tH} = K_1^t(F_{-v})^* = \int_\Omega \exp\left\{-\int_0^t V \circ X_s\, ds\right\} dM_1^t. \quad (4.2)$$

The last equality follows from the identity $M_1^t(A) = K_1^t(\chi_A)^*$, $A \in S_t$, by monotone convergence for operator valued measures. Equation (4.2) is therefore a reformulation of the Feynman-Kac formula in which the operator $e^{-tH}$ is represented directly in terms of an integral with respect to an operator valued measure.

A similar argument still works if $G$ is the generator of a $C_0$-semigroup $S$ and the $(S, Q)$-process $\left(\Omega, S, \langle M^t \rangle_{t \geq 0}, \langle S_t \rangle_{t \geq 0}; \langle X_t \rangle_{t \geq 0}\right)$ is $\sigma$-additive in the sense that each set function $M_t$ is actually an operator valued measure, $\sigma$-additive for the strong operator topology of $\mathcal{L}(\mathcal{X})$. The equality

$$e^{t(G+Q(V))} = \int_\Omega \exp\left\{\int_0^t V \circ X_s\, ds\right\} dM^t \quad (4.3)$$

then holds for all bounded measurable potentials $V$ under weak measurability assumptions on the process $X$. The equality holds in the limit if the left hand side converges in the strong operator topology to a $C_0$-semigroup and the integrand converges to an $M^t$-integrable function for all $t > 0$ as $V$ is approximated by bounded cutoff potentials. Conditions for this to obtain are considered in [28, Chapter 3].

Formulae of the type (4.3) for operator valued measures $M^t$ are also termed Feynman–Kac formulae in [28], although the associated semigroup $S$ need not be associated with the heat equation or any parabolic equation. Comparison of (4.3) with formula (2.4) suggests that we may be able to write

$$e^{-it(H_0+Q(V))} = \int_\Omega \exp\left\{-i \int_0^t V \circ X_s \, ds\right\} dM_{-i}^t \qquad (4.4)$$

for a suitable class of real valued potentials $V$, following the idea of Feynman's formula (1.1). However, as mentioned in Section 3, if $S_t$ denotes the algebra generated by cylinder sets based before time $t$, then the collection $\{M_{-i}^t(A) : A \in S_t\}$ of bounded linear operators is unbounded in the operator norm of $\mathcal{L}(L^2(\mathbb{R}^d))$. Worse, the total variation of the scalar set function

$$\langle M_\lambda^t \psi, \phi \rangle : A \longmapsto \langle M_\lambda^t(A)\psi, \phi \rangle, \quad A \in S_t,$$

is either zero or infinite for $\phi, \psi \in L^2(\mathbb{R}^d)$ and $\lambda \in \mathbb{C}_+$ with $\Im\lambda \neq 0$ (see Example 5.1 below), so the right hand side of (4.4) would have to be an integral with respect to a very singular object indeed. Nevertheless, integration with respect to unbounded set functions is a hidden feature of many problems in analysis and a few of these are mentioned in passing in Section 8.

## 5 Boundedness of processes

Before embarking on the study of particular examples of evolution processes, it is worth looking at conditions on a semigroup $S$ which guarantee the boundedness of the associated set functions $M^t$ defined by formula (3.8) on the *algebra* generated by cylinder sets (3.7). In this section, $S$ acts on some $L^p$-space and the spectral measure $Q$ is multiplication by characteristic functions. Such boundedness is needed to prove the Feynman–Kac formula (4.3), in which the right-hand side of the equation is the usual integral with respect to an operator valued measure.

It has been mentioned several times already that for $\lambda \in \mathbb{C}_+$ and $\Im\lambda \neq 0$, the collection $\{M_\lambda^t(A) : A \in S_t\}$ of bounded linear operators associated with the $(S_\lambda, Q)$-process is *unbounded* in the operator norm as sets $A$ range over the *algebra* $S_t$ generated by cylinder sets (3.4). It is worthwhile doing the calculation explicitly as a cautionary example prior to looking at the general situation.

**Example 5.1.** Suppose that $t > 0$ and $C^t$ is the collection of all continuous functions $\omega : [0, t] \to \mathbb{R}^d$. Set $X_s(\omega) = \omega(s)$ for every $\omega \in C^t$ and $0 \le s \le t$. Let $\phi, \psi \in L^2(\mathbb{R}^d)$ and $\lambda \in \mathbb{C} \setminus \{0\}$ with $\Re\lambda > 0$. Then for every cylinder set

$$E = \{X_0 \in B_0, \ X_{t_1} \in B_1, \ldots, X_{t_n} \in B_n, \ X_t \in B \} \qquad (5.1)$$

with $0 < t_1 < \cdots < t_n < t$, we define

$$\mu_{\lambda,\phi,\psi}^t(E) = C_{n+1}\lambda^{\frac{d(n+1)}{2}} \int_B \int_{B_n} \cdots \int_{B_1} \int_{B_0} \overline{\psi}(x_{n+1}) e^{-\frac{\lambda|x_{n+1}-x_n|^2}{2(t-t_n)}} e^{-\frac{\lambda|x_n-x_{n-1}|^2}{2(t_n-t_{n-1})}} \cdots$$

$$(5.2)$$

$$\cdots e^{-\frac{\lambda|x_2-x_1|^2}{2(t_2-t_1)}} e^{-\frac{\lambda|x_1-x_0|^2}{2t_1}} \phi(x_0) \, dx_0 dx_1 \cdots dx_n dx_{n+1}.$$

Here $C_{n+1} = (2\pi(t - t_n))^{-d/2} \ldots (2\pi t_1)^{-d/2}$. Then $\mu^t_{\lambda,\phi,\psi}$ defines an additive set function on the algebra $S_t$ generated by all cylinder sets $E$ of the form (5.1) as the times $0 < t_1 < \cdots < t_n < t$ vary, the Borel subsets $B_0, \ldots, B_n, B$ of $\mathbb{R}$ vary, and the index $n = 1, 2, \ldots$ varies. In the limiting case with $\Re\lambda = 0$, the integrals (5.2) converge as improper iterated integrals. Comparison of equation (5.2) with equations (3.1–3.5) shows that $\mu^t_{\lambda,\phi,\psi} = \langle M^t_\lambda \phi, \psi \rangle$ for the scalar set function $\langle M^t_\lambda \phi, \psi \rangle : A \longmapsto \langle M^t_\lambda(A)\phi, \psi \rangle$, $A \in S_t$.

Now fix the times $t_1, \ldots, t_n$ and consider the algebra $S_{t_1,\ldots,t_n}$ generated by the sets of the form (5.1), just as the Borel subsets $B_0, \ldots, B_n, B$ of $\mathbb{R}$ vary. A calculation shows that the total variation $\|\mu^t_{\lambda,\phi,\psi}\|_{S_{t_1,\ldots,t_n}}$ of $\mu^t_{\lambda,\phi,\psi}$ over the algebra $S_{t_1,\ldots,t_n}$ is given by

$$\|\mu^t_{\lambda,\phi,\psi}\|_{S_{t_1,\ldots,t_n}} = \left(\frac{\Re\lambda}{2\pi t}\right)^{d/2} \left(\frac{|\lambda|}{\Re\lambda}\right)^{\frac{d(n+1)}{2}} \int_{\mathbb{R}^d} \int_{\mathbb{R}^d} |\psi(x_1)| e^{-\frac{\Re\lambda|x_1-x_0|^2}{2t}} |\phi(x_0)| \, dx_0 dx_1$$

$$= \left(\frac{|\lambda|}{\Re\lambda}\right)^{\frac{d(n+1)}{2}} \langle S_{\Re\lambda}(t)|\phi|, |\psi|\rangle.$$

(5.3)

The total variation $\|\mu^t_{\lambda,\phi,\psi}\|$ of $\mu^t_{\lambda,\phi,\psi}$ over the whole algebra $S^t$ is necessarily greater than the total variation $\|\mu^t_{\lambda,\phi,\psi}\|_{S_{t_1,\ldots,t_n}}$ for any choice of times $0 < t_1 < \cdots < t_n < t$ and any $n = 1, 2, \ldots$. Hence, if $|\lambda| > \Re\lambda$, then either $\|\mu^t_{\lambda,\phi,\psi}\| = +\infty$, or $\|\mu^t_{\lambda,\phi,\psi}\| = 0$, the last case occurring when either $\phi$ or $\psi$ is zero almost everywhere. So, in the case $\Im\lambda \neq 0$, the additive set function $\mu^t_{\lambda,\phi,\psi}$ is highly singular.

The key element of the argument above is that $S_\lambda$ is a semigroup of convolution operators for each $\lambda \in \mathbb{C}_+$. If $k_\xi : \mathbb{R}^d \to \mathbb{C}$ is the function defined by

$$k_\xi(x) = \left(\frac{\xi}{2\pi}\right)^{d/2} e^{-\frac{\xi}{2}|x|^2}, \quad x \in \mathbb{R}^d, \, \xi \in \overline{\mathbb{C}_+}, \quad (5.4)$$

then $S_\lambda(t)f = k_{\frac{\lambda}{t}} * f$ for all $f \in L^2(\mathbb{R}^d)$, $t > 0$ and $\lambda \in \overline{\mathbb{C}_+} \setminus \{0\}$. If $\Re\lambda > 0$, then $S_\lambda(t)$ is bounded on $L^1(\mathbb{R}^d)$ and $L^\infty(\mathbb{R}^d)$ and its operator norms are given by

$$\|S_\lambda(t)\|_{\mathcal{L}(L^1(\mathbb{R}^d))} = \|S_\lambda(t)\|_{\mathcal{L}(L^\infty(\mathbb{R}^d))} = \|k_{\frac{\lambda}{t}}\|_1 = \frac{|\lambda|}{\Re\lambda}.$$

Although $S_\lambda(t) = e^{t\Delta/(2\lambda)}$ is a $C_0$-contraction semigroup on $L^2(\mathbb{R}^d)$, it is not a contraction (or quasicontraction) semigroup on $L^1(\mathbb{R}^d)$ or $L^\infty(\mathbb{R}^d)$ when $\Im\lambda \neq 0$ and $\Re\lambda > 0$. If $\Re\lambda = 0$ and $\lambda \neq 0$, then $S_\lambda(t)$ is only bounded on $L^2(\mathbb{R}^d)$. Because we end up taking infinite operator products in the formula (3.5) for $M^t_\lambda$, this seemingly innocuous deficiency of the $C_0$-semigroup $S_\lambda$ assumes catastrophic significance for the boundedness of the set functions $M^t_\lambda$ of an $(S_\lambda, Q)$-process when $\Im\lambda \neq 0$.

Let $(\Sigma, \mathcal{E}, \mu)$ be a $\sigma$-finite measure space, $N = 1, 2, \ldots$ and $1 \leq p < \infty$. The space of all $\mu$-equivalence classes of strongly measurable functions $f : \Sigma \to \mathbb{C}^N$ such that $|f_j|^p$ is $\mu$-integrable for each $j = 1, \ldots, N$ is written as $L^p(\mu, \mathbb{C}^N)$. Then

$L^p(\mu, \mathbb{C}^N)$ is a Banach space with the norm $\|f\|_p = \left(\sum_{j=1}^N \int_\Sigma |f_j|^p \, d\mu\right)^{1/p}$. For $p = \infty$, the functions $|f_j|$, $j = 1, \ldots, N$ are assumed to be $\mu$-essentially bounded with $\|f\|_\infty = \max_j \{\text{ess. sup } |f_j|\}$. Then $L^2(\mu, \mathbb{C}^N)$ is itself a Hilbert space with inner product

$$\langle f, g \rangle = \sum_{j=1}^N \int_\Sigma f_j(\sigma)\overline{g_j(\sigma)} \, d\mu(\sigma), \quad f, g \in L^2(\mu, \mathbb{C}^N).$$

Let $1 \leq p < \infty$. The spectral measure $Q : \mathcal{E} \to L^p(\mu, \mathbb{C}^N)$ is given by multiplication by characteristic functions, that is, for each $B \in \mathcal{E}$ and $f \in L^p(\mu, \mathbb{C}^N)$, the equality

$$(Q(B)f)(\sigma) = \chi_B(\sigma)f(\sigma) \tag{5.5}$$

holds for $\mu$-almost all $\sigma \in \Sigma$. For $p = \infty$, the set function $Q$ is defined by the same formula, but it is only $\sigma$-additive for the weak*-topology $\sigma\left(L^\infty(\mu, \mathbb{C}^N), L^1(\mu, \mathbb{C}^N)\right)$.

**Theorem 5.2.** *Let $S$ be a semigroup of operators acting on $L^\infty(\mu, \mathbb{C}^N)$. Suppose that there exists an invertible linear map $T : \mathbb{C}^N \to \mathbb{C}^N$ and $\alpha \geq 0$ such that*

$$\|T \circ (S(t)(T^{-1} \circ f))\|_\infty \leq e^{\alpha t}\|f\|_\infty \tag{5.6}$$

*for all $f \in L^\infty(\mu, \mathbb{C}^N)$ and $t \geq 0$.*

*Then for every $t \geq 0$ and $1 \leq p \leq \infty$, the set function $M^t$ of an $(S, Q)$-process is bounded in $\mathcal{L}(L^\infty(\mu, \mathbb{C}^N))$ on the algebra $\mathcal{S}_t$ generated by cylinder sets (3.7) based before time $t$. In fact, there exists $C \geq 0$ such that the inequality*

$$\|M^t(A)f\|_\infty \leq Ce^{\alpha t}\|f\|_\infty, \quad A \in \mathcal{S}_t, \ f \in L^\infty(\mu, \mathbb{C}^N), \ t \geq 0, \tag{5.7}$$

*is valid.*

*Proof.* By abusing notation, if $V : \mathbb{C}^N \to \mathbb{C}^N$ is a mapping, then the mapping $f \longmapsto V \circ f$, $f \in L^p(\mu, \mathbb{C}^N)$, is written just as $Vf$; the notation $(I_{L^p} \otimes V)f$ with $I_{L^p}$ the identity map on $L^p(\Sigma, \mathcal{E}, \mu)$ would be more appropriate.

Now $t \longmapsto TS(t)T^{-1}$, $t \geq 0$ is itself a semigroup $TST^{-1}$ of operators. Because the linear map $T$ and the spectral measure $Q$ commute, if $M^t$ denotes the $(S, Q)$-set function on $\mathcal{S}_t$, then the $(TST^{-1}, Q)$-set function is just $TM^tT^{-1}$.

By a small modification of the proof of [28, Corollary 2.3.4], we have

$$\|TM^t(A)T^{-1}\|_{\mathcal{L}(L^\infty(\mu, \mathbb{C}^N))} \leq e^{\alpha t} \tag{5.8}$$

for all $A \in \mathcal{S}_t$, $t \geq 0$. The essential idea in the proof of [28, Corollary 2.3.4] is that the spectral measure $Q$ has the property that

$$\left\|\sum_{j=1}^n Q(B_j)f_j\right\|_\infty \leq 1$$

for all pairwise disjoint sets $B_j \in \mathcal{E}$ and elements $f_j$ of the closed unit ball of $L^\infty(\mu, \mathbb{C}^N)$, for $j = 1, \ldots, n$ and $n = 1, 2, \ldots$.

Let $\mathbb{C}^N_\infty$ denote $\mathbb{C}^N$ equipped with the max norm. The conclusion now follows by setting $C = \|T\|_{\mathcal{L}(\mathbb{C}^N_\infty)} \cdot \|T^{-1}\|_{\mathcal{L}(\mathbb{C}^N_\infty)}$.     □

It is usually a semigroup on a Hilbert space of states that arises in physical problems rather than a semigroup on $L^\infty$. However, as Example 5.1 shows, it is not enough to have a continuous unitary group of operators. The following result suffices for most of the applications considered in these notes.

**Corollary 5.3.** *Let $S$ be a semigroup of selfadjoint operators or a group of unitary operators acting on $L^2(\mu, \mathbb{C}^N)$. Suppose that there exists an invertible linear map $T : \mathbb{C}^N \to \mathbb{C}^N$ and $\alpha \geq 0$ such that*

$$\|T \circ (S(t)(T^{-1} \circ f))\|_\infty \leq e^{\alpha t} \|f\|_\infty \tag{5.9}$$

*for all $f \in L^2(\mu, \mathbb{C}^N) \cap L^\infty(\mu, \mathbb{C}^N)$ and $t \geq 0$ (or $t \in \mathbb{R}$ in the case that $S$ is a group of unitaries).*

*Then for every $t \geq 0$ and $1 \leq p \leq \infty$, the set function $M^t$ of an $(S, Q)$-process is bounded in $\mathcal{L}(L^p(\mu, \mathbb{C}^N))$ on the algebra $\mathcal{S}_t$ generated by cylinder sets (3.7) based before time $t$. In fact, there exists $C \geq 0$ such that for each $1 \leq p \leq \infty$, the inequality*

$$\|M^t(A)f\|_p \leq C e^{\alpha t} \|f\|_p, \quad A \in \mathcal{S}_t, \ f \in L^2 \cap L^p(\mu, \mathbb{C}^N), \ t \geq 0, \tag{5.10}$$

*is valid.*

**Remark 5.4.** If $H$ is a selfadjoint operator defined on $L^2(\mu, \mathbb{C}^N)$, then we have in mind the semigroups $S(t) = e^{-tH}$, $t \geq 0$ in the case that $H$ is bounded below (by $-\alpha I$) or $S(t) = e^{-itH}$, $t \in \mathbb{R}$. Continuity assumptions for the semigroup $S$ are irrelevant for the consideration of boundedness in Corollary 5.3.

*Proof.* The spaces $L^\infty(\mu, \mathbb{C}^N)$ and $L^1(\mu, \mathbb{C}^N)$ are in duality via the bilinear pairing

$$\langle f, g \rangle_{J_1} = \sum_{j=1}^N \int_\Sigma f_j(\sigma) g_j(\sigma) \, d\mu(\sigma), \quad f \in L^\infty(\mu, \mathbb{C}^N), \ g \in L^1(\mu, \mathbb{C}^N),$$

giving rise to the isometry $L^1(\mu, \mathbb{C}^N)' \equiv L^\infty(\mu, \mathbb{C}^N)$. The antilinear conjugation $x \longmapsto \bar{x}, x \in \mathbb{C}^N$, is denoted by $J$. Let $S(t)^*$ be the adjoint of $S(t)$ on the Hilbert space $L^2(\mu, \mathbb{C}^N)$ and $T^*$ the adjoint of $T$ on $\mathbb{C}^N$. The operator $J(T^*)^{-1} S(t)^* T^* J$ has the property that

$$\langle T S(t) T^{-1} f, g \rangle_{J_1} = \langle f, J(T^*)^{-1} S(t)^* T^* J g \rangle_{J_1}$$

for every $f \in L^\infty(\mu, \mathbb{C}^N) \cap L^2(\mu, \mathbb{C}^N)$ and $g \in L^1(\mu, \mathbb{C}^N) \cap L^2(\mu, \mathbb{C}^N)$. Hence, the dual operator of $T S(t) T^{-1}$ with respect to the bilinear pairing $\langle f, g \rangle_{J_1}$ is a bounded linear operator on $L^1(\mu, \mathbb{C}^N)$ equal to $J(T^*)^{-1} S(t)^* T^* J$ on the subspace $L^1(\mu, \mathbb{C}^N) \cap L^2(\mu, \mathbb{C}^N)$. It has $\mathcal{L}(L^1(\mu, \mathbb{C}^N))$-operator norm bounded by $e^{\alpha t}$.

If $S(t)$ is selfadjoint, then $S(t)^* = S(t)$ for all $t \geq 0$. On the other hand, if $S$ is a unitary group of operators, then $S(t)^* = S(-t)$ for all $t > 0$. In either case, it follows from the bound (5.9), Theorem 5.2 and duality, that

$$\|J(T^*)^{-1}M^t(A)T^*J\|_{\mathcal{L}(L^1(\mu,\mathbb{C}^N))} \le e^{\alpha t} \tag{5.11}$$

for all $A \in \mathcal{S}_t$, $t \ge 0$. If we set $C = \|T\|_{\mathcal{L}(\mathbb{C}_\infty^N)} \cdot \|T^{-1}\|_{\mathcal{L}(\mathbb{C}_\infty^N)}$, then the conclusion follows from the vector valued Riesz–Thorin interpolation theorem [39, Theorem XX].  $\square$

In the case that $1 \le p \le \infty$, $\mathbb{C}^N = \mathbb{C}$, $Q$ has multiplication by characteristic functions on $L^p(\mu)$ and $S$ is a semigroup acting on $L^p(\mu)$, necessary and sufficient conditions have recently been obtained for the $(S, Q)$-measures $M^t$ to be bounded for all $t > 0$, see [31].

Although the conditions of Corollary 5.3 are sufficient in applications, they are not the whole story. Suppose for example, that $S$ is a semigroup acting on $L^p(\mu)$, as just mentioned, and $S(t)f \ge 0$ $\mu$-a.e. whenever $t > 0$ and $f \in L^p(\mu)$ is nonnegative $\mu$-a.e., that is, $S(t)$ is a *positive operator* in the sense of Banach lattices. Each of the operators (3.8) defining the $(S, Q)$-measure $M^t$ is positive. The additivity of $M^t$ ensures that

$$0 \le M^t(A)f \le M^t(A)f + M^t(A^c)f = M^t(\Omega)f = S(t)f \quad \mu\text{-a.e.}$$

whenever $f \ge 0$ $\mu$-a.e. and $A$ is an element of the algebra $\mathcal{S}_t$ generated by cylinder sets (3.4). In particular, $\|M^t(A)f\|_p \le \|S(t)f\|_p$ for all $A \in \mathcal{S}_t$. There is no guarantee that $S(t)$ even defines a bounded linear operator on $L^\infty(\mu)$. It turns out that all $(S, Q)$-set functions $M^t$, $t \ge 0$, are bounded if and only if they are dominated by positive $(|S|, Q)$-set functions $\tilde{M}^t$, $t \ge 0$, that is,

$$|M^t(A)f| \le \tilde{M}^t(A)f, \quad A \in \mathcal{S}_t, \ t \ge 0,$$

for all nonnegative $f \in L^p(\mu)$. The positive semigroup $|S|$ is the modulus semigroup of $S$.

**Example 5.5.** There are positive semigroups on $L^2(\mathbb{R}^3)$ of physical interest which do not satisfy the conditions of Corollary 5.3. The $\delta$-interaction at zero for the Laplacian $\Delta$ in $L^2(\mathbb{R}^3)$ is an operator of the form $-\Delta + \delta\frac{\partial}{\partial r}r$ defined by choosing a particular selfadjoint extension $-\Delta_\tau$ of $-\Delta$ with domain $C_0^\infty(\mathbb{R}^3 \setminus \{0\})$ [10, p. 255]. The resolvent of $-\Delta_\tau$ is given by

$$((\lambda I + \Delta_\tau)^{-1}f)(x) = K_\lambda \frac{e^{-\lambda|x|}}{4\pi|x|} \int_{\mathbb{R}^3} \frac{e^{-\lambda|y|}}{4\pi|y|} f(y)\,dy, \quad \text{a.e.} \tag{5.12}$$

for some positive number $K_\lambda$ [10, p. 264]. The resolvent is bounded on $L^p(\mathbb{R}^3)$ only for $\frac{3}{2} < p < 3$ and for this range of values, the operator $\Delta_\tau$ is the generator of an analytic semigroup $S$. Therefore, the $(S, Q)$-measures $M^t$ are bounded on $L^p(\mathbb{R}^3)$ for all $\frac{3}{2} < p < 3$, but $S$ is not bounded on $L^\infty(\mathbb{R}^3)$, nor on $L^1(\mathbb{R}^3)$.

A semigroup $T$ is said to *dominate* $S$ if

$$T(t)|x| \ge |S(t)x|, \quad x \in L^p(\mu), \tag{5.13}$$

for every $t \ge 0$. A semigroup $S$ is said to be *dominated* if such a dominating semigroup $T$ exists. When $S$ is dominated and $|S|$ is a semigroup of positive operators with the property that if $T$ dominates $S$ then $|S|(t) \leq T(t)$ for all $t \geq 0$, we call $|S|$ the *modulus semigroup* of $S$. The following result [31, Theorem 3.1] characterizes bounded $(S, Q)$-processes on $L^p$-spaces.

**Theorem 5.6.** *Let $1 \leq p \leq \infty$, and let $S$ be a semigroup of continuous linear operators on $L^p(\mu)$ and in the case $p = \infty$, assume the extra condition that $S(t) \in \mathcal{L}((L^\infty(\mu), weak^*))$ for each $t \geq 0$.*

*Then the operator valued, additive $(S, Q)$-set function $M^t : \mathcal{S}_t \to \mathcal{L}(L^p(\mu))$ has uniformly bounded range for each $t \geq 0$ if and only if $S$ is dominated. In this case, the modulus semigroup $|S|$ of $S$ exists.*

*Sketch of Proof.* Suppose that $S$ is dominated by a positive semigroup $T$ and let $\tilde{M}^t$ be a $(T, Q)$-set function. The previous discussion shows that the $(S, Q)$-set functions $M^t$ have bounded range because $|M^t(A)f| \leq \tilde{M}^t(A)f$ for all $A \in \mathcal{S}_t$, $t \geq 0$ and nonnegative $f \in L^p(\mu)$.

In the other direction, suppose that $M^t$ has bounded range on $\mathcal{S}_t$. Let $p'$ be the dual index to $p$ and take positive functions $f \in L^p(\mu)$ and $g \in L^{p'}(\mu)$. It turns out that $\langle |S|(t)f, g \rangle = |\langle M^t f, g \rangle|(\Omega)$. On the right-hand side of this equation, $|\langle M^t f, g \rangle| : \mathcal{S}_t \to \mathbb{R}_+$ is the variation of the additive scalar set function $\langle M^t f, g \rangle : A \longmapsto \langle M^t(A)f, g \rangle$, $A \in \mathcal{S}_t$. The proof that $|S|$ is the required dominating semigroup is given in [31, Theorem 3.1]. $\qquad\square$

# 6 Path integrals on finite sets

One of the simplest situations in which the conditions of Theorem 5.2 are satisfied is where $A$ is a bounded linear operator acting on $L^\infty(\mu)$ for some $\sigma$-finite measure $\mu$ and $S(t) = e^{tA}$. Because $A$ is bounded, the exponential is given by

$$e^{tA} = I + \sum_{n=1}^{\infty} \frac{t^n}{n!} A^n. \tag{6.1}$$

The convergence of the sum of operators is in the uniform operator topology of $\mathcal{L}(L^\infty(\mu))$ and the estimate

$$\|e^{tA}\|_{\mathcal{L}(L^\infty(\mu))} \leq 1 + \sum_{n=1}^{\infty} \frac{t^n}{n!} \|A\|_{\mathcal{L}(L^\infty(\mu))}^n = e^{t\|A\|_{\mathcal{L}(L^\infty(\mu))}} \tag{6.2}$$

holds, so the conditions of Theorem 5.2 are valid with $T$ equal to the identity map on $\mathbb{C}$ and $\alpha = \|A\|_{\mathcal{L}(L^\infty(\mu))}$. If $\mu$ is counting measure on the finite set $\{1, \ldots, N\}$, then $L^\infty(\mu)$ can be identified with Euclidean space $\mathbb{C}^N$ endowed with the max-norm. In this case, $A$ can be the linear map defined by *any* $(N \times N)$ matrix with complex entries. The corresponding matrix valued measures have been studied by E. Thomas [47] and integrals with respect to them have been called *path integrals on finite sets*.

Let $\Sigma$ be a Hausdorff topological space with Borel $\sigma$-algebra $\mathcal{B}(\Sigma)$ and let $\mu : \mathcal{B}(\Sigma) \to [0, \infty]$ be a Radon measure. Let $Q$ be the spectral measure on $E = \langle L^\infty(\mu), \text{weak*} \rangle$ of multiplication by characteristic functions. We shall use 'weak*' to denote the topology $\sigma(L^\infty(\mu), L^1(\mu))$.

Let $\Gamma$ be collection of all functions $\omega : [0, \infty) \to \Sigma$ such that in each time interval $[0, T]$, there exist finitely many times $0 < t_1 < t_2 < \ldots < t_k < T, k = 1, 2, \ldots$, such that $\omega(t) = \omega(t_{j-1})$ for each $j = 1, \ldots, k$ and $t \in [t_{j-1}, t_j)$, and $\omega(t) = \omega(t_k)$ for all $t_k \leq t < T$.

Let $F$ be a locally convex space, and let $(\Xi, \mathcal{R})$ be a measurable space. A vector measure $m : \mathcal{R} \to F$ is said to be *concentrated* on a subset $U$ of $\Xi$ if $m(R) = 0$ for every set $R \in \mathcal{R}$ disjoint from $U$. If $m$ is concentrated on $U$, then the set function $m_U : \mathcal{R} \cap U \to F$ defined by $m_U(A \cap U) = m(A)$, $A \in \mathcal{R}$, is $\sigma$-additive in $F$. The following result appears in [28, Theorem 6.3.1].

**Theorem 6.1.** *Let $\Sigma$ be a Hausdorff topological space with Borel $\sigma$-algebra $\mathcal{B}(\Sigma)$ and let $\mu : \mathcal{B}(\Sigma) \to [0, \infty]$ be a Radon measure. Let $A$ be a weak\*-continuous operator on $L^\infty(\mu)$, $S(s) = e^{As}$, $s \geq 0$, and let $Q$ be the spectral measure of multiplication by characteristic functions, acting on the space $\langle L^\infty(\mu), \text{weak*} \rangle$.*

*Let $(\Omega, \langle S_t \rangle_{t \geq 0}, \langle M_t \rangle_{t \geq 0}; \langle X_t \rangle_{t \geq 0})$ be the $(S, Q)$-process with $\Omega$ equal to the set all paths $\omega : [0, \infty) \to \Sigma$ and $X_t(\omega) = \omega(t)$ for all $t \geq 0$.*

*Then the $(S, Q, t)$-set function $M_t$ is the restriction to the semi-algebra $\mathcal{S}_t$ of an operator valued measure $\widetilde{M}_t$ defined on the $\sigma$-algebra $\sigma(\mathcal{S}_t)$ generated by $\mathcal{S}_t$ and acting on $\langle L^\infty(\mu), \text{weak*} \rangle$. Furthermore, $\widetilde{M}_t$ is concentrated on the space $\Gamma$.*

On taking $\Sigma = \{1, \ldots, N\}$ and $\mu$ counting measure, we have

**Corollary 6.2.** *Let $A$ be an $(N \times N)$ matrix, $S(s) = e^{As}$, $s \geq 0$, and for each set $B \subseteq \{1, \ldots, N\}$, let $Q(B) \in \mathcal{L}(\mathbb{C}^N)$ be the projection onto the coordinates belonging to $B$.*

*Let $(\Omega, \langle S_t \rangle_{t \geq 0}, \langle M_t \rangle_{t \geq 0}; \langle X_t \rangle_{t \geq 0})$ be the $(S, Q)$-process with $\Omega$ equal to the set of all paths $\omega : [0, \infty) \to \Sigma$ and $X_t(\omega) = \omega(t)$ for all $t \geq 0$.*

*Then the $(S, Q, t)$-set function $M_t$ is the restriction to the semi-algebra $\mathcal{S}_t$ of a matrix valued measure $\widetilde{M}_t$ defined on the $\sigma$-algebra $\sigma(\mathcal{S}_t)$ generated by $\mathcal{S}_t$ and acting on $\mathbb{C}^N$. Furthermore, $\widetilde{M}_t$ is concentrated on the space $\Gamma$.*

**Remark 6.3.** A proof similar to that of Theorem 6.1 works if $\langle L^\infty(\mu), \text{weak*} \rangle$ is replaced by $L^p(\mu)$ for $1 \leq p < \infty$ and we suppose that $A$ is a *regular operator* acting on $L^p(\mu)$ in the sense of Banach lattices [42]. The proof appears in [41, Theorem 7.3.1, p. 136]. A bounded linear operator on $L^\infty(\mu)$ is automatically regular [42, Theorem IV.1.5 (i)].

## 6.1 The associated Markov jump process

The constant function identically equal to one is denoted by **1**. By Theorems 6.1 and 5.6, if $A : L^\infty(\mu) \to L^\infty(\mu)$ is a weak\*-continuous linear operator, then the semigroup $S(s) = e^{As}$, $s \geq 0$, has a modulus semigroup $|S|$ for which

$$\| \, |S|(s) \, \|_{\mathcal{L}(L^\infty(\mu))} \leq e^{s\lambda}, \quad \text{for all } s \geq 0.$$

with $\lambda = \|A\|_{\mathcal{L}(L^\infty(\mu))}$. If $\lambda$ now denotes the smallest such number, then $P_s = e^{-\lambda s}|S|(s), s \geq 0$ is a sub-Markovian semigroup of operators acting on $L^\infty(\mu)$, that is, $P_s f \geq 0$ if $f \in L^\infty(\mu)$ and $f \geq 0$ and $P_s \mathbf{1} \leq 1$ for all $s \geq 0$. Actually, the semigroup $P_s$ should act on the space $\mathcal{L}^\infty(\mu)$ of bounded $\mu$-measurable functions, but this can be obtained from a multiplicative lifting in general, or computed explicitly in many cases. Then

$$p_s(x, A) = (P_s \chi_A)(x), \quad x \in \Sigma, \ A \in \mathcal{E}, s \geq 0,$$

are the transition functions associated with a sub-Markovian process. By appending a point at infinity, if necessary, a Markov process with state space $\Sigma \cup \{\infty\}$ can be constructed. Because the modulus semigroup $|S|$ and hence, the transition functions $p_s$, $s \geq 0$, are constructed from the semivariation of the $(S, Q)$-measures, the associated sub-Markovian process has the space $\Gamma$ as its state space.

The following example is taken from [47, Example 10.2].

**Example 6.4.** Let $\mu$ be counting measure on $\Sigma = \{0, 1\}$ and identify $L^\infty(\mu)$ with $\mathbb{C}^2$. Let $H_0 = \begin{pmatrix} 0 & i \\ -i & 0 \end{pmatrix}$. Then

$$S(s) = e^{-is H_0} = \begin{pmatrix} \cos s & \sin s \\ -\sin s & \cos s \end{pmatrix}, \quad s \in \mathbb{R}.$$

The modulus semigroup $|S|$ is given by

$$|S|(s) = \lim_{n \to \infty} \begin{pmatrix} |\cos(s/n)| & |\sin(s/n)| \\ |\sin(s/n)| & |\cos(s/n)| \end{pmatrix}^n = \begin{pmatrix} \cosh(s) & \sinh(s) \\ \sinh(s) & \cosh(s) \end{pmatrix}$$

The stochastic matrix $P_s$ is therefore given by

$$P_s = e^{-s}|S|(s) = \frac{1}{2} \begin{pmatrix} 1 + e^{-2s} & 1 - e^{-2s} \\ 1 - e^{-2s} & 1 + e^{-2s} \end{pmatrix}, \quad s \geq 0.$$

# 7 The Dirac equation in one space dimension

After jump processes associated with a bounded linear operator on an $L^\infty$-space, the next simplest (nontrivial) process is associated with the free Dirac equation for a particle of mass $m \geq 0$ with motion restricted to the line. There is no difficulty with the unboundedness of the operator valued set functions associated with the evolution of the system like there is with the Schrödinger equation as described in Section 5. Moreover, when singular interactions are considered, the support of the associated measures accommodate the possibility of particle-pair production. Unfortunately, the propagator of the Dirac equation in four space–time dimensions is a distribution of order one, so the associated evolution process is unbounded—the finite dimensional distributions are not even Radon polymeasures (see Remark 8.7 (iii) below), as they are for the Schrödinger equation.

The Dirac equation in $\mathbb{R}^2$ is given by

$$i\hbar \frac{\partial}{\partial t}\psi(x, t) = \left(-i\hbar c\alpha\frac{\partial}{\partial x} + mc^2\beta\right)\psi(x, t) = (H_0\psi)(x, t)$$

with $\psi(x, t) \in \mathbb{C}^2$ for $x \in \mathbb{R}$ and $t \geq 0$. The $(2 \times 2)$ matrices $\alpha$ and $\beta$ satisfy $\alpha^2 = \beta^2 = Id$ and $\alpha\beta + \beta\alpha = 0$. With an interaction represented by a potential $V : \mathbb{R} \to \mathbb{R}$, the Dirac equation becomes

$$i\hbar\frac{\partial}{\partial t}\psi(x, t) = (H_0 + V)\psi(x, t), \tag{7.1}$$

so in a coordinate system in which $\hbar = c = 1$, solutions of the initial value problem associated with equation (7.1) define a group $e^{-it(H_0+V)}$, $t \in \mathbb{R}$, of operators for suitable potentials $V$. The symbol $V$ in the expression $H_0 + V$ is interpreted as a multiplication operator acting in $L^2(\mathbb{R}, \mathbb{C}^2)$.

## 7.1 Boundedness of the associated process

Now suppose that $S(t) = e^{-itH_0}$ for $t \in \mathbb{R}$. Let $Q$ be multiplication by characteristic functions acting on $L^2(\mathbb{R}, \mathbb{C}^2)$ and let $S_t$ be the collection of cylinder sets (3.7) before time $t$. Then the $(S, Q)$-set function $M^t$ on the semi-algebra $S_t$ is actually the restriction to $S_t$ of an operator valued measure $\tilde{M}^t : \sigma(S_t) \to \mathcal{L}(L^2(\mathbb{R}, \mathbb{C}^2))$ that is $\sigma$-additive for the strong operator topology of $\mathcal{L}(L^2(\mathbb{R}, \mathbb{C}^2))$ for each $t \geq 0$. One way to see this is to take $\alpha$ and $\beta$ to be the Pauli matrices

$$\alpha = \sigma_3 = \begin{pmatrix} 1 & 0 \\ 0 & -1 \end{pmatrix}, \quad \beta = -\sigma_1 = \begin{pmatrix} 0 & -1 \\ -1 & 0 \end{pmatrix}. \tag{7.2}$$

Other possible representations are unitarily equivalent to this choice of matrices. Then $im\sigma_1$ is a bounded perturbation of the operator $\sigma_3\frac{\partial}{\partial x}$, which is the generator of a direct sum of translations acting on $L^2(\mathbb{R}, \mathbb{C}^2)$ and $L^p(\mathbb{R}, \mathbb{C}^2)$, for each $1 \leq p < \infty$.

It follows that for each $1 \leq p \leq \infty$, the inequality $\|S(t)\psi\|_p \leq e^{m|t|}\|\psi\|_p$ holds for all $t \in \mathbb{R}$ and $\psi \in L^p(\mathbb{R}, \mathbb{C}^2) \cap L^2(\mathbb{R}, \mathbb{C}^2)$. The $(S, Q)$-set functions $M^t$ are therefore bounded on the algebra $S_t$ in operator norm by $e^{mt}$ for all $1 \leq p \leq \infty$ by Corollary 5.3.

Further analysis shows that the operator valued measures $M^t$ so defined are actually supported on the space $\Omega$ of all continuous paths $\omega : [0, \infty) \to \mathbb{R}$ with velocity equal to $\pm 1$ and only finitely many changes of direction in each bounded time interval, see [23] or Subsection 7.5 below.

A similar argument shows that the operator valued measures $M^t$, $t \geq 0$, also act on the space $\mathcal{M}(\mathbb{R}, \mathbb{C}^2)$ of $\mathbb{C}^2$-valued Borel measures equipped with the total variation norm and on the space $\mathcal{L}^\infty(\mathbb{R}, \mathbb{C}^2)$ of bounded Borel measurable functions with the weak topology $\sigma\left(\mathcal{L}^\infty(\mathbb{R}, \mathbb{C}^2), \mathcal{M}(\mathbb{R}, \mathbb{C}^2)\right)$ defined by the duality between $\mathcal{L}^\infty(\mathbb{R}, \mathbb{C}^2)$ and $\mathcal{M}(\mathbb{R}, \mathbb{C}^2)$.

Denote the constant function equal to one on $\mathbb{R}$ by **1**. The space of all $n \times n$ matrices over $\mathbb{C}$ is denoted by $\mathbf{M}_n(\mathbb{C})$. The matrix valued measures $\nu_{t,x} : \sigma(S_t) \to \mathbf{M}_2(\mathbb{C})$ defined by

$$\nu_{t,x}(E)v = \left(M^t(E)(\mathbf{1}v)\right)(x), \quad E \in \sigma(S_t), \ v \in \mathbb{C}^2, \ x \in \mathbb{R}, \ t > 0, \tag{7.3}$$

were constructed by T. Ichinose [22] using the method of E. Nelson [37]. Representations of $\nu_{t,x}$ using a Poisson process have been obtained by Ph. Blanchard et al. [7] and T. Zastawniak [50], see also [23] for a survey. The sample space properties mentioned above follow immediately from the Poisson process representations, see Proposition 7.8 below.

## 7.2 Locally integrable potentials

For suitable functions $V : \mathbb{R} \to \mathbb{R}$, we obtain the Feynman–Kac formula

$$e^{-it(H_0+V)} = \int_\Omega e^{-i\int_0^t V(\omega(s))\, ds}\, dM^t(\omega). \tag{7.4}$$

In order that $\int_0^t |V(\omega(s))|\, ds < \infty$ for all $\omega \in \Omega$ and $t > 0$, the function $V$ must be *locally integrable* on $\mathbb{R}$. The set of paths with no changes in direction and crossing a given point has nonzero $M^t$-measure, so the same conclusion holds if $\int_0^t |V(\omega(s))|\, ds < \infty$ for $M^t$-almost all $\omega \in \Omega$ and all $t > 0$.

Now suppose that $V$ is locally integrable on $\mathbb{R}$. Because $H_0 + V$ is in the limit point case at $\pm\infty$ [48, Theorem 6.8], the minimal operator [48, p41] associated with the differential expression $H_0 + V$ is essentially selfadjoint [48, Theorem 5.7]. The functional calculus for selfadjoint operators makes sense of the left hand side of equation (7.4). Note that the symbol '$H_0+V$' must be interpreted in this special sense, because we are not assuming that $H_0 + V$ is densely defined as the sum of two selfadjoint operators or as a quadratic form sum. Finally, the equality (7.4) follows after approximating $V$ by cutoff functions $V_k = \chi_{\{|V|\le k\}} V, k = 1, 2, \ldots$, as in the Feynman–Kac formula (4.3).

## 7.3 The Coulomb potential

We are concerned in this section with the potential $V(x) = \frac{\gamma}{|x|}, x \in \mathbb{R} \setminus \{0\}$, which is *not* a locally integrable function on $\mathbb{R}$. The operator $H_0 + V$ with domain $C_0^\infty(\mathbb{R} \setminus \{0\}, \mathbb{C}^2)$ of all smooth functions $\mathbb{C}^2$-valued with compact support in $\mathbb{R} \setminus \{0\}$ has a four parameter family of selfadjoint extensions studied by S. Benvegnù [6]. For every path $\omega \in \Omega$ passing through zero, $\int_0^t |V(\omega(s))|\, ds = \infty$ and equation (7.4) now makes no sense. In particular, there is no uniquely determined selfadjoint operator $H_0 + V$ associated with the left hand side of (7.4).

By analogy with the three dimensional case, $V$ is called a *Coulomb potential* and $H_0 + V$ is sometimes called the Hamiltonian of a one dimensional hydrogen atom. Models of this type arise in the investigation of 'quantum wires' and in the description of electrons near the surface of liquid helium, see [6] for the references.

Another motivation for investigating selfadjoint extensions of $H_0 + V$ is that it is a particularly simple example of a singular perturbation of the free Hamiltonian $H_0$, in the sense that the differential expression $H_0 + V$ does not have an essentially selfadjoint minimal operator [48, p41]. Other examples include point interactions $H_0 + c\delta$ and $H_0 + c\delta'$ which have been the subject of recent investigations in both the nonrelativistic [3] and relativistic settings [19].

Singular perturbations also arise in the perturbative approach to quantum field theory where 'renormalization' of the interaction terms involving the subtraction of divergences is necessary, see Section 10 below. In order to make sense of the Feynman–Kac formula (7.4) for the Coulomb potential $V$, we find an analogous subtraction of divergences from the expression $\int_0^t V(\omega(s))\,ds$. The exponential functional $e^{-i\int_0^t V(\omega(s))\,ds}$ is thereby modified in the limit by terms whose modulus is one. By adjusting the subtraction of the divergence, we obtain a two-parameter family of selfadjoint extensions of $(H_0 + V)|C_0^\infty(\mathbb{R}\setminus\{0\},\mathbb{C}^2)$.

The appropriate modification of the Feynman–Kac formula (7.4) is given by

$$e^{-it(H_0 + \Gamma V)} = \int_\Omega F_t^\Gamma(\omega)\,dM^t(\omega). \tag{7.5}$$

Here $H_0 +_\Gamma V$ is an element of the two-parameter family of selfadjoint extensions of $H_0 + V|C_0^\infty(\mathbb{R}\setminus\{0\},\mathbb{C}^2)$ corresponding to certain boundary conditions denoted by $\Gamma$, and $F_t^\Gamma$ is a *multiplicative functional* of modulus one associated with $\Gamma$.

The term 'multiplicative functional' is borrowed from the theory of Markov processes. To say that $F_t, t \geq 0$ is a multiplicative functional means that for each $t \geq 0$, the function $F_t : \Omega \to \mathbb{C}$ is measurable with respect to the $\sigma$-algebra $\sigma(S_t)$ (or perhaps, an appropriate completion) and $F_{s+t} = F_t \circ \theta_s F_s$, $M^{t+s}$-a.e., where $\theta_s : \Omega \to \Omega$ is the shift map defined by $\theta_s\omega = \omega(s + \cdot)$, for all $s \geq 0$ and $\omega \in \Omega$.

It follows from the properties of the operator valued measures $M^t$ that $t \longmapsto \int_\Omega F_t\,dM^t$, $t > 0$, is necessarily a *semigroup of operators* if $F_t$ is $M^t$-integrable for each $t > 0$. The point of the Feynman–Kac formula is to represent the dynamical group of an interacting quantum system in terms of the integral of such a multiplicative functional with respect to the measures associated with a free system. In the present context, there is no need to take the analytic continuation of the dynamical group to imaginary time in order to obtain such a representation, because the evolution process associated with the free system is bounded.

The usual example of a multiplicative functional is the Feynman–Kac functional $F_t(\omega) = e^{-i\int_0^t V(\omega(s))\,ds}$. The multiplicative functional $F_t^\Gamma$ mentioned above is obtained by subtracting divergences from the expression $\int_0^t V(\omega(s))\,ds$ in a way prescribed by the boundary conditions $\Gamma$. In the case of the nonrelativistic hydrogen atom in one dimension, W. Fischer, H. Leschke and P. Müller [16] show that the multiplicative functional

$$F_t(\omega) = \chi_{\{\int_0^t |V(\omega(s))|\,ds < \infty\}} e^{-\int_0^t V(\omega(s))\,ds}$$

is associated with the Dirichlet boundary condition at zero via integration with respect to Wiener measure. We shall lead up to equation (7.5) by a number of steps starting with the free system with zero mass.

## 7.4 The translation measures

The path space measures associated with translation on $\mathbb{R}$ are particularly simple. We shall construct the operator valued measures associated with the one-dimensional Dirac equation out of these, so we give the explicit calculation of this special case here.

Let $Q : \mathcal{B}(\mathbb{R}) \to \mathcal{L}(L^2(\mathbb{R}))$ be the spectral measure of multiplication by characteristic functions of Borel subsets of $\mathbb{R}$. Let $p$ be the selfadjoint operator $\frac{1}{i} \frac{\partial}{\partial x}$ acting in $L^2(\mathbb{R})$. We can explicitly calculate the $(S, Q)$-measures in the case that $S(t) = e^{ipt}$, $t \in \mathbb{R}$.

Then for $n$ times $0 < t_1 < \cdots < t_n < t$ and $n$ Borel subsets $B_1, \ldots, B_n$ of $\mathbb{R}$, and $f \in L^2(\mathbb{R})$, we have

$$
\begin{aligned}
\left(e^{ip(t-t_n)} Q(B_n) e^{ip(t_n - t_{n-1})} \cdots Q(B_1) e^{ipt_1} f\right)(x) & \\
&= \mathcal{X}_{B_n}(x + t - t_n)\left(e^{ip(t_n - t_{n-1})} Q(B_{n-1}) \cdots Q(B_1) e^{ipt_1} f\right)(x + t - t_n) \\
&= \mathcal{X}_{B_n}(x + t - t_n)\mathcal{X}_{B_{n-1}}(x + t - t_{n-1}) \cdots \mathcal{X}_{B_1}(x + t - t_1)f(x + t) \\
&= \left(e^{ipt} Q \circ \gamma_+^{-1}(E) f\right)(x),
\end{aligned}
$$

$$(7.6)$$

for almost all $x \in \mathbb{R}$. Here $E \subset \Omega$ is the cylinder set (3.7) and $\Omega$ is the collection of all paths $\omega : [0, \infty) \to \mathbb{R}$ for which there exists $x \in \mathbb{R}$ such that either

a) $\omega(s) = x + s$, for all $s \geq 0$, or,
b) $\omega(s) = x - s$, for all $s \geq 0$.

The mapping $\gamma_+ : \mathbb{R} \to \Omega$ is defined by $\gamma_+(x)(s) = x - s$ for $s \geq 0$ and analogously, $\gamma_- : \mathbb{R} \to \Omega$ is defined by $\gamma_-(x)(s) = x + s$ for $x \geq 0$ and $s \geq 0$. The set $\Omega$ has the finest topology for which the maps $\gamma_\pm$ are continuous. Then

$$
M_t^+ := e^{ipt} Q \circ \gamma_+^{-1}
$$

is an $\mathcal{L}(L^2(\mathbb{R}))$-valued Borel measure on $\Omega$. Similarly, set

$$
M_t^- := e^{-ipt} Q \circ \gamma_-^{-1}.
$$

The range of $\gamma_\pm$ is the set of characteristic lines of the equation

$$
\frac{\partial u}{\partial t} = \pm i \, pu = \pm \frac{\partial u}{\partial x}
$$

defining translations on $\mathbb{R}$.

**Remark 7.1.** The same argument as above applies to a classical dynamical system with the maps $\gamma_\pm$ replaced by the flow of the system.

Now suppose that $A$ is the selfadjoint differential operator

$$
A = \sigma_3 \frac{1}{i} \frac{\partial}{\partial x} = \frac{1}{i} \begin{pmatrix} 1 & 0 \\ 0 & -1 \end{pmatrix} \frac{\partial}{\partial x}
$$

$$(7.7)$$

acting in $L^2(\mathbb{R}, \mathbb{C}^2)$. Set $S(t) = e^{-iAt}$ for all $t \in \mathbb{R}$ and $Q : \mathcal{B}(\mathbb{R}) \to \mathcal{L}(L^2(\mathbb{R}, \mathbb{C}^2))$ is multiplication by characteristic functions. Let $M^t$ be the $(S, Q)$-measure. Then $A = p \oplus (-p)$ in $L^2(\mathbb{R}, \mathbb{C}^2)$ and $M^t$ defines an $\mathcal{L}(L^2(\mathbb{R}, \mathbb{C}^2))$-valued Borel measure on $\Omega$ given by

$$M^t = M_-^t \oplus M_+^t := \left( e^{-ipt} Q \circ \gamma_-^{-1} \right) \oplus \left( e^{ipt} Q \circ \gamma_+^{-1} \right). \tag{7.8}$$

## 7.5 The Feynman–Kac formula for nonsingular potentials

Let $V : \mathbb{R} \to \mathbb{R}$ be a locally integrable function. In the present context, this is what is meant by a 'nonsingular' interaction. The spectral measure $Q : \mathcal{B}(\mathbb{R}) \to \mathcal{L}(L^2(\mathbb{R}, \mathbb{C}^2))$ of multiplication by characteristic functions is the spectral resolution of the position operator for a Dirac particle on the line. Then $Q(V) := \int_{\mathbb{R}} V \, dQ$ is the selfadjoint operator of multiplication by $V$ acting in $L^2(\mathbb{R}, \mathbb{C}^2)$. The symbol '$\int_{\mathbb{R}} V \, dQ$' is interpreted literally because

$$\mathcal{D}(Q(V)) = \{ f \in L^2(\mathbb{R}, \mathbb{C}^2) : V \text{ is } (Qf)\text{-integrable} \}$$

and $Q(V)f := \int_{\mathbb{R}} V \, d(Qf)$ for all $f \in \mathcal{D}(Q(V))$.

As mentioned in Subsection 7.2, the operator $A + Q(V)$ whose domain is the set of all functions $u \in L^2(\mathbb{R}, \mathbb{C}^2)$, absolutely continuous on bounded intervals such that $Au + Vu \in L^2(\mathbb{R}, \mathbb{C}^2)$ is selfadjoint. Therefore, $e^{-it(A+Q(V))}$, $t \in \mathbb{R}$, is a continuous unitary group of operators. The following *Feynman–Kac* formula represents the semigroup of operators as a functional integral. In this special case, both sides of the formula can be calculated explicitly. The proof below sets the stage for the more involved calculations later. For each $\omega \in \Omega$, set $X_s(\omega) = \omega(s)$, $s \geq 0$.

According to the notation mentioned above, the value of the operator valued measure

$$e^{-ipt} Q \circ \gamma_-^{-1} : \mathcal{B}(\Omega) \to \mathcal{L}(L^2(\mathbb{R}, \mathbb{C}^2))$$

on the Borel subset $E$ of $\Omega$ is the bounded linear operator $e^{-ipt} Q(\gamma_-^{-1}(E))$.

**Proposition 7.2.** *The function* $\omega \longmapsto e^{-i \int_0^t V \circ X_s(\omega) \, ds}$, $\omega \in \Omega$, *exists and is* $M^t$-*integrable for each* $t \geq 0$. *Furthermore,*

$$e^{-it(A+Q(V))} = \int_\Omega e^{-i \int_0^t V \circ X_s \, ds} \, dM^t. \tag{7.9}$$

*Proof.* Because $V$ is locally integrable, $\int_0^t |V(x-s)| \, ds < \infty$ and $\int_0^t |V(x+s)| \, ds < \infty$ for every $x \in \mathbb{R}$. Hence, $\int_0^t |V(\omega(s))| \, ds < \infty$ for every $\omega \in \Omega$. The measurable function $\omega \longmapsto e^{-i \int_0^t V \circ X_s(\omega) \, ds}$, $\omega \in \Omega$, has absolute value one, so it is integrable with respect to the operator valued measure $M^t$. We calculate its integral from equation (7.8) by observing that

$$\int_\Omega e^{-i \int_0^t V \circ X_s \, ds} \, dM^t = \int_\Omega e^{-i \int_0^t V \circ X_s \, ds} \, d(M_-^t \oplus M_+^t)$$

$$= \left( \int_\Omega e^{-i \int_0^t V \circ X_s \, ds} \, dM_-^t \right) \oplus \left( \int_\Omega e^{-i \int_0^t V \circ X_s \, ds} \, dM_+^t \right)$$

$$= \left( \int_\Omega e^{-i \int_0^t V \circ X_s \, ds} \, d(e^{-ipt} Q \circ \gamma_-^{-1}) \right) \oplus \left( \int_\Omega e^{-i \int_0^t V \circ X_s \, ds} \, d(e^{ipt} Q \circ \gamma_+^{-1}) \right)$$

$$= \left( e^{-ipt} \int_\Omega e^{-i \int_0^t V \circ X_s \, ds} \, d(Q \circ \gamma_-^{-1}) \right) \oplus \left( e^{-ipt} \int_\Omega e^{-i \int_0^t V \circ X_s \, ds} \, d(Q \circ \gamma_+^{-1}) \right)$$

$$= \left( e^{-ipt} Q \left( e^{-i \int_0^t V \circ X_s \circ \gamma_- \, ds} \right) \right) \oplus \left( e^{ipt} Q \left( e^{-i \int_0^t V \circ X_s \circ \gamma_+ \, ds} \right) \right). \qquad (7.10)$$

Now let $v \in L^2(\mathbb{R})$. Then for almost all $x \in \mathbb{R}$, we have

$$\left( Q \left( e^{-i \int_0^t V \circ X_s \circ \gamma_+ \, ds} \right) v \right)(x) = e^{-i \int_0^t V \circ X_s \circ \gamma_+(x) \, ds} v(x)$$

$$= e^{-i \int_0^t V(x-s) \, ds} v(x)$$

so we have

$$\left( e^{ipt} Q \left( e^{-i \int_0^t V \circ X_s \circ \gamma_+ \, ds} \right) v \right)(x) = e^{-i \int_0^t V \circ X_s \circ \gamma_+(x+t) \, ds} v(x+t)$$

$$= e^{-i \int_0^t V(x+t-s) \, ds} v(x+t)$$

$$= e^{-i \int_0^t V(x+s) \, ds} v(x+t)$$

$$= e^{-i \int_x^{x+t} V(s) \, ds} v(x+t)$$

Similarly,

$$\left( e^{-ipt} Q \left( e^{-i \int_0^t V \circ X_s \circ \gamma_- \, ds} \right) u \right)(x) = e^{-i \int_{x-t}^x V(s) \, ds} u(x-t).$$

Integration by parts verifies that for $\phi = \begin{pmatrix} \phi_1 \\ \phi_2 \end{pmatrix} \in L^2(\mathbb{R}, \mathbb{C}^2)$, the equality

$$\left( e^{-it(A+Q(V))} \phi \right)(x) = \begin{pmatrix} e^{-i \int_{x-t}^x V(s) \, ds} \phi_1(x-t) \\ e^{-i \int_x^{x+t} V(s) \, ds} \phi_2(x+t) \end{pmatrix}$$

holds for almost all $x \in \mathbb{R}$, so now equality (7.9) follows from (7.10). $\qquad \square$

## 7.6 Point interactions: zero mass

Let $A_\Gamma$ be the operator (7.7) with the boundary condition

$$\begin{pmatrix} \phi_1(0+) \\ \phi_2(0-) \end{pmatrix} = \Gamma \begin{pmatrix} \phi_1(0-) \\ \phi_2(0+) \end{pmatrix} \qquad (7.11)$$

at zero with respect to the unitary matrix

$$\Gamma = \eta \begin{pmatrix} \alpha & -\overline{\beta} \\ \beta & \overline{\alpha} \end{pmatrix} \tag{7.12}$$

with $\alpha, \beta, \eta \in \mathbb{C}$ satisfying $|\alpha|^2 + |\beta|^2 = 1$ and $|\eta| = 1$.

In this simple case, a calculation shows that the matrix $\Gamma$ corresponds to a unitary map from one deficiency subspace of $A|C_0^\infty(\mathbb{R} \setminus \{0\}, \mathbb{C}^2)$ onto another, so $A_\Gamma$ is self-adjoint and every selfadjoint extension of $A|C_0^\infty(\mathbb{R} \setminus \{0\}, \mathbb{C}^2)$ may be obtained in this way [48, Theorem 4.4]; see also [8] for further discussion of this point.

A careful analysis of the selfadjoint operator $A_\Gamma - m\sigma_1$ (or rather, a selfadjoint operator unitarily equivalent to this one) is given in [21]. Certain choices of the boundary condition $\Gamma$ are associated with point interactions.

The present Chiral representation is better adapted to the calculation of functional integrals, but the Dirac representation of [6] and [21] leads to a simpler determination of the boundary conditions that arise from taking the nonrelativistic limit.

For suitable $\phi \in L^2(\mathbb{R}, \mathbb{C}^2)$ the function

$$\begin{pmatrix} u(x, t) \\ v(x, t) \end{pmatrix} = \left( e^{-itA_\Gamma} \phi \right)(x), \quad x \in \mathbb{R}, \tag{7.13}$$

is a solution $\psi(x, t) = \begin{pmatrix} u(x, t) \\ v(x, t) \end{pmatrix}$ of the equation (cf. (7.7) for $\sigma_3$)

$$\frac{\partial \psi}{\partial t} + \sigma_3 \frac{\partial \psi}{\partial x} = 0, \quad \psi(\cdot, t) \in L^2(\mathbb{R}, \mathbb{C}^2)$$

satisfying the initial condition $\psi(x, 0) = \phi(x)$, $x \in \mathbb{R}$, and satisfying the boundary condition (7.11).

Let $t > 0$. If $x > 0$ or $x < -t$, then $v(x, t) = \phi_2(x + t)$ and if $x < 0$ or $x > t$, then $u(x, t) = \phi_1(x - t)$. In the region $-t < x < 0$, there exists another function $\Phi$ such that $v(x, t) = \Phi(x + t)$. To ensure that the boundary condition (7.11) is satisfied, we must have

$$\Phi(t-) = \eta \left[ \beta \phi_1((-t)-) + \overline{\alpha} \phi_2(t+) \right]$$

for all $t > 0$. Therefore, if $\phi_1, \phi_2$ are absolutely continuous on subintervals of $\mathbb{R} \setminus \{0\}$ and satisfy the boundary condition (7.11), it follows that

$$v(x, t) = \eta \left[ \beta \phi_1((-(x + t)) + \overline{\alpha} \phi_2(x + t) \right], \quad -t < x < 0.$$

Similarly, we have

$$u(x, t) = \eta \left[ \alpha \phi_1(x - t) - \overline{\beta} \phi_2(-(x - t)) \right], \quad 0 < x < t.$$

It follows that for any $\phi \in L^2(\mathbb{R}, \mathbb{C}^2)$ equation (7.13) holds with

$$u(x, t) = \begin{cases} \phi_1(x - t), & \text{for all } x > t, \ x < 0, \\ \eta \left[ \alpha \phi_1(x - t) - \overline{\beta} \phi_2(-(x - t)) \right], & \text{for all } 0 < x < t. \end{cases} \tag{7.14}$$

$$v(x, t) = \begin{cases} \phi_2(x + t), & \text{for all } x < -t, \ x > 0, \\ \eta \left[ \beta \phi_1(-(x + t)) + \overline{\alpha} \phi_2(x + t) \right], & \text{for all } -t < x < 0. \end{cases} \tag{7.15}$$

Let $S(t) = e^{-itA_\Gamma}$ for all $t \geq 0$ and let $M_\Gamma^t$ be $(S, Q)$-set function. Define the mapping $\zeta_\pm : \mathbb{R} \to \Omega$ onto paths which reflect at $x = 0$ by the formulae

$$\zeta_+(x)(s) = \text{sgn}(x)|x - s|, \quad s \geq 0, \ x \in \mathbb{R}, \tag{7.16}$$

$$\zeta_-(x)(s) = \text{sgn}(x)|x + s|, \quad s \geq 0, \ x \in \mathbb{R}. \tag{7.17}$$

Here we enlarge $\Omega$ so as to include not only the ranges of the characteristic mappings $\gamma_\pm$, but also of the reflected paths $\zeta_\pm(x)$, $x \in \mathbb{R}$, that is, set $\Omega_0 = \gamma_+(\mathbb{R}) \cup \gamma_-(\mathbb{R})$, $\Omega_1 = \zeta_+((0, \infty)) \cup \zeta_-((-\infty, 0))$ and $\Omega = \Omega_0 \cup \Omega_1$. Then $\Omega_0$ is the set of paths with no reflection at zero and $\Omega_1$ is the set of paths with a reflection at zero for some positive time.

As the following representation suggests, paths belonging to $\Omega_1$ are associated with the creation of 'antiparticles' or reflected waves.

**Theorem 7.3.** *Let* $S(t) = e^{-iA_\Gamma t}$ *for all* $t \in \mathbb{R}$ *and for each* $t \geq 0$ *let* $M_\Gamma^t$ *be the* $(S, Q)$-*measure acting on* $L^2(\mathbb{R}, \mathbb{C}^2)$.

*Then there exist operator valued measures* $M_0^t$ *and* $M_{\Gamma,1}^t$ *such that* $M_\Gamma^t = M_0^t + M_{\Gamma,1}^t$, *where* $M_0^t$ *is concentrated on paths* $\omega \in \Omega$ *which do not hit zero in the interval* $(0, t)$ *and* $M_{\Gamma,1}^t$ *is concentrated on those paths* $\omega \in \Omega$ *that do.*

*More precisely, set* $(Rf)(x) = f(-x)$ *for* $f \in L^2(\mathbb{R}, \mathbb{C})$, $x \in \mathbb{R}$. *Then*

$$M_0^t = \chi_{\{X_0 X_t > 0\}} \cdot \left[ \left( e^{-ipt} Q \circ \gamma_-^{-1} \right) \oplus \left( e^{ipt} Q \circ \gamma_+^{-1} \right) \right] \text{ and} \tag{7.18}$$

$$M_{\Gamma,1}^t = \eta \begin{pmatrix} \alpha \chi_{\{X_0 X_t < 0\}} \cdot \left( e^{-ipt} Q \circ \gamma_-^{-1} \right) & -\bar{\beta}(\chi_{(0,t)} \cdot [e^{-ipt} QR]) \circ \zeta_-^{-1} \\ \beta(\chi_{(-t,0)} \cdot [e^{ipt} QR]) \circ \zeta_+^{-1} & \bar{\alpha} \chi_{\{X_0 X_t < 0\}} \cdot \left( e^{ipt} Q \circ \gamma_+^{-1} \right) \end{pmatrix}. \tag{7.19}$$

*The operator valued measure* $M_0^t$ *is concentrated on the Borel set* $\Omega_0 \cap \{X_0 X_t > 0\}$ *and* $M_{\Gamma,1}^t$ *is concentrated on* $(\Omega_0 \cap \{X_0 X_t < 0\}) \cup \zeta_+((0, t)) \cup \zeta_-((-t, 0))$.

*Proof.* Let $E \subset \Omega$ be a nonempty cylinder set (3.7) and $f \in L^2(\mathbb{R}, \mathbb{C}^2)$. Then for almost all $x$ outside the interval $[-t, t]$, a calculation similar to (7.6) holds because the formula for $e^{-iAt}$ applies, so that

$$(M_\Gamma^t(E) f)(x) = \left( \left[ \left( e^{-ipt} Q \circ \gamma_-^{-1}(E) \right) \oplus \left( e^{ipt} Q \circ \gamma_+^{-1}(E) \right) \right] f \right)(x).$$

Then neither path $\gamma_\pm(x \pm t)$ hits zero in the interval $[0, t]$. Hence, $\left( M_\Gamma^t(E) f \right)(x)$ is equal to

$$\left( \left( \chi_{\{X_0 X_t > 0\}} \cdot \left[ \left( e^{-ipt} Q \circ \gamma_-^{-1} \right) \oplus \left( e^{ipt} Q \circ \gamma_+^{-1} \right) \right] \right) (E) f \right)(x) +$$
$$\eta \begin{pmatrix} \alpha \chi_{\{X_0 X_t < 0\}} \cdot \left( e^{-ipt} Q \circ \gamma_-^{-1} \right)(E) & -\bar{\beta}(\chi_{(0,t)} \cdot [e^{-ipt} QR]) \circ \zeta_-^{-1}(E) \\ \beta(\chi_{(-t,0)} \cdot [e^{ipt} QR]) \circ \zeta_+^{-1}(E) & \bar{\alpha} \chi_{\{X_0 X_t < 0\}} \cdot \left( e^{ipt} Q \circ \gamma_+^{-1} \right)(E) \end{pmatrix} f(x). \tag{7.20}$$

for almost all of these $x$ because the second term is zero.

Now suppose that $-t < x < 0$ and $E$ is a nonempty cylinder set (3.7). Then there exists $k = 0, \ldots, n$ such that $x + t - t_k \geq 0$ and $x + t - t_{k+1} < 0$, where we have set $t_0 = 0$ and $t_{n+1} = t$. Then we have

$$(M_\Gamma^t(E)f)_2(x) = \left(e^{-i(t-t_n)A_\Gamma} Q(B_n)e^{-i(t_n-t_{n-1})A_\Gamma} \cdots Q(B_1)e^{-it_1 A_\Gamma} f\right)_2(x)$$
$$= \chi_{B_n}(x + t - t_n)\left(e^{-i(t_n-t_{n-1})A_\Gamma} Q(B_{n-1}) \cdots Q(B_1)e^{-it_1 A_\Gamma} f\right)_2(x + t - t_n)$$

$$\vdots$$

$$= \chi_{B_n}(x + t - t_n) \cdots \chi_{B_{k+1}}(x + t - t_{k+1}) \times$$
$$\left(e^{-i(t_{k+1}-t_k)A_\Gamma} Q(B_k) \cdots Q(B_1)e^{-it_1 A_\Gamma} f\right)_2(x + t - t_{k+1})$$
$$= \eta \chi_{B_n}(x + t - t_n) \cdots \chi_{B_{k+1}}(x + t - t_{k+1}) \times$$
$$[\beta \chi_{B_k}(-(x + t - t_k))\left(e^{-i(t_k-t_{k-1})A_\Gamma} Q(B_{k-1}) \cdots Q(B_1)e^{-it_1 A_\Gamma} f\right)_1(-(x + t - t_k))$$
$$+\overline{\alpha} \chi_{B_k}(x + t - t_k)\left(e^{-i(t_k-t_{k-1})A_\Gamma} Q(B_{k-1}) \cdots Q(B_1)e^{-it_1 A_\Gamma} f\right)_2(x + t - t_k)]$$
$$\text{(by equation (7.15))}$$

$$\vdots$$

$$= \eta \chi_{B_n}(x + t - t_n) \cdots \chi_{B_{k+1}}(x + t - t_{k+1}) \times$$
$$\left[\beta \chi_{B_k}(-(x + t - t_k)) \cdots \chi_{B_1}(-(x + t - t_1)) f_1(-(x + t))\right.$$
$$\left. +\overline{\alpha} \chi_{B_k}(x + t - t_k) \cdots \chi_{B_1}(x + t - t_1) f_2(x + t)\right]$$
$$= \eta\left(\beta e^{itp} Q \circ \zeta_+^{-1}(E)Rf_1 + \overline{\alpha} e^{itp} Q \circ \gamma_+^{-1}(E) f_2\right)(x).$$

These equalities hold true for almost all $-(t - t_k) < x < -(t - t_{k+1})$.

Because $x < 0$, we have $x - (t - t_j) < 0$ for all $j = 0, \ldots, n + 1$. It follows that

$$(M_\Gamma^t(E)f)_1(x) = \left(e^{-i(t-t_n)A_\Gamma} Q(B_n)e^{-i(t_n-t_{n-1})A_\Gamma} \cdots Q(B_1)e^{-it_1 A_\Gamma} f\right)_1(x)$$
$$= \chi_{B_n}(x - (t - t_n))\left(e^{-ip(t_n-t_{n-1})} Q(B_{n-1}) \cdots Q(B_1)e^{-ipt_1} f\right)_1(x - (t - t_n))$$
$$= \chi_{B_n}(x - (t - t_n))\chi_{B_{n-1}}(x - (t - t_{n-1})) \cdots \chi_{B_1}(x - (t - t_1)) f_1(x - t)$$
$$= \left(e^{-ipt} Q \circ \gamma_-^{-1}(E) f_1\right)(x).$$

Hence, the equality

$$(M_\Gamma^t(E)f)(x) = \begin{pmatrix} e^{-ipt} Q \circ \gamma_-^{-1}(E) f_1 \\ \eta\left(\beta e^{itp} Q \circ \zeta_+^{-1}(E)Rf_1 + \overline{\alpha} e^{itp} Q \circ \gamma_+^{-1}(E) f_2\right) \end{pmatrix}(x) \quad (7.21)$$

holds for almost all $-(t - t_k) < x < -(t - t_{k+1})$. As $k$ varies from 0 to $n$, we obtain the representation (7.21) almost everywhere on the interval $(-t, 0)$ for the given cylinder set $E$. Because $E$ is any cylinder set, the representation (7.21) holds on every cylinder set for almost all $x \in (-t, 0)$. We need to check that the right-hand side of equation (7.21) is given by the expression (7.20) for almost all $x \in (-t, 0)$.

Now the element

$$\left(\left(\chi_{\{X_0 X_t > 0\}} \cdot \left[\left(e^{-ipt} Q \circ \gamma_-^{-1}\right) \oplus \left(e^{ipt} Q \circ \gamma_+^{-1}\right)\right]\right)(E)f\right)_1$$

of $L^2(\mathbb{R})$ is equal to $e^{-ipt} \chi_{\gamma_-^{-1}(\{X_0 X_t > 0\} \cap E)} f_1$, which, at $x \in (-t, 0) = \chi_{\gamma_-^{-1}(\{X_0 X_t > 0\} \cap E)}(x - t) f_1(x - t)$. But $X_0(\gamma_-(x - t))X_t(\gamma_-(x - t)) > 0$ for $x < 0$, so this is just

$$\chi_{\gamma_-^{-1}(E)}(x-t)f_1(x-t) = \left(e^{-ipt}Q\circ\gamma_-^{-1}(E)f_1\right)(x),$$

corresponding to the first element of (7.21). Because $X_0(\gamma_-(x-t))X_t(\gamma_-(x-t)) > 0$ for $x < 0$, no other contribution is made by the first elements of (7.20).

On the other hand,

$$\left(\left(\chi_{\{X_0X_t>0\}}\cdot\left[\left(e^{-ipt}Q\circ\gamma_-^{-1}\right)\oplus\left(e^{ipt}Q\circ\gamma_+^{-1}\right)\right]\right)(E)f\right)_2(x) = 0$$

for $-t < x < 0$, because then $X_0(\gamma_+(x+t))X_t(\gamma_+(x+t)) < 0$. It follows that $\left(M_\Gamma^t(E)f\right)(x)$ is equal to the expression (7.20)

A similar argument applies to the interval $(0, t)$, for then $X_0(\gamma_+(x+t))X_t(\gamma_+(x+t)) > 0$.

Let $\phi$ be an integrable $\mathbb{C}^2$-valued simple function. Replacing $f$ by $\phi$, the formulae above make sense for *each* $x \in \mathbb{R}$. But, for each $x \in \mathbb{R}$ and $t > 0$, there is at most one path $\omega \in \Omega$ such that $\omega = \gamma_\pm(x \pm t)$ or $\omega = \zeta_\pm(x \pm t)$ and $X_0(\omega)X_t(\omega) = 0$.

Then for almost all $\propto x \in \mathbb{R}$, $(M_\Gamma^t(E \cap \{X_0X_t = 0\})\phi)(x) = 0$ for all $E \in S_t$. The image $[\phi]$ of $\phi$ in $L^2(\mathbb{R}, \mathbb{C}^2)$ has the property that $M_\Gamma^t(E \cap \{X_0X_t = 0\})[\phi] = 0$ as an element of $L^2(\mathbb{R}, \mathbb{C}^2)$ for every $E \in S_t$. Integrable simple functions are dense in $L^2(\mathbb{R}, \mathbb{C}^2)$, so the cylinder set $\{X_0X_t = 0\}$ is an $M_\Gamma^t$-null set. This establishes the representation $M_\Gamma^t(E) = M_0^t(E) + M_{\Gamma,1}^t(E)$ on all cylinder sets $E$. Both sides of the equation are $\sigma$-additive, so we have equality on all Borel subsets $E$ of $\Omega$.    □

**Remark 7.4.** a) According to Theorem 7.3, relativistic point interactions in one dimension are also associated with $\mathcal{L}(L^2(\mathbb{R}, \mathbb{C}^2))$-valued measures on the path space $\Omega$. Nevertheless, the smallest number $C > 0$ for which $\|e^{-itA_\Gamma}f\|_\infty \le C\|f\|_\infty$ for all $t \in \mathbb{R}$ and $f \in L^2(\mathbb{R}, \mathbb{C}^2) \cap L^\infty(\mathbb{R}, \mathbb{C}^2)$ is $|\alpha| + |\beta|$. Because

$$\sup\{|a| + |b| : a, b \in \mathbb{C}, |a|^2 + |b|^2 = 1\} = \sqrt{2},$$

the group $e^{-itA_\Gamma}$ need not be similar to a group of contractions on $L^\infty(\mathbb{R}, \mathbb{C}^2)$ which is the usual condition for constructing path space measures, see Corollary 5.3. In the case of point interactions $\delta$ associated with the Laplacian $\Delta$ in $\mathbb{R}^d$, the positive operators $e^{t(\Delta-\delta)}$, $t > 0$, are not even *bounded* on $L^\infty(\mathbb{R}^d)$ for $d = 1, 2, 3$ ([10], [4]).

b)  The effect of off-diagonal terms in the unitary matrix $\Gamma$ is to introduce paths $\zeta_\pm(x)$, $x \in \mathbb{R}$, that scatter off the singular interaction at the origin. The off-diagonal terms of $\Gamma$ are the reflection coefficients and the diagonal terms are the transmission coefficients of the transmitted path. An operator valued measure $M_{\Gamma,1}^t$ concentrated on the associated reflected or transmitted paths is associated with each of these coefficients via formula (7.19). The singularity represented by the off-diagonal terms in the unitary matrix $\Gamma$ therefore gives rise to paths associated with particle-pair production. Of course, we are only dealing with a 'toy model', so the idea is only suggestive.

## 7.7 The Feynman–Kac formula for singular potentials : zero mass

Let $\gamma \in \mathbb{R}$ and $V(x) = \frac{\gamma}{|x|}$ for all $x \in \mathbb{R}$ with $x \neq 0$. Let $\Gamma$ be a $(2 \times 2)$ unitary matrix—any such matrix can be expressed in the form (7.12). Then the operator

$$\frac{1}{i}\begin{pmatrix} 1 & 0 \\ 0 & -1 \end{pmatrix}\frac{\partial}{\partial x} + \begin{pmatrix} 1 & 0 \\ 0 & 1 \end{pmatrix}\frac{\gamma}{|x|} \tag{7.22}$$

satisfying the boundary conditions

$$\begin{pmatrix} \lim_{x \to 0+} |x|^{i\gamma} u(x) \\ \lim_{x \to 0-} |x|^{i\gamma} v(x) \end{pmatrix} = \Gamma \begin{pmatrix} \lim_{x \to 0-} |x|^{-i\gamma} u(x) \\ \lim_{x \to 0+} |x|^{-i\gamma} v(x) \end{pmatrix} \tag{7.23}$$

is selfadjoint and written as $A +_\Gamma Q(V)$. As mentioned above in the case $\gamma = 0$, the collection of all unitary matrices $\Gamma$ is in one-to-one correspondence with isometries from one deficiency subspace of $(A + Q(V))|C_0^\infty(\mathbb{R} \setminus \{0\}, \mathbb{C}^2)$ onto another.

Then $e^{-it(A+_\Gamma Q(V))}$, $t \in \mathbb{R}$, is a continuous unitary group of operators. Calculations similar to those of Subsection 7.6 show that for $\phi \in L^2(\mathbb{R}, \mathbb{C}^2)$, the function

$$\begin{pmatrix} u(x, t) \\ v(x, t) \end{pmatrix} = \left( e^{-it(A+_\Gamma Q(V))}\phi \right)(x), \quad x \in \mathbb{R}, \tag{7.24}$$

is given by

$$u(x, t) = \begin{cases} e^{-i\int_{x-t}^{x} V(s)\,ds}\phi_1(x - t), & x > t,\ x < 0, \\ \eta e^{-i\gamma(\ln|x|+\ln|x-t|)}[\alpha\phi_1(x - t) - \bar{\beta}\phi_2(-(x - t))], & 0 < x < t. \end{cases}$$

$$v(x, t) = \begin{cases} e^{-i\int_{x}^{x+t} V(s)\,ds}\phi_2(x + t), & x < -t,\ x > 0, \\ \eta e^{-i\gamma(\ln|x|+\ln(x+t))}[\beta\phi_1(-(x + t)) + \bar{\alpha}\phi_2(x + t)], & -t < x < 0. \end{cases}$$

Suppose that $\beta = 0$ and $\eta\alpha = e^{-i\kappa_1}$, $\eta\bar{\alpha} = e^{-i\kappa_2}$ for numbers $0 \leq \kappa_j < 2\pi$.

For the moment, we take $\Omega$ to be the collection of all paths that lie in the range of the maps $\gamma_\pm$, that is, there exists $x \in \mathbb{R}$ such that either $\omega(s) = x + s$ for all $s \geq 0$, or $\omega(s) = x - s$. Suppose that $\omega \in \Omega$ is a path for which $X_0(\omega)X_t(\omega) < 0$, which is to say that the path $\omega$ hits the origin at some time in the open interval $(0, t)$. Set

$$\left( \int_0^t V \circ X_s(\omega)\,ds \right)_\Gamma = \begin{cases} \gamma(\ln|\omega(0)| + \ln|\omega(t)|) + \kappa_1 & \text{if } \omega'(s) = 1,\ s > 0, \\ \gamma(\ln|\omega(0)| + \ln|\omega(t)|) + \kappa_2 & \text{if } \omega'(s) = -1,\ s > 0. \end{cases} \tag{7.25}$$

Then the measurable function $F_t^\Gamma : \Omega \to \mathbb{C}$ is defined by

$$F_t^\Gamma = X_{\{X_0 X_t > 0\}} \cdot e^{-i\int_0^t V \circ X_s\,ds} + X_{\{X_0 X_t < 0\}} \cdot e^{-i(\int_0^t V \circ X_s\,ds)_\Gamma} \tag{7.26}$$

Here $F_t^\Gamma$ is a *multiplicative functional*. The multiplicative property is borrowed from the theory of Markov processes

$$F_{s+t}^\Gamma(\omega) = F_t^\Gamma(\omega_s)F_s^\Gamma(\omega) \quad \text{a.e.,} \tag{7.27}$$

with $\omega_s(r) = \omega(s+r)$. It is natural to view

$$\left\langle \int_0^t V \circ X_s \, ds \right\rangle_\Gamma$$

as a 'renormalization' of the expression '$\int_0^t V \circ X_s \, ds$' on the set of all paths $\omega \in \Omega$ such that $X_0(\omega)X_t(\omega) < 0$. For example, suppose that $0 < x < t$ and $\omega(s) = x - s$. Then

$$\int_{[0,t]\cap\{|\omega(s)|>\epsilon\}} V(\omega(s)) \, ds = \int_{x+\epsilon}^t \frac{\gamma}{|x-s|} \, ds + \int_0^{x-\epsilon} \frac{\gamma}{|x-s|} \, ds$$
$$= \gamma\,(\ln|\omega(0)| + \ln|\omega(t)|) - 2\gamma \ln \epsilon,$$

as $\epsilon \to 0+$, so we are subtracting a logarithmic divergence. Nevertheless, for each $\epsilon > 0$ sufficiently small, $e^{-i\int_{[0,t]\cap\{|\omega(s)|>\epsilon\}} V(\omega(s))\,ds}$ and $e^{-i\langle\int_0^t V\circ X_s\,ds\rangle_\Gamma}$ differ by a complex factor with modulus one.

**Lemma 7.5.** $F_t^\Gamma$ *is a continuous multiplicative functional, that is,* $s \longmapsto F_s^\Gamma(\omega)$, $s \in [0, t]$ *is continuous for $M^t$-almost all $\omega \in \Omega$.*

*Proof.* We look at the case $\omega'(s) = -1$ for all $s > 0$. The argument is similar for the other type of path. We have

$$F_s^\Gamma(\omega) = e^{-i\int_0^s V(\omega(r))\,dr} \chi_{\{X_s>0\}\cup\{X_s<-s\}}(\omega) + e^{-i\langle\int_0^s V(\omega(r))\,dr\rangle_\Gamma} \chi_{\{-s<X_s<0\}}(\omega)$$
$$F_t^\Gamma(\omega_s) = e^{-i\int_0^t V(\omega_s(r))\,dr} \chi_{\{X_t>0\}\cup\{X_t<-t\}}(\omega_s)$$
$$\qquad e^{-i\langle\int_0^t V(\omega_s(r))\,dr\rangle_\Gamma} \chi_{\{-t<X_t<0\}}(\omega_s)$$
$$= e^{-i\int_s^{s+t} V(\omega(r))\,dr} \chi_{\{X_{s+t}>0\}\cup\{X_{s+t}<-t\}}(\omega)$$
$$\qquad + e^{-i\langle\int_0^t V(\omega_s(r))\,dr\rangle_\Gamma} \chi_{\{-t<X_{s+t}<0\}}(\omega)$$
$$= e^{-i\int_s^{s+t} V(\omega(r))\,dr} \chi_{\{X_s>t\}\cup\{X_s<0\}}(\omega)$$
$$\qquad + e^{-i\langle\int_0^t V(\omega_s(r))\,dr\rangle_\Gamma} \chi_{\{0<X_s<t\}}(\omega).$$

The last line follows from the observation $X_r(\omega) = X_0(\omega) - r$ for all $r \geq 0$. Then

$$F_t^\Gamma(\omega_s)F_s^\Gamma(\omega) = e^{-i\int_0^{s+t} V(\omega(r))\,dr} \chi_{\{X_s>t\}\cup\{X_s<-s\}}(\omega)$$
$$\qquad + e^{-i\int_s^{s+t} V(\omega(r))\,dr} e^{-i\langle\int_0^s V(\omega(r))\,dr\rangle_\Gamma} \chi_{\{-s<X_s<0\}}(\omega)$$
$$\qquad + e^{-i\int_0^s V(\omega(r))\,dr} e^{-i\langle\int_0^t V(\omega_s(r))\,dr\rangle_\Gamma} \chi_{\{0<X_s<t\}}(\omega)$$
$$= e^{-i\int_0^{s+t} V(\omega(r))\,dr} \chi_{\{X_{s+t}>0\}\cup\{X_{s+t}<-s-t\}}(\omega)$$
$$\qquad + e^{-i\int_s^{s+t} V(\omega(r))\,dr} e^{-i\langle\int_0^s V(\omega(r))\,dr\rangle_\Gamma} \chi_{\{-s-t<X_{s+t}<-t\}}(\omega)$$
$$\qquad + e^{-i\int_0^s V(\omega(r))\,dr} e^{-i\langle\int_0^t V(\omega_s(r))\,dr\rangle_\Gamma} \chi_{\{-t<X_{s+t}<0\}}(\omega)$$

But from equation (7.25),

$$\left\langle \int_0^t V(\omega_s(r)) \, dr \right\rangle_\Gamma = \gamma\,(\ln(\omega_s(0)) + \ln|\omega_s(t)|) + \kappa_2$$

$$= \gamma \left( \ln(\omega(s)) + \ln |\omega(s+t)| \right) + \kappa_2$$

and for $s < \omega(0) < s + t$, we have $\int_0^s V(\omega(r)) \, dr = \gamma \left( \ln(\omega(0)) - \ln(\omega(s)) \right)$. Hence,

$$e^{-i \int_0^s V(\omega(r)) \, dr} e^{-i \left( \int_0^t V(\omega_s(r)) \, dr \right) \Gamma} = e^{-i \gamma (\ln(\omega(0)) + \ln(\omega(s+t))) - i \kappa_2}$$
$$= e^{-i \left( \int_0^{s+t} V(\omega(r)) \, dr \right) \Gamma}.$$

For $0 < \omega(0) < s$, we have $\int_s^{s+t} V(\omega(r)) \, dr = \gamma \left( \ln(|\omega(s+t)|) - \ln(|\omega(s)|) \right)$. Hence,

$$e^{-i \int_s^{s+t} V(\omega(r)) \, dr} e^{-i \left( \int_0^s V(\omega(r)) \, dr \right) \Gamma} = e^{-i \gamma (\ln(|\omega(0)|) + \ln(|\omega(s+t)|)) - i \kappa_2}$$
$$= e^{-i \left( \int_0^{s+t} V(\omega(r)) \, dr \right) \Gamma}.$$

The equality $F_{s+t}^\Gamma(\omega) = F_t^\Gamma(\omega_s) F_s^\Gamma(\omega)$ therefore holds unless $\omega(0)$ belongs to the finite set $\{0, s, s+t\}$. However, the set of all such $\omega$ has $M_{t+s}$-measure zero.

Continuity of $s \longmapsto F_s^\Gamma(\omega)$ at $s_0 \in [0, t]$ follows from formulae (7.25) and (7.26) and the fact that the set of all $\omega$ with $\omega(0) = 0$ or $\omega(s_0) = 0$ has $M^t$-measure zero.    □

The following result represents the operator $e^{it(A + \Gamma Q(V))}$ as an integral with respect to the free Dirac measure $M^t$ for each $t > 0$.

**Theorem 7.6.** *The function $F_t^\Gamma$ is $M^t$-integrable for each $t \geq 0$. Furthermore,*

$$e^{-it(A + \Gamma Q(V))} = \int_\Omega F_t^\Gamma \, dM^t. \tag{7.28}$$

*Proof.* As in the proof of Theorem 2.1,

$$\int_\Omega F_t^\Gamma \, dM^t = \left( e^{-ipt} Q \left( F_t^\Gamma \circ \gamma_- \right) \right) \oplus \left( e^{ipt} Q \left( F_t^\Gamma \circ \gamma_+ \right) \right).$$

Now let $\phi_2 \in L^2(\mathbb{R})$. Then for almost all $x \in \mathbb{R}$, we have

$$\left( Q \left( F_t^\Gamma \circ \gamma_+ \right) \phi_2 \right)(x) = F_t^\Gamma \circ \gamma_+(x) \phi_2(x)$$
$$= X_{\{X_0 X_t > 0\}} (\gamma_+(x)) . e^{-i \int_0^t V(x-s) \, ds} \phi_2(x)$$
$$+ X_{\{X_0 X_t < 0\}} (\gamma_+(x)) . e^{-i \left( \int_0^t V \circ X_s \circ \gamma_+(x) \, ds \right) \Gamma} \phi_2(x)$$
$$= X_{\{x' > t\} \cup \{x' < 0\}} (x) . e^{-i \int_0^t V(x-s) \, ds} \phi_2(x)$$
$$+ X_{\{0 < x' < t\}} (x) . e^{-i \left( \int_0^t V \circ X_s \circ \gamma_+(x) \, ds \right) \Gamma} \phi_2(x)$$

so we have

$$\left( e^{ipt} Q \left( F_t^\Gamma \circ \gamma_+ \right) \phi_2 \right)(x) = F_t^\Gamma \circ \gamma_+(x+t) \phi_2(x+t)$$
$$= X_{\{x' > t\} \cup \{x' < 0\}} (x+t) . e^{-i \int_0^t V(x+t-s) \, ds} \phi_2(x+t)$$
$$+ X_{\{0 < x' < t\}} (x+t) . e^{-i \left( \int_0^t V \circ X_s \circ \gamma_+(x+t) \, ds \right) \Gamma} \phi_2(x+t)$$
$$= X_{\{x' > 0\} \cup \{x' < -t\}} (x) . e^{-i \int_x^{x+t} V(s) \, ds} \phi_2(x+t)$$
$$+ X_{\{-t < x' < 0\}} (x) . e^{-i \left( \int_0^t V \circ X_s \circ \gamma_+(x+t) \, ds \right) \Gamma} \phi_2(x+t)$$
$$= X_{\{x' > 0\} \cup \{x' < -t\}} (x) . e^{-i \int_0^t V(x+s) \, ds} \phi_2(x+t)$$
$$+ X_{\{-t < x' < 0\}} (x) . e^{-i \gamma (\ln |x| + \ln |x+t|) - i \kappa_2} \phi_2(x+t)$$

Similarly,

$$\left(e^{-ipt} Q \left(F_t^\Gamma \circ \gamma_-\right) \phi_1\right)(x) = \chi_{\{x'<0\}\cup\{x'>t\}}(x).e^{-i\int_{x-t}^{x} V(s)\,ds} \phi_1(x-t)$$

$$+\chi_{\{0<x'<t\}}(x).e^{-i\gamma(\ln|x|+\ln|x-t|)-i\kappa_1} \phi_1(x-t).$$

Comparison with formula (7.24) for $e^{-it(A+\Gamma Q(V))}\phi$ for $\phi \in L^2(\mathbb{R}, \mathbb{C}^2)$ establishes the result.                                                                                              □

Suppose $\Gamma$ is *any* $(2 \times 2)$ unitary matrix. Another representation of $e^{-it(A+\Gamma Q(V))}$ is possible by using the measures $M_\Gamma^t$ associated with point interactions (7.11). Take the path space $\Omega$ to be the union of the ranges of the functions $\gamma_\pm$ and $\zeta_\pm$. For each $\omega \in \Omega$, set $\tau(\omega) = \inf\{s \geq 0 : \omega(s) = 0\}$, where we allow the possibility that $\tau(\omega) = \infty$ if $\omega$ never hits the origin.

The integral $\int_0^t V \circ X_s\,ds$ does not converge on the set $\{0 < \tau < t\}$ of paths, so we use the following 'renormalization' or principal value. Let

$$\left\langle\!\!\left\langle \int_0^t V \circ X_s(\omega)\,ds \right\rangle\!\!\right\rangle = \gamma(\ln|\omega(0)| + \ln|\omega(t)|) \tag{7.29}$$

for all $\omega \in \Omega$ for which $0 < \tau(\omega) < t$ and set

$$F_t = \chi_{\{\tau>t\}}.e^{-i\int_0^t V\circ X_s\,ds} + \chi_{\{0<\tau<t\}}.e^{-i\langle\!\langle\int_0^t V\circ X_s\,ds\rangle\!\rangle}. \tag{7.30}$$

Note that according to formula (7.18–7.19), the sets $\{\tau = 0\}$ and $\{\tau = t\}$ are $M_\Gamma^t$-null. The proof of the following statement is similar to that above.

**Theorem 7.7.** *Let $\Gamma$ be any $(2 \times 2)$ unitary matrix. The function $F_t$, $t > 0$, is a multiplicative functional and $F_t$ is $M_\Gamma^t$-integrable for each $t \geq 0$. Furthermore,*

$$e^{-it(A+\Gamma Q(V))} = \int_\Omega F_t\,dM_\Gamma^t. \tag{7.31}$$

### 7.8 The Feynman–Kac formula for singular potentials : nonzero mass

Let $A$ be the selfadjoint operator (7.7) and let $\beta$ be the hermitian matrix $\begin{pmatrix} 0 & -1 \\ -1 & 0 \end{pmatrix}$ and $m > 0$. Then $A + m\beta$ is a selfadjoint operator acting in $L^2(\mathbb{R}, \mathbb{C}^2)$. Let $S(t) = e^{-it(A+m\beta)}$ and suppose that $M_m^t$ is an $(S, Q)$-measure.

The matrix $m\beta$ is a bounded perturbation of $A$, so the Dyson series expansion [34, Theorem IX.2.1]

$$S(t) = e^{-it(A+m\beta)} = e^{-itA} + \sum_{n=1}^{\infty}(-im)^n R_n(t) \tag{7.32}$$

$$R_n(t) = \int_0^t \cdots \int_0^{s_2} e^{-i(t-s_n)A}\beta e^{-i(s_n-s_{n-1})A} \cdots \beta e^{-i(s_2-s_1)A}\beta e^{-is_1 A}\,ds_1 \ldots ds_n$$

converges absolutely in the operator norm of $\mathcal{L}(L^2(\mathbb{R}, \mathbb{C}^2))$. Put $R_0(t) = e^{-itA}$ for $t \geq 0$.

Denote the algebra generated by cylinder sets (3.7) by $\mathcal{Z}_t(\Omega)$. For each set $E$ of the form (3.7), set

$$M^t_{(n)}(E) = \sum_{\substack{n_0 + \cdots + n_k = n \\ n_0, \ldots, n_k \geq 0}} R_{n_k}(t - t_k) Q(B_k) R_{n_{k-1}}(t_k - t_{k-1}) \cdots Q(B_1) R_{n_0}(t_1). \quad (7.33)$$

The identities

$$\sum_{\substack{l + m = n \\ l, m \geq 0}} R_l(s) R_m(t) = R_n(s + t), \quad n = 0, 1, 2, \ldots, \quad s, t \geq 0,$$

ensure that (7.33) defines an additive operator valued set function $E \longmapsto M^t_{(n)}(E)$, $E \in \mathcal{Z}_t(\Omega)$. Furthermore, the sum $M^t_m(E) = \sum_{n=0}^{\infty} (-im)^n M^t_{(n)}(E)$ converges uniformly in the operator norm of $\mathcal{L}(L^2(\mathbb{R}, \mathbb{C}^2))$ for each $E \in \mathcal{Z}_t(\Omega)$. For each $n = 0, 1, 2 \ldots$, the operator valued measure $M^t_{(n)}$ is supported on the set of those paths belonging to $\Omega$ with exactly $n$ changes in direction in the interval $[0, t]$. This follows immediately from the Poisson process representation of T. Zastawniak [50] to which we now turn.

As mentioned above, the operator valued measures $M^t_m$ and $M^t_{(n)}$ may be viewed as acting on the space $\mathcal{M}(\mathbb{R}, \mathbb{C}^2)$ of $\mathbb{C}^2$-valued measures defined on the Borel $\sigma$-algebra $\mathcal{B}(\mathbb{R})$ of $\mathbb{R}$ with the total variation norm. We shall identify this action more carefully now.

In the case that $\mu \in \mathcal{M}(\mathbb{R}, \mathbb{C}^2)$ has a density $f : \mathbb{R} \to \mathbb{C}^2$ with respect to Lebesgue measure on $\mathbb{R}$, then according to formula (7.7), we have $(e^{-iAs} f)(x) = (f_1(x - s), f_2(x + s))$ for almost all $x \in \mathbb{R}$, because the expression is valid on the dense subspace $L^1 \cap L^2(\mathbb{R}, \mathbb{C}^2)$ of $L^1(\mathbb{R}, \mathbb{C}^2)$. For any continuous function $\phi : \mathbb{R} \to \mathbb{C}^2$ with compact support,

$$\langle \phi, e^{-iAs} f \rangle = \int_{\mathbb{R}} \phi_1(x) (e^{-iAs} f)_1(x) \, dx + \int_{\mathbb{R}} \phi_2(x) (e^{-iAs} f)_2(x) \, dx$$

$$= \int_{\mathbb{R}} \phi_1(x + s) f_1(x) \, dx + \int_{\mathbb{R}} \phi_2(x - s) f_2(x) \, dx = \langle e^{iAs} \phi, f \rangle,$$

with respect to the duality between $L^\infty(\mathbb{R}, \mathbb{C}^2)$ and $L^1(\mathbb{R}, \mathbb{C}^2)$.

For cylinder sets (3.7), it is therefore consistent to interpret formulae (3.8) and (7.32) with $S(s) = e^{-iAs}$ acting on a measure $\mu \in \mathcal{M}(\mathbb{R}, \mathbb{C}^2)$ for each $s \in \mathbb{R}$ by the formula

$$\langle \phi, S(s)\mu \rangle = \int_{\mathbb{R}} \phi_1(x) \, d[(S(s)\mu)_1](x) + \int_{\mathbb{R}} \phi_2(x) \, d[(S(s)\mu)_2](x)$$

$$= \int_{\mathbb{R}} \phi_1(x + s) \, d\mu_1(x) + \int_{\mathbb{R}} \phi_2(x - s) \, d\mu_2(x) = \langle S(-s)\phi, \mu \rangle \quad (7.34)$$

for all continuous functions $\phi : \mathbb{R} \to \mathbb{C}^2$ with compact support. The embedding of $L^1(\mathbb{R}, \mathbb{C}^2)$ in $\mathcal{M}(\mathbb{R}, \mathbb{C}^2)$ induces the direct sum of translations on $L^1(\mathbb{R}, \mathbb{C}^2)$ given above. Note that the group $S$ of operators is not actually a $C_0$-group acting on $\mathcal{M}(\mathbb{R}, \mathbb{C}^2)$.

Let $\delta_x$ denote the unit point mass at $x \in \mathbb{R}$. In the same spirit as above, the operator $e^{ipy} : L^1(\mathbb{R}) \to L^1(\mathbb{R})$ defined for all $f \in L^1(\mathbb{R})$ by $(e^{ipy}f)(x) = f(x+y)$, $x \in \mathbb{R}$, is translation by $y \in \mathbb{R}$. Then denoting the induced operator on $\mathcal{M}(\mathbb{R})$ by the same symbol, we have $e^{ipy}\delta_x = \delta_{x-y}$ for all $x \in \mathbb{R}$. The standard basis vectors of $\mathbb{C}^2$ are written as $e_1, e_2$.

**Proposition 7.8.** (Zastawniak [50]) *Let* $(\Xi, \mathcal{E}, P, \langle N_t \rangle_{t \geq 0})$ *be the standard Poisson process with intensity one. Let* $\tau_k(\xi) = \inf\{t \geq 0 : N_t(\xi) = k\}$ *be the* $k^{th}$ *jump time of* $\xi \in \Xi$. *For each* $t \geq 0$, $j = 1, 2$ *and* $x \in \mathbb{R}$, *let* $Y_t^{(x,j)} : \Xi \to \mathbb{R}$ *be the random variable defined by*

$$Y_t^{(x,j)} = x - (-1)^j \int_0^t (-1)^{N_s}\, ds.$$

*Then for each* $x \in \mathbb{R}$ *and* $j = 1, 2$ *and cylinder set (3.7), the* $\mathbb{C}^2$-*valued Borel measure* $M_{(n)}^t(X_{t_1} \in B_1, \ldots, X_{t_k} \in B_k)(\delta_x e_j)$ *equals*

$$B \longmapsto e^t \sum_{\substack{n_0 + \cdots + n_k = n \\ n_0, \ldots, n_k \geq 0}} \int_{E_{n_0, \ldots, n_k}(t_1, \ldots, t_k)} \chi_{B_k} \circ Y_{t_k}^{(x,j)} \cdots \chi_{B_1} \circ Y_{t_1}^{(x,j)} \times$$
$$\left[ e^{i(t - \tau_n)A} \beta e^{-i(\tau_n - \tau_{n-1})A} \cdots \beta e^{-i\tau_1 A}(\delta_x e_j) \right](B)\, dP, \quad B \in \mathcal{B}(\mathbb{R}).$$

*Here* $E_{n_0, \ldots, n_k}(t_1, \ldots, t_k)$ *is the set of* $\xi \in \Xi$ *with* $n_{j-1}$ *jumps at times greater than* $t_{j-1}$ *but less than or equal to* $t_j$ *for each* $j = 1, \ldots, k+1$.

Let $\Phi_{x,j} : \Xi \to \Omega$ denote the map sending $\xi \in \Xi$ onto the path $\omega : [0, \infty) \to \mathbb{R}$ defined by $\omega(s) = Y_s^{(x,j)}(\xi)$ for all $s \geq 0$. Let $P_{(n)}^t = \chi_{\{N_t = n\}}P$. Now let $\tau_k(\omega)$, $k = 1, \ldots, K(\omega)$ be the consecutive times where $\omega \in \Omega$ changes direction in the time interval $[0, t]$. It follows from Proposition 7.8 that

$$M_{(n)}^t(E)(\delta_x e_j) = e^t \int_E e^{i(t - \tau_n)A} \beta e^{-i(\tau_n - \tau_{n-1})A} \cdots \beta e^{-i\tau_1 A}(\delta_x e_j)\, d\left( P_{(n)}^t \circ \Phi_{x,j}^{-1} \right)$$

for all elements $E$ of the $\sigma$-algebra $\sigma(\mathcal{Z}_t(\Omega))$ generated by the algebra $\mathcal{Z}_t(\Omega)$ of cylinder sets. In other words, the $\mathcal{M}(\mathbb{R}, \mathbb{C}^2)$-valued measure $E \longmapsto M_{(n)}^t(E)(\delta_x e_j)$ has an $\mathcal{M}(\mathbb{R}, \mathbb{C}^2)$-valued density

$$e^t e^{-i(t - \tau_n)A} \beta e^{-i(\tau_n - \tau_{n-1})A} \cdots \beta e^{-i\tau_1 A}(\delta_x e_j) \tag{7.35}$$

with respect to the finite measure $P_{(n)}^t \circ \Phi_{x,j}^{-1}$.

We may take $\Omega = \bigcup_{x \in \mathbb{R}, j = 1,2} \Phi_{x,j}(\Xi)$. For each $v \in \mathbb{C}^2$, the $\mathcal{M}(\mathbb{R}, \mathbb{C}^2)$-valued measure $M_{(n)}^t(\,\cdot\,)[\delta_x v]$ is therefore concentrated on all paths $\omega$ with $\omega(0) = x$, velocity $\pm 1$ and exactly $n$ changes of direction in the interval $[0, t]$. Each operator $M_{(n)}^t(E)$ is also continuous for the weak topology of the duality between $\mathcal{M}(\mathbb{R}, \mathbb{C}^2)$ and the space $\mathcal{L}^\infty(\mathbb{R}, \mathbb{C}^2)$ of bounded Borel measurable functions. Because the set of all measures

$\delta_x v$ for $x \in \mathbb{R}$ and $v \in \mathbb{C}^2$ separates the vector space $\mathcal{L}^\infty(\mathbb{R}, \mathbb{C}^2)$, it follows that $M^t_{(n)}$ is concentrated on all paths $\omega$ with velocity $\pm 1$ and exactly $n$ changes of direction in the interval $[0, t]$, as mentioned above. The proof of the next lemma is somewhat technical and may be found in [9, Lemma 7.2].

**Lemma 7.9.** *Suppose that* $G^{(j)}_{u,v}$, $j = 0, \ldots, n$ *are bounded random variables such that* $G^{(j)}_{u,v}$ *is measurable with respect to* $\sigma\{X_s : u \leq s \leq v\}$ *for every* $0 \leq u < v \leq t$. *Let* $\omega_{-r}(s) = \omega(s - r)$ *for all* $s \geq r$.
*Then* $G^{(0)}_{0,\tau_1} \cdots G^{(n-1)}_{\tau_{n-1},\tau_n} G^{(n)}_{\tau_n,t}$ *is* $M^t_{(n)}$*-integrable and*

$$
\int_\Omega G^{(0)}_{0,\tau_1} \cdots G^{(n-1)}_{\tau_{n-1},\tau_n} G^{(n)}_{\tau_n,t} \, dM^t_{(n)}
$$
$$
= \int_0^t \cdots \int_0^{s_2} \left( \int_\Omega G^{(n)}_{s_n,t}(\omega_{-s_n}) \, dM^{t-s_n}(\omega) \right)
$$
$$
\times \beta \left( \int_\Omega G^{(n-1)}_{s_{n-1},s_n}(\omega_{-s_{n-1}}) \, dM^{s_n-s_{n-1}}(\omega) \right) \cdots
$$
$$
\beta \left( \int_\Omega G^{(1)}_{s_1,s_2}(\omega_{-s_1}) \, dM^{s_2-s_1}(\omega) \right) \beta \left( \int_\Omega G^{(0)}_{0,s_1}(\omega) \, dM^{s_1}(\omega) \right) ds_1 \cdots ds_n
$$

Let $\Gamma$ be the $(2 \times 2)$ unitary matrix (7.12). Suppose that $\beta = 0$ and $\eta\alpha = e^{-i\kappa_1}$, $\eta\bar\alpha = e^{-i\kappa_2}$ for numbers $0 \leq \kappa_j < 2\pi$. As above, denote the consecutive times where $\omega \in \Omega$ changes direction by $\tau_k(\omega)$, $k = 1, \ldots, K(\omega)$ with $(\tau_{K+1})(\omega) = t$. The shift operator $\theta_u$ maps $\omega \in \Omega$ into the path $\theta_u(\omega)(s) = \omega(s + u)$ defined for all $s \geq -u$.

**Lemma 7.10.** *There is essentially only one right continuous multiplicative functional* $F^\Gamma_t$ *on* $\Omega$ *satisfying (7.25) and (7.26) on* $\{\tau_1 > t\}$. *It is given* $M^t$*-almost everywhere by*

$$
F^\Gamma_t = \exp\left[ -i \sum_{k=0}^n \chi_{\{X_{\tau_k} X_{\tau_{k+1}} > 0\}} \left( \int_{\tau_k}^{\tau_{k+1}} V \circ X_s \, ds \right) \right. \tag{7.36}
$$
$$
\left. + \chi_{\{X_{\tau_k} X_{\tau_{k+1}} < 0\}} \left\langle \int_{\tau_k}^{\tau_{k+1}} V \circ X_s \, ds \right\rangle_\Gamma \right],
$$

*on* $\{N_t = n\}$ *where* $\tau_0 = 0$ *and* $\tau_{n+1} = t$. *The expression* $\left\langle \int_{\tau_k}^{\tau_{k+1}} V \circ X_s \, ds \right\rangle_\Gamma$ *is given by*

$$
\begin{cases} \gamma \left( \ln |\omega(\tau_k)| + \ln |\omega(\tau_{k+1})| \right) + \kappa_1, & \text{if } \omega'(s) = 1, \; \tau_k(\omega) < s < \tau_{k+1}(\omega), \\ \gamma \left( \ln |\omega(\tau_k)| + \ln |\omega(\tau_{k+1})| \right) + \kappa_2, & \text{if } \omega'(s) = -1, \; \tau_k(\omega) < s < \tau_{k+1}(\omega). \end{cases} \tag{7.37}
$$

*Furthermore, the multiplicative functional* $F^\Gamma_t$ *satisfies*

$$
F^\Gamma_t = F^\Gamma_{\tau_1}.(F^\Gamma_{\tau_2-\tau_1} \circ \theta_{\tau_1}) \cdots (F^\Gamma_{t-\tau_n} \circ \theta_{\tau_n}) \tag{7.38}
$$

$M^t$*-almost everywhere on* $\{N_t = n\}$.

*Proof.* Suppose first that $F_t^\Gamma$ is a multiplicative functional satisfying equations (7.25) and (7.26) on the set $\{\tau_1 > t\}$ of all paths with no change in direction before time $t$.

Then for every $0 < t_1 < \cdots < t_n < t$

$$F_t^\Gamma = F_{t_1}^\Gamma.(F_{t_2-t_1}^\Gamma \circ \theta_{t_1}) \cdots (F_{t-t_n}^\Gamma \circ \theta_{t_n})$$

a.e. on $\{N_t = n\}$, because $F_t^\Gamma$ is a multiplicative functional. Right-continuity ensures that we can replace $t_1, \ldots, t_n$ by the jump times $\tau_1, \ldots, \tau_n$ so that equation (7.38) holds.

Now suppose that $\omega \in \Omega$ and $\tau_k(\omega) = t_k$, $\tau_{k+1}(\omega) = t_{k+1}$ with $0 < t_k < t_{k+1} < t$, $k = 0, \ldots, n-1$, $\tau_{n+1} = t$. Then on the interval $[0, t_{k+1} - t_k]$, the sample path $\theta_{t_k}(\omega)$ is equal to the restriction of an element of $\{\tau_1 > t_{k+1} - t_k\}$ to $[0, t_{k+1} - t_k]$, so that by equations (7.25) and (7.26) we have

$$(F_{\tau_{k+1}-\tau_k}^\Gamma \circ \theta_{\tau_1})(\omega) = (F_{t_{k+1}-t_k}^\Gamma \circ \theta_{t_k})(\omega)$$

$$= X_{\{X_0 X_{t_{k+1}-t_k} > 0\}}(\theta_{t_k}(\omega)).e^{-i \int_0^{t_{k+1}-t_k} V \circ X_s(\theta_{t_k}(\omega))\,ds}$$

$$+ X_{\{X_0 X_{t_{k+1}-t_k} < 0\}}(\theta_{t_k}(\omega)).e^{-i\langle \int_0^{t_{k+1}-t_k} V \circ X_s(\theta_{t_k}(\omega))\,ds\rangle_\Gamma}$$

$$= X_{\{X_{t_k} X_{t_{k+1}} > 0\}}(\omega).e^{-i \int_{t_k}^{t_{k+1}} V \circ X_s(\omega)\,ds}$$

$$+ X_{\{X_{t_k} X_{t_{k+1}} < 0\}}(\omega).e^{-i\langle \int_{t_k}^{t_{k+1}} V \circ X_s(\omega)\,ds\rangle_\Gamma}$$

$$= X_{\{X_{\tau_k} X_{\tau_{k+1}} > 0\}}(\omega).e^{-i \int_{\tau_k}^{\tau_{k+1}} V \circ X_s(\omega)\,ds}$$

$$+ X_{\{X_{\tau_k} X_{\tau_{k+1}} < 0\}}(\omega).e^{-i\langle \int_{\tau_k}^{\tau_{k+1}} V \circ X_s(\omega)\,ds\rangle_\Gamma}.$$

Hence equation (7.36) follows by applying equation (7.38).

On the other hand, (7.38) follows from the definition (7.36). To check that (7.36) defines a multiplicative functional, it is enough to verify that equation (7.27) holds for almost all $\omega \in \{N_{t+s} = n\}$, for each $n = 1, 2, \ldots$ and on each set $\{\tau_{k-1} \le s < \tau_k\}$, $k = 1, \ldots, n$.

According to formula (7.36) and Lemma 5.1, we have

$$F_{\tau_k-\tau_{k-1}} \circ \theta_{\tau_k} = (F_{s-\tau_{k-1}} \circ \theta_{\tau_{k-1}}).(F_{\tau_k-s} \circ \theta_s)$$
$$F_t \circ \theta_s = (F_{\tau_k-s} \circ \theta_s).(F_{s+t-\tau_k} \circ \theta_{\tau_k})$$

on the set $\{N_{t+s} = n, \tau_{k-1} \le s < \tau_k\}$. Then

$$F_{s+t}^\Gamma = F_{\tau_1}^\Gamma.(F_{\tau_2-\tau_1}^\Gamma \circ \theta_{\tau_1}) \cdots (F_{s+t-\tau_n}^\Gamma \circ \theta_{\tau_n})$$
$$= F_{\tau_1}^\Gamma.(F_{\tau_2-\tau_1}^\Gamma \circ \theta_{\tau_1}) \cdots (F_{s-\tau_{k-1}} \circ \theta_{\tau_{k-1}})(F_{\tau_k-s} \circ \theta_s) \cdots (F_{s+t-\tau_n}^\Gamma \circ \theta_{\tau_n})$$
$$= F_s.(F_{\tau_k-s} \circ \theta_s).(F_{s+t-\tau_k}^\Gamma \circ \theta_{\tau_k})$$
$$= F_s.F_t \circ \theta_s.$$

Hence, formula (7.36) does indeed define a multiplicative functional.    □

The operator $H_0 +_\Gamma Q(V) := A +_\Gamma Q(V) + m\beta$ is interpreted as a bounded perturbation of the selfadjoint operator $A +_\Gamma Q(V)$, so it is itself selfadjoint. A careful study of the operator $H_0 +_\Gamma Q(V)$ in the Dirac representation is given in [6].

**Theorem 7.11.** *For every $t \geq 0$, the equality*

$$e^{-it(H_0+\Gamma Q(V))} = \int_{\Omega} F_t^{\Gamma}\, dM_m^t \tag{7.39}$$

*holds.*

*Proof.* It suffices to establish that for each $n = 1, 2, \ldots,$ the equality

$$\int_{\Omega} F_t^{\Gamma}\, dM_{(n)}^t = \int_0^t \cdots \int_0^{s_2} e^{-i(t-s_n)(A+\Gamma Q(V))} \beta e^{-i(s_n-s_{n-1})(A+\Gamma Q(V))} \cdots$$
$$\beta e^{-i(s_2-s_1)(A+\Gamma Q(V))} \beta e^{-is_1(A+\Gamma Q(V))}\, ds_1 \cdots ds_n \tag{7.40}$$

is valid, for then the operator coefficients of the Taylor expansion in powers of $m$ of the right and left-hand sides of equation (7.39) agree. The equality

$$\int_{\Omega} F_t^{\Gamma}\, dM_{(0)}^t = \int_{\Omega} F_t^{\Gamma}\, dM^t = e^{-it(A+\Gamma Q(V))}$$

corresponding to $n = 0$ is proved in Theorem 7.6.

Appealing to Theorem 7.6 again, we can write the right hand side of equation (7.40) as

$$\int_0^t \cdots \int_0^{s_2} \left( \int_{\Omega} F_{t-s_n}^{\Gamma}\, dM^{t-s_n} \right) \beta \left( \int_{\Omega} F_{s_n-s_{n-1}}^{\Gamma}\, dM^{s_n-s_{n-1}} \right) \cdots$$
$$\beta \left( \int_{\Omega} F_{s_2-s_1}^{\Gamma}\, dM^{s_2-s_1} \right) \beta F_{s_1}^{\Gamma}\, dM^{s_1}\, ds_1 \cdots ds_n.$$

By equation (7.38), we have

$$F_t^{\Gamma} = F_{\tau_1}^{\Gamma}.(F_{\tau_2-\tau_1}^{\Gamma} \circ \theta_{\tau_1}) \cdots (F_{t-\tau_n}^{\Gamma} \circ \theta_{\tau_n}),$$

so Lemma 7.9 shows that equation (7.40) holds. $\qquad\square$

We also state the following Feynman–Kac formula with respect to the measure $M_m^{t,\Gamma}$ associated with point interactions. The proof is similar to that above, except we replace the appeal to Theorem 7.6 by Theorem 7.7.

Let $\Gamma$ be an arbitrary $(2 \times 2)$ unitary matrix (4.2), $A_{\Gamma}$ the operator (7.7) with the boundary condition (7.11), $S(s) = e^{-is(A_{\Gamma}+m\beta)}$ for every $s \in \mathbb{R}$ and suppose that $M_m^{t,\Gamma}$ is the measure defined by formula (3.8). The multiplicative functional $F_t$ is defined by formulae (7.36) and (7.37), but with $\kappa_1 = \kappa_2 = 0$.

**Theorem 7.12.** *For every $t \geq 0$, the equality*

$$e^{-it(H_0+\Gamma Q(V))} = \int_{\Omega} F_t\, dM_m^{t,\Gamma} \tag{7.41}$$

*holds.*

**Remark 7.13.** a)  In the case that $\Gamma$ is a unitary diagonal matrix, the equality

$$F_t^\Gamma . M_m^t = F_t . M_m^{t,\Gamma}$$

holds for all $t \geq 0$.

b)  One point of view of 'Feynmanism' is that the evolution of the states of an interacting quantum system can be written as integrals $t \longmapsto \int_\Omega F_t \, dM^t, t \geq 0$, over path space $\Omega$. The process of 'renormalization' in perturbative quantum field theory is required to construct the multiplicative functional $\langle F_t \rangle_{t \geq 0}$, just as it is required to obtain the representations (7.39) and (7.41) for the Dirac equation with a Coulomb interaction in one space dimension. Because interactions in quantum field theory are never defined at a point, the process of renormalization is necessarily more involved. Nevertheless, the construction of a multiplicative functional with modulus one, as befits quantum physics, just requires the notion of *measurability*, see Section 10 below.

# 8  Integration with respect to unbounded set functions

Before moving on to integration with respect to the set functions $\langle M_\lambda^t \rangle_{\lambda \in \mathbb{C}_+}$ defined in Section 3, it is worthwhile looking at integration with respect to unbounded set functions which are less singular than the operator valued set function $M_\lambda^t$ with $\Im \lambda \neq 0$ and $t > 0$. In Subsection 8.2 below, the class of *regular* unbounded set functions is described. Additive set functions of this type arise as the finite dimensional distributions $M_{-i}^t \circ (X_{t_1} \otimes \cdots \otimes X_{t_n})^{-1}$ of the Schrödinger process for fixed times $0 \leq t_1 < \cdots < t_n \leq t$ and any $n = 1, 2, \ldots$, see Example 8.2 below.

The situation we are in with the set functions $\langle M_\lambda^t \rangle_{\lambda \in \mathbb{C}_+}$ is not unprecedented. Even to interpret the SDE

$$dX_t = f(X_t, t)dt + g(X_t, t)db_t,$$

it should be rewritten as an integral equation

$$X_t = X_0 + \int_0^t f(X_s, s) \, ds + \int_0^t g(X_t, t) \, db_t.$$

As is well known for Brownian motion $\langle b_t \rangle_{t \geq 0}$, with probability one, the additive set function defined by $[r, s) \longmapsto b_s - b_r$, for all $0 \leq r < s$ has infinite variation on every subinterval of $\mathbb{R}_+$. Nevertheless, stochastic integrals with respect to $db_t$ are now well-understood and may be viewed as integrals with respect to unbounded set functions.

The essential point is that for, say, a suitable square integrable semimartingale $X$, there exists a constant $C > 0$ and a $\sigma$-finite measure $\beta$ constructed from $X$ such that the bound

$$\left\| \int_0^t f \, dX \right\|_2 \leq \left( \int_{[0,t] \times \Omega} |f|^2 \, d\beta \right)^{1/2}$$

holds for all predictable simple functions $f$, see [38, Section 6.2] for an explanation of these terms and a far-reaching generalization of this idea. In the language of Subsection

8.1 below, the functional $\rho$ defined by $\rho(f) = \left(\int_{[0,t]\times\Omega} |f|^2 \, d\beta\right)^{1/2}$ is an *integrating gauge* for the $L^2$-valued integral $f \longmapsto \int_0^t f \, dX$ defined on the space of predictable simple functions $f$. As pointed out in [38, Remark 6.2.11, p. 471], the actual construction of the integrating gauge $\rho$ is generally nontrivial.

In this section, by an *unbounded* set function, we mean an additive scalar or vector valued set function $m$ defined on a semi-algebra $S$ of subsets of a set $\Sigma$ such that the unique additive extension $\tilde{m}$ of $m$ to the algebra $[S]$ generated by $S$ has unbounded range, that is, $\sup\{\|\tilde{m}(E)\| : E \in [S]\} = \infty$. Here $\|\cdot\|$ denotes the norm of the range space.

The simplest example of this type is provided by the product of two noncommuting spectral measures.

**Example 8.1.** Let $Q : \mathcal{B}(\mathbb{R}) \to \mathcal{L}(L^2(\mathbb{R}))$ be the spectral measure of multiplication by characteristic functions and let $\mathcal{F} : L^2(\mathbb{R}) \to L^2(\mathbb{R})$ be the Fourier transform defined for each $f \in L^1 \cap L^2(\mathbb{R})$ by $(\mathcal{F}f)(\xi) = \hat{f}(\xi) := \int_{\mathbb{R}} e^{-ix\xi} f(x) \, dx$, for almost all $\xi \in \mathbb{R}$. Let $P : \mathcal{B}(\mathbb{R}) \to \mathcal{L}(L^2(\mathbb{R}))$ be the spectral measure defined by $P = \mathcal{F}^{-1} Q \mathcal{F}$. Then $Q$ is the spectral measure associated with the position operator for a quantum particle on the line and $P$ is the spectral measure of the momentum operator.

The operator valued set function $m : A \times B \longmapsto Q(A)P(B)$, $A, B \in \mathcal{B}(\mathbb{R})$ is unbounded in the above sense: the collection of all product sets $A \times B$ with $A, B \in \mathcal{B}(\mathbb{R})$ is a semi-algebra $S$ of subsets of $\mathbb{R}^2$ and the set function $m$ is unbounded on the algebra $[S]$ generated by $S$. For $\phi, \psi \in L^2(\mathbb{R})$, we have

$$\langle m(A \times B)\phi, \psi \rangle = (2\pi)^{-1} \lim_{n\to\infty} \int_{A\cap[-n,n]} \left( \int_{B\cap[-n,n]} e^{ixy} \hat{\phi}(y) \, dy \right) \overline{\psi}(x) \, dx.$$

Let $m_{\phi,\psi}(A \times B) = \langle m(A \times B)\phi, \psi \rangle$ for all Borel subsets $A, B$ of $\mathbb{R}$. Note that if the Borel set $B$ is kept fixed, then the set function $A \longmapsto m_{\phi,\psi}(A \times B)$, $A \in \mathcal{B}(\mathbb{R})$, is $\sigma$-additive and the analogous property holds if $A$ is kept fixed, that is, $m_{\phi,\psi}$ is *separately* $\sigma$-additive.

The total variation $\|m_{\phi,\psi}\|$ of the set function $m_{\phi,\psi}$ on the algebra $S$ is $\|m_{\phi,\psi}\| = \|\hat{\phi}\|_1 \|\psi\|_1/(2\pi)$. The functions $\phi$ and $\psi$ are just elements of $L^2(\mathbb{R})$, so $\|m_{\phi,\psi}\|$ may be infinite. The variation of $m_{\phi,\psi}$ is finite on compact product sets. Nevertheless, $m$-integrable functions $a : \mathbb{R}^2 \to \mathbb{C}$ correspond to bounded pseudodifferential operators $a(x, D)$ acting on $L^2(\mathbb{R})$. The subject of harmonic analysis on phase space and the Weyl functional calculus [17] is intimately connected with the mathematics of quantum theory.

**Example 8.2.** For another related example, let $S(t) = e^{it\Delta/2}$ for all $t \in \mathbb{R}$ on $L^2(\mathbb{R})$. Let $g_s(x) = e^{-isx^2/2}$ for all $x, s \in \mathbb{R}$. Then the $(S, Q)$-set function on $L^2(\mathbb{R})$ is given by

$$M^t(E) = S(t - t_n)Q(B_n)S(t_n - t_{n-1}) \cdots Q(B_1)S(t_1)$$
$$= P(g_{t-t_n})Q(B_n)P(g_{t_n-t_{n-1}}) \cdots Q(B_1)P(g_{t_1})$$

for each cylinder set $E$ of the form (3.7) and with the spectral measure $P$ as defined in Example 8.1. Then the set function

$$(B_1 \times \cdots \times B_n) \longmapsto M^t(\{X_{t_1} \in B_1, \ldots, X_{t_n} \in B_n\}), \quad B_1, \ldots, B_n \in \mathcal{B}(\mathbb{R}),$$

is separately $\sigma$-additive in the strong operator topology, but unbounded on the algebra generated by product sets. Unbounded set functions of this form have been studied in [32]. Integration with respect to them uses the property that restricted to the intersection $S_K$ of products of Borel sets to a fixed compact product set $K$, the additive set function is bounded on the algebra $[S_K]$, so there is no difficulty defining a natural class of integrable cylinder functions, including products of finitely many bounded functions.

By contrast, the variation of the additive set function $\mu^t_{\lambda, \phi, \psi}$ considered in Example 5.1 takes only the values $+\infty$ and 0.

## 8.1 Integration structures

The notion of an *integration structure* was introduced in [36] to encompass integration with respect to both measures and certain examples of unbounded set functions. The slight generalisation of this scheme we need is as follows. Let $\mathcal{H}$ be a collection of functions defined on a nonempty set $\Omega$. We suppose that $\mathcal{H}$ at least contains the function 0 identically zero on $\Omega$. A *gauge* on $\mathcal{H}$ is a function $\rho : \mathcal{H} \to [0, \infty)$ such that $\rho(0) = 0$. A family $\Gamma$ of gauges on $\mathcal{H}$ is said to be *collectively integrating* if the following condition holds: if $f \in \mathcal{H}$, $c_i \in \mathbb{C}$ and $f_i \in \mathcal{H}$, $i = 1, 2, \ldots$, have the property that

$$\sum_{i=1}^{\infty} |c_i| \rho(f_i) < \infty \quad \text{for all} \quad \rho \in \Gamma, \tag{8.1}$$

and for every $\omega \in \Omega$ such that

$$\sum_{i=1}^{\infty} |c_i| |f_i(\omega)| < \infty \tag{8.2}$$

it follows that $f(\omega) = \sum_{i=1}^{\infty} c_i f_i(\omega)$, then satisfies

$$\rho(f) \le \sum_{i=1}^{\infty} |c_i| \rho(f_i) \quad \text{for all} \quad \rho \in \Gamma. \tag{8.3}$$

If $\Gamma$ is just a singleton set $\{\rho\}$, then $\rho$ is called an *integrating gauge*.

Following the theory of [36], suppose that $\Gamma$ is a collectively integrating family of gauges. Let us denote by $\mathcal{L}(\mathcal{H}, \Gamma)$ the vector space of all functions $f : \Omega \to \mathbb{C}$ for which there exist numbers $c_i \in \mathbb{C}$, and functions $f_i \in \mathcal{H}$, $i = 1, 2, \ldots$ such that (8.1) holds, and for every $\omega \in \Omega$ such that (8.2) holds, it follows that $f(\omega) = \sum_{i=1}^{\infty} c_i f_i(\omega)$. A function belonging to $\mathcal{L}(\mathcal{H}, \Gamma)$ is said to be $(\mathcal{H}, \Gamma)$-*integrable*. If the collection of functions $\mathcal{H}$ is understood from the context, we shall merely say that a function is $\Gamma$-*integrable*. The vector space $\mathbf{sim}(\mathcal{H})$ of all finite linear combinations of functions belonging to $\mathcal{H}$ is clearly contained in $\mathcal{L}(\mathcal{H}, \Gamma)$.

For each $\tau \in \Gamma$, and $f \in \mathcal{L}(\mathcal{H}, \Gamma)$, the number $q_\tau(f)$ is defined by

$$q_\tau(f) = \inf \sum_{i=1}^{\infty} |c_i|\tau(f_i); \tag{8.4}$$

the infimum is taken over all numbers $c_i \subset \mathbb{C}$, and functions $f_i \in \mathcal{H}$, $i = 1, 2, \ldots$ such that (8.1) holds, and with the property that for every $\omega \in \Omega$ such that (8.2) holds, the equality $f(\omega) = \sum_{i=1}^{\infty} c_i f_i(\omega)$ is true. The condition that a family $\Gamma$ of gauges on $\mathcal{H}$ is collectively integrating may be reformulated in terms of the condition $q_\tau(f) = \tau(f)$ for all $f \in \mathcal{H}$ and $\tau \in \Gamma$. It follows that for each $\tau \in \Gamma$, the functional $q_\tau : \mathcal{L}(\mathcal{H}, \Gamma) \to [0, \infty)$ is a seminorm on the vector space $\mathcal{L}(\mathcal{H}, \Gamma)$.

If $\rho : \mathcal{H} \to [0, \infty)$ is any gauge on $\mathcal{H}$, not necessarily an integrating gauge, then the gauge $q_\rho$ defined by (8.4) is an *integrating gauge* defined on the vector space $\mathcal{L}(\mathcal{H}, \rho)$ of all functions $f$ such that $q_\rho(f) < \infty$: if $f$ is a function defined on $\Omega$, and if $c_i \in \mathbb{C}$ and $f_i \in \mathcal{L}(\mathcal{H}, \rho)$, $i = 1, 2, \ldots$ have the property that

$$\sum_{i=1}^{\infty} |c_i|q_\rho(f_i) < \infty, \tag{8.5}$$

and for every $\omega \in \Omega$ such that (8.2) holds, it follows that $f(\omega) = \sum_{i=1}^{\infty} c_i f_i(\omega)$, then $f \in \mathcal{L}(\mathcal{H}, \rho)$ and

$$q_\rho(f) \leq \sum_{i=1}^{\infty} |c_i|q_\rho(f_i). \tag{8.6}$$

A function $f$ for which $q_\rho(f) = 0$ for each $\rho \in \Gamma$ is termed a $\Gamma$-*null function*. It is then possible to form the quotient space $L^1(\mathcal{H}, \Gamma)$ of $\mathcal{L}(\mathcal{H}, \Gamma)$ with the vector space of all null functions so that each of the seminorms $q_\rho$, $\rho \in \Gamma$ induces a corresponding seminorm, also denoted by $q_\rho$, on the quotient space $L^1(\mathcal{H}, \Gamma)$. The collection of seminorms $q_\rho$, $\rho \in \Gamma$ defines a locally convex Hausdorff topology $\tau_\Gamma$ on $L^1(\mathcal{H}, \Gamma)$. It turns out that the image of $\mathbf{sim}(\mathcal{H})$ via the quotient map is dense in the locally convex space $L^1(\mathcal{H}, \Gamma)$.

The following statement provides a convenient condition for guaranteeing that a function is $\rho$-null for an integrating gauge $\rho$. The proof is given in [36, Proposition 2.2], which is reproduced here to give a flavor of how the passage from measures to integrating gauges works.

**Proposition 8.3.** *Let $\rho$ be an integrating gauge on $\mathcal{H}$. A function $f$ is $\rho$-null if and only if there exists functions $h_j \in \mathcal{L}(\mathcal{H}, \rho)$, $j = 1, 2, \ldots$ such that $\sum_{j=1}^{\infty} q_\rho(h_j) < \infty$ and $\sum_{j=1}^{\infty} |h_j(\omega)| = \infty$ for all $\omega \in \Omega$ such that $f(\omega) \neq 0$.*

*Proof.* First, suppose that $q_\rho(f) = 0$. Let $X = \{\omega \in \Omega : f(\omega) \neq 0\}$. Then condition (8.5) holds for the functions $f_i = f$ and the numbers $c_i = 1$, $i = 1, 2, \ldots$; the sum is zero. Moreover $\chi_X(\omega) = \sum_{i=1}^{\infty} f_i(\omega)$ for all $\omega \in \Omega$ for which (8.2) holds, namely, for all $\omega \notin X$. It follows by the inequality (8.6) that $\chi_X \in \mathcal{L}(\mathcal{H}, \rho)$ and $\chi_X$ is $\Gamma$-null. The functions $h_j = \chi_X$, $j = 1, 2, \ldots$ have the required properties.

Now let $h_j \in \mathcal{L}(\mathcal{H}, \rho)$, $j = 1, 2, \ldots$ be functions with the property mentioned above and set $f_{2j} = h_j$ and $f_{2j-1} = -h_j$ for all $j = 1, 2 \ldots$. Then $\sum_{j=1}^{\infty} q_\rho(f_j) < \infty$, and for every $n = 1, 2, \ldots$, $f(\omega) = \frac{1}{n}\sum_{j=1}^{\infty} f_j(\omega) = 0$ for all $\omega \in \Omega$ such that $\sum_{j=1}^{\infty} |f_j(\omega)| < \infty$. It follows from (8.6), that the inequality $q_\rho(f) \leq \frac{1}{n}\sum_{j=1}^{\infty} q_\rho(f_j)$ is valid. This is true for all $n = 1, 2, \ldots$ only if $q_\rho(f) = 0$.     $\square$

It follows that $L^1(\mathcal{H}, \rho)$ is complete. It turns out that the problem of the quasi-completeness of the space $\langle L^1(\mathcal{H}, \Gamma), \tau_\Gamma \rangle$ under more general conditions is more subtle (see, for example, [35]).

The collectively integrating family of gauges $\Gamma$ is said to be *integrating* for a linear map $m : \mathbf{sim}(\mathcal{H}) \to \mathbb{C}$ if there exists a number $C > 0$, and gauge $\rho \in \Gamma$ such that for every $f \in \mathbf{sim}(\mathcal{H})$, $|m(f)| \leq Cq_\rho(f)$. If $\Gamma$ is integrating for $m$, then it follows that $m$ is the restriction to $\mathbf{sim}(\mathcal{H})$ of a $\tau_\Gamma$-continuous linear functional $\tilde{m} : L^1(\mathcal{H}, \Gamma) \to \mathbb{C}$. When there is no danger of confusion, $\tilde{m}(f)$ will be denoted by $m(f)$ for any $f \in L^1(\mathcal{H}, \Gamma)$. We will also write

$$\int_\Omega f \, dm, \quad \int_\Omega f(\omega) \, dm(\omega)$$

for $m(f)$. The triple $\langle L^1(\mathcal{H}, \Gamma), \Gamma, m \rangle$ is called an *integration structure*.

The following example shows that the treatment of Feynman path integrals in [1], or more generally, oscillatory integrals, may be viewed as an integration structure. This is already apparent from [1], but is worth mentioning here. Integration considered in the next section leads to a larger class of integrable functions.

**Example 8.4.** Let $H$ be a separable Hilbert space with inner product $\langle \cdot, \cdot \rangle$. The Fourier-Stieltjes transform of a signed Borel measure $\mu$ on $H$ is the continuous function $\hat{\mu} : H \to \mathbb{C}$ defined by

$$\hat{\mu}(\xi) := \int_H e^{i\langle \xi, h \rangle} \, d\mu(h), \quad \xi \in H.$$

Let $\mathcal{F}(H)$ denote the commutative Banach algebra of all such Fourier-Stieltjes transforms $\hat{\mu}$ with the norm $\|\hat{\mu}\|_{\mathcal{F}(H)}$ of the element $\hat{\mu}$ of $\mathcal{F}(H)$ equal to the total variation $\|\mu\|$ of the signed Borel measure $\mu$. Multiplication in $\mathcal{F}(H)$ is pointwise so that $\hat{\mu}.\hat{\nu} = \widehat{(\mu * \nu)}$.

The *Fresnel integral* $\mathcal{F}(\hat{\mu})$ of $\hat{\mu} \in \mathcal{F}(H)$ is defined by

$$\mathcal{F}(\hat{\mu}) := \int_H e^{-\frac{i}{2}\|h\|^2} \, d\mu(h).$$

Clearly $|\mathcal{F}(\hat{\mu})| \leq \|\mu\| = \|\hat{\mu}\|_{\mathcal{F}(H)}$, so the Banach algebra norm $\| \cdot \|_{\mathcal{F}(H)}$ of $\mathcal{F}(H)$ is integrating for the Fresnel integral $\mathcal{F} : \mathcal{F}(H) \to \mathbb{C}$.

To check that $\| \cdot \|_{\mathcal{F}(H)}$ is actually an integrating gauge, suppose that (8.1) holds for $f_j = \hat{\mu}_j \in \mathcal{F}(H)$, $c_j \in \mathbb{C}$, $j = 1, 2, \ldots$ and $\rho = \| \cdot \|_{\mathcal{F}(H)}$. Set $\mu = \sum_{j=1}^{\infty} c_j \mu_j$. The sum converges in the total variation norm and $\|\mu\| \leq \sum_{j=1}^{\infty} |c_j| \|\mu_j\|$. Then $\hat{\mu} = \sum_{j=1}^{\infty} c_j \hat{\mu}_j$ uniformly on $H$, so that (8.2) holds *everywhere* and $\|\hat{\mu}\|_{\mathcal{F}(H)} \leq \sum_{j=1}^{\infty} |c_j| \|\hat{\mu}_j\|_{\mathcal{F}(H)}$.

The connection between the Fresnel integral as described above and the Feynman path integral of functions defined on the space of continuous paths is treated in [33, Section 20.1, pp. 613–636].

**Example 8.5.** Suppose that $\mathcal{E}$ is a nonempty semi-ring of subsets of a set $\Sigma$. We shall also denote the collection of characteristic functions of elements of $\mathcal{E}$ by the same symbol. If $\mu : \mathcal{E} \to \mathbb{C}$ is an additive set function such that the variation $|\mu|(E)$ of $\mu$ on every set $E \in \mathcal{E}$ is finite, then $\rho(f) = \int_\Sigma |f| \, d|\mu|$ defines an integrating gauge for the linear mapping $f \longmapsto \int_\Sigma f \, d\mu$, $f \in \mathrm{sim}(\mathcal{E})$ if and only if $\mu$ (or, equivalently, $|\mu|$) is $\sigma$-additive on $\mathcal{E}$ [36, Proposition 2.13]. In this case, $L^1(\mathcal{E}, \rho) = L^1(|\mu|)$. The unique $\sigma$-additive extension of the variation $|\mu|$ of $\mu$ to the $\sigma$-ring generated by $\mathcal{E}$ has also been written as $|\mu|$.

The set function $\mu = m_{\phi, \psi}$ of Example 8.1 is actually $\sigma$-additive on the semi-ring $\mathcal{E}$ of all bounded product sets in $\mathbb{R}^2$. Then $|\mu| = (2\pi)^{-1}(|\hat{\phi}| . |\psi|)\lambda$ with $\lambda$ Lebesgue measure on $\mathbb{R}^2$. A less restrictive integration structure sufficient to encompass pseudodifferential operators $a(x, D)$ is given in Definition 8.10 below.

An efficient method for constructing an integrating gauge for additive set functions by using an auxiliary measure follows. Let $S$ be a semi-algebra of subsets of a set $\Omega$ and suppose that $m : S \to \mathbb{C}$ is a bounded additive set function. Let $ba(S)$ be the family of bounded additive set functions on the semi-algebra $S$ endowed with the uniform norm over $S$. Let $\mathrm{sim}(S)$ be the collection of all finite linear combinations of characteristic functions of elements of $S$. Then the bounded additive set function $fm : S \to \mathbb{C}$ is defined by linearity, in the obvious way, for each $f \in \mathrm{sim}(S)$. The set function $m$ is said to be *closable* with respect to a finite measure $\mu$ defined on the $\sigma$-algebra $\sigma(S)$ generated by $S$ if the closure of the graph $\{(f, fm) : f \in \mathrm{sim}(S)\}$ of the integration map $f \longmapsto fm$, $f \in \mathrm{sim}(S)$ in the product space $L^1(\mu) \times ba(S)$ is the graph of a function. Then $\rho(f) = \mu(|f|) + \sup_{A \in S} |fm(A)|$, $f \in \mathrm{sim}(S)$ is integrating for $m$. In the next subsection, this idea is developed in the situation where there is a natural family of closing measures associated with $m$.

## 8.2 Regular set functions

Let $S$ be a semi-algebra of subsets of a Hausdorff topological space $\Omega$. We also make the technical assumption that for each compact set $K$ belonging to the algebra $a(S)$ generated by $S$, there exists a dense subspace $X_K$ of the space $C(K)$ of continuous functions such that each function $f \in X_K$ is measurable with respect to the $\sigma$-algebra $\sigma(S)$ generated by $S$. This condition is satisfied, for example, when $S$ is the semi-algebra of all products of Borel sets in the Cartesian product of finitely many Hausdorff topological spaces. Take $X_K$ to be the vector space of all linear combinations of continuous product functions defined on $K$; the algebra $X_K$ is dense in $C(K)$ by the Stone-Weierstrass theorem. This is the principal environment for the class of unbounded set functions considered in this section. The assumption takes care of the fact that the Borel $\sigma$-algebra of a product space needs to be the product of the Borel $\sigma$-algebras of the component spaces. A Radon measure on a Hausdorff topological space is taken in the sense of L. Schwartz [43].

**Definition 8.6.** Let $\mathcal{K}$ be a nonempty family of compact subsets of $\Omega$ belonging to $a(S)$. A bounded additive set function $m : S \to \mathbb{C}$ is said to be $\mathcal{K}$-*regular* if the following two conditions are satisfied:

(i) for all $K \in \mathcal{K}$, the set function $A \longmapsto m(A)$, for $A \in \mathcal{S}$ and $A \subseteq K$ is the restriction of a complex valued Radon measure $m_K$ on $K$, and,

(ii) for all $A \in \mathcal{S}$ and $\epsilon > 0$, there exists $K \in \mathcal{K}$ such that $K \subseteq A$ and $|m(A) - m(K)| < \epsilon$.

The complex Radon measure $m_K$ in (i) is unique because it is determined by the subspace $X_K$ mentioned above. Moreover, if $K_1, K_2 \in \mathcal{K}$ and $A \subseteq K_1 \cap K_2$ is a Borel set, then $m_{K_1}(A) = m_{K_2}(A)$. Clearly, linear combinations of $\mathcal{K}$-regular set functions are $\mathcal{K}$-regular. If $m$ is $\mathcal{K}$-regular and $m(K) = 0$ for all $K \in \mathcal{K}$, then $m(A) = 0$ for all $A \in \mathcal{S}$.

**Remark 8.7.** (i) If $m$ is a $\mathcal{K}$-regular set function and $m(\mathcal{S}) \subseteq [0, \infty)$, then $m$ is actually the restriction to $\mathcal{S}$ of a nonnegative measure defined on $\sigma(\mathcal{S})$.

(ii) If $m$ is a $\mathcal{K}$-regular set function and $m$ is bounded on the algebra $a(\mathcal{S})$ of sets generated by the family $\mathcal{S}$, then $m$ is actually the restriction to $\mathcal{S}$ of a complex measure defined on $\sigma(\mathcal{S})$, for, if $|m|$ denotes the variation of $m$ on $a(\mathcal{S})$, then $|m|(\Omega) < \infty$ and for any set $A \in a(\mathcal{S})$ and any $\epsilon > 0$, there exist pairwise disjoint subsets $B_j \in \mathcal{S}$ of $A$ such that $0 \leq |m|(A) - \sum_{j=1}^{n} |m|(B_j) < \epsilon/2$. Now choose $K_j \in \mathcal{K}$ such that $K_j \subseteq B_j$ for $j = 1, \ldots, n$ and $\sum_{j=1}^{n} |m(B_j) - m(K_j)| < \epsilon/2$. Then $0 \leq |m|(A) - \sum_{j=1}^{n} |m|(K_j) < \epsilon$, so the set $\cup_{j=1}^{n} K_j$ is a compact set approximating $|m|(A)$ from the interior. Then $|m|$ is $\sigma$-additive [43, p. 51], so $m$ is also $\sigma$-additive.

(iii) If $X_1, \ldots, X_n$ are locally compact Hausdorff spaces, $\mathcal{S} = \mathcal{B}(X_1) \times \cdots \times \mathcal{B}(X_n)$ and $\mathcal{K}$ is the family of all compact product sets, then an additive set function $m : \mathcal{S} \to \mathbb{C}$ with bounded range is $\mathcal{K}$-regular if and only if it is a *Radon polymeasure* [32]. In particular, the set function $m_{\phi, \psi}$ of Example 8.2 is a Radon bimeasure.

Let $m$ be a $\mathcal{K}$-regular set function and let $\Omega_m$ be the collection of all points $x \in \Omega$ for which there exists an open neighborhood $V_x$ of $x$ such that

(i) for all compact $C \subseteq V_x$, there exists $K \in \mathcal{K}$ such that $C \subseteq K \subseteq V_x$, and
(ii) $\sup\{|m_K|(K) : K \in \mathcal{K}, K \subseteq V_x\} < \infty$.

It follows that $\Omega_m$ is an open subset of $\Omega$. The complement $\Omega \setminus \Omega_m$ of $\Omega_m$ may be viewed as the set of singularities of the additive set function $m$. It is worth noting that if $\Omega_m \neq \emptyset$, then there is a natural measure $\mu$ associated with $\mathcal{K}$-regular set function $m$.

**Proposition 8.8.** *If $\Omega_m \neq \emptyset$, then there exists a unique Radon measure $\mu$ on $\Omega_m$ such that $\mu(A) = |m|(A)$ for all $A \in \mathcal{S}$ such that $A \subseteq \Omega_m$.*

*Proof.* For every $K \in \mathcal{K}$ such that $K \subseteq \Omega_m$, let $j_K : K \to \Omega$ be the inclusion of $K$ in $\Omega$ and let $\mu_K$ be the variation of the signed Radon measure $m_K$. Then the image $\mu_K \circ j_K^{-1}$ of $\mu_K$ in $\Omega$ is a Radon measure. Let $\mu$ be the least upper bound of the family $\mathcal{R}$ of all Radon measures $\mu_K \circ j_K^{-1}$ with $K \in \mathcal{K}$ such that $K \subseteq \Omega_m$. By virtue of conditions (i) and (ii) and [43, Proposition 7, p56], $\mathcal{R}$ is bounded above, so that the least upper bound $\mu$ exists.

If $A \in \mathcal{S}$ and $A \subseteq \Omega$, then for all $\epsilon > 0$, there exists $K \in \mathcal{K}$ such that $K \subseteq A$ and $|m(A) - m(K)| < \epsilon$. Moreover, if $|m|(A) < \infty$, then there exist pairwise disjoint sets $K_j \in \mathcal{K}$, $j = 1, \ldots, n$ such that $K_j \subseteq A$ and $|m|(A) - |m|(\bigcup_{j=1}^n K_j) < \epsilon$. If $|m|(A) = \infty$, then the sets $K_j \in \mathcal{K}$, $j = 1, \ldots, n$ may be chosen so that the number $|m|(\bigcup_{j=1}^n K_j)$ is arbitrarily large. To prove the equality $\mu(A) = |m|(A)$ for all $A \in \mathcal{S}$ such that $A \subseteq \Omega_m$, it is therefore sufficient to establish that $\mu(K) = |m|(K)$ for all $K \in \mathcal{K}$ such that $K \subseteq \Omega_m$.

Let $K \in \mathcal{K}$. The variation $\mu_K$ of the signed Radon measure $m_K$ satisfies the equality $\mu_K(S) = |m|(S)$ for all $S \in \mathcal{S}$ such that $S \subset K$. Moreover, $\sup\{\mu_J(K \cap J) : J \in \mathcal{K}\} = \mu_K(K)$ by virtue of the consistency of the Radon measures $m_K$, $K \in \mathcal{K}$. By [43, Proposition 7, p56], $\mu(K) = \mu_K(K)$, so $\mu(K) = |m|(K)$.

To prove uniqueness, suppose that $\mu_1, \mu_2$ are two Radon measures on $\Omega_m$ such that $\mu_1(A) = \mu_2(A) = |m|(A)$ for all $A \in \mathcal{S}$ such that $A \subset \Omega_m$. It is enough to show that for every compact subset $K$ of $\Omega_m$, $\mu_1(K) = \mu_2(K)$. Now each compact subset $K$ of $\Omega_m$ is contained in a finite union of open sets $V_1, \ldots, V_n$ satisfying (i) and (ii), so by the inner-regularity of $\mu_1$ and $\mu_2$, for every $\epsilon > 0$, there exist compact subsets $C_1 \subseteq V_1, \ldots, C_n \subseteq V_n$ such that $\mu_1(V_j \backslash C_j) < \epsilon/(2n)$ and $\mu_2(V_j \backslash C_j) < \epsilon/(2n)$ for all $j = 1, \ldots, n$. According to condition (i), there exist sets $K_j \in \mathcal{K}$ such that $C_j \subseteq K_j \subseteq V_j$, $j = 1, \ldots, n$. It follows that $\mu_1(K \backslash (\bigcup_{j=1}^n K_j)) < \epsilon/2$ and $\mu_2(K \backslash (\bigcup_{k=1}^n K_j)) < \epsilon/2$. The quality $\mu_1(A) = \mu_2(A)$ holds for all $A \in \mathcal{S}$ such that $A \subseteq \bigcup_{j=1}^n K_j$, so $\mu_1(f) = \mu_2(f)$ for all continuous functions $f : \bigcup_{j=1}^n K_j \to \mathbb{R}$ and so $\mu_1(A) = \mu_2(A)$ for all Borel sets $A \subseteq \bigcup_{j=1}^n K_j$. In particular, $\mu_1((\bigcup_{j=1}^n K_j) \cap K) = \mu_2((\bigcup_{j=1}^n K_j) \cap K)$. Then $|\mu_1(K) - \mu_2(K)| \leq \mu_1(K \backslash (\bigcup_{j=1}^n K_j)) + \mu_2(K \backslash (\bigcup_{j=1}^n K_j)) < \epsilon$. It follows that $\mu_1(K) = \mu_2(K)$. □

The Banach space $L^1(\mu)$ of the Lebesgue integration theory for the measure $\mu$ defined above is often too restrictive to take as class of $m$-integrable functions. For example, uniformly bounded, Borel measurable product functions $f_1 \otimes f_2$ may not be integrable with respect to the variation $|m_{\phi,\psi}|$ of the Radon bimeasure $m_{\phi,\psi}$ of Example 8.1 although $m_{\phi,\psi}(f_1 \otimes f_2) = (Q(f_1)P(f_2)\phi, \psi)$ makes perfect sense.

If $\Omega_m = \Omega$, then the singularities of our unbounded set function are 'at infinity'. The following example shows that a bimeasure can be $\mathcal{K}$-regular with respect to some family $\mathcal{K}$ of compact sets without being a Radon bimeasure and that singularities can easily be concentrated on some proper closed subset of $\Omega$.

**Example 8.9.** Let $D = \frac{1}{i}\frac{d}{dx}$ be the self-adjoint operator associated with the group of translations in $L^2(\mathbb{R})$. Then $\mathrm{sgn}(D)$ is the operator defined by the operational calculus for self-adjoint operators in $L^2(\mathbb{R})$ associated with the signum function sgn. The operator $\mathrm{sgn}(D)$ is precisely the *Hilbert transform* $H$ given by

$$(H\phi)(x) = \frac{i}{\pi} \lim_{\epsilon \to 0} \int\limits_{|x-y|>\epsilon} \frac{\phi(y)}{x - y} dy$$

for all $\phi \in L^2(\mathbb{R})$ and almost all $x \in \mathbb{R}$ [45, p54].

Let $\mathcal{B}(\mathbb{R})$ be the family of all Borel sets in $\mathbb{R}$. As noted previously the collection of all sets $A \times B$ with $A \in \mathcal{B}(\mathbb{R})$ and $B \in \mathcal{B}(\mathbb{R})$ is denoted by $\mathcal{B}(\mathbb{R}) \times \mathcal{B}(\mathbb{R})$. The spectral measure $Q : \mathcal{B}(\mathbb{R}) \to L^2(\mathbb{R})$ is defined by $Q(B)\phi = \chi_B \phi$ for all $B \in \mathcal{B}(\mathbb{R})$ and $\phi \in L^2(\mathbb{R})$. Let $\phi \in L^2(\mathbb{R})$ be nonzero and define the map $m : \mathcal{B}(\mathbb{R}) \times \mathcal{B}(\mathbb{R}) \to \mathbb{C}$ by $m(A \times B) = (Q(B)HQ(A)\phi, \phi)$ for every $A \in \mathcal{B}(\mathbb{R})$ and $B \in \mathcal{B}(\mathbb{R})$. The map $m$ has a unique additive extension to the algebra generated by $\mathcal{B}(\mathbb{R}) \times \mathcal{B}(\mathbb{R})$. Then $m$ is separately $\sigma$-additive but the variation $|m|$ of $m$ is not the restriction of a Radon measure on the product space $\mathbb{R} \times \mathbb{R}$. It is not difficult to see that

$$|m|(A \times B) = \frac{1}{\pi} \int\limits_{A \times B} \frac{|\phi(x)\phi(y)|}{|x - y|} dx \, dy$$

for all $A, B \in \mathcal{B}(\mathbb{R})$. It follows that if $A$ and $B$ are closed intervals such that the interior of $A \times B$ intersects $diag = \{(x, x) : x \in \mathbb{R}\}$, then $|m|(A \times B) = \infty$ and if $A \times B$ is a positive distance from $diag$, then $|m|(A \times B) < \infty$. The diagonal in $\mathbb{R}^2$ is the set of singularities mentioned in Proposition 8.8, that is, $\Omega_m = \mathbb{R}^2 \setminus diag$.

Let $\mathcal{K}$ be the collection of all finite unions of compact product sets disjoint from $diag$. The variation of $m$ on each set $K \in \mathcal{K}$ is the restriction of the indefinite integral of the function $(x, y) \longmapsto \frac{1}{\pi}|\phi(x)\phi(y)|/|x - y|$, $(x, y) \in K$ with respect to Lebesgue measure, so condition (i) of Definition 8.6 is satisfied by $m$.

To verify condition (ii), we need to show that for every $A, B \in \mathcal{B}(\mathbb{R})$, $\epsilon > 0$, there exists $K \in \mathcal{K}$ such that $|m(A \times B) - m(K)| < \epsilon$. The separate $\sigma$-additivity of $m$ ensures that there exists some compact set $K = K_1 \times K_2$ such that $m(K_1 \times K_2)$ approximates $m(A \times B)$, but the special nature of $m$ ensures that $K$ may be chosen from the class $\mathcal{K}$.

For each $\delta > 0$, set

$$(H_\delta \psi)(x) = \frac{i}{\pi} \int_{|x-y|>\delta} \frac{\psi(y)}{x - y} dy$$

for all $\psi \in L^2(\mathbb{R})$ and almost all $x \in \mathbb{R}$, and set $m_\delta(A \times B) = (Q(B)H_\delta Q(A)\phi, \phi)$ for every $A \in \mathcal{B}(\mathbb{R})$ and $B \in \mathcal{B}(\mathbb{R})$. The variation of $m_\delta$ is the restriction to $\mathcal{B}(\mathbb{R}) \times \mathcal{B}(\mathbb{R})$ of a Radon measure on $\mathbb{R}^2$ denoted by $|m_\delta|$. Given $A, B \in \mathcal{B}(\mathbb{R})$ and $\epsilon > 0$, choose $\delta > 0$ such that $|m(A \times B) - m_\delta(A \times B)| < \frac{\epsilon}{3}$ and choose compact sets $K_1 \subset A$ and $K_2 \subset B$ such that $|m_\delta(A \times B) - m_\delta(K_1 \times K_2)| < \frac{\epsilon}{3}$. Such a choice is possible by the separate $\sigma$-additivity of $m_\delta$. Now the set $W = \{(x, y) \in K_1 \times K_2 : |x - y| \leq \delta\}$ is compact with $|m_\delta|$-measure zero. There exists a finite open cover $\mathcal{U}$ of $W$ by product sets, such that $|m_\delta|(\cup \mathcal{U}) < \frac{\epsilon}{3}$. Then $K = (K_1 \times K_2) \setminus \cup \mathcal{U} \in \mathcal{K}$ and $|m_\delta(K_1 \times K_2) - m_\delta(K)| = |m_\delta((K_1 \times K_2) \cap (\cup \mathcal{U}))| \leq |m_\delta|(\cup \mathcal{U}) < \frac{\epsilon}{3}$. Because the set $K$ is disjoint from $W$, $m_\delta(K) = m(K)$, so combining the estimates, $|m(A \times B) - m(K)| < \epsilon$. It follows that $m$ is a $\mathcal{K}$-regular set function on the semi-algebra $\mathcal{B}(\mathbb{R}) \times \mathcal{B}(\mathbb{R})$.   □

It is clear that similar examples can be manufactured by replacing the Hilbert transform $H$ in the definition of the set function $m$ by some other singular integral operator.

A $\mathcal{K}$-regular set function admits a natural integration structure as follows.

**Definition 8.10.** Let $\mathcal{K}$ be a family of compact subsets of $\Omega$ belonging to the algebra $a(\mathcal{S})$ of sets generated by the family $\mathcal{S}$. Let $m : \mathcal{S} \to \mathbb{C}$ be a bounded additive $\mathcal{K}$-regular set function. A function $f : \Omega \to \mathbb{C}$ is said to be *integrable with respect to m* if for each $K \in \mathcal{K}$ the restriction $f_K$ of $f$ to the set $K$ is $m_K$-integrable and there exists a $\mathcal{K}$-regular set function $fm : \mathcal{S} \to \mathbb{C}$ such that $fm(K) = \int_K f_K dm_K$ for all $K \in \mathcal{K}$.

The set function $fm$ is uniquely defined among the family of all $\mathcal{K}$-regular set functions. The number $fm(A)$ is sometimes denoted by $\int_A f dm$ and $m(f)$ is used to denote $fm(\Omega)$. Let $L^1(m)$ be the space of (equivalence classes of) $m$-integrable functions with the family of norms

$$\rho_K : f \longmapsto \sup_{A \in \mathcal{S}} |fm(A)| + |m_K|(|f_K|), \qquad f \in L^1(m).$$

defined for each $K \in \mathcal{K}$.

The set function $m$ may actually be bounded on the algebra $a(\mathcal{S})$ generated by $\mathcal{S}$ so that it is the restriction to $\mathcal{S}$ of a complex measure $\tilde{m}$ defined on the $\sigma$-algebra $\sigma(\mathcal{S})$ generated by $\mathcal{S}$. In this case, $L^1(\tilde{m})$ is, in general, a proper subspace of $L^1(m)$ (see [29, Example 3.7]).

**Proposition 8.11.** *Let $m$ be a $\mathcal{K}$-regular set function. The family $\Gamma = \{\rho_K : K \in \mathcal{K}\}$ of gauges is integrating for the integration map $f \longmapsto m(f)$, $f \in L^1(m)$.*

*Proof.* Let $f, f_i \in L^1(m)$, $i = 1, 2, \ldots$ be functions such that (8.1) holds, and for every $\omega \in \Omega$ such that (8.2) is true, it follows that $f(\omega) = \sum_{i=1}^{\infty} f_i(\omega)$. Here we may take $c_i = 1$ for all $i = 1, 2, \ldots$, because each gauge $\rho_K$, $K \in \mathcal{K}$ is a seminorm on the vector space $L^1(m)$.

To prove that the inequality (8.3) holds, it is enough to show that for each $K \in \mathcal{K}$, $\rho_K(f - \sum_{i=1}^{n} f_i) \to 0$ as $n \to \infty$ because $\rho_K(f) \le \rho_K(f - \sum_{i=1}^{n} f_i) + \sum_{i=1}^{n} \rho_K(f_i)$ for all $i = 1, 2, \ldots$. By the Beppo–Levi convergence theorem, $|m_K|(|f_K - \sum_{i=1}^{n} (f_i)_K|) \to 0$ as $n \to \infty$, so for each compact set $K \in \mathcal{K}$,

$$|fm(K) - \sum_{i=1}^{n} f_i m(K)| \le \sum_{i=n+1}^{\infty} |f_i m(K)| \le \sum_{i=n+1}^{\infty} |m_K|(|(f_i)_K|) \to 0$$

as $n \to \infty$.

Given $\epsilon > 0$, by (8.1) there exists $N = 1, 2, \ldots$ such that $\sum_{i=n}^{\infty} |f_i m(A)| < \epsilon$ for all $n > N$ and all $A \in \mathcal{S}$. It follows that $|fm(K) - \sum_{i=1}^{N} f_i m(K)| < \epsilon$ for all $K \in \mathcal{K}$. The regularity of $fm$ and $f_i m$, $i = 1, 2, \ldots$ shows that $\sup_{A \in \mathcal{S}} |fm(A) - \sum_{i=1}^{N} f_i m(A)| \le \epsilon$, proving that for each $K \in \mathcal{K}$, $\rho_K(f - \sum_{i=1}^{n} f_i) \to 0$ as $n \to \infty$. $\square$

In practice, it is useful to have a single gauge which is integrating for $m$.

**Proposition 8.12.** *Let $m$ be a $\mathcal{K}$-regular set function. Let $\mu : \sigma(\mathcal{S}) \to [0, \infty)$ be a measure such that for every $K \in \mathcal{K}$ there exists $b_K > 0$ such that $|m_K|(A) \le b_K \mu(A)$ for all $A \in \sigma(\mathcal{S})$ such that $A \subseteq K$. Then the graph $\{(f, fm) : f \in L^1(m)\}$ of the integration map $f \longmapsto fm$, $f \in L^1(m) \cap L^1(\mu)$ is closed in the product space $L^1(\mu) \times ba(\mathcal{S})$.*

*Proof.* Suppose that $f_n \to f$ in $L^1(\mu)$, $f_n$ belongs to $L^1(m) \cap L^1(\mu)$ and $f_n m$ converges in $ba(\mathcal{S})$. Now $f_n \chi_K \to f \chi_K$ in $L^1(\mu)$ as $n \to \infty$ for each $K \in \mathcal{K}$, so $|m_K|(|f_K -$

$(f_n)_K|) \leq b_K \mu(|f_n \chi_K - f \chi_K|) \leq b_K \mu(|f - f_n|) \to 0$ as $n \to \infty$. In particular, the restriction $f_K$ of $f$ to $K$ is $m_K$-integrable.

Because the set functions $f_n m, n = 1, 2, \ldots$ converge uniformly on $\mathcal{S}$, the limit $r$ of $f_n m, n = 1, 2, \ldots$ in $ba(\mathcal{S})$ is an additive $\mathcal{K}$-regular set function. Moreover, $r(K) = \lim\limits_{n \to \infty} f_n m(K) = \lim\limits_{n \to \infty} \int_K (f_n)_K dm_K = \int_K f_K dm_K$, so it follows that $f$ is $m$-integrable and $r = fm$.   □

**Corollary 8.13.** *Let $m, \mu$ be as above. Let $\rho_\mu(f) = \sup_{A \in \mathcal{S}} |fm(A)| + \mu(|f|)$ for all $f \in L^1(m) \cap L^1(\mu)$. Then $\left(L^1(m) \cap L^1(\mu), \rho_\mu, m\right)$ is an integration structure.*

So, as above, a $\mathcal{K}$-regular set function is closable with respect to a suitable measure. It can happen, however, that a bimeasure is not closable with respect to any measure (see the example in [29, Section 4]) because the regularity assumption fails.

**Remark 8.14.** For the bimeasure $m_{\phi, \psi}$ considered in Example 8.1 we have what might be called the *Lebesgue integration structure* described in Example 8.5 and also the less restrictive integration structure determined by Definition 8.10 with $\mathcal{K}$ being the family of compact product sets in $\mathbb{R}^2$, so that $m_{\phi, \psi}$ is treated as a Radon bimeasure. In the latter case, if $a$ is the symbol of a bounded pseudodifferential operator acting on $L^2(\mathbb{R})$, then $m_{\phi, \psi}(a) = \langle a(x, D)\phi, \psi \rangle$.

The appropriate integration structure is determined by the problem at hand. The same remark applies to the Fresnel integrals described in Example 8.4 compared to the more complicated construction of the next section.

# 9 The Feynman integral with singular potentials

The integration structure associated with regular set functions is sufficient to deal with the integrals of cylinder functions $f(X_{t_1}, \ldots, X_{t_n})$ with respect to $M_{-i}^t$ for $0 \leq t_1 < \cdots < t_n \leq t$ and $n = 1, 2, \ldots$, as remarked upon in Example 8.2. As shown in Section 5, the total variation of the set functions $\mu_{\lambda, \phi, \psi}^t$ is either 0 or $+\infty$ if $t > 0$ and $\Im\lambda \neq 0$, so a different approach is needed to deal with the integration of functions depending on the values of the process at infinitely many distinct times.

Analytic continuation is a feature of the discussion in Sections 2 and 3 above. Adopting a similar approach, integration with respect to the family $\langle M_\lambda^t \rangle_{\lambda > 0}$ of operator valued measures may be used to control the convergence of integrals with respect to the set functions $\langle M_\lambda^t \rangle_{\lambda \in \mathbb{C}_+}$. First we have to make precise the idea of integrating with respect to a *family* of operator valued measures.

The space $L^1(\langle M_\lambda^t \rangle_{\lambda > 0})$ of equivalence classes of functions integrable with respect to each operator valued measure $M_\lambda^t$, $\lambda > 0$, is equipped with a natural locally convex topology given by the seminorms (9.1) below, with respect to which it is a sequentially complete lcs. The seminorms defining the topology of $L^1(\langle M_\lambda^t \rangle_{\lambda > 0})$ are given by

$$f \longmapsto \sup \left\{ \int_{C^t} |f| \, d\mu_{\lambda, |\phi|, |\psi|}^t : \psi \in L^2(\mathbb{R}^d), \ \|\psi\|_2 \leq 1 \right\} \qquad (9.1)$$

for every $\phi \in L^2(\mathbb{R}^d)$ and $\lambda > 0$.

The completeness is a consequence of the operator valued measures $M_\lambda^t$ and $M_\nu^t$ having disjoint support for all $\lambda > 0$ and $\nu > 0$ such that $\lambda \neq \nu$, that is, the operator valued measures live on spaces of paths with distinct quadratic variation according to a result of P. Lévy. The measurability of functions belonging to $L^1(\langle M_\lambda^t \rangle_{\lambda>0})$ is closely related to the scale-invariant measurability studied in [33, Sections 4.2–4.4], except we do not require our paths $\omega$ to satisfy $\omega(0) = 0$. Define $\mu_{\lambda,\phi,\psi}^t = \langle M_\lambda^t \phi, \psi \rangle$ by equation (5.2).

As is usual in integration theory, in order to integrate with respect to the operator valued set functions $M_\lambda^t : S_t \to \mathcal{L}(L^2(\mathbb{R}^d))$ in the case that $\Im \lambda \neq 0$, one starts with *simple functions*, in this case, a finite linear combination $s = \sum_{j=1}^k c_k \chi_{E_j}$ of characteristic functions of sets $E_j \in S_t$, for $j = 1, \ldots, k$. Then linearity gives

$$\int_{C^t} s \, dM_\lambda^t = \sum_{j=1}^k c_k M_\lambda^t(E_k).$$

Let $\mathbf{sim}(S_t)$ be the linear space of simple functions $s$ based on $S_t$.

One idea is to give the topology on $\mathbf{sim}(S_t)$ so that a net $\langle s_\alpha \rangle_{\alpha \in A}$ of simple functions converges to a function $f$ if and only if it converges to $f$ in the quasicomplete space $L^1(\langle M_\lambda^t \rangle_{\lambda>0})$ and the net is also Cauchy with respect to the seminorms

$$p_{E,K,\phi} : s \longmapsto \sup_{\lambda \in K} \left\| \int_E s(\omega) \, (M_\lambda^t \phi)(d\omega) \right\|_2 \tag{9.2}$$

as $E$ varies over cylinder sets (3.4), the function $\phi$ varies over $L^2(\mathbb{R}^d)$ and $K$ varies over compact subsets of $\mathbb{C}_+$. Then we can *define*

$$\int_E f(\omega) \, (M_\lambda^t \phi)(d\omega) := \lim_{\alpha \in A} \int_E s_\alpha(\omega) \, (M_\lambda^t \phi)(d\omega)$$

so that the convergence is uniform in the strong operator topology as $\lambda$ varies over compact subsets of $\mathbb{C}_+$. It follows that the operator valued function

$$(E, \lambda) \longmapsto \int_E f(\omega) \, (M_\lambda^t \phi)(d\omega)$$

is finitely additive in $E \in S_t$ and analytic in $\lambda \in \mathbb{C}_+$.

It is easy to see by analytic continuation from positive values of the parameter $\lambda$, that for $\phi \in L^2(\mathbb{R}^d)$ fixed, $\Gamma_\phi = \{p_{E,K,\phi}\}$ is a collectively integrating family of seminorms defined on $\mathbf{sim}(S_t)$ as $E$ and $K$ vary. Moreover, for each $\lambda \in \mathbb{C}_+$, the inequality $\|(M_\lambda^t \phi)(s)\|_2 \leq p_{\Omega,K,\phi}(s)$ holds for all $s \in \mathbf{sim}(S_t)$, if we choose $K$ any compact subset of $\mathbb{C}_+$ containing $\lambda$. The scheme considered here fits into the general framework of Section 8.

For the situation of interest—quantum mechanics—$\lambda$ is interpreted as $-i$ times a mass parameter $m$. It is not unreasonable to expect that the dynamics of a quantum system should exhibit continuous dependence upon nonzero (and positive) mass. Then analytic continuation in $\lambda$ from the boundary values on $(i\mathbb{R}) \setminus \{0\}$ to $\mathbb{C}_+$ can be achieved by the Poisson integral formula.

Let us look at two possibilities for going from the set $\mathbb{C}_+$ to the boundary $\partial\mathbb{C}_+ = i\mathbb{R}$ of $\mathbb{C}_+$. Let $H(\mathbb{C}_+)$ denote the space of all functions which are analytic in $\mathbb{C}_+$ and continuous on $\overline{\mathbb{C}}_+ \setminus \{0\}$, endowed with the topology of uniform convergence on compact subsets of $\mathbb{C}_+$. The space $H(\mathbb{C}_+)$ endowed with the topology of uniform convergence on compact subsets of $\overline{\mathbb{C}}_+ \setminus \{0\}$ is written as $\overline{H}(\mathbb{C}_+)$. The two locally convex spaces $\overline{H}(\mathbb{C}_+)$ and $H(\mathbb{C}_+)$ have the same underlying sets, only the topologies differ. The space $\overline{H}(\mathbb{C}_+)$ is complete and metrizable. Although $H(\mathbb{C}_+)$ is metrizable, it is not a complete locally convex space.

Of course, we could equally use $\lambda = -im$ with some smaller interval $I$ of the mass parameter $m$, say, all positive real values. Then we would look at analytic functions in $\mathbb{C}_+$ with continuous boundary values on $-iI$.

We consider two types of integrability for a function $f : C^t \to \mathbb{C}$.

$[H(\mathbb{C}_+)]$:    *the function* $\lambda \longmapsto \int_E f(\omega)\,(M_\lambda^t \phi)(d\omega)$ *defined above belongs to* $H(\mathbb{C}_+)$
     *for each* $E \in \mathcal{S}_t$ *and all* $\phi \in L^2(\mathbb{R}^d)$;

$[\overline{H}(\mathbb{C}_+)]$:    *the net* $\langle s_\alpha \rangle_{\alpha \in A}$ *of* $\mathcal{S}_t$-*simple functions mentioned above converges to* $f$ *in the quasicomplete space* $L^1(\langle M_\lambda^t \rangle_{\lambda>0})$ *and the net is also Cauchy with respect to the seminorms* $p_{E,K,\phi}$ *defined in formula* (9.2) *as* $E$ *varies over cylinder sets* (3.4), *the function* $\phi$ *varies over* $L^2(\mathbb{R}^d)$ *and* $K$ *varies over compact subsets of* $\overline{\mathbb{C}}_+ \setminus \{0\}$.

We now look at the distinction between $H(\mathbb{C}_+)$-integrability and $\overline{H}(\mathbb{C}_+)$-integrability. The second definition of integrability fits better into the considerations of Section 8 because for $\phi \in L^2(\mathbb{R}^d)$ fixed, $\Gamma_\phi = \{p_{E,K,\phi}\}$ is a collectively integrating family of seminorms defined on $\mathbf{sim}(\mathcal{S}_t)$ as $E$ and $K$ vary and the inequality $\|(M_{-i}^t \phi)(s)\|_2 \leq p_{\Omega,K,\phi}(s)$ holds for all $s \in \mathbf{sim}(\mathcal{S}_t)$, if we choose $K$ any compact subset of $\overline{\mathbb{C}}_+ \setminus \{0\}$ containing $-i$.

Every $\overline{H}(\mathbb{C}_+)$-integrable function is $H(\mathbb{C}_+)$-integrable because the function

$$\lambda \longmapsto \int_E f(\omega)\,(M_\lambda^t \phi)(d\omega) \tag{9.3}$$

is the uniform limit of functions

$$\lambda \longmapsto \int_E s_\alpha(\omega)\,(M_\lambda^t \phi)(d\omega), \quad \alpha \in A,$$

on compact subsets of $\overline{\mathbb{C}}_+ \setminus \{0\}$, with $\langle s_\alpha \rangle_{\alpha \in A}$ a net of simple functions. Hence, (9.3) is analytic in $\mathbb{C}_+$ and continuous on $\overline{\mathbb{C}}_+ \setminus \{0\}$. A few examples of functions integrable in the above senses follow.

## 9.1  $\overline{H}(\mathbb{C}_+)$-integrable functions

i) $F : \omega \longmapsto f_1(\omega(t_1)) \cdots f_n(\omega(t_n))$, $\omega \in C^t$, with $f_1, \ldots, f_n$ bounded and Borel measurable on $\mathbb{R}^d$.

ii) $e^{-i \int_0^t V \circ X_s \, ds}$ with $V \in L^p(\mathbb{R}^d) + L^\infty(\mathbb{R}^d)$, $p > d/2$ for $d \geq 3$.

In Example ii), we have the representation

$$
\begin{aligned}
e^{-itH(m)} &= \int_{C^t} e^{-i \int_0^t V \circ X_s \, ds} \, dM_\lambda^t \\
&= K_{-\lambda}^t \left( e^{i \int_0^t V \circ X_s \, ds} \right)^*, \quad \lambda = -im, \; m \in \mathbb{R}, \; m \neq 0, \quad (9.4)
\end{aligned}
$$

relative to the operator $H(m) = -\Delta/(2m) + V$. Here $X_s : \omega \longmapsto \omega(s)$ for all $\omega \in C^t$ and $0 \leq s \leq t$.

To check what is involved in proving $\overline{H}(\mathbb{C}_+)$-integrability for the function i) and ii) above, the calculations are given below.

*Proof of $\overline{H}(\mathbb{C}_+)$-integrability.* i) Let $s_{k,j} : \mathbb{R}^d \to \mathbb{C}$ be Borel measurable simple functions defined for all $k = 1, 2, \ldots$ and $j = 1, \ldots, n$, such that $\|s_{k,j}\|_\infty \leq \|f_j\|_\infty$ for all $k = 1, 2 \ldots$, and with the property that for each $x \in \mathbb{R}^d$ and $j = 1, \ldots, n$, we have $s_{k,j}(x) \to f_j(x)$ as $k \to \infty$. Then by dominated convergence $(s_{k,1} \circ X_{t_1}) \cdots (s_{k,n} \circ X_{t_n}) \to F$ in $L^1((\langle M_\lambda^t \rangle_{\lambda > 0})$ as $k \to \infty$.

We need to look at the convergence in $L^2(\mathbb{R}^d)$ of

$$
\int_E (s_{k,1} \circ X_{t_1}) \cdots (s_{k,n} \circ X_{t_n}) \, d(M_\lambda^t \phi) \tag{9.5}
$$

as $k \to \infty$, $E$ varies over cylinder sets and $\lambda$ varies over compact subsets of $\overline{\mathbb{C}}_+ \setminus \{0\}$. The proof of convergence for $E = C^t$ gives the idea, for then (9.5) is equal to

$$
S_\lambda(t - t_n) Q(s_{k,n}) S_\lambda(t_n - t_{n-1}) \cdots Q(s_{k,1}) S_\lambda(t_1) \phi \tag{9.6}
$$

for all $\lambda \in \overline{\mathbb{C}}_+ \setminus \{0\}$. The operator of multiplication by a Borel measurable function $f$ on $\mathbb{R}^d$ is written as $Q(f)$. Because $Q(s_{k,j}) \to Q(f_j)$ in the strong operator topology of $\mathcal{L}(L^2(\mathbb{R}^d))$ as $k \to \infty$, convergence also holds for the topology of precompact convergence, that is, uniform convergence on precompact sets of $L^2(\mathbb{R}^d)$. Now for $\tau \geq 0$ fixed, $K$ a compact subset of $\overline{\mathbb{C}}_+ \setminus \{0\}$ and $C$ a precompact subset of $L^2(\mathbb{R}^d)$, the set

$$
\{S_\lambda(\tau)g : \lambda \in K, \; g \in C\}
$$

is a precompact subset of $L^2(\mathbb{R}^d)$ because the mapping $(\lambda, g) \longmapsto S_\lambda(\tau)g$ is continuous from $(\overline{\mathbb{C}}_+ \setminus \{0\}) \times L^2(\mathbb{R}^d)$ into $L^2(\mathbb{R}^d)$. It follows that (9.6) converges to

$$
S_\lambda(t - t_n) Q(f_n) S_\lambda(t_n - t_{n-1}) \cdots Q(f_1) S_\lambda(t_1) \phi \tag{9.7}
$$

in $L^2(\mathbb{R}^d)$ uniformly for $\lambda \in K$ as $k \to \infty$.

Repeating the argument with $C^t$ replaced by a cylinder set $E$, we see that $F$ is $\overline{H}(\mathbb{C}_+)$-integrable and

$$
\int_{C^t} F \, dM_\lambda^t = S_\lambda(t - t_n) Q(f_n) S_\lambda(t_n - t_{n-1}) \cdots Q(f_1) S_\lambda(t_1) \tag{9.8}
$$

for all $\lambda \in \overline{\mathbb{C}}_+ \setminus \{0\}$.

ii) The argument here appears in [26], [28]. We first consider the case of bounded continuous functions $V : \mathbb{R}^d \to \mathbb{R}$ for any $d = 1, 2, \ldots$. Let

$$F_n = \prod_{j=1}^{n} e^{-it(V \circ X_{jt/n})/n}$$

for each $n = 1, 2, \ldots$. Then $F_n : C^t \to \mathbb{C}$ is a uniformly bounded function of the form i), so $F_n$ is $\overline{H}(\mathbb{C}_+)$-integrable and from (9.8) we have

$$\int_{C^t} F_n \, d(M_\lambda^t \phi) = \prod_{j=1}^{n} \left[ e^{-itQ(V)/n} S_\lambda(t/n) \right] \phi \qquad (9.9)$$

for all $\lambda \in \overline{\mathbb{C}}_+ \setminus \{0\}$ and $\phi \in L^2(\mathbb{R}^d)$. The operator $Q(e^{-itV/n})$ of multiplication by the function $e^{-itV/n}$ is just the unitary operator $e^{-itQ(V)/n} = \sum_{k=0}^{\infty} (-itQ(V)/n)^k / k!$.

By the continuity of $V$, the continuity of paths in the sample space $C^t$ and dominated convergence, it follows that $F_n \to e^{-i \int_0^t V \circ X_s \, ds}$ in $L^1(\langle M_\lambda^t \rangle_{\lambda > 0})$ as $n \to \infty$. We need to show that (9.9) converges in $L^2(\mathbb{R}^d)$ uniformly for $\lambda$ in compact subsets of $\overline{\mathbb{C}}_+ \setminus \{0\}$ as $n \to \infty$. The argument with $C^t$ replaced by a cylinder set $E$ is similar.

Clearly, we are looking at convergence of the type of the Lie–Kato–Trotter product formula. The infinitesimal generator of $S_\lambda$ is $\Delta/(2\lambda)$ for all $\lambda \in \overline{\mathbb{C}}_+ \setminus \{0\}$. Let $K$ be a compact subset of $\overline{\mathbb{C}}_+ \setminus \{0\}$ and let $C(K, L^2(\mathbb{R}^d))$ be the Banach space of all continuous functions $f : K \to L^2(\mathbb{R}^d)$ with the uniform norm $\|f\|_\infty = \sup_{\lambda \in K} \|f(\lambda)\|_2$. Then

$$S_\lambda(t + s)f(\lambda) = S_\lambda(t)(S_\lambda(s)f(\lambda)), \quad \text{for all } \lambda \in K,$$

so the operator $f \longmapsto S_{(\cdot)}(t)(f(\cdot))$, $f \in C(K, L^2(\mathbb{R}^d))$, defines a contraction semi-group $\tilde{S}$ of bounded linear operators for each $t \geq 0$. Because the linear subspace $C(K) \otimes L^2(\mathbb{R}^d)$ is dense in $C(K, L^2(\mathbb{R}^d))$, the semigroup $\tilde{S}$ is continuous at zero and its generator is the application of the operator $\Delta/(2\lambda)$ to $f(\lambda)$ for a dense set of functions $f \in C(K, L^2(\mathbb{R}^d))$. But $-iQ(V)$ is a bounded perturbation of this generator, so appealing to the elementary version of the Lie–Kato–Trotter product formula applied to the Banach space $C(K, L^2(\mathbb{R}^d))$, the operators (9.9) converge in $L^2(\mathbb{R}^d)$ uniformly for $\lambda \in K$ as $n \to \infty$. It follows that $e^{-i \int_0^t V \circ X_s \, ds}$ is $\overline{H}(\mathbb{C}_+)$-integrable and

$$\int_{C^t} e^{-i \int_0^t V \circ X_s \, ds} \, dM_\lambda^t = e^{\frac{t}{2\lambda} \Delta - itQ(V)} \qquad (9.10)$$

for all $\lambda \in \overline{\mathbb{C}}_+ \setminus \{0\}$.

Next, approximate $V \in L^\infty(\mathbb{R}^d)$ almost everywhere by continuous functions $V_\epsilon$ such that $\|V_\epsilon\|_\infty \leq \|V\|_\infty$ for all $\epsilon > 0$. Again, convergence in the Banach space $C(K, L^2(\mathbb{R}^d))$ yields $\overline{H}(\mathbb{C}_+)$-integrability and equation (9.10). For the general case $V \in L^p(\mathbb{R}^d) + L^\infty(\mathbb{R}^d)$, $p > d/2$ for $d \geq 3$, the selfadjoint operator $Q(V)$ is a small perturbation of $\Delta$, that is, the operator $Q(V)$ is relatively bounded with respect to $\Delta$ and the relative bound is zero. Hence, we can approximate $Q(V)$ by bounded operators

$Q(V_n)$ associated with cutoff potentials $V_n = V\chi_{\{|V|\le n\}}$, $n = 1, 2, \ldots$, acting in the Banach space $C(K, L^2(\mathbb{R}^d))$. Again, we get uniform convergence in $\lambda \in K$ and equation (9.10) holds. $\qquad\qquad\qquad\qquad\qquad\qquad\qquad\qquad\qquad\qquad\qquad\qquad$ $\square$

A critical part of the proof above is the condition that $Q(V)$ is a small perturbation of $\Delta$. If this fails, it looks like we have to relax the conditions of $\overline{H}(\mathbb{C}_+)$-integrability.

## 9.2  $H(\mathbb{C}_+)$-integrable functions

iii) $e^{-i\int_0^t V\circ X_s\, ds}$ with $V \in L^{d/2}(\mathbb{R}^d) + L^\infty(\mathbb{R}^d)$, $d \ge 3$.

iv) $e^{-i\int_0^t V\circ X_s\, ds}$ with $V(x) = -c/|x|^2$, $c > 0$ in $\mathbb{R}^3$.

v) $f^t = e^{-i\alpha_t}$ with $\langle\alpha_t\rangle_{t\ge 0}$ a positive continuous additive functional of Brownian motion [33, Section 13.7].

*Proof of $H(\mathbb{C}_+)$-integrability.* We just look at how to verify the conditions $[H(\mathbb{C}_+)]$ in the case that $t > 0$ and the cylinder set $E$ is equal to the whole space $C^t$. For all of these functionals $f^t$, the approximation of $\int_{C^t} f^t\, dM_\lambda^t$ for $\lambda > 0$ converges uniformly for $\lambda$ belonging to compact subsets of $\mathbb{C}_+$ by the operator version of Vitali's convergence theorem. The proofs of $H(\mathbb{C}_+)$-integrability of $f^t$ in the cases above merely require the proof of continuity of $\lambda \longmapsto \int_{C^t} f^t\, dM_\lambda^t$ in the strong operator topology for all $\lambda \in \partial\mathbb{C}_+ \setminus \{0\}$ and the identification of the function $\lambda \longmapsto \int_{C^t} f^t\, dM_\lambda^t$, $\lambda \in \partial\mathbb{C}_+ \setminus \{0\}$, as the continuous boundary value of the operator valued analytic function $\lambda \longmapsto \int_{C^t} f^t\, dM_\lambda^t$, $\lambda \in \mathbb{C}_+ \setminus \{0\}$. This sort of property is known from perturbation theory. Example iii) is treated in [34, Lemma VI.4.8b, Remark 4.9a], Example iv) in [37] and a suitable modification of [5], [33, Section 13.7] gives Example v). $\qquad$ $\square$

More generally, Example iii) could be any measurable function $V : \mathbb{R}^d \to \mathbb{R}$ such that the form sum $-\Delta + aV$ is bounded below for all $a \in \mathbb{R}$ [34, Theorem IV.3.6], for example, if $V$ is a small (zero relative bound) *form* perturbation of $-\Delta$.

The functionals $e^{-it(V\circ X_{jt/n})/n}$ cannot converge in the topology of $\overline{H}(\mathbb{C}_+)$-integrable functions to $e^{-i\int_0^t V\circ X_s\, ds}$ in Example iv), otherwise $t \longmapsto \int_{C^t} e^{-i\int_0^t V\circ X_s\, ds}\, dM_\lambda^t$ would be a unitary group for purely imaginary $\lambda$. For each $c > 0$, this is known not to be the case for sufficiently large positive values of the mass parameter $-\Im\lambda$ [37]. However, to establish that $\left(\prod_{j=1}^n e^{-it(V\circ X_{\frac{jt}{n}})/n}\right)$ is *not* $\overline{H}(\mathbb{C}_+)$-integrable, we would need to show that it is not the limit in the seminorms (9.2) of *any* net of simple functions, rather than just this particular sequence of approximations. This looks hard to prove.

The space of all $\overline{H}(\mathbb{C}_+)$-integrable functions may be given a locally convex topology under which all Cauchy *sequences* converge in the lcs. This is not possible for the space of all $H(\mathbb{C}_+)$-integrable functions, because $H(\mathbb{C}_+)$ is not itself complete.

The notions of integrability just described are sufficient to treat most potentials $V$ of physical interest in quantum mechanics. The question of the *integrability* of the multiplicative functional $e^{-i\int_0^t V\circ X_s\, ds}$, $t > 0$, is intimately connected with the *existence and uniqueness of the dynamics* of the quantum system, so it is not surprising that operator theoretic arguments should feature in the proofs sketched above. In the next section, even the existence of an appropriate multiplicative functional of the process is an issue.

## 10 Quantum field theory

The Feynman representation (1.1) has an analogue in quantum field theory discussed at the heuristic level in [33, Section 20.2], especially with regards to knot theory and low dimensional topology. The Feynman–Kac formula is also a tool in the construction of quantum fields with polynomial self-interactions in Minkowski space with two and three space–time dimensions [20]. An overview of constructive quantum field theory relevant to the discussion below appears in [25].

The question arises of what is the evolution process $\left(\Omega, \langle S_t \rangle_{t \geq 0}, \langle M^t \rangle_{t \geq 0}; \langle \Phi_t \rangle_{t \geq 0}\right)$ associated with a free quantum field and how can we represent the dynamical group of an interacting field in the form

$$e^{-itH} = \int_\Omega F_t \, dM^t, \quad t \geq 0, \tag{10.1}$$

with $t \longmapsto F_t$, $t \geq 0$, some multiplicative functional. This is just what we have been considering in the context of quantum mechanics.

The situation is profoundly different in quantum field theory. The Hamiltonian operator $H$ is not constructed directly by perturbation theory and the physically realistic multiplicative functional $F_t$ is not simply a Feynman–Kac functional $F_t = e^{-i \int_0^t V \circ \Phi_s \, ds}$—a process of *renormalization* is needed to construct $F_t$, $t \geq 0$. The detailed analysis in Section 7 of the Dirac equation on the line illustrates the point, while avoiding technical difficulties associated with integration with respect to unbounded set functions. The Coulomb-like potential is singular for the Dirac operator on the line. Equation (7.41) shows that a dynamical group for the Dirac equation on the line with a given boundary condition has a Feynman representation, if we construct the Feynman–Kac functional by renormalizing the interaction at the singularity of the potential, and take into account particle-pair production at the singularity for all paths belonging to the support of the associated measure $M_m^{t, \Gamma}$.

In this section, the evolution process $\left(\Omega, \langle S_t \rangle_{t \geq 0}, \langle M^t \rangle_{t \geq 0}; \langle X_t \rangle_{t \geq 0}\right)$ associated with a free quantum field is constructed along the lines suggested in [20]. As in quantum mechanics, there is actually an associated family $\left(\Omega, \langle S_t \rangle_{t \geq 0}, \langle M_\lambda^t \rangle_{t \geq 0}; \langle X_t \rangle_{t \geq 0}\right)$ of evolution processes defined for all $\lambda \in \overline{\mathbb{C}}_+ \setminus \{0\}$, so that for each $t \geq 0$, the equality $M_{-i}^t = M^t$ holds, the function $\lambda \longmapsto M_\lambda^t(E)$, $\lambda \in \mathbb{C}_+ \setminus \{0\}$, is analytic and continuous in $\overline{\mathbb{C}}_+ \setminus \{0\}$ for each $E \in \mathcal{S}_t$, and $M_\lambda^t$ is associated with the free Euclidean field for each $\lambda > 0$, in the same way that the operator valued measures (3.5) are associated with Wiener measure by scaling. The section concludes with some comments about the construction of the multiplicative functional $F_t$, $t \geq 0$, for which the representation (10.1) is valid, although more work needs to be done even in two space–time dimensions, for which the Euclidean field theory for polynomial interactions is well-understood.

## 10.1 The free Euclidean field

The starting point for $d$-dimensional Euclidean field theory is a probability measure $\mu$ defined on the Borel $\sigma$-algebra of the space $\mathcal{D}'(\mathbb{R}^d)$ of Schwartz distributions defined on $\mathbb{R}^d$. The space of fields $\phi \in \mathcal{D}'(\mathbb{R}^d)$ plays the role of the paths $\omega \in C([0, \infty), \mathbb{R}^n)$ in quantum mechanics and for the free field, $\mu$ is analogous to Wiener measure.

The inverse Fourier transform $S\mu : \mathcal{D}(\mathbb{R}^d) \to \mathbb{C}$ of $\mu$ is defined by

$$S\mu(f) = \int_{\mathcal{D}'(\mathbb{R}^d)} e^{i\langle f, \phi \rangle}\, d\mu(\phi), \quad f \in \mathcal{D}(\mathbb{R}^d). \tag{10.2}$$

In this section, the notation $\langle f, \phi \rangle = \phi(f)$ for $\phi \in \mathcal{D}'(\mathbb{R}^d)$ and $f \in \mathcal{D}(\mathbb{R}^d)$ is used to represent the duality between $\mathcal{D}(\mathbb{R}^d)$ and $\mathcal{D}'(\mathbb{R}^d)$. The inner product of a Hilbert space $\mathcal{H}$ is written as $\langle \cdot, \cdot \rangle_{\mathcal{H}}$.

The properties that the probability measure $\mu$ possesses (*Osterwalder–Schrader axioms*) are formulated in terms of the functional $S\mu$ defined on $\mathcal{D}(\mathbb{R}^d)$; these are listed in [20, pp. 89–90]. Properties of the quantum field theory, such as its time evolution are derived from the probability measure $\mu$. Because our discussion is at the most basic level, we shall go directly to the objects of interest for the free quantum field of mass $m = 1$.

For each $d = 1, 2, \ldots$, let $\Delta_d$ be the selfadjoint Laplacian in $L^2(\mathbb{R}^d)$ and let $H^{-1}(\mathbb{R}^d)$ be the Sobolev space of order $-1$, defined as the completion of $L^2(\mathbb{R}^d)$ with respect to the Hilbert space norm $f \longmapsto \langle (-\Delta_d + I)^{-1} f, f \rangle_{L^2(\mathbb{R}^d)}$, $f \in L^2(\mathbb{R}^d)$. The Hilbert space $H^{-1}(\mathbb{R}^d)$ can be identified with tempered distributions $T \in \mathcal{S}'(\mathbb{R}^d)$ for which

$$\int_{\mathbb{R}^d} |\hat{T}(\xi)|^2 (1 + |\xi|^2)^{-1}\, d\xi < \infty.$$

The norm of $H^{-1}(\mathbb{R}^d)$ is denoted by $\| \cdot \|_{H^{-1}(\mathbb{R}^d)}$ and inner product by $\langle \cdot, \cdot \rangle_{H^{-1}(\mathbb{R}^d)}$. The subscript is dropped from $\Delta_d$ if it is clear from the context.

Now suppose that $d \geq 2$. We take the free Euclidean field of unit mass to be the canonical Gaussian process over $H^{-1}(\mathbb{R}^d)$, that is, a continuous linear map $f \longmapsto \Phi_f$ from the Hilbert space $H^{-1}(\mathbb{R}^d)$ into the space $L^0(\Omega, \mathcal{F}, \mu)$ of random variables with respect to a probability measure $\mu$ such that

$$\int_\Omega \Phi_f \Phi_g\, d\mu = \langle f, g \rangle_{H^{-1}(\mathbb{R}^d)}, \quad f, g \in H^{-1}(\mathbb{R}^d). \tag{10.3}$$

Just as Brownian motion is usually represented by Wiener measure on the space of continuous functions, we take the probability measure $\mu$ to be the unique Gaussian measure defined on the Borel $\sigma$-algebra $\mathcal{F} = \mathcal{B}(\mathcal{D}'(\mathbb{R}^d))$ of $\Omega = \mathcal{D}'(\mathbb{R}^d)$ with mean zero and variance $\int_\Omega |\phi(f)|^2\, d\mu(\phi) = \|f\|^2_{H^{-1}(\mathbb{R}^d)}$ for all $f \in \mathcal{D}(\mathbb{R}^d)$. The measure $\mu$ exists by the Bochner–Minlos theorem [18, Chap. IV Section 3].

Let $\mathcal{B}_f(\mathbb{R}^n)$ denote the collection of all Borel sets of finite Lebesgue measure in $\mathbb{R}^n$, $n = 1, 2, \ldots$. In keeping with the spirit of these notes, by restricting the map $f \longmapsto \Phi_f$ to the collection of all characteristic functions of sets of $A \in \mathcal{B}_f(\mathbb{R}^d)$, the field may also be viewed as an $L^2(\mu)$-valued measure $A \longmapsto \Phi_A$, $A \in \mathcal{B}_f(\mathbb{R}^d)$, defined on the $\delta$-ring

$B_f(\mathbb{R}^d)$. In a similar way to the properties of Brownian motion, the set of all $\omega \in \Omega$ for which there exists an open subset $U$ of $\mathbb{R}^d$ on which $f \longmapsto \langle f, \omega \rangle$, $f \in \mathcal{D}(\mathbb{R}^d)$, is a distribution of order zero, has $\mu$-measure zero [11, Proposition 3.1].

It is easy to see that every measurable function $f$ such that $|f| \in H^{-1}(\mathbb{R}^d)$ is $\Phi$-integrable on $\mathbb{R}^d$ and $\Phi_f = \int_{\mathbb{R}^d} f \, d\Phi$. In particular, if $f \in L^2(\mathbb{R}^{d-1})$ and $B$ is a Borel subset of $\mathbb{R}$ with finite Lebesgue measure, the function $f \otimes B : (x, t) \longmapsto f(x)\chi_B(t)$, $x \in \mathbb{R}^{d-1}, t \in \mathbb{R}$, belongs to $H^{-1}(\mathbb{R}^d)$.

The following proposition is a slight reformulation of the definition of sharp time fields.

**Proposition 10.1.** *The process $\Phi$ admits a continuous disintegration $X : \mathbb{R} \times \Omega \to S'(\mathbb{R}^{d-1})$ in the sense that $X$ is $\mathcal{B}(\mathbb{R}) \otimes \mathcal{F}$-measurable and*

(i) *for each $\omega \in \Omega$, the $S'(\mathbb{R}^{d-1})$-valued function $t \longmapsto X_t(\omega) := X(t, \omega), t \in \mathbb{R}$, is continuous and*

(ii) *for each $f \in S(\mathbb{R}^{d-1})$, the $L^2(\mu)$-valued function $t \longmapsto \langle f, X_t(\cdot) \rangle$ is locally weakly Lebesgue integrable in $L^2(\mu)$ and the equality*

$$\Phi_{f \otimes B} = \int_B \langle f, X_t(\cdot) \rangle \, dt, \tag{10.4}$$

*holds in $L^2(\mu)$ for every $B \in \mathcal{B}_f(\mathbb{R})$.*

*Furthermore, $X$ is a version of the Ornstein–Uhlenbeck process, that is, $X$ is Gaussian with mean zero and covariance*

$$\int_\Omega \langle f, X_s(\omega) \rangle \langle g, X_t(\omega) \rangle \, d\mu(\omega)$$

$$= \frac{1}{2}\left\langle (-\Delta_{d-1} + I)^{-1/2} e^{-|t-s|(-\Delta_{d-1}+I)^{-1/2}} f, g \right\rangle, \quad f, g \in S(\mathbb{R}^{d-1}), \ t, s \in \mathbb{R}. \tag{10.5}$$

*Any two such continuous disintegrations are indistinguishable.*

*Proof.* Let $\delta_t$ denote the unit point mass at $t \in \mathbb{R}$. A calculation shows that for each $f \in S(\mathbb{R}^{d-1})$, the distribution $f \otimes \delta_t \in S'(\mathbb{R}^d)$ belongs to $H^{-1}(\mathbb{R}^d)$ and

$$\int_\Omega \Phi_{f \otimes \delta_s} \cdot \Phi_{g \otimes \delta_t} \, d\mu = \langle f \otimes \delta_s, g \otimes \delta_t \rangle_{H^{-1}(\mathbb{R}^d)} \tag{10.6}$$

equals the right-hand side of equation (10.5) for each $f, g \in S(\mathbb{R}^{d-1})$ [20, Proposition 6.2.5]. From equations (10.6) and (10.5), we have

$$\int_\Omega |\Phi_{f \otimes \delta_s} - \Phi_{f \otimes \delta_t}|^2 \, d\mu = \left\langle (I - e^{-|t-s|(-\Delta_{d-1}+I)^{-1/2}})(-\Delta_{d-1} + I)^{-1/2} f, f \right\rangle \tag{10.7}$$

for each $f \in S(\mathbb{R}^{d-1})$ and $s, t \in \mathbb{R}$, so the $L^2(\mu)$-valued function $t \longmapsto \Phi_{f \otimes \delta_t}$ is continuous and so uniformly bounded on compact subsets of $\mathbb{R}$. The equality

$$\Phi_{f \otimes B} = \lim_{n \to \infty} \int_{B \cap [-n,n]} \Phi_{f \otimes \delta_t} \, dt, \quad B \in \mathcal{B}_f(\mathbb{R}),$$

ensures that $t \longmapsto \Phi_{f \otimes \delta_t}$ is locally weakly Lebesgue integrable in $L^2(\mu)$ and $\Phi_{f \otimes B} = \int_B \Phi_{f \otimes \delta_t} \, dt$, for all $B \in \mathcal{B}_f(\mathbb{R})$, so it is enough to show that there exists $X$ satisfying (i) such that $\langle f, X_t(\cdot) \rangle = \Phi_{f \otimes \delta_t}$ $\mu$-almost everywhere for each $f \in S(\mathbb{R}^{d-1})$. A version of Kolmogorov's criterion for the continuity of random processes gives this: see [13, Theorem 3.4.1] and [11, Theorem 2.2] for explicit proofs.    $\square$

It is possible to realize the process $X$ in a space much smaller that $S'(\mathbb{R}^{d-1})$ [40]. If $\theta_s : \Omega \to \Omega$ denotes the time shift map

$$\langle f, \theta_s \omega \rangle = \langle \theta_{-s} f, \omega \rangle, \quad \omega \in \Omega, \ f \in \mathcal{D}(\mathbb{R}^d), \tag{10.8}$$

where $\theta_{-s} f(x_1, \ldots, x_d) = f(x_1 \ldots, x_{d-1}, x_d - s)$, then $X_t \circ \theta_s = X_{t+s}$ for all $s, t \geq 0$.

The Hilbert space of the free field is $\mathcal{H} = L^2(\mu \circ X_0^{-1})$. By formula (10.5), the image measure $\mu \circ X_0^{-1}$ is the unique Gaussian measure on $S'(\mathbb{R}^{d-1})$ with mean zero and covariance $(f, g) \longmapsto \frac{1}{2} \langle (-\Delta_{d-1} + I)^{-1/2} f, g \rangle$, $f, g \in S(\mathbb{R}^{d-1})$. According to [20, Corollary 6.2.8], the Hilbert space $\mathcal{H}$ can be identified with the state space constructed from the Osterwalder–Schrader axioms [20, pp. 89–92].

The mapping $f \longmapsto f \circ X_0$, $f \in \mathcal{H}$, is an isometry between $\mathcal{H}$ and $L^2(\Omega, \mathcal{F}_0, \mu|_{\mathcal{F}_0})$, where $\mathcal{F}_0$ is the $\sigma$-algebra generated by the random variables

$$\{ \Phi_f : f \in H^{-1}(\mathbb{R}^d), \ \text{supp } f \subset \mathbb{R}^{d-1} \times \{0\} \}.$$

In particular, for each $u \in \mathcal{H}$ and $t \geq 0$, there exists a unique element $e^{-t H_0} u \in \mathcal{H}$ such that

$$(e^{-t H_0} u) \circ X_0 = E(u \circ X_t | \mathcal{F}_0),$$

where the right-hand side of the equation is the $\mu$-conditional expectation of $u \circ X_t$ with respect to $\mathcal{F}_0$. The Markov property of the process $\langle X_t \rangle_{t \geq 0}$ ensures that $t \longmapsto e^{-t H_0}$, $t \geq 0$, is a Markov semigroup acting on $\mathcal{H}$ and its generator $H_0$ is a positive selfadjoint operator—the free Hamiltonian of the quantum field [2, Section 5.1]. The general proof given in [20, Theorem 6.1.3] for the construction of the Hamiltonian operator avoids the use of sharp-time fields, whose existence is problematic in more general situations.

## 10.2 Evolution processes associated with the free field

Let $\mathcal{H}$ be the Hilbert space of Section 10.1 and $H_0$ the free Hamiltonian defined in $\mathcal{H}$. The spectral measure $Q$ of multiplication by characteristic functions of Borel subsets of $S'(\mathbb{R}^{d-1})$ acts on $\mathcal{H}$. It is the spectral measure associated with the position operators $q(f)$, $f \in S(\mathbb{R}^{d-1})$, of the quantum field mapping $F \in \mathcal{H}$ to the function $\phi \longmapsto \phi(f) F(\phi)$, $\phi \in S'(\mathbb{R}^{d-1})$, that is, for each $f \in S(\mathbb{R}^{d-1})$, the spectral measure of the selfadjoint operator $q(f)$ is $Q \circ \langle f, \cdot \rangle^{-1}$.

As in the consideration of quantum mechanics in Section 3, we set

$$S_\lambda(t) = e^{-(t/\lambda) H_0}, \quad t \geq 0, \ \lambda \in \overline{\mathbb{C}}_+ \setminus \{0\}. \tag{10.9}$$

The operator is defined by the functional calculus for selfadjoint operators.

Let $(X_t)_{t\geq 0}$ be the Ornstein–Uhlenbeck process defined in Proposition 10.1 and let $S_t$ be the algebra defined by all cylinder sets

$$E = \{X_{t_1} \in B_1, \ldots, X_{t_n} \in B_n\} \tag{10.10}$$

for $0 \leq t_1 < \cdots < t_n \leq t$, $B_1, \ldots, B_n \in \mathcal{B}(\mathcal{S}(\mathbb{R}^{d-1}))$ and $n = 1, 2, \ldots$. Then for each $\lambda \in \overline{\mathbb{C}}_+ \setminus \{0\}$, we have an associated $(S_\lambda, Q)$-process

$$\left(\Omega, \langle S_t\rangle_{t\geq 0}, \langle M_\lambda^t\rangle_{t\geq 0}; \langle X_t\rangle_{t\geq 0}\right).$$

For $\lambda > 0$ and $E$ defined by (10.10), the Markov property for $\langle X_t\rangle_{t\geq 0}$ ensures that

$$\langle M_\lambda^t(E)u, v\rangle = \mu(\overline{v} \circ X_{t/\lambda}\{X_{t_1/\lambda} \in B_1, \ldots, X_{t_n/\lambda} \in B_n\}u \circ X_0), \ u, v \in \mathcal{H}, \tag{10.11}$$

so by analytic continuation from $\lambda > 0$, the operator valued set functions $\langle M_\lambda^t\rangle_{t\geq 0}$ are defined independently of the version of $\langle X_t\rangle_{t\geq 0}$ chosen for every $\lambda \in \overline{\mathbb{C}}_+ \setminus \{0\}$.

It follows from the representation (10.11) that $M_\lambda^t$ is the restriction to $S_t$ of an operator valued measure, denoted again by the same symbol, for each $\lambda > 0$.

Following the argument of Section 9 for the case of quantum mechanics, a net $\langle s_\alpha\rangle_{\alpha \in A}$ of simple functions converges to a function $f$ if and only if it converges to $f$ in the quasicomplete space $L^1(\langle M_\lambda^t\rangle_{\lambda>0})$ and the net is also Cauchy with respect to the seminorms

$$p_{E,K,\phi} : s \longmapsto \sup_{\lambda \in K} \left\| \int_E s(\omega)\,(M_\lambda^t\phi)(d\omega) \right\|_{\mathcal{H}} \tag{10.12}$$

as $E$ varies over cylinder sets (10.10), the function $\phi$ varies over $\mathcal{H}$ and $K$ varies over compact subsets of $\overline{\mathbb{C}}_+ \setminus \{0\}$. Then we can *define*

$$\int_E f(\omega)\,(M_\lambda^t\phi)(d\omega) := \lim_{\alpha \in A} \int_E s_\alpha(\omega)\,(M_\lambda^t\phi)(d\omega)$$

so that the convergence is uniform in the strong operator topology as $\lambda$ varies over compact subsets of $\overline{\mathbb{C}}_+ \setminus \{0\}$. It follows that the operator valued function

$$(E, \lambda) \longmapsto \int_E f(\omega)\,(M_\lambda^t\phi)(d\omega)$$

is additive in $E \in S_t$, analytic in $\lambda \in \mathbb{C}_+$ and continuous on $\overline{\mathbb{C}}_+ \setminus \{0\}$. The appropriate notions of measurability and null sets here is with respect to the family $\langle M_\lambda^t\rangle_{\lambda>0}$ of operator valued measures.

## 10.3 Multiplicative functionals of the free field and renormalization

In the present context, to say that $F_t$, $t \geq 0$, is a multiplicative functional means that for each $t \geq 0$, the function $F_t : \Omega \to \mathbb{C}$ is measurable with respect to the $\sigma$-algebra $\sigma(S_t)$ (or perhaps, an appropriate completion) and

$$F_{s+t} = F_t \circ \theta_s F_s, \quad (M_\lambda^{t+s})_{\lambda>0}\text{-a.e.,} \tag{10.13}$$

where $\theta_s : \Omega \to \Omega$ is the shift map given by formula (10.8) for all $s \geq 0$ and $\omega \in \Omega$.

If $F_t, t \geq 0$, is a multiplicative functional which is integrable in the sense of Section 10.2, then the operators

$$S_\lambda^F(t) = \int_\Omega F_t \, dM_\lambda^t, \quad t \geq 0,$$

have the semigroup property for all $\lambda \in \overline{\mathbb{C}}_+ \setminus \{0\}$ because the equality

$$\int_\Omega F_t \circ \theta_s F_s \, dM_\lambda^{t+s} = \int_\Omega F_t \, dM_\lambda^t \int_\Omega F_s \, dM_\lambda^s$$

holds for all $\lambda > 0$, and so for all $\lambda \in \overline{\mathbb{C}}_+ \setminus \{0\}$ by analytic continuation. We are seeking an integrable multiplicative functional $F_t : \Omega \to \mathbb{T}$ (with $\mathbb{T}$ the unit circle in $\mathbb{C}$) such that

$$e^{-itH} = \int_\Omega F_t \, dM_{-i}^t, \quad t \geq 0,$$

represents the dynamics of an interacting quantum field for some selfadjoint operator $H$. We saw in Section 7 how this was achieved in the elementary example of a Dirac particle on the line by suitably subtracting logarithmic divergences from the Feynman–Kac functional as paths crossed the simple singularity in the Coulomb-like potential at the origin, thereby adjusting the phase of the evolving state as a sum over paths. *Any measurable multiplicative functional $F_t : \Omega \to \mathbb{T}$ is $M_\lambda^t$-integrable for $\lambda > 0$. The existence of the dynamics for an interacting field is determined by the $M_\lambda^t$-integrability of $F_t$ for all $\lambda \in \overline{\mathbb{C}}_+ \setminus \{0\}$ and $t > 0$.*

According to [45, pp. 132-133], the kernel $G_{\frac{1}{2}}$ of the operator $(-\Delta_{d-1} + 1)^{-1/2}$ has the properties

$$G_{\frac{1}{2}}(x) = c|x|^{-(d-1)+1/2} + o(|x|^{-(d-1)+1/2}) \quad \text{as } x \to 0, \tag{10.14}$$

$$G_{\frac{1}{2}}(x) = O(e^{-|x|/2}) \quad \text{as } x \to \infty.$$

Because $\int_\Omega \langle f, X_t \rangle^2 \, d\mu = \int_{\mathbb{R}^{2(d-1)}} f(x) G_{\frac{1}{2}}(x-y) f(y) \, dx dy$ for every $f \in \mathcal{S}(\mathbb{R}^{d-1})$, it follows from the estimate (10.14) that $\int_\Omega \langle f_n, X_t \rangle^2 \, d\mu$ diverges as $f_n \to \delta_x$ weakly in the sense of measures as $n \to \infty$, for each $x \in \mathbb{R}^{d-1}$. This is unfortunate, because polynomials in the random field are just the type of interactions that need to be represented, for example, in the quantization of the classical $\phi^4$ field satisfying

$$-\Box\phi + \phi + 4\gamma\phi^3 = 0$$

in Minkowski space [20, p. 112].

For the purpose of discussing the most basic type of renormalization in the context of the construction of integrable multiplicative functionals of a random field, attention is restricted to $d = 2$. The 1-dimensional Laplacian $\Delta_1$ is actually the operator $d^2/dx^2$ with the domain of all functions $f \in L^2(\mathbb{R})$ such that $\int_{\mathbb{R}} \xi^2 |\hat{f}(\xi)|^2 \, d\xi < \infty$.

For $n = 1, 2, \ldots,$ let $\mathcal{P}^{\leq n}$ be the closed linear span in $\mathcal{H}$ of all monomials

$$\xi \longmapsto \prod_{j=1}^{k} \langle f_j, \xi \rangle, \quad \xi \in \mathcal{S}'(\mathbb{R}),$$

for all $f_j \in \mathcal{S}(\mathbb{R})$, $j = 1, \ldots, k$ and $k = 1, \ldots, n$. The union of all spaces $\mathcal{P}^{\leq n}$ is dense in $\mathcal{H}$ and the Hilbert space $\mathcal{H}$ can be represented as the direct sum of the orthogonal complements $\mathcal{P}^{(n)} = \mathcal{P}^{\leq n} \ominus \mathcal{P}^{\leq (n-1)}$ of $\mathcal{P}^{\leq (n-1)}$ in $\mathcal{P}^{\leq n}$ for $n = 1, 2, \ldots.$

Then for each $f \in \mathcal{S}'(\mathbb{R})$ and $n = 1, 2, \ldots,$ there exists a unique function $\Xi_f^n \in \mathcal{P}^{(n)}$ such that

$$\left( \Xi_f^n, \prod_{j=1}^{n} \langle f_j, \cdot \rangle \right)_{\mathcal{H}}$$

$$= \frac{n!}{2^n} \int_{\mathbb{R}} \cdots \int_{\mathbb{R}} \prod_{j=1}^{n} \left( (-\Delta_1 + 1)^{-1/2} (y_j - x) f_j(y_j) \, dy_j \right) f(x) \, dx$$

for all $f_1, \ldots, f_n \in \mathcal{S}(\mathbb{R})$. In particular, $f \longmapsto \Xi_f^n$, $f \in \mathcal{S}(\mathbb{R})$, is a linear map from $\mathcal{S}(\mathbb{R})$ to $\mathcal{H}$. The *Wick monomial* $\Xi_f^n$ is the orthogonal projection of the monomial function $\xi \longmapsto \langle f, \xi \rangle^n, \xi \in \mathcal{S}'(\mathbb{R})$ in $\mathcal{P}^{(n)}$. Denoting the Hermite polynomial of degree $n$ with leading coefficient one by $H_n$ and setting $c(f) = \frac{1}{\sqrt{2}} \left( (-\Delta_1 + I)^{-1/2} f, f \right)^{1/2}$, it follows that $\Xi_f^n$ is just $\xi \longmapsto c(f)^n H_n(c(f)^{-1} \langle f, \xi \rangle), \xi \in \mathcal{S}'(\mathbb{R})$ [20, Section 6.3]. Because the $\mathcal{H}$-valued function $f \longmapsto \Xi_f^n$ is uniformly continuous with respect to the norm $f \longmapsto c(f)$ on $\mathcal{S}(\mathbb{R})$, the random variable $\Xi_\Lambda^n$ may be defined by continuity as the element $\Xi_{\chi_\Lambda}^n$ of $\mathcal{H}$ for every $\Lambda \in \mathcal{B}_f(\mathbb{R})$.

Our process $X$ takes values in $\mathcal{S}'(\mathbb{R})$, so the random variable $\Xi_\Lambda^n \circ X_t$ over the probability space $(\Omega, \mathcal{F}, \mu)$ makes sense and is traditionally written as

$$\int_\Lambda : X_t^n :(x) \, dx.$$

The mapping $\Lambda \longmapsto \int_\Lambda : X_t^n :(x) \, dx$, $\Lambda \in \mathcal{B}_f(\mathbb{R})$, is an $L^2(\mu)$-valued measure but $: X_t^n :(x)$ is not defined as a random variable, that is, there is no $L^2(\mu)$-valued density with respect to Lebesgue measure on $\mathbb{R}$. Nevertheless, the multiplicative functional

$$F_t^{(\Lambda)}(\omega) := \exp\left[ -i \int_0^t \left( \int_\Lambda : X_s^n : (x) \, dx \right) (\omega) ds \right], \quad \Lambda \in \mathcal{B}_f(\mathbb{R}), \ t \geq 0,$$

$$\tag{10.15}$$

has the property that $F_t^{(\Lambda)}$ is measurable with respect to the $\sigma$-algebra generated by all random variables $\Phi_A$ with $A \in \mathcal{B}(\mathbb{R}^2)$ and $A \subset \Lambda \times [0, t], t \geq 0$. It is also possible to express the multiplicative functional (10.15) as the limit of multiplicative functionals of regularized processes $X^{(\epsilon)}$ defined at points of $\mathbb{R}$ [20, Proposition 8.5.1].

It is clear that the function $F_t^{(\Lambda)}$ is $M_\lambda^t$-integrable for each $\lambda > 0$ because it is $M_\lambda^t$-measurable with absolute value one. That $F_t^{(\Lambda)}$ is $M_\lambda^t$-integrable for $\lambda \in \overline{\mathbb{C}}_+ \setminus \{0\}$

and $t \geq 0$ follows from the fact that the closure of the operator $- \left( \frac{1}{\lambda} H_0 + i \, Q(\Xi_\Lambda^n) \right)$ is the generator of a contraction semigroup on $\mathcal{H}$ for every $\lambda \in \overline{\mathbb{C}}_+ \setminus \{0\}$, see [2, Section 5.3].

The treatment of the $P(\phi)_2$ Euclidean fields in [20] is not sufficient to determine the limiting behavior of the multiplicative functionals $F_t^{(\Lambda)}$ as $\Lambda \nearrow \mathbb{R}$ simply because $F_t^{(\Lambda)}$ is a random variable with absolute value one for each $\Lambda \in \mathcal{B}_f(\mathbb{R})$. The oscillatory nature of the expression (10.15) is a new feature and needs to be taken into account in order to make sense of formula (10.1). This is the subject of future work.

# References

[1] S. Albeverio and R. Høegh-Krohn, *Feynman Path Integrals*, Springer, Berlin, Heidelberg, New York, 1975.

[2] _____, Dirichlet forms and diffusion processes on rigged Hilbert spaces, *Z. Wahrsch. Verw. Gebiete* **40** (1977), 1–57.

[3] S. Albeverio, F. Gesztesy, R. Høegh-Krohn, H. Holden, *Solvable Models in Quantum Mechanics*, Springer-Verlag, Berlin,Heidelberg,New York, 1988.

[4] S. Albeverio, Z. Brzeźniak, L. Dąbrowski, Fundamental solution of the heat and Schrödinger equations with point interaction, *J. Funct. Anal.* **130** (1995), 220–254.

[5] S. Albeverio, G.W. Johnson and Z.M. Ma, The analytic operator valued Feynman integral via additive functionals of Brownian motion, *Acta Appl. Math.* **42** (1996), 267–295.

[6] S. Benvegnù, Relativistic point interaction with Coulomb potential in one dimension, *J. Math. Phys.* **38** (1997), 556–570.

[7] Ph. Blanchard, Ph. Combe, M. Sirugue, and M. Sirugue-Collin, Probabilistic solution of the Dirac equation, path integral representation for the solution of the Dirac equation in the presence of an electromagnetic field, *Bielefeld BiBoS Preprint Nos.* **44, 66** (1985).

[8] Z. Brzeźniak and B. Jefferies, Characterization of one-dimensional point interactions for the Schrödinger operator by means of boundary conditions, *J. Phys. A: Math. Gen.* **34** (2001), 2977–2983.

[9] _____, Renormalization of Coulomb interactions for the 1-D Dirac equation, *J. Math. Phys.* **44** (2003), 1638–1659.

[10] W. Caspers and Ph. Clément, Point interactions in $L^p$, *Semigroup Forum* **46** (1994), 253–265.

[11] P. Colella and O. Lanford III, Sample field behaviour for free Markov random fields, in *Constructive Quantum field Theory*, Lecture Notes in Physics **25**, Springer-Verlag, Berlin, Heidelberg, New York, 1973, 44–70.

[12] J. Diestel and J.J. Uhl Jr., *Vector Measures*, Math. Surveys No. 15, Amer. Math. Soc., Providence, 1977.

[13] R.L. Dobrushin and R.A. Minlos, An investigation of the properties of generalized Gaussian random fields, *Selecta Math. Soviet* **1** (1981), 215–263; originally published in *Zadachy Mekh. Mat. Fiz.* Moscow: Nauka, 1976, 117–165.

[14] A. Dynin, Feynman integral for functional Schrödinger equations, *Amer. Math. Soc. Transl.* Ser. 2 **206** (2002), 65–80.

[15] R. Feynman, space–time approach to non-relativistic quantum mechanics, *Rev. Mod. Phys.* **20** (1948), 367–387.

[16] W. Fischer, H. Leschke, P. Müller, The functional-analytic versus the functional-integral approach to quantum Hamiltonians: The one-dimensional hydrogen atom, *J. Math. Phys.* **36** (1995), 2313–2323.

[17] G. Folland, *Harmonic Analysis in Phase Space* , Princeton University press, Princeton, 1989.

[18] I.M. Gelfand and N.J. Vilenkin, *Generalized Functions* Vol. 4, Academic Press, New York, 1964.

[19] F. Gesztesy, P. Šeba, New analytically solvable models of relativistic point interactions, *Lett. Math. Phys.* **13** (1987), 345–358.

[20] J. Glimm and A. Jaffe, *Quantum Physics: A Functional integral Point of View*, Springer-Verlag, Berlin, Heidelberg, New York, 1981.

[21] R. Hughes, Renormalization of the relativistic delta potential in one dimension, *Lett. Math. Phys.* **34** (1995), 395–406.

[22] T. Ichinose, Path integral for the Dirac equation in two space–time dimensions, *Proc. Japan Acad. Ser. A Math. Sci* **58** (1982), 290–293.

[23] ———, Path integral for the Dirac equation, *Sugaku* **6** (1993), 15–31.

[24] T. Ichinose and H. Tamura, Zitterbewegung of a Dirac particle in two-dimensional space–time *J. Math. Phys.* **29** (1988), 103–109.

[25] A. Jaffe, Constructive quantum field theory. *Mathematical Physics 2000*, Imp. Coll. Press, London, 2000, 111–127.

[26] B. Jefferies, Remarks on the Feynman representation, *Publ. Res. Inst. Math. Sci., Kyoto U.* **21** (1985), 1311–1323.

[27] ———, Integration with respect to closable set functions, *J. Funct. Anal.* **66** (1986), 381–405.

[28] ———, *Evolution Processes and the Feynman–Kac Formula*, Kluwer Academic Publishers, Dordrecht, Boston,London, 1995.

[29] ———, Regular unbounded set functions, *J. d'Anal. Math.* **65** (1995), 125–144.

[30] ———, Integration structures for the operator valued Feynman integral, *J. Korean Math. Soc.* **38** (2001), 349–363.

[31] B. Jefferies and S. Okada, Dominated semigroups of operators and evolution processes, *Hokkaido Math. J.* **33** (2004), 127–151.

[32] B. Jefferies and W. Ricker, Integration with respect to vector valued Radon polymeasures, *J. Austral. Math. Soc.* (Ser. A) **56** (1994), 17–40.

[33] G.W. Johnson and M.L. Lapidus, *The Feynman Integral and Feynman's Operational Calculus*, Oxford U. Press, Oxford Mathematical Monograph, 2000.

[34] T. Kato, *Perturbation Theory for Linear Operators, 2nd Ed.*, Springer-Verlag, Berlin, Heidelberg, New York, 1980.

[35] I. Kluvánek and G. Knowles, *Vector Measures and Control Systems*, North Holland, Amsterdam, 1976

[36] I. Kluvánek, *Integration Structures*, Proc. Centre for Mathematical Analysis **18**, Australian Nat. Univ., Canberra, 1988.

[37] E. Nelson, Feynman integrals and the Schrödinger equation *J. Math. Phys.* **5** (1964), 332–343.

[38] M.M. Rao, *Stochastic Processes: General Theory*, Kluwer Academic Publishers, Dordrecht/Boston/London, 1995.

[39] M. Reed and B. Simon, *Methods of Modern Mathematical Physics* I–II, Academic Press, New York, 1973

[40] M. Röckner, Traces of harmonic functions and a new path space for the free quantum field, *J. Funct. Anal.* **79** (1988), 211–249.

[41] P. Rothnie, *Bilinear Integrals and the Feynman–Kac Formula*, PhD. thesis, UNSW, 2000.

[42] H.H. Schaefer, *Banach Lattices and Positive Operators*, Springer-Verlag, Berlin, 1974.

[43] L. Schwartz, *Radon Measures on Arbitrary Topological Spaces and Cylindrical Measures*, Tata Institute Publications, Oxford University Press, Bombay, 1973.

[44] O.G. Smolyanov, A.G. Tokarev and A. Truman, Hamiltonian Feynman path integrals via the Chernoff formula, *J. Math. Phys.* **43** (2002), 5161–5171.

[45] E. Stein, *Singular Integrals and the Differentiability Properties of Functions*, Princeton Univ. Press, Princeton, NJ, 1970.

[46] B. Thaller, *The Dirac Equation*, Springer-Verlag, Amsterdam, 1992.

[47] E. Thomas, Finite path integrals, *Acta. Applic. Math.* **43** (1996), 191–232.

[48] J. Weidmann, *Spectral Theory of Ordinary Differential Operators*, Lecture Notes in Mathematics 1258, Springer-Verlag, Berlin, Heidelberg, New York, 1987.

[49] E. Witten, Physical law and the quest for mathematical understanding, *Bull. Amer. Math. Soc.* **40** (2003), 21–29.

[50] T. Zastawniak, *Path integrals for the Dirac equation — some recent developments in the mathematical theory* in "Stochastic Analysis, Path Integration and Dynamics", K.D. Elworthy and J-C Zambrini,eds. Pitman Res. Notes in Math. **200**, Longman Scientific and Technical, Harlow, 1989, 243–263.

# Stochastic Differential Equations Based on Lévy Processes and Stochastic Flows of Diffeomorphisms

Hiroshi Kunita

Department of Mathematical Sciences, Nanzan University, Seirei-cho, Seto, 489-0863, Japan
kunita@ms.nanzan-u.ac.jp

## Introduction

Continuous stochastic differential equations (SDE) based on Brownian motions have been studied a lot. Among them, pathwise properties of the solution such as the continuity, the differentiability and the diffeomorphic properties of the solution with respect to the initial state were studied in detail in the past two decades. Some of these results can be found in the author's book [13].

In the mean time, similar problems have been studied for SDEs of jump type based on Lévy processes or semimartingales with jumps. Results are not parallel to those of continuous SDEs. In some cases the diffeomorphic property of the solution may fail.

The purpose of this chapter is to expose basic facts about the solution of a certain SDE with jumps. It will be shown in Section 3 that the solution is differentiable with respect to the initial state if coefficients of the equation are smooth. However, the homeomorphic property or the diffeomorphic property is not always satisfied owing to the behavior of jumps. In Section 3.4, we will show that the solution defines a stochastic flow of homeomorphisms, if it makes a "homeomorphic" jump.

For the study of SDE with jumps, we need stochastic analysis of semimartingales with jumps. In Section 1, we discuss briefly stochastic integrals based on semimartingales and establish Itô's formula for semimartingales with jumps. These could be considered as a basis for stochastic analysis of processes with jumps. For more details, see Meyer [13], Jacod-Shiryaev [12] and Protter [20]. In Section 2, we study Lévy processes by applying results of Section 1. Among Lévy processes, Brownian motions and Poisson random measures play important roles in this work. We study these two with details in Section 2. For related problems, we refer to Ikeda–Watanabe [10] and Sato [21].

Section 3 is the main part of this chapter. We introduce a SDE with jumps. In order to make the discussion simple, we will restrict our attention to a SDE based on a Brownian motion and a Poisson random measure, though more general SDEs based on semimartingales (with spatial parameter) are studied in the literature (e.g., Fujiwara–Kunita [7,8,9], Carmona–Nualart [3], Applebaum–Tang [2]). We study the pathwise properties of the solutions such as the differentiability and the diffeomorphic property of the solution with respect to the initial state. For this purpose we obtain various types of $L^p$ estimates of the solution by applying Burkholder's inequality for stochastic integrals and then we apply Kolmogorov's criterion on the continuity of random fields. In Section 4 (Appendix), we discuss the Kolmogorov criterion or Kolmogorov–Totoki's theorem.

Most material of Section 3 is chosen from the joint works with Fujiwara, though some improvements are given here. The author expresses his gratitude to T. Fujiwara for his cooperative work on stochastic flows with jumps. Also he thanks D. Applebaum for pointing out errors in the first version of this article.

# 1 Stochastic integrals for semimartingales

## 1.1 Martingales, localmartingales and semimartingales

Let $(\Omega, \mathcal{F}, P)$ be a complete probability space. Suppose that we are given a family of sub $\sigma$-fields $\{\mathcal{F}_t\}, t \in [0, T]$ of $\mathcal{F}$ satisfying the following properties.
1) (Increasing) $\mathcal{F}_s \subset \mathcal{F}_t$ for any $s < t$.
2) (Right continuous) $\cap_{h>0}\mathcal{F}_{t+h} = \mathcal{F}_t$ holds for any $t$.
3) (Complete) Each $\mathcal{F}_t$ contains null sets of $\mathcal{F}$.
Then $\{\mathcal{F}_t\}, t \in [0, T]$ is called a (standard) filtration.

Let $X(t), t \in [0, T]$ be a real stochastic process. If $X(t)$ is $\mathcal{F}_t$-measurable for any $t \in [0, T]$, the process is called *adapted*. An adapted process $X(t)$ is called a *martingale* if $X(t)$ is integrable for any $t$ and equalities

$$E[X(t)|\mathcal{F}_s] = X(s), \quad a.s. \ \forall s < t$$

hold. If equality signs are replaced by $\geq$ in the above, it is called a *submartingale*. Further, if $-X(t)$ is a submartingale $X(t)$ is called a *supermartingale*. It is known that any submartingale $X(t)$ has a modification $\tilde{X}(t)$ whose sample paths are cadlag (right continuous with the left hand limits) a.s. Further if its sample paths are continuous a.s., it is called a *continuous martingale*. In the following we always consider cadlag martingales or submartingales.

Here we quote a useful inequality for martingales, which may be found in text books discussing martingale theory. See, e.g., Dellacherie–Meyer [4], Ikeda–Watanabe [10].

**Theorem 1.1 (Doob's inequality).** *Let $p > 1$ be any number. Let $X(t)$ be a martingale such that $E[|X(t)|^p] < \infty$. Then it holds*

$$E[\sup_{r \leq t} |X(r)|^p] \leq q^p E[|X(t)|^p], \quad \forall t \tag{1.1}$$

*where $q$ is a positive number such that $q^{-1} = 1 - p^{-1}$.*

A random variable $\tau$ with values in $[0, T]$ is called a *stopping time* if $\{\omega; \tau(\omega) \leq t\} \in \mathcal{F}_t$ holds for any $t$. We denote by $\mathcal{T}$ the set of all stopping times.

An adapted cadlag process $X(t)$ is called a *localmartingale* if there exists an increasing sequence of stopping times $\tau_n$ such that $P(\tau_n < T) \to 0$ as $n \to \infty$ and each stopped process $X^n(t) = X(t \wedge \tau_n)$ is a martingale. The corresponding sequence of stopping times $\{\tau_n\}$ is called the *reducing stopping times*. Further if each $X^n(t)$ is a square integrable martingale, it is called a *locally square integrable martingale*.

An adapted cadlag process $A(t)$ with $A(0) = 0$ is called an *increasing process* if it is increasing with respect to $t$ a.s. Further a process $A(t)$ is called a *process of finite variation* if it is written as the difference of two increasing processes. An adapted cadlag process $X(t)$ is called a *semimartingale* if it is written as the sum of a locally square integrable martingale and a process of finite variation.

The *predictable $\sigma$-field* $\mathcal{P}$ is the $\sigma$-field on $\Omega \times [0, T]$ generated by left continuous adapted processes. A $\mathcal{P}$-measurable stochastic process is called a *predictable process*.

**Theorem 1.2 (Doob–Meyer decomposition).** *Let $X(t)$ be a supermartingale such that the class of random variables $\{X(\tau); \tau \in \mathcal{T}\}$ is uniformly integrable. Then there exists a unique martingale $M(t)$ and an integrable predictable increasing process $A(t)$ such that*

$$X(t) = M(t) - A(t), \quad A(0) = 0.$$

For the proof, see Dellacherie–Meyer [4], Ikeda–Watanabe [10], Protter [20].

## 1.2 Stochastic integrals

Let $X(t)$ be a square integrable martingale. Then $X(t)^2$ is a nonnegative submartingale. Further, $\sup_{0 \leq t \leq T} X(t)^2$ is integrable by Doob's inequality. Then the class of random variables $\{X(\tau)^2; \tau \in \mathcal{T}\}$ is uniformly integrable. Therefore there exists a unique integrable predictable increasing process $A(t)$ such that $X(t)^2 - A(t)$ is a martingale by Doob–Meyer's decomposition theorem. We denote $A(t)$ by $\langle X \rangle_t$. Thus $X(t)^2 - \langle X \rangle_t$ is a martingale. Then it holds for any $s < t$,

$$E[(X(t) - X(s))^2 | \mathcal{F}_s] = E[\langle X \rangle_t - \langle X \rangle_s | \mathcal{F}_s], \quad a.s. \tag{1.2}$$

Conversely, suppose that for a given square integrable martingale $X(t)$, there exists a predictable increasing process $A(t)$ such that

$$E[(X(t) - X(s))^2 | \mathcal{F}_s] = E[A(t) - A(s) | \mathcal{F}_s], \quad a.s. \ \forall s < t.$$

Then $X(t)^2 - A(t)$ is a martingale, so that $A(t) = \langle X \rangle_t$ holds by the uniqueness of the Doob–Meyer decomposition of a supermartingale.

We set

$$L^2(\langle X \rangle) = \left\{ f(s); \text{predictable and } E\left[ \int_0^T |f(s)|^2 d\langle X \rangle_s \right] < \infty \right\}.$$

It is a Hilbert space with the norm $\|f\| = E\left[ \int_0^T |f(s)|^2 d\langle X \rangle_s \right]^{1/2}$. Let $f(s)$ be an adapted process. It is called a *simple predictable process* if it is written as

$$f(s) = \sum_i f_i 1_{(s_i, s_{i+1}]}(s),$$

where $0 = s_0 < s_1 < \cdots < s_n = T$ with $f_i$ bounded and $\mathcal{F}_{s_i}$-measurable. The set of all simple predictable processes $S$ is dense in $L^2(\langle X \rangle)$.

We shall define the stochastic integral based on a square integrable martingale $X(t)$. Let $f(s)$ be a simple predictable process. We define

$$M(t) = \sum_{i=1}^{n} f_i(X(s_{i+1} \wedge t) - X(s_i \wedge t)),$$

and call it the stochastic integral of $f$ by $X$ and denote it by $\int_0^t f(s) dX(s)$.

**Lemma 1.3.** *The stochastic integral is a square integrable martingale. Further*

$$\left\langle \int_0^{\cdot} f(s) dX(s) \right\rangle_t = \int_0^t f(s)^2 d\langle X \rangle_s, \tag{1.3}$$

$$E\left[ \left| \int_0^t f(s) dX(s) \right|^2 \right] = E\left[ \int_0^t |f(s)|^2 d\langle X \rangle_s \right]. \tag{1.4}$$

*Proof.* It can be verified directly that the above $M(t)$ is a square integrable martingale. If $s_{n_0} = s$ and $s_{n_1} = t$, we have the equality

$$E[M(t)^2|\mathcal{F}_s] - M(s)^2 = \sum_{i=n_0}^{n_1-1} E[E[(M(s_{i+1}) - M(s_i))^2|\mathcal{F}_{s_i}]|\mathcal{F}_s]$$

$$+2 \sum_{n_0 \le i < j \le n_1-1} E[E[f_i f_j(X(s_{j+1}) - X(s_j))(X(s_{i+1}) - X(s_i))|\mathcal{F}_{s_i}]|\mathcal{F}_s].$$

It holds

$$E[(M(s_{i+1}) - M(s_i))^2|\mathcal{F}_{s_i}] = f_i^2 E[(X(s_{i+1}) - X(s_i))^2|\mathcal{F}_{s_i}]$$
$$= f_i^2 E[\langle X \rangle_{s_{i+1}} - \langle X \rangle_{s_i}|\mathcal{F}_{s_i}],$$

and for $i < j$

$$E[f_i f_j(X(s_{j+1}) - X(s_j))(X(s_{i+1}) - X(s_i))|\mathcal{F}_{s_i}] =$$
$$E[f_i f_j E[X(s_{j+1}) - X(s_j)|\mathcal{F}_{s_j}](X(s_{i+1}) - X(s_i))|\mathcal{F}_{s_i}] = 0.$$

Therefore,

$$E[(M(t) - M(s))^2|\mathcal{F}_s] = \sum_{n_0 \le i < n_1-1} E[f_i^2(\langle X \rangle_{s_{i+1}} - \langle X \rangle_{s_i})|\mathcal{F}_s]$$

$$= E[\int_s^t |f(u)|^2 d\langle X \rangle_u|\mathcal{F}_s],$$

for any $s < t$. This implies $\langle M \rangle_t = \int_0^t f(u)^2 d\langle X \rangle_u$, proving the first equality of the lemma. The second equality is immediate from the first equality.

Next consider any $f \in L^2(\langle X \rangle)$. We can choose a sequence of simple predictable processes $\{f_n(s)\}$ such that $E[\int_0^T |f(s) - f_n(s)|^2 d\langle X \rangle_s] \to 0$. Let $M_n(t)$ and $M_m(t)$ be stochastic integrals of $f_n$ and $f_m$, respectively. Then we have by Doob's inequality

$$E[\sup_{t \leq T} |M_n(t) - M_m(t)|^2] \leq 4E[|M_n(T) - M_m(T)|^2]$$

$$= 4E[\int_0^T |f_n(s) - f_m(s)|^2 d\langle X \rangle_s].$$

It converges to 0 as $n, m \to \infty$. Then the sequence of martingales $\{M_n(t)\}$ converges uniformly in $t$ in $L^2$-sense. The $L^2$-limit $M(t)$ is a cadlag martingale. It holds

$$E[(M(t) - M(s))^2|\mathcal{F}_s] = \lim_{n \to \infty} E[(M_n(t) - M_n(s))^2|\mathcal{F}_s]$$

$$- \lim_{n \to \infty} E[\int_s^t f_n(u)^2 d\langle X \rangle_u|\mathcal{F}_s] = E[\int_s^t f(u)^2 d\langle X \rangle_u|\mathcal{F}_s].$$

Therefore we have $\langle M \rangle_t = \int_0^t f(u)^2 d\langle X \rangle_u$.

We denote the above $M(t)$ by $\int_0^t f(s)dX(s)$ and call it the *stochastic integral of $f(s)$ based on $X(t)$*. We have isometric properties (1.3) and (1.4).

Next suppose that $X(t)$ is a locally square integrable martingale and $\{\tau_n\}$ is a sequence of reducing stopping times. Then for each $X^n(t) = X(t \wedge \tau_n)$, there exists a unique predictable increasing process $\langle X^n \rangle_t$ such that $X^n(t)^2 - \langle X^n \rangle_t$ is a martingale. If $m < n$, we have $X^n(t \wedge \tau_m) = X^m(t)$. Then we have $\langle X^n \rangle_{t \wedge \tau_m} = \langle X^m \rangle_t$. Therefore there exists a predictable increasing process $\langle X \rangle_t$ such that $\langle X \rangle_{t \wedge \tau_n} = \langle X^n \rangle_t$. It follows that the process $X(t)^2 - \langle X \rangle_t$ is a localmartingale.

We shall extend the definition of the stochastic integrals to locally square integrable martingales. Let $X(t)$ be a locally square integrable martingale with a sequence of reducing stopping times $\{\tau_n\}$. Let $f(s)$ be a predictable process with the square integrability condition $\int_0^T |f(s)|^2 d\langle X \rangle_s < \infty$, a.s. Define

$$\sigma_n = \inf\{t \in [0, T]; \int_0^t |f(s)|^2 ds \geq n\} \wedge \tau_n,$$

and set $f^n(s) = f(s)1_{\{s < \sigma_n\}}$. Then it holds $P(\sigma_n < T) \to 0$ as $n \to \infty$ and $E[\int_0^T |f^n(s)|^2 d\langle X^n \rangle_s] < \infty$ for any $n$. Now we can define stochastic integrals $M^n(t) := \int_0^t f^n(s)dX^n(s)$ as square integrable martingales. Further it holds for $m < n$ $M^n(t \wedge \sigma_m) = M^m(t)$. Then there exists a locally square integrable martingale $M(t)$ such that $M(t \wedge \sigma_n) = M^n(t)$. We denote the localmartingale $M(t)$ by $\int_0^t f(s)dX(s)$ and call it the stochastic integral of $f(s)$ by $X(t)$.

Now let $X(t)$ be a semimartingale decomposed as $X(t) = M(t) + A(t)$, where $M(t)$ is a locally square integrable martingale and $A(t)$ is a process of finite variation. The total variation process of $A(t)$ is denoted by $|A|(t)$. Suppose that $f(t)$ is a predictable process satisfying the integrability condition, $\int_0^T |f(s)|^2 d\langle M \rangle_s + \int_0^T |f(s)|d|A|(s) < \infty$. We may define the stochastic integral of $f$ by $X(t)$ by

$$\int_0^t f(s)dX(s) := \int_0^t f(s)dM(s) + \int_0^t f(s)dA(s),$$

where the last integral $\int_0^t f(s)dA(s)$ is the usual Stieltjes integral by the function of finite variation. Then the stochastic integral $\int_0^t f(s)dX(s)$ is a semimartingale for any $f$ with the above integrability condition.

Now let $X(t)$ be a semimartingale decomposed as $X(t) = M(t) + A(t)$, where $M(t)$ is a locally square integrable martingale and $A(t)$ is a process of finite variation, and let $f(t)$ be a cadlag adapted process. Then the left limit $f(t-) = \lim_{h \downarrow 0} f(t-h)$ is a predictable process and it satisfies the integrability condition mentioned above. Therefore the stochastic integral $\int_0^t f(s-)dX(s)$ is well defined. We shall approximate it by a sequence of finite sums. Let us denote by $\Pi$ a partition of $[0, T]$; $\Pi = \{0 = t_0 < t_1 < \cdots < t_{n-1} < t_n = T\}$. We set $|\Pi| = \max_i |t_{i+1} - t_i|$, and define

$$Y^\Pi(t) = \sum_i f(t_i \wedge t)(X(t_{i+1} \wedge t) - X(t_i \wedge t)). \tag{1.5}$$

**Theorem 1.4.** *Let $\{\Pi_k\}$ be a sequence of partitions such that $|\Pi_k| \to 0$. Then $\{Y^{\Pi_k}(t)\}$ converges to $\int_0^t f(s-)dX(s)$ uniformly in $t$ in probability, i.e.,*

$$\lim_{k \to \infty} P\left( \sup_{0 \le t \le T} \left| Y^{\Pi_k}(t) - \int_0^t f(s-)dX(s) \right| > \epsilon \right) = 0$$

*is valid for any $\epsilon > 0$.*

*Proof.* It is sufficient to prove the theorem in the case where $X(t)$ is a locally square integrable martingale and the case where $X(t)$ is a process of finite variation. In the latter case, the stochastic integral is just equal to the pathwise Stieltjes integral. Then $Y^{\Pi_k}(t)$ converges to the stochastic integral $\int_0^t f(s-)dX(s)$ uniformly in $t$ a.s. as $k \to \infty$. So we consider the case where $X(t)$ is a locally square integrable martingale. Let $\{\tau_n\}$ be a sequence of reducing stopping times of $X(t)$. Set $\sigma_n = \inf\{t;\ |f(t)|^2 > n\} \wedge \tau_n$. Then $\{\sigma_n\}$ is a sequence of the reducing stopping times of $\int_0^t f(s-)dX(s)$. It holds

$$E\left[ \sup_{0 \le t \le T} \left| Y^{\Pi_k}(t \wedge \sigma_n) - \int_0^{t \wedge \sigma_n} f(s-)dX(s) \right|^2 \right]$$

$$\le 4E\left[ \int_0^{T \wedge \sigma_n} |f^{\Pi_k}(s-) - f(s-)|^2 d\langle X \rangle_s \right],$$

where $f^{\Pi_k}(s) = \sum_i f(s_i)1_{[s_i, s_{i+1})}(s)$. Then the above expectation converges to 0 as $k \to \infty$. On the other hand, it holds $P(\sigma_n < T) \to 0$ as $n \to \infty$. For any $\delta > 0$, choose $n$ such that $P(\sigma_n < T) < \delta$. Hence we have

$$P\left( \sup_{0 \le t \le T} \left| Y^{\Pi_k}(t) - \int_0^t f(s-)dX(s) \right| > \epsilon \right)$$

$$\le P\left( \sup_{0 \le t \le T} \left| Y^{\Pi_k}(t \wedge \sigma_n) - \int_0^{t \wedge \sigma_n} f(s-)dX(s) \right| > \epsilon, \sigma_n = T \right) + \delta.$$

The first term of the right hand side converges to 0 as $k \to \infty$ by Chebychev's inequality. Since $\delta$ is arbitrary, we get the assertion of the theorem.

## 1.3 Orthogonal martingales

For two square integrable martingales $X(t)$ and $Y(t)$, we define

$$\langle X, Y \rangle_t = \frac{1}{4}\Big\{\langle X+Y \rangle_t - \langle X-Y \rangle_t\Big\}.$$

Then $X(t)Y(t) - \langle X, Y \rangle_t$ is a martingale. We have obviously $\langle X, X \rangle_t = \langle X \rangle_t$. The process $\langle X, Y \rangle_t$ is called the bracket of $X$ and $Y$. The bracket satisfies

$$E[(X(t) - X(s))(Y(t) - Y(s))|\mathcal{F}_s] = E[\langle X, Y \rangle_t - \langle X, Y \rangle_s|\mathcal{F}_s], \quad \forall s < t.$$

We give a characterization of stochastic integrals by means of the bracket. We first show

**Lemma 1.5.** *Let $X, Y$ be square integrable martingales. Let $f, g$ be predictable processes belonging to $L^2(\langle X \rangle)$ and $L^2(\langle Y \rangle)$, respectively. Then $fg$ is integrable with respect to $\langle X, Y \rangle$. Further, we have*

$$\left|\int_0^t f(s)g(s)d\langle X, Y \rangle_s\right| \le \left(\int_0^t |f(s)|^2 d\langle X \rangle_s\right)^{1/2} \left(\int_0^t |g(s)|^2 d\langle Y \rangle_s\right)^{1/2}. \quad (1.6)$$

*Proof.* We first observe that for any $t > s$, the bracket $\langle X, Y \rangle_t - \langle X, Y \rangle_s$ is a positive bilinear form a.s. Then we have by Schwarz's inequality,

$$|\langle X, Y \rangle_t - \langle X, Y \rangle_s| \le (\langle X \rangle_t - \langle X \rangle_s)^{1/2}(\langle Y \rangle_t - \langle Y \rangle_s)^{1/2}.$$

Now suppose that both $f, g$ are simple predictable processes of the form $\sum_i f_i 1_{(s_i, s_{i+1}]}(s)$ and $\sum_i g_i 1_{(s_i, s_{i+1}]}(s)$, respectively, where $f_i, g_i$ are bounded $\mathcal{F}_{s_i}$-measurable random variables. Then we have

$$\left|\sum_i f_i g_i(\langle X, Y \rangle_{s_{i+1}} - \langle X, Y \rangle_{s_i})\right|$$
$$\le \sum_i (f_i^2(\langle X \rangle_{s_{i+1}} - \langle X \rangle_{s_i}))^{1/2}(g_i^2(\langle Y \rangle_{s_{i+1}} - \langle Y \rangle_{s_i}))^{1/2}$$
$$\le \{\sum_i f_i^2(\langle X \rangle_{s_{i+1}} - \langle X \rangle_{s_i})\}^{1/2}\{\sum_i g_i^2(\langle Y \rangle_{s_{i+1}} - \langle Y \rangle_{s_i})\}^{1/2}.$$

This proves the inequality (1.6) in the case where $f, g$ are simple predictable processes. We can show the inequality for general $f, g$ by approximating them by sequences of simple predictable processes.

**Theorem 1.6.** *Let* $X, Y$ *be square integrable martingales and let* $f \in L^2(\langle X \rangle)$. *Then we have*

$$\left\langle \int f dX, Y \right\rangle_t = \int_0^t f(s) d\langle X, Y \rangle_s. \tag{1.7}$$

*Conversely suppose that a square integrable martingale* $N(t)$ *satisfies*

$$\langle N, Y \rangle_t = \int_0^t f(s) d\langle X, Y \rangle_s$$

*for any square integrable martingale* $Y$. *Then* $N(t) = N(0) + \int_0^t f(s) dX(s)$ *holds valid.*

*Proof.* Suppose that $f$ is a simple process given by $f = \sum_i f_i 1_{(s_i, s_{i+1}]}$, where $s_{n_0} = s$ and $s_{n_1} = t$. Set $M(t) = \int_0^t f(s) dX(s)$. Then we have

$$E[(M(t) - M(s))(Y(t) - Y(s))|\mathcal{F}_s]$$

$$= \sum_{n_0 \leq k < n_1} E[E[(M(s_{k+1}) - M(s_k))(Y(s_{k+1}) - Y(s_k))|\mathcal{F}_{s_k}]|\mathcal{F}_s]$$

$$= \sum_{n_0 \leq k < n_1} E[f_k E[(X(s_{k+1}) - X(s_k))(Y(s_{k+1}) - Y(s_k))|\mathcal{F}_{s_k}]|\mathcal{F}_s]$$

$$= \sum_{n_0 \leq k < n_1} E[f_k E[\langle X, Y \rangle_{s_{k+1}} - \langle X, Y \rangle_{s_k}|\mathcal{F}_{s_k}]|\mathcal{F}_s]$$

$$= E[\sum_{n_0 \leq k < n_1} f_k(\langle X, Y \rangle_{s_{k+1}} - \langle X, Y \rangle_{s_k})|\mathcal{F}_s]$$

$$= E[\int_s^t f(u) d\langle X, Y \rangle_u|\mathcal{F}_s].$$

This proves (1.7) in the case where $f$ is a simple predictable process. The equality can be extended to any $f \in L^2(\langle X \rangle)$ by using Lemma 1.5, since simple predictable processes are dense in $L^2(\langle X \rangle)$.

For the proof of the latter assertion, observe that $M - N$ satisfies $\langle M - N, Y \rangle_s = 0$, for all square integrable martingales $Y$. Then setting $Y = M - N$, we find that $\langle M - N \rangle = 0$, proving $M - N = $ constant. The proof is complete.

We denote by $\mathcal{M}$ the set of all square integrable martingales $X(t)$ such that $X(0) = 0$. For each $X \in \mathcal{M}$, we define an $L^2$ norm by $\|X\| = E[|X(T)|^2]^{1/2}$. It is a Hilbert space.

Let $\mathcal{N}$ be a subset of $\mathcal{M}$. It is called a *stable subspace* of $\mathcal{M}$ if it satisfies the following.
1) $\mathcal{N}$ is a closed subspace of $\mathcal{M}$ as a vector space.
2) For any $X \in \mathcal{N}$ and $f \in L^2(\langle X \rangle)$, the stochastic integral $\int f dX$ belongs to $\mathcal{N}$.

For a given $X \in \mathcal{M}$, we set

$$\mathcal{L}(X) = \left\{ \int_0^t f(s) dX(s) : \ f \in L^2(\langle X \rangle) \right\}.$$

Then it is a stable subspace of $\mathcal{M}$. The set $\mathcal{L}(X)$ is called the *stable subspace generated by X*.

Let $X, Y$ be two square integrable martingales. These are called *orthogonal* if $\langle X, Y \rangle_t \equiv 0$ or equivalently the product $X(t)Y(t)$ is a martingale. For a given stable subspace $\mathcal{N}$, we set

$$\mathcal{N}^{\perp} = \{Y \in \mathcal{M} : \langle Y, X \rangle = 0 \ \forall X \in \mathcal{N}\}.$$

It is clearly a closed vector space. Further it is also stable. Indeed, if $Y \in \mathcal{N}^{\perp}$ and $g \in L^2(\langle Y \rangle)$, then $\langle \int g dY, X \rangle = \int g(s) d \langle Y, X \rangle_s = 0$ holds for any $X \in \mathcal{N}$. Therefore $\int g dY \in \mathcal{N}^{\perp}$.

We will prove that an arbitrary given element $Y$ of $\mathcal{M}$ is decomposed uniquely to the sum of $Y_1$ and $Y_2$, where $Y_1 \in \mathcal{L}(X)$ and $Y_2$ is orthogonal to $\mathcal{L}(X)$. We first prepare a lemma:

**Lemma 1.7.** *For any given* $X, Y \in \mathcal{M}$, *there exists a unique* $f \in L^2(\langle X \rangle)$ *satisfying*

$$\langle X, Y \rangle = \int f d \langle X \rangle.$$

*Proof.* Let $A$ be a set in $\mathcal{B}([0, T]) \times \mathcal{F}$ such that the indicator function $1_A$ is predictable. Suppose that $\int 1_A d \langle X \rangle = 0$ a.s. Then we have $\int 1_A d \langle X, Y \rangle = 0$ by Lemma 1.5. Therefore, the process of finite variation $\langle X, Y \rangle_t(\omega)$ is absolutely continuous with respect to the increasing function $\langle X \rangle_t(\omega)$, for almost all $\omega$. The Radon–Nikodym density $f(t, \omega)$ can be chosen as a predictable process. We will prove that the above $f(t)$ belongs to $L^2(\langle X \rangle)$. Let $c$ be a positive constant. We set $f^c(s) = f(s)$ if $|f(s)| \le c$ and $f^c(s) = 0$ if $|f(s)| > c$. Since $|f^c(s)|^2 = f^c(s)f(s)$ holds, we have

$$\int |f^c(s)|^2 d \langle X \rangle_s = \int f^c(s) \langle X, Y \rangle_s \le \left( \int |f^c(s)|^2 d \langle X \rangle_s \right)^{1/2} \langle Y \rangle^{1/2}.$$

Therefore we get $\int |f^c(s)|^2 d \langle X \rangle_s \le \langle Y \rangle$. Since $c$ is arbitrary, we have $\int |f(s)|^2 d \langle X \rangle_s \le \langle Y \rangle$, showing $f \in L^2(\langle X \rangle)$. The uniqueness of $f$ will be obvious.

**Proposition 1.8.** *Let* $X, Y$ *be any elements of* $\mathcal{M}$. *Then* $Y$ *is decomposed uniquely to the sum of* $Y_1 \in \mathcal{L}(X)$ *and* $Y_2$ *which is orthogonal to* $\mathcal{L}(X)$.

*Proof.* Let $f(s)$ be the predictable process of Lemma 1.7. Set $Y_1 = \int_0^t f(s) dX(s)$ and $Y_2 = Y - Y_1$. Then $Y_1 \in \mathcal{L}(X)$. Further, $\langle Y_1, X \rangle = \int f(s) d \langle X \rangle = \langle Y, X \rangle$. Therefore we have $\langle Y_2, X \rangle = 0$. We have thus shown the existence of the orthogonal decomposition. The uniqueness will be obvious.

We denote $Y_1$ of the proposition by $P_{\mathcal{L}(X)} Y$ and call it the *orthogonal projection* of $Y$ to $\mathcal{L}(X)$.

Now suppose that we are given a sequence of martingales $X_1, ..., X_n$ of $\mathcal{M}$. We define a sequence of the orthogonal martingales by the Gram–Schmidt's orthogonalization method;

$$Y_1(t) = X_1(t), \quad Y_2(t) = X_2(t) - P_{\mathcal{L}(Y_1)}X_2, ...,$$

$$Y_n(t) = X_n(t) - \sum_{i=1}^{n-1} P_{\mathcal{L}(Y_i)}X_n.$$

Then these $Y_i, i = 1, ..., n$ are orthogonal martingales.

An orthogonal system $\{Y_n, n = 1, 2, ...\}$ is called an orthogonal base of $\mathcal{M}$ if $Y_n \neq 0$ for any $n$ and every $X$ is represented by $X = \sum_n P_{\mathcal{L}(Y_n)}X$.

For an arbitrary stable subspace $\mathcal{N}$, there exists a (at most) countable orthogonal basis $\{Y_n\}$. This can be shown similarly to the existence of orthogonal bases of $\mathcal{M}$. Now let $X$ be an arbitrary element of $\mathcal{M}$. We set $X_1 = \sum_n P_{\mathcal{L}(Y_n)}X$ and $X_2 = X - X_1$. Then $X_1 \in \mathcal{N}$ and $X_2 \in \mathcal{N}^{\perp}$. Therefore $X$ has the orthogonal decomposition $X = X_1 + X_2$. We can write the orthogonal decomposition as $\mathcal{M} = \mathcal{N} \oplus \mathcal{N}^{\perp}$.

Let $\mathcal{M}_c$ be the set of all continuous martingales $X(t)$ in $\mathcal{M}$. Then stochastic integrals $\int_0^t f(s)dX(s)$ for $X \in \mathcal{M}_c$ and $f \in L^2(\langle X \rangle)$ are continuous martingales. Therefore $\mathcal{M}_c$ is a stable subspace of $\mathcal{M}$. We set $\mathcal{M}_d = \mathcal{M}_c^{\perp}$ and call elements of $\mathcal{M}_d$ *purely discontinuous martingales*. Thus we have the orthogonal decomposition

$$\mathcal{M} = \mathcal{M}_c \oplus \mathcal{M}_d,$$

and any element $X$ of $\mathcal{M}$ is written as the sum of a continuous martingale $X_c \in \mathcal{M}_c$ and a purely discontinuous one $X_d \in \mathcal{M}_d$. Such a decomposition is unique.

## 1.4  Quadratic variations and Stratonovich integrals

Let $X(t)$ be a semimartingale. Then $X(t)$ is a cadlag process and $X(s-) = \lim_{h \downarrow 0} X(s - h)$ exists. Then the stochastic integral $\int_0^t X(s-)dX(s)$ is well defined as a semimartingale. We define the quadratic variation of the semimartingale $X(t)$ by

$$[X]_t = X(t)^2 - 2 \int_0^t X(s-)dX(s) - X(0)^2. \tag{1.8}$$

Co-quadratic variation of two semimartingales $X, Y$ is defined by

$$[X, Y]_t = \frac{1}{4}\Big\{ [X + Y] - [X - Y] \Big\}. \tag{1.9}$$

For a partition $\Pi = \{0 = t_0 < t_1 < \cdots < t_n = T\}$, we set

$$[X, Y]_t^{\Pi} = \sum_{i=1}^{n-1} (X(t_{i+1} \wedge t) - X(t_i \wedge t))(Y(t_{i+1} \wedge t) - Y(t_i \wedge t)). \tag{1.10}$$

**Lemma 1.9.** *Let $\{\Pi_n\}$ be a sequence of partitions such that $|\Pi_n| \to 0$. Then we have*

$$\lim_{n \to \infty} [X, Y]_t^{\Pi_n} = [X, Y]_t$$

*uniformly in $t \in [0, T]$ in probability.*

*Proof.* We give the proof in the case where $X = Y$ only. We will apply Theorem 1.4. Let $\{Y^{\Pi_n}(t)\}$ be a sequence of stochastic integrals which approximates the integral $\int_0^t X(s-)dX(s)$. It holds $X(t)^2 - X(0)^2 - 2Y^{\Pi_n}(t) = [X]_t^{\Pi_n}$. Let $n$ tend to infinity. Then we find that $[X]_t^{\Pi_n}$ converges uniformly in probability and the limit is equal to $X(t)^2 - X(0)^2 - 2\int_0^t X(s-)dX(s)$. Therefore, $\lim_{n\to\infty}[X]_t^{\Pi_n}$ exists in probability and it coincides with $[X]$.

From the above lemma, the quadratic variation $[X]_t$ is an increasing process and the quadratic co-variation $[X, Y]$ is a process of finite variation. Then the latter is written as the sum of a continuous process of finite variation denoted by $[X, Y]_t^c$ and a purely discontinuous process of finite variation denoted by $[X, Y]_t^d$. The decomposition is unique. $[X, Y]_t^c$ is called the *continuous part* of $[X, Y]_t$.

**Proposition 1.10.** *Let $X(t)$ be a square integrable martingale. If $X(t)$ is a continuous martingale, then we have $[X]_t = \langle X \rangle_t$. If $X(t)$ is purely discontinuous, then we have $[X]_t = \sum_{s \leq t}(\Delta X(s))^2$. Generally, let $X(t) = X_c(t) + X_d(t)$ be the orthogonal decomposition such that $X_c \in \mathcal{M}_c$ and $X_d \in \mathcal{M}_d$. Then it holds*

$$[X]_t = \langle X_c \rangle_t + \sum_{s \leq t}(\Delta X(s))^2. \tag{1.11}$$

*Proof.* If $X(t)$ is a continuous martingale, then the stochastic integral $\int_0^t X(s-)dX(s)$ is a continuous martingale. Therefore the quadratic variation $[X]_t$ is a continuous increasing process. Further, since $X(t)^2 - [X]_t$ is a martingale, $\langle X \rangle_t - [X]_t$ is also a martingale. However a continuous martingale with finite variation is a constant, proving $\langle X \rangle_t - [X]_t \equiv 0$. We omit the case where $X(t)$ is purely discontinuous. See Lemma I.4.51 in Jacod–Shiryaev [12].

We will prove (1.11). Since $\langle X_c, X_d \rangle_t = 0$, we have $[X_c, X_d]_t = 0$. Then we get

$$[X]_t = [X_c]_t + [X_d]_t = \langle X_c \rangle_t + \sum_{s \leq t}(\Delta X(s))^2.$$

**Remark.** If $X(t)$ and $Y(t)$ are continuous locally square integrable martingales, then these two are orthogonal if and only if the quadratic co-variation $[X, Y]$ is 0. However it is not the case if both of $X, Y$ have jumps. Indeed, if both $X$ and $Y$ are purely discontinuous, $[X, Y] = 0$ holds if and only if $X, Y$ have no common jumps, i.e., $\Delta X(s)\Delta Y(s) = 0$ a.s. for all $s$. In this case $X$ and $Y$ are orthogonal. However the orthogonality of $X, Y$ does not imply that they have no common jumps.

Let $X(t)$ and $f(t)$ be semimartingales. We define the *Stratonovich integral* of $f(t)$ based on $X(t)$ by

$$\int_0^t f(s) \circ dX(s) := \int_0^t f(s-)dX(s) + \frac{1}{2}[f, X]_t.$$

For a partition $\Pi = \{0 = t_0 < t_1 < \cdots < t_n = T\}$, set

$$Z^{\Pi}(t) = \sum_i \frac{1}{2}\left(f(t_{i+1} \wedge t) + f(t_{t_i} \wedge t)\right)(X(t_{i+1} \wedge t) - X(t_i \wedge t)).$$

**Theorem 1.11.** *Let $\{\Pi_n\}$ be a sequence of partitions of $[0, T]$ such that $|\Pi_n| \to 0$. Then it holds*

$$\exists \lim_{n \to \infty} Z^{\Pi_n}(t) = \int_0^t f(s) \circ dX(s).$$

*Proof.* Let $Y^{\Pi_n}(t)$ be the process defined by (1.5). Then we have the relation $Z^{\Pi_n}(t) = Y^{\Pi_n}(t) + \frac{1}{2}[f, X]_t^{\Pi_n}$. Let $|\Pi_n| \to 0$. Then we obtain

$$\lim_{n \to \infty} Z^{\Pi_n}(t) = \int_0^t f(s-)dX(s) + \frac{1}{2}[f, X]_t.$$

by Theorem 1.4 and Lemma 1.9.

## 1.5 Itô's formula I

**Theorem 1.12 (Itô's formula).** *Let $X(t) = (X^1(t), ..., X^d(t))$ be a d-dimensional semimartingale and let $F(x_1, ..., x_d)$ be a $C^2$ function. Then $F(X^1(t), ..., X^d(t))$ is again a semimartingale, and the following formula holds.*

$$F(X(t)) - F(X(0)) = \tag{1.12}$$
$$\sum_i \int_0^t \frac{\partial F}{\partial x_i}(X(s-))dX_s^i + \frac{1}{2}\sum_{i,j}\int_0^t \frac{\partial^2 F}{\partial x_i \partial x_j}(X(s-))d[X^i, X^j]_s^c$$
$$+ \sum_{0 < s \leq t}\left\{F(X(s)) - F(X(s-)) - \sum_i \frac{\partial F}{\partial x_i}(X(s-))\Delta X_s^i\right\}.$$

*Here $[X^i, X^j]_t^c$ is the continuous part of $[X^i, X^j]_t$.*

**Remark.** 1) The infinite sum of the last term of Itô's formula is absolutely convergent, because

$$\sum_{0 < s \leq t}\left|F(X(s)) - F(X(s-)) - \sum_i \frac{\partial F}{\partial x_i}(X(s-))\Delta X_s^i\right|$$
$$\leq \frac{1}{2}\sum_{ij}\sum_{0 \leq s \leq t}\left|\int_0^1 (1-\theta)\frac{\partial^2 F}{\partial x_i \partial x_j}(X(s-) + \theta \Delta X(s)d\theta)\right|$$
$$|\Delta X^i(s)\Delta X^j(s)| < \infty.$$

2) Here is another expression of Itô's formula:

$$F(X(t)) - F(X(0)) = \tag{1.13}$$
$$\sum_i \int_0^t \frac{\partial F}{\partial x_i}(X(s-))dX_s^i + \frac{1}{2}\sum_{i,j}\int_0^t \frac{\partial^2 F}{\partial x_i \partial x_j}(X(s-))d[X^i, X^j]_s$$
$$+ \sum_{0 < s \leq t}\left\{F(X(s)) - F(X(s-)) - \sum_i \frac{\partial F}{\partial x_i}(X(s-))\Delta X_s^i\right.$$
$$\left.- \frac{1}{2}\sum_{i,j}\sum_{s \leq t}\frac{\partial^2 F}{\partial x_i \partial x_j}(X(s-))\Delta X^i(s)\Delta X^j(s)\right\}.$$

In fact, the last infinite sum of (1.13) can be divided into two parts. One is the same one as in (1.12). The other is the term involving $\Delta X^i(s)\Delta X^j(s)$. Sum up this term with the second term involving $[X^i, X^j]_t$. Then we get the second term on the right hand side of (1.12).

*Proof.* We give the proof in the case $d = 1$. We will prove (1.13). We assume $F''$ is bounded and uniformly continuous in $\mathbf{R}$. (If $F''$ is not bounded, consider the semi-martingales $X(t)1_{[0,\tau_n)}(t)$, where $\tau_n = \inf\{t; |X(t)| \geq n\}$.)

Let $\{\Pi_n = \{0 = t_1^n < \cdots < t_{k_n}^n = t\}\}$ be a sequence of partitions of the interval $[0, t]$ such that $|\Pi_n| \to 0$ as $n \to \infty$. We have

$$F(X(t)) - F(X(0)) = \sum_i \{F(X(t_{i+1}^n)) - F(X(t_i^n))\}$$

$$= \sum_i F'(X(t_i^n))(X(t_{i+1}^n) - X(t_i^n))$$

$$+ \frac{1}{2}\sum_i F''(X(t_i^n))(X(t_{i+1}^n) - X(t_i^n))^2$$

$$+ \sum_i R(X(t_{i+1}^n), X(t_i^n)),$$

where we have used Taylor's formula

$$F(y) - F(x) = F'(x)(y - x) + \frac{1}{2}F''(x)(y - x)^2 + R(x, y),$$

$$R(x, y) = \left(\int_0^1 (1 - \theta)F''(x + \theta(y - x))d\theta\right)(y - x)^2$$

$$- \frac{1}{2}F''(x)(y - x)^2.$$

It holds

$$\lim_{n\to\infty}\sum_i F'(X(t_i^n))(X(t_{i+1}^n) - X(t_i^n)) = \int_0^t F'(X(s-))dX(s),$$

by Theorem 1.4 and

$$\lim_{n\to\infty}\frac{1}{2}\sum_i F''(X(t_i^n))(X(t_{i+1}^n) - X(t_i^n))^2 = \frac{1}{2}\int_0^t F''(X(s-))d[X]_s,$$

by Lemma 1.9. We will prove

$$\lim_{n\to\infty}\sum_i R(X(t_{i+1}^n), X(t_i^n)) = \sum_{0\leq s\leq t} r(X(s), X(s-))(\Delta X(s))^2, \tag{1.14}$$

where $r(x, y) = \frac{1}{2}\{\int_0^1 F''(x + \theta(y - x))d\theta - F''(x)\}$. Given $\epsilon > 0$, we set $J(\epsilon) = \{s \in [0, t]; |\Delta X(s)| > \epsilon\}$. It is a finite set a.s. Then we have

$$\lim_{n \to \infty} \sum_{i; J(\epsilon) \cap (t_i^n, t_{i+1}^n] \neq \phi} R(X(t_{i+1}^n), X(t_{t_i}^n)) = \sum_{s \in J(\epsilon)} r(X(s), X(s-)) \Delta X(s)^2.$$

On the other hand, observe the inequality

$$\left| \sum_{i; J(\epsilon) \cap (t_i^n, t_{i+1}^n] = \phi} R(X(t_{i+1}^n), X(t_{t_i}^n)) \right|$$

$$\leq \left( \sup_{i; J(\epsilon) \cap (t_i^n, t_{i+1}^n] = \phi} |r(X(t_{i+1}^n), X(t_i^n))| \right) \sum_i (X(t_{i+1}^n) - X(t_i^n))^2.$$

Therefore,

$$\limsup_{n \to \infty} \left| \sum_{i; J(\epsilon) \cap (t_i^n, t_{i+1}^n] = \phi} R(X(t_{i+1}^n), X(t_{t_i}^n)) \right| \leq \sup_{|x-y| \leq \epsilon} |r(x, y)| [X]_t.$$

It tends to 0 as $\epsilon \to 0$. Therefore we get (1.14).

Now we have

$$r(x, y)(y - x)^2 = F(y) - F(x) - F'(x)(y - x) - \frac{1}{2} F''(x)(y - x)^2.$$

Then we get

$$\sum_{0 < s \leq t} (X(s), X(s-))(\Delta X(s))^2$$

$$= \sum_{0 < s \leq t} \left\{ F(X(s)) - F(X(s-)) - F'(X(s-)) \Delta X(s) - \right.$$

$$\left. \frac{1}{2} F''(X(s-))(\Delta X(s))^2 \right\}.$$

Therefore we get the formula (1.13).

**Complex martingales.**    Finally in this section we shall briefly discuss complex martingales. Let $Z(t)$ be an adapted cadlag process with values in $\mathbf{C}$, the set of complex numbers. Let $X(t)$ and $Y(t)$ be the real part and the imaginary part of $Z(t)$, respectively. These are real adapted cadlag processes such that $Z(t) = X(t) + iY(t)$. $Z(t)$ is called a complex martingale if both of $X(t)$ and $Y(t)$ are martingales. Further $Z(t)$ is called a square integrable martingale if $|Z(t)|^2 = X(t)^2 + Y(t)^2$ is integrable. Complex localmartingales and complex semimartingales are defined similarly.

Now let $Z(t)$ and $\tilde{Z}(t) = \tilde{X}(t) + i\tilde{Y}(t)$ be square integrable complex martingales. The bracket of $Z(t)$ and $\tilde{Z}(t)$ is defined by

$$\langle Z, \tilde{Z} \rangle_t = \langle X, \tilde{X} \rangle_t + i \langle X, \tilde{Y} \rangle_t + i \langle Y, \tilde{X} \rangle_t - \langle Y, \tilde{Y} \rangle_t.$$

Then $Z(t)\tilde{Z}(t) - \langle Z, \tilde{Z} \rangle_t$ is a complex martingale. Note that $\langle Z, \bar{Z} \rangle_t = \langle X \rangle_t + \langle Y \rangle_t$ holds, where $\bar{Z}_t = X(t) - iY(t)$ is the complex conjugate of $Z(t)$. Let $f(t)$ be a complex predictable process such that $\int_0^T |f(s)|^2 d\langle Z, \bar{Z} \rangle_t < \infty$. Then we can define the stochastic integral $\int_0^t f(s) dZ(s)$ as a complex localmartingale. Theorem 1.6 is valid for complex stochastic integrals.

We can define the quadratic co-variation $[Z, \tilde{Z}]_t$ of two complex semimartingales $Z(t)$ and $\tilde{Z}(t)$ similarly to the bracket $\langle Z, \tilde{Z} \rangle_t$.

We can extend Itô's formula to complex $C^2$ functions and to complex semimartingales. The formula (1.12) is valid for a complex $C^2$ function $F$ and for a complex semimartingale $X(t)$.

# 2 Stochastic analysis of Lévy processes

## 2.1 Lévy processes and Poisson random measures

Let $Z(t)$, $t \in [0, T]$ be an $m$-dimensional cadlag process defined on a probability space $(\Omega, \mathcal{F}, P)$. It is called a *Lévy process* if it has the following three properties.
1) (*Independent increments*) For any $0 \le t_0 < t_1 < \cdots < t_n \le T$, random variables $Z(t_i) - Z(t_{i-1})$, $i = 1, ..., n$, are independent.
2) (*Time homogeneous*) Laws of $Z(t + h) - Z(s + h)$ do not depend on $h$.
3) (*Continuous in probability*) For any $t$ it holds $\lim_{h \to 0} P(|Z(t+h) - Z(t)| > \epsilon) = 0$ for any $\epsilon > 0$.

In this work, we assume $Z(0) = 0$ for simplicity.

Given a Lévy process $Z(t)$, we define a family of sub $\sigma$-fields of $\mathcal{F}$ by

$$\mathcal{F}'_t := \cap_{\epsilon > 0} \sigma(Z(s) : 0 \le s \le t + \epsilon).$$

It is called the *filtration generated by the Lévy process* $Z(t)$.

Suppose that a Lévy process $Z(t)$ is integrable for any $t$. Its mean vector is proportional to $t$, which we denote by $tm$. Then $M(t) = Z(t) - tm$ is a (vector) martingale with respect to the filtration $\{\mathcal{F}'_t\}$. In fact, $M(t) - M(s)$ is independent of $\mathcal{F}'_s$ and its expectation is 0. Suppose further that $Z(t)$ is square integrable. Its covariance matrix is proportional to $t$. It is written by $tV$, where $V = (v_{ij})$ is a nonnegative definite symmetric matrix. In this case $M(t) = (M^1(t), ..., M^m(t))$ is a square integrable martingale. The bracket of $M^i(t)$ and $M^j(t)$ is given by $\langle M^i, M^j \rangle_t = t v_{ij}$. In fact, $(M^i(t) - M^i(s))(M^j(t) - M^j(s))$ is independent of $\mathcal{F}'_s$ and its expectation is $(t-s) v_{ij}$. Generally, however, Lévy processes are not always integrable. We will see in Section 2.4 that any Lévy process is a semimartingale.

In some applications, it is convenient to deal with another filtration $\{\mathcal{F}_t\}$, larger than the filtration $\{\mathcal{F}'_t\}$ generated by the Lévy process. Suppose that we are given a filtration $\{\mathcal{F}_t\}$, an $\{\mathcal{F}_t\}$ adapted cadlag process $Z(t)$ is called a $\{\mathcal{F}_t\}$-*Lévy process* if it is time homogeneous, continuous in probability and $Z(t) - Z(s)$ is independent of $\mathcal{F}_s$ for any $s$. Any $\{\mathcal{F}_t\}$-Lévy process is a Lévy process. Conversely any Lévy process is an $\{\mathcal{F}'_t\}$-Lévy process.

One of the most important Lévy processes is a Brownian motion. A continuous $\{\mathcal{F}_t\}$-Lévy process $Z(t)$ is called a $\{\mathcal{F}_t\}$-*Brownian motion*. It will be shown in Section

2.4 that any $\{\mathcal{F}_t\}$-Brownian motion $Z(t)$ is square integrable and in fact $Z(t)$ is Gaussian distributed with mean $tm$ and covariance $tV$. In particular if $m = 0$ and $V$ is the identity matrix $I$, the process is called a standard $\{\mathcal{F}_t\}$-Brownian motion.

Another important $\{\mathcal{F}_t\}$-Lévy process is a Poisson random measure. Let $(\mathcal{Z}, \mathcal{B})$ be a measurable space. A mapping $p : \mathbf{D}_p \to \mathcal{Z}$ is said to be a *point function* if its domain $\mathbf{D}_p$ is a countable subset of the time interval $(0, T]$. $p$ defines a counting measure on $(0, T] \times \mathcal{Z}$ by

$$N_p((s, t] \times U) = \#\{r \in \mathbf{D}_p; s < r \le t, p(r) \in U\}, \quad U \in \mathcal{B}. \tag{2.1}$$

Let $\Pi_{\mathcal{Z}}$ be the totality of point functions. A point process is a measurable mapping $\Omega \to \Pi_{\mathcal{Z}}$. It induces a random counting measure $N(dtdz) \equiv N_{p(\omega)}(dtdz)$. A point process $p$ is called a $\{\mathcal{F}_t\}$-*Poisson point process* if for any $E_1, ..., E_n \in \mathcal{B}$ such that $N((0, T] \times E_i) < \infty$ a.s. $i = 1, 2, ...,$ the process $X(t) = (N((0, t] \times E_1), ..., N((0, t] \times E_n))$ is a $\{\mathcal{F}_t\}$-Lévy process. The random measure $N$ is called a $\{\mathcal{F}_t\}$-*Poisson random measure*.

**Theorem 2.1.** *Let $N(dtdz)$ be a $\{\mathcal{F}_t\}$-Poisson random measure. Set $N_t(E_0) = N((0, t] \times E_0)$. If $N_t(E_0) < \infty$ holds for any $t$, it is integrable and is Poisson distributed with intensity $v(E_0) = E[N_1(E_0)]$.*

*Let $E_1, ..., E_n$ be disjoint measurable subsets of $\mathcal{Z}$ such that $v(E_i) < \infty, i = 1, ..., n$. Then $N_t(E_1), ..., N_t(E_n)$ are independent $\{\mathcal{F}_t\}$-Lévy processes.*

*Proof.* Suppose that $N_t(E_0) < \infty$ a.s. We shall compute its characteristic function $\varphi_t(\alpha) = E[e^{i\alpha N_t(E_0)}]$, $\alpha \in \mathbf{R}$. Since the distribution of $N_t(E_0)$ is infinitely divisible and is time homogeneous, it is known that the characteristic function is written as $\varphi_t(\alpha) = e^{t\psi(\alpha)}$ (Sato [21]). Set $M^\alpha(t) = e^{i\alpha X(t)} e^{-t\psi(\alpha)}$. Since $M^\alpha(t)M^\alpha(s)^{-1}$ is independent of $\mathcal{F}_s$, we have

$$E[M^\alpha(t)M^\alpha(s)^{-1}|\mathcal{F}_s] = E[M^\alpha(t)M^\alpha(s)^{-1}] = 1.$$

Therefore $M^\alpha(t)$ is a bounded complex martingale. We consider the stochastic integral

$$Y^\alpha(t) = \int_0^t \frac{1}{M^\alpha(s-)} dM^\alpha(s). \tag{2.2}$$

It is a square integrable complex martingale, since the integrand is a bounded process. It holds

$$Y_t^\alpha = \sum_{0 < s \le t} \frac{\Delta M^\alpha(s)}{M^\alpha(s-)} + \int_0^t (-\psi(\alpha))ds = (e^{i\alpha} - 1)N_t(E_0) - t\psi(\alpha).$$

Taking its expectation, we obtain $(e^{i\alpha} - 1)E(N_t(E_0)) - t\psi(\alpha) = 0$. Therefore $E(N_t(E_0)) = v_t(E_0)$ is finite and $t\psi(\alpha) = (e^{i\alpha} - 1)v_t(E_0)$. This yields $v_t(E_0) = tv(E_0)$ and

$$E[e^{i\alpha N_t(E_0)}] = \exp\{t(e^{i\alpha} - 1)v(E_0)\}.$$

The above formula is valid for any $\alpha \in \mathbf{R}$. It indicates that $N_t(E)$ is Poisson distributed with intensity $tv(E)$.

We next consider the $n$ vector process $X(t) = (N_t(E_1), \ldots, N_t(E_n))$. Consider its characteristic function $\varphi_t(\alpha) = E[\exp(i \sum_k \alpha_k N_t(E_k))]$, where $\alpha = (\alpha_1, \ldots, \alpha_n)$. It is represented by $\exp(t\psi(\alpha))$ as before. We define a complex martingale with parameter $\alpha$ by

$$M^\alpha(t) = \exp(i \sum_k \alpha_k N_t(E_k)) \exp(-t\psi(\alpha))$$

and we define $Y^\alpha(t)$ by (2.2). It is a square integrable complex martingale. Note that $N_t^\alpha(E_1), \ldots, N_t^\alpha(E_n)$ have no common jumps, since $E_1, \ldots, E_n$ are disjoint. Then we have

$$Y^\alpha(t) = \sum_{k=1}^n (e^{i\alpha_k} - 1) N_t(E_k) - t\psi(\alpha).$$

Taking its expectation in the case $t = 1$, we have $\sum_{k=1}^n (e^{i\alpha} - 1)v(E_k) - \psi(\alpha) = 0$. Therefore we have

$$\varphi_t(\alpha) = \exp(t \sum_k (e^{i\alpha_k} - 1)v(E_k)) = \prod_k \exp(t(e^{i\alpha_k} - 1)v(E_k)).$$

This proves

$$E\Big[ \prod_{k=1}^n e^{i\alpha_k N_t(E_k)} \Big] = \prod_{k=1}^n E[e^{i\alpha_k N_t(E_k)}].$$

The above formula is valid for any $\alpha_1, \ldots, \alpha_n$. Thus, $N_t(E_1), \ldots, N_t(E_n)$ are independent.

Now consider $N_t(E_k) - N_s(E_k), k = 1, \ldots, n$, in place of $N_t(E_k), k = 1, \ldots, n$. Then we can show that these random variables are also independent if $E_k, k = 1, \ldots, n$, are disjoint. Further these are independent of $\mathcal{F}_s$, since the $N_t(E_k)$ are $\{\mathcal{F}_t\}$-Lévy processes. Consequently the stochastic processes $N_t(E_k), k = 1, \ldots, n$, are independent of each other.

We set

$$\tilde{N}(dtdz) = N(dtdz) - dsv(dz), \tag{2.3}$$

and call it the *compensated* Poisson random measure.

**Corollary 2.2.** *If $v(E) < \infty$, then $\tilde{N}_t(E) := \tilde{N}((0, t] \times E)$ is a square integrable martingale with $\langle \tilde{N}(E) \rangle_t - tv(E)$.*

*Let $F$ be a set such that $v(F) < \infty$. Then we have*

$$\langle \tilde{N}(E), \tilde{N}(F) \rangle_t = tv(E \cap F).$$

*Let $\phi(z), \psi(z)$ be measurable on $\mathcal{Z}$ such that $\int [|\phi(z)|^2 + |\psi(z)|^2]v(dz) < \infty$. Then both $X(t) = \int \phi(z)\tilde{N}_t(dz)$ and $Y(t) = \int \psi(z)\tilde{N}_t(dz)$ are square integrable martingales and satisfy*

$$\langle X, Y \rangle_t = t \int \phi(z)\psi(z)v(dz).$$

*Proof.* Set $M(t) = \tilde{N}_t(E)$. It is a martingale such that $\Delta M(s)$ is 0 or 1. Then we have $[M]_t = N_t(E)$. Therefore $[M]_t$ and $M(t)^2$ are integrable. Further, since $M(t)^2 - [M]_t$ and $[M]_t - t\nu(E)$ are martingales, $M(t)^2 - t\nu(E)$ is a martingale. This proves $\langle M \rangle_t = t\nu(E)$.

Let $E, F$ be measurable subsets of $\mathcal{Z}$ such that $\nu(E) < \infty$ and $\nu(F) < \infty$. We split $E$ and $F$ as $E = (E \cap F) \cup (E - F)$ and $F = (E \cap F) \cup (F - E)$. Then we have

$$\langle \tilde{N}(E), \tilde{N}(F) \rangle_t = \langle \tilde{N}(E \cap F) + \tilde{N}(E - F), \tilde{N}(E \cap F) + \tilde{N}(F - E) \rangle_t$$
$$= \langle \tilde{N}(E \cap F), \tilde{N}(E \cap F) \rangle_t = t\nu(E \cap E).$$

In fact, the pair of $\tilde{N}_t(E - F)$ and $\tilde{N}_t(E \cap F)$ or the pair of $\tilde{N}_t(F - E)$ and $\tilde{N}_t(E \cap F)$ are orthogonal since they have no common jumps.

Suppose that $\phi, \psi$ are simple functions of the form $\phi = \sum_i c_i 1_{E_i}$ and $\psi = \sum_j d_j 1_{E_j}$, where $E_1, ..., E_n$ are disjoint subsets of $\mathcal{Z}$ such that $\nu(E_i) < \infty$ for any $i$. Then we have

$$\langle X, Y \rangle_t = t \sum_{ij} c_i d_j \nu(E_i \cap E_j) = t \sum_i c_i d_i \nu(E_i)$$
$$= t \int \phi(z) \psi(z) \nu(dz).$$

It will be obvious that the above equality can be extended to any square integrable functions $\phi, \psi$ with respect to $\nu$.

We set $\hat{N}(dsdz) = ds\nu(dz)$ and call it the *compensator* of the Poisson random measure $N(dsdz)$.

## 2.2 Stochastic integrals based on compensated Poisson random measure

We shall define the stochastic integral based on the compensated Poisson random measure. Let $g(s, z)$ be a simple process written as $g(s, z) = \sum_{i=1}^n \psi_i(z) 1_{(s_i, s_{i+1}]}(s)$, where $\psi_i(z)$ is a $\mathcal{F}_{s_i} \times \mathcal{B}$-measurable function with the square integrability condition $E[\int |\psi_i(z)|^2 \nu(dz)] < \infty$. It is called a simple predictable process with square integrability condition.

A functional $g(s, z, \omega)$, $(s, z, \omega) \in [0, T] \times \mathcal{Z} \times \Omega$ is called a predictable process if it is $\mathcal{P} \times \mathcal{B}$-measurable, where $\mathcal{P}$ is the predictable $\sigma$-field on $[0, T] \times \Omega$. We denote by $L^2(\hat{N})$ the set of all predictable functionals $g(s, z)$ such that

$$E[\int_0^T \int_{\mathcal{Z}} |g(s, z)|^2 \hat{N}(dsdz)] < \infty.$$

The following fact can be verified by a standard argument.

**Lemma 2.3.** *Simple predictable processes $g$ with the square integrability conditions are dense in $L^2(\hat{N})$.*

We will define the stochastic integral of the form $\int\int g(s,z)\tilde{N}(dsdz)$. We first remark that

$$E[\tilde{N}((s,t]\times E_1)\tilde{N}((s,t]\times E_2)|\mathcal{F}_s] = \hat{N}((s,t]\times(E_1\cap E_2)) = (t-s)\nu(E_1\cap E_2)$$

is valid because of Proposition 2.1. Then if $\psi(z)$ is $\mathcal{F}_s$-measurable random variable such that $E[\int|\psi|^2\nu(dz)] < \infty$, then

$$E[\left(\int\psi(z)\tilde{N}((s,t],dz)\right)^2|\mathcal{F}_s] = \int\psi(z)^2\hat{N}((s,t],dz) = (t-s)\int\psi(z)^2\nu(dz).$$

Let $g(t,z)$ be a simple predictable process defined by $\sum\psi_i(z)1_{[s_i,s_{i+1})}(t)$. We set

$$X(t) = \sum_i\psi_i(z)\Big(\tilde{N}(s_{i+1}\wedge t,dz) - \tilde{N}(s_i\wedge t,dz)\Big).$$

We denote it by $\int_0^t\int_Z g(s,z)\tilde{N}(dsdz)$ and call it the *stochastic integral of $g$ by the compensated Poisson random measure*.

**Lemma 2.4.** *The stochastic integral is a square integrable martingale. Further it holds*

$$\langle\int\int_Z g(s,z)\tilde{N}(dsdz)\rangle_t = \int_0^t\int_Z g(s,z)^2\hat{N}(dsdz), \qquad (2.4)$$

$$E[|\int_0^t\int_Z g(s,z)\tilde{N}(dsdz)|^2] = E[\int_0^t\int_Z g(s,z)^2\hat{N}(dsdz)]. \qquad (2.5)$$

*Proof.* For simplicity we only consider the case $s = s_{n_0}$ and $t = s_{n_1}$. Since $X(t)$ is a square integrable martingale, we have

$$E[(X(t) - X(s))^2|\mathcal{F}_s] = \sum_{n_0\le i<n_1} E[(X(s_{i+1}) - X(s_i))^2|\mathcal{F}_s]$$

$$= \sum_i E[\left(\int_Z\psi_i(z)\tilde{N}((s_i,s_{i+1}],dz)\right)^2|\mathcal{F}_s]$$

$$= \sum_i E[E[\left(\int_Z\psi_i(z)\tilde{N}((s_i,s_{i+1}],dz)\right)^2|\mathcal{F}_{s_i}|\mathcal{F}_s]$$

$$= \sum_i E[E[\int_Z\psi_i(z)^2\hat{N}((s_i,s_{i+1}],dz)|\mathcal{F}_s]]$$

$$= E[\int_s^t\int_Z|g(r,z)|^2\hat{N}(drdz)|\mathcal{F}_s].$$

This proves the first equality of the lemma. The second equality follows immediately if we take the expectation of the first one.

Now let $g$ be any element of $L^2(\hat{N})$ and let $\{g_n\}$ be a sequence of simple predictable processes converging to $g$ with respect to the norm of $L^2(\hat{N})$. Then the sequence of stochastic integrals $\{\int_0^t g_n(s, z)\tilde{N}(dsdz)\}$ converges in $L^2$ in view of (2.5). The limit is denoted by $\int_0^t \int_Z g(s, z)\tilde{N}(dsdz)$. It is called the stochastic integral of $g(s, z)$ by the compensated Poisson random measure.

**Remark.** Suppose that $E[\int_0^T \int_Z |g(r, z)|dr\nu(dz)] < \infty$. We may define

$$\int_0^t \int_Z g(s, z)N(dsdz), \quad \int_0^t \int_Z g(s, z)ds\nu(dz),$$

as integrable processes of finite variation. Further,

$$\int_0^t \int_Z g(s, z)\tilde{N}(dsdz) := \int_0^t \int_Z g(s, z)N(dsdz) - \int_0^t \int_Z g(s, z)dr\nu(dz)$$

is a martingale.

### 2.3 Itô's formula II

We shall rewrite Itô's formula introduced in Section 1.

**Theorem 2.5 (Kunita–Watanabe).** [16]  *Let* $X(t) = (X^1(t), ..., X^d(t))$ *be a d-dimensional semimartingale represented by*

$$X^i(t) = X^i(0) + A^i(t) + M^i(t) \tag{2.6}$$
$$+ \int_0^t \int_Z g^i(s, z)\tilde{N}(dsdz) + \int_0^t \int_Z h^i(s, z)N(dsdz), \quad i = 1, ..., d.$$

*Here* $A^i(t)$ *are continuous adapted processes of finite variation,* $M^i(t)$ *are continuous localmartingales and* $h, g$ *satisfy* $|g||h| = 0$. *Let* $F(x_1, ..., x_d)$ *be a* $C^2$ *function. Then we have*

$$F(X^1(t), ..., X^d(t)) \tag{2.7}$$

$$= F(X^1(0), ..., X^d(0)) + \sum_{i=1}^d \int_0^t \frac{\partial F}{\partial x_i}(X(s-))dA^i(s)$$

$$+ \sum_{i=1}^d \int_0^t \frac{\partial F}{\partial x_i}(X(s-))dM^i(s)$$

$$+ \frac{1}{2} \sum_{i,j=1}^d \int_0^t \frac{\partial^2 F}{\partial x_i \partial x_j}(X(s))d\langle M^i, M^j \rangle_s$$

$$+ \int_0^t \int_Z \{F(X(s-) + g(s, z)) - F(X(s-))\}\tilde{N}(dsdz)$$

$$+ \int_0^t \int_Z \{F(X(s) + h(s, x)) - F(X(s-))\}N(dsdx)$$

$$+ \int_0^t \int_Z \left\{ F(X(s-) + g(s, z)) - F(X(s-)) \right.$$

$$\left. - \sum_{i=1}^d g^i(s, z) \frac{\partial F}{\partial x_i} (X(s-)) \right\} \hat{N}(dsdz).$$

*Proof.* We will use Itô's formula I. Denote the sum of the two discontinuous processes in (2.6) by $X_d(t)$. Then the semimartingale $X(t)$ is decomposed as $X(t) = X(0) + A(t) + M(t) + X_d(t)$. In Itô's formula I, we split the integral $\int \frac{\partial F}{\partial x_i}(X(s))dX^i(s)$ as the sum of the integral based on $A^i(t) + M^i(t)$ and that based on $X_d^i(t)$. We have

$$\int_0^t \frac{\partial F}{\partial x_i}(X(s-))(dA^i(s) + dM^i(s))$$

$$= \int_0^t \frac{\partial F}{\partial x_i}(X(s-))dA^i(s) + \int_0^t \frac{\partial F}{\partial x_i}(X(s-))dM^i(s).$$

It holds $[X^i, X^j]_t^c = \langle M^i, M^j \rangle_t$. The remaining parts of Itô's formula I are the two jump parts. These are

$$\sum_i \int_0^t \frac{\partial F}{\partial x_i}(X(s-))dX_d^i(s)$$

$$= \sum_i \int_0^t \int_Z \frac{\partial F}{\partial x_i}(X(s-))g^i(s, z)\tilde{N}(dsdz)$$

$$+ \sum_i \int_0^t \int_Z \frac{\partial F}{\partial x_i}(X(s-))h^i(s, z)N(dsdz),$$

and

$$\sum_{0 < s \le t} \left\{ F(X(s)) - F(X(s-)) - \sum_i \frac{\partial F}{\partial x_i} \Delta X_s^i \right\}$$

$$= \int_0^t \int_Z \{ F(X(s-) + g(s, z)) - F(X(s-))$$

$$- \sum_i \frac{\partial F}{\partial x_i}(X(s-))g^i(s, z) \} N(dsdx)$$

$$+ \int_0^t \int_Z \{ F(X(s-) + h(s, z)) - F(X(s-))$$

$$- \sum_i \frac{\partial F}{\partial x_i}(X(s-))h^i(s, z) \} N(dsdx).$$

The sum of the above two is written as

$$\int_0^t \int_Z \{F(X(s-) + g(s, z)) - F(X(s-))\}\tilde{N}(dsdz)$$

$$+ \int_0^t \int_Z \{F(X(s-) + g(s, z)) - F(X(s-))$$

$$- \sum_{i=1}^d g^i(s, z)\frac{\partial F}{\partial x_i}(X(s-))\bigg\} \hat{N}(dsdx)$$

$$+ \int_0^t \int_Z \{F(X(s-) + h(s, z)) - F(X(s-))\}N(dsdz).$$

Therefore we get Itô's formula II.

**Remark.** The condition $|g(t, z)||h(t, z)| = 0$ means that two processes $\tilde{N}_t(g)$ and $N_t(h)$ do not jump simultaneously. In fact, setting $\Delta \tilde{N}_t(g) = \tilde{N}_t(g) - \tilde{N}_{t-}(g)$, we have

$$\sum_{s \leq t} |\Delta \tilde{N}_s(g)||\Delta N_s(h)| = N_t(|g||h|).$$

The above is 0 if and only if $|g||h| = 0$ a.e. $(t, x, \omega)$.

As an application of Itô's formula, we will give a proposition concerning distributions of $\{\mathcal{F}_t\}$-Brownian motion.

**Proposition 2.6.** *Let $Y(t)$ be a square integrable $\{\mathcal{F}_t\}$-Brownian motion. Then it is Gaussian distributed. Further it is independent of any $\{\mathcal{F}_t\}$-Poisson random measure.*

*Proof.* We consider the case where $Y(t)$ is a one-dimensional $\{\mathcal{F}_t\}$-Brownian motion with mean 0. Set $\sigma^2 t = E[Y(t)^2]$. Then it holds $\langle Y \rangle_t = \sigma^2 t$. Let $X(t) = N_t(E_0)$, where $v(E_0) < \infty$. We have by Itô's formula II,

$$U(t) := e^{i\alpha Y(t)}e^{i\beta X(t)}$$

$$= i\alpha \int_0^t U(s-)Y(ds) - \frac{1}{2}\alpha^2 \int_0^t U(s-)\sigma^2 ds$$

$$+ \int_0^t \int_Z U(s-)(e^{i\beta} - 1)1_{E_0}(z)\tilde{N}(dsdz)$$

$$+ \int_0^t \int_Z (e^{i\beta} - 1)1_{E_0}(z)dsv(dz).$$

Now taking its expectation, since $E[U(s)] = E[U(s-)]$, we have

$$E[U(t)] = -\frac{1}{2}\alpha^2\sigma^2 \int_0^t E[U(s)]ds + (e^{i\beta} - 1)v(E_0) \int_0^t E[U(s)]ds.$$

Therefore we have

$$E[U(t)] = \exp\left[t\left\{-\frac{1}{2}\sigma^2\alpha^2 + (e^{i\beta} - 1)v(E_0)\right\}\right].$$

This means

$$E[e^{i\alpha Y(t)}e^{i\beta X(t)}] = \exp\left\{-\frac{1}{2}t\sigma^2\alpha^2\right\}\exp\left[t\left\{(e^{i\beta-1}-1)\nu(E_0)\right\}\right]$$
$$= E[e^{i\alpha Y(t)}]E[e^{i\beta X(t)}].$$

Consequently the processes $Y(t)$ and $X(t)$ are independent. Further, $Y(t)$ is Gaussian distributed.

## 2.4 Lévy–Itô's decomposition and representation of martingales

Let $Z(t)$ be an $m$-dimensional $\{\mathcal{F}_t\}$-Lévy process. We study jumps of the Lévy process $Z(t)$. Let $\mathcal{B}(\mathbf{R}^m - \{0\})$ be the Borel field of $\mathbf{R}^m - \{0\}$. We define a random measure $N((s, t] \times E)$ on $\mathcal{B}$ by

$$N((s, t] \times E) = \sum_{r \in (s,t]} \#\{r; \Delta Z(r) \in E\}.$$

Then $\{N((s, t] \times E_0); E_0 \in \mathcal{B}(\mathbf{R}^m - \{0\})\}$ is independent of $\mathcal{F}_s$. Also it is continuous in probability and is time homogeneous. Therefore it is an $\{\mathcal{F}_t\}$-Poisson random measure on $\mathcal{B}(\mathbf{R}^m - \{0\})$ with the intensity $\nu(E_0) = E[N((0, 1] \times E_0)]$. Note that $N_t(E_0)$ is finite for any $t$ a.s. if $E_0 \cap \{z; |z| < \epsilon\}$ is empty for some $\epsilon > 0$. Then $\nu(E_0) < \infty$ holds by Theorem 2.1. Consequently it holds $\nu(\{z : |z| \geq \epsilon\}) < \infty$ for any $\epsilon > 0$. However it may occur that $\nu(\mathbf{R}^m - \{0\}) = \infty$.

**Theorem 2.7.** *1) (Lévy–Itô's decomposition) The $\{\mathcal{F}_t\}$-Lévy process $Z(t)$ is represented as*

$$Z(t) = W(t) + bt + \int_{(0,t]}\int_{|z|>1} xN(dsdz) + \int_{(0,t]}\int_{0<|z|\leq 1} x\tilde{N}(dsdz)\}, \quad (2.8)$$

*where $W(t) = (W^1(t), ..., W^m(t))$ is an m-dimensional square integrable $\{\mathcal{F}_t\}$-Brownian motion with mean 0 and covariance $tV$. $N(dsdz)$ is a $\{\mathcal{F}_t\}$-Poisson random measure on $[0, T] \times (\mathbf{R}^m - \{0\})$ with intensity measure $\hat{N}(dsdz) = dsd\nu(z)$, which is independent of $W(t)$. Further the Lévy measure satisfies $\int \frac{|z|^2}{1+|z|^2}\nu(dz) < \infty$.*

*The representation is unique, i.e., suppose there exists another $\{\mathcal{F}_t\}$-Brownian motion $W'(t)$ and another Poisson random measure $N'$ and $Z(t)$ is represented by (2.8). Then we have $W = W'$ and $N = N'$.*

*2) (Lévy–Khintchin) The characteristic function $\varphi_t(\alpha) = E[e^{i(\alpha, Z(t))}]$ admits the representation*

$$\varphi_t(\alpha) = \exp\left[t\left\{i(b, \alpha) - \frac{1}{2}(\alpha, V\alpha) + \int_{\mathbf{R}^m - \{0\}}(e^{i(\alpha,z)} - 1 - i(\alpha, z)1_{|z|\leq 1})\nu(dz)\right\}\right].$$
$$(2.9)$$

*Proof.* We give the proof in the one dimensional case only. We adopt a method used in the proof of Theorem 2.1. The characteristic function of $Z(t)$ can be written as $e^{t\psi(\alpha)}$, since the law of $Z(t)$ is infinitely divisible. Set $M^\alpha(t) = e^{i(\alpha, Z(t))} e^{-t\psi(\alpha)}$. Then for any $\alpha$, $M^\alpha(t)$ is a bounded complex martingale. We define a process $Y^\alpha(t)$ by (2.2). It is a square integrable complex martingale. We will show that it is a complex $\{\mathcal{F}_t\}$-Lévy process. For a partition $\Pi = \{0 = t_0 < t_1 < \cdots < t_n = T\}$, consider a stochastic process

$$Y^\Pi(t) := \sum_k \frac{1}{M^\alpha(t_{k-1} \wedge t)} (M^\alpha(t_k \wedge t) - M^\alpha(t_{k-1} \wedge t))$$

$$= \sum_k e^{i\alpha(Z(t_k \wedge t) - Z(t_{k-1} \wedge t))} e^{-(t_k \wedge t - t_{k-1} \wedge t)\psi(\alpha)}.$$

Then $Y^\Pi(t_k) - Y^\Pi(t_{k-1})$ are independent of $\mathcal{F}_{t_k-1}$ for $k = 1, 2, ..., n$, since $Z(t)$ is a $\mathcal{F}_t$-Lévy process. Let $\{\Pi_n\}$ be a sequence of partitions such that $\Pi_n \subset \Pi_{n+1} \subset \cdots$ and $|\Pi_n| \to 0$. Then the sequence $\{Y^{\Pi_n}(t)\}$ converges to $Y^\alpha(t)$, and the limit $Y^\alpha(t)$ is a process such that $Y(t) - Y(s)$ are independent of $\mathcal{F}_s$ for any $s < t$ such that $s, t \in \cup_n \Pi$. The process is continuous in probability and is time homogeneous. Therefore it is a complex $\{\mathcal{F}_t\}$-Lévy process.

The process $Y^\alpha(t)$ is then decomposed as the sum of a continuous (complex) $\{\mathcal{F}_t\}$-Brownian motion $Y_c^\alpha(t)$ and a purely discontinuous $\{\mathcal{F}_t\}$-Lévy process $Y_d^\alpha(t)$. The jumps of $Y^\alpha(t)$ are $\Delta Y^\alpha(t) = \frac{1}{M_{t-}^\alpha} \Delta M^\alpha(t) = e^{i\alpha \Delta Y(t)} - 1$. Therefore $Y_d^\alpha(t)$ is represented by

$$Y_d^\alpha(t) = \int_0^t \int_{0<|z|<\infty} (e^{i\alpha z} - 1) \tilde{N}(ds\,dz).$$

We have

$$\langle Y_d^\alpha, \bar{Y}_d^\alpha \rangle_t = t \int_{0<|z|<\infty} |e^{i\alpha z} - 1|^2 \nu(dz) < \infty.$$

Then the Lévy measure $\nu$ should satisfy $\int \frac{|z|^2}{1+|z|^2} \nu(dz) < \infty$.

Now the complex martingale $M^\alpha(t)$ can be regarded as the solution of a linear SDE

$$M^\alpha(t) = 1 + \int_0^t M_{s-}^\alpha \, dY^\alpha(s). \tag{2.10}$$

Then, $M^\alpha(t)$ is written as

$$M^\alpha(t) = \exp\left( Y_c^\alpha(t) + i\alpha \left( \int_0^t \int_{0<|z|\leq 1} z\tilde{N}(dr\,dz) + \int_0^t \int_{|z|>1} zN(dr\,dz) \right) \right.$$
$$- \frac{1}{2}\langle Y_c^\alpha, Y_c^\alpha \rangle_t - t \int_{0<|z|\leq 1} (e^{i\alpha z} - 1 - i\alpha z)\nu(dz)$$
$$\left. - t \int_{|z|>1} (e^{i\alpha z} - 1)\nu(dz) \right).$$

In fact, denote the right hand side of the above by $M'(t)$. Then applying Itô's formula II to the exponential function $F(x) = e^x$, we can show that $M'(t)$ satisfies the linear SDE (2.10). Since the solution of SDE (2.10) is unique, we get $M^\alpha(t) = M'(t)$.

We define a (complex) $\{\mathcal{F}_t\}$-Lévy process $Y(t)$ by

$$Y(t) = (i\alpha)^{-1} Y_c^{\alpha}(t) + \int_0^t \int_{0 < |x| \le 1} x\tilde{N}(dsdz) + \int_0^t \int_{|x| > 1} xN(dsdz).$$

Since $M^{\alpha}(t) = e^{i\alpha Z(t)} e^{-t\psi(\alpha)}$, the above equation is rewritten as

$$e^{i\alpha(Z(t) - Y(t))}$$
$$= e^{t\psi(\alpha)} \exp\left( -\frac{1}{2} \langle Y_c^{\alpha}, Y_c^{\alpha} \rangle_t - t \int (e^{i\alpha z} - 1 - i\alpha z 1_{|z| \le 1}) \nu(dz) \right).$$

Then $Z(t) - Y(t)$ is a deterministic process. It is differentiable with respect to $t$. The derivative does not depend on $t$, since $Z(t) - Y(t)$ is time homogeneous. We denote it by $b$. Then we get $Y(t) = Z(t) - bt$, and the process $Y(t)$ is a real process and $b$ is a real constant.

Now $W(t) = (i\alpha)^{-1} Y_c^{\alpha}(t)$ is a real continuous square integrable $\{\mathcal{F}_t\}$-Lévy process. Hence it is an $\{\mathcal{F}_t\}$-Brownian motion. Further the $\{\mathcal{F}_t\}$-Brownian motion $W(t)$ and $\{\mathcal{F}_t\}$-Poisson random measure are independent by Proposition 2.6. Consequently we get the Lévy–Itô decomposition of $Z(t)$.

Finally we have $\langle Y_c^{\alpha}, Y_c^{\alpha} \rangle_t = -t\alpha^2 \sigma^2$. Therefore $\exp(t\psi(\alpha))$ is represented by (2.9). Thus the Lévy–Khintchin formula is established.

**Corollary 2.8.** *Any $\{\mathcal{F}_t\}$-Brownian motion is square integrable and is Gaussian distributed.*

**Remark.** If $\bar{b} := \int_{|x| \le 1} |x| \nu(dx) < \infty$ (this includes the case where $\nu$ is a finite measure), the last integral of equation (2.8) can be decomposed as the difference of two integrals based on $N(dsdx)$ and $\hat{N}(dsdx)$. Then equation (2.8) can be written as

$$Z(t) = W(t) + b't + \int_{(0,t]} \int_{0 < |z| < \infty} zN(dsdz), \qquad (2.11)$$

where $b' = b - \int_{|z| \le 1} x\nu(dz)$. However if $\bar{b} = \infty$, we cannot decompose the last integral in (2.8) as the difference of two integrals. It may happen that both integrals $\int_0^t \int_{0 < |z| \le 1} xN(dsdz)$ and $\int_0^t \int_{0 < |z| \le 1} x\hat{N}(dsdz)$ may diverge, but the last integral of (2.8) can converge.

Now we shall return to $\{\mathcal{F}_t'\}$-martingale, where $\{\mathcal{F}_t'\}$ is the filtration generated by a given Lévy process $Z(t)$.

**Theorem 2.9 (Kunita–Watanabe).** [16] *Let $M(t)$ be a square integrable $\{\mathcal{F}_t'\}$-martingale. Then there exist a constant $h$, predictable processes $f(s) = (f_1(s), ..., f_m(s))$ and $g(s, x)$ such that*

$$\int_0^T |f(s)|^2 ds < \infty, \qquad \int_0^T \int_{\mathbf{R}^m - \{0\}} |g(s, z)|^2 ds\nu(dz) < \infty, \quad a.s. \qquad (2.12)$$

*and $M(t)$ is represented by*

$$M(t) = h + \sum_{i=1}^{m} \int_0^t f_i(s) dW^i(s) + \int_0^t \int_{\mathbf{R}^m - \{0\}} g(s, z) \tilde{N}(ds dz). \qquad (2.13)$$

*The triple $(h, f(s), g(s, z))$ is uniquely determined from $M(t)$, i.e., if $M(t)$ is represented by (2.13) with another $(h', f'(s), g'(s, z))$ satisfying (2.12), then we have $h = h'$, $f(s) = f'(s)$ a.e. $\lambda \otimes P$ and $g(s, z) = g'(s, z)$ a.e. $\lambda \otimes \nu \otimes P$, where $\lambda$ is the Lebesgue measure on $[0, T]$ and $\nu$ is the Lévy measure on $\mathbf{R}^m - \{0\}$.*

*Proof.* (Taken from [15]) For simplicity we prove the theorem in the case $m = 1$ only. Let $\mathbf{Z} = (Z(t))$ be a one dimensional Lévy process and let (2.8) be the Lévy–Itô decomposition. We introduce a random measure $M(E_1)$ on $[0, T] \times \mathbf{R}$ by

$$M(E_1) = \int_{E_1(0)} dW(t) + \int_{E_1 - E_1(0)} \frac{z}{1 + |z|} \tilde{N}(dt dz), \qquad (2.14)$$

where $E_1(0) = \{(t, 0) : (t, 0) \in E_1\}$. Then $\int E[M(E_1)M(E_2)] dP = \mu(E_1 \cap E_2)$, where

$$\mu(E) = |E(0)| + \int_{E - E(0)} \left( \frac{z}{1 + |z|} \right)^2 dt \nu(dz).$$

For each positive integer $p$, we define the multiple Wiener integral by

$$I_p(f) = \int \cdots \int f(\xi_1, ..., \xi_p) dM(\xi_1) \cdots dM(\xi_p). \qquad (2.15)$$

Let $\mathbf{H_Z}$ be the $L^2$ space over $(\Omega, \mathcal{F}'_T, P)$ and let $\mathbf{H_Z}^{(p)}$ be the closed linear manifold of $\{I_p(f); f \in L_p^2\}$, where $L_p^2$ is the $L^2$ space on $\mathbf{R}^p$ with the product measure of $\mu$. Then it is shown in Itô [11] that one has the direct sum expansion: $\mathbf{H_Z} = \sum_{p \geq 0} \oplus \mathbf{H_Z}^{(p)}$. Note that each $I_p(f)$ is written as the sum of the following terms

$$\int \cdots \int_{0 \leq t_1 < \cdots < t_p \leq T, (z_1, ..., z_p) \in \mathbf{R}^p} f((t_1, z_1), ..., (t_p, z_p)) dM(t_1 z_1) \cdots dM(t_p z_p)$$

$$\qquad (2.16)$$

$$= \int_0^T \int_{\mathbf{R}} \varphi(t_p, z_p) dM(t_p z_p),$$

where

$$\varphi(t_p, z_p) = \int \cdots \int_{\Lambda(t_p, z_p)} f((t_1, z_1), ..., (t_p, z_p)) dM(t_1 z_1) \cdots dM(t_{p-1} z_{p-1})$$

and
$\Lambda(t_p, z_p) = \{0 < t_1 < \cdots < t_{p-1} < t_p, (z_1, ..., z_{p-1}, z_p) \in \mathbf{R}^p\}$. Setting $\phi(t) = \varphi(t, 0)$ and $\psi(t, z) = \varphi(t, z) \frac{1 + |z|}{z}$ ($|z| > 0$), we find that the above is written as

$$\int_0^T \phi(t) dW(t) + \int_0^T \int_{\mathbf{R} - \{0\}} \psi(t, z) \tilde{N}(dt dz). \qquad (2.17)$$

Therefore any element of $\mathbf{H_Z}^{(p)}$ and hence any element $X$ of $\mathbf{H_Z}$ with mean 0 is written as the above. Now taking the conditional expectation of (2.17), we obtain the representation (2.13) for the square integrable martingale $M(t) = E[X|\mathcal{F}_t]$.

**Corollary 2.10.** *The spaces of continuous martingales $\mathcal{M}_c$ and that of purely discontinuous martingales $\mathcal{M}_d$ are characterized as follows.*

$$\mathcal{M}_c = \left\{ \sum_i \int_0^t f^i(s)dW^i(s) : E\left[ \int_0^T \sum_{i,j} f^i(s)f^j(s)v_{ij}ds \right] < \infty \right\}, \quad (2.18)$$

$$\mathcal{M}_d = \left\{ \int_0^t \int_{\mathbf{R}^m - \{0\}} g(s,z)\tilde{N}(dsdz) : \; g \in L^2(\hat{N}) \right\}. \quad (2.19)$$

Let $W(t) = (W^1(t), ..., W^m(t))$ be a standard $\{\mathcal{F}_t\}$-Brownian motion. Let $f(s) = (f_1(s), ..., f_m(s))$ be a predictable process with the square integrability condition. Then the stochastic integrals $\sum_i \int_0^t f_i(s)dW^i(s)$ are continuous square integrable martingales. These martingales are orthogonal to martingales $\int \int g\tilde{N}(dsdz)$. Indeed, the quadratic co-variation of the Brownian motion $W(t)$ and Poisson random measure $N(dsdz)$ is 0, since the former is a continuous martingale and the latter is a process of finite variation. Then their stochastic integrals are orthogonal.

Let $M, N$ be square integrable martingales represented by

$$M(t) = \sum_{i=1}^m \int_0^t f_i(s)dW^i(s) + \int_0^t \int_Z g(s,z)\tilde{N}(dsdz),$$

$$N(t) = \sum_{i=1}^m \int_0^t \phi_i(s)dW^i(s) + \int_0^t \int_Z \psi(s,z)\tilde{N}(dsdz),$$

then we have the formula

$$\langle M, N \rangle_t = \sum_{i=1}^m \int_0^t f_i(s)\phi_i(s)ds + \int_0^t \int_Z g(s,z)\psi(s,z)ds\nu(dz),$$

$$[M, N]_t = \sum_{i=1}^m \int_0^t f_i(s)\phi_i(s)ds + \int_0^t \int_Z g(s,z)\psi(s,z)N(dsdz).$$

## 2.5 $L^p$ estimates of stochastic integrals

In this section we are concerned with $L^p$ estimates of semimartingales $X(t)$. The $L^p$ estimate plays an important role for the study of the solution of SDEs. In particular, for the study of regularity of solutions with respect to the initial conditions, we will apply Kolmogorov's criterion for the continuity of random fields, where the $L^p$ estimates of solutions are required (see Appendix). In this section we study $L^p$ estimates for general semimartingales and in the next section we apply these estimates to the solution of a SDE.

We assume that we are given an $m$ dimensional standard $\{\mathcal{F}_t\}$-Brownian motion $W(t) = (W^1(t), ..., W^m(t))$ and a $\{\mathcal{F}_t\}$-Poisson random measure $N$ on a measurable space $\mathcal{Z}$. Let us consider a $d$-dimensional semimartingale $X(t) = (X^1(t), ..., X^d(t))$ represented by.

$$X^i(t) = x_i + \int_0^t b^i(r)dr + \sum_{j=1}^m \int_0^t f^{ij}(r)dW^j(r) + \int_0^t \int_{\mathcal{Z}} g^i(r, z)\tilde{N}(drdz), \quad (2.20)$$

for $i = 1, ..., d$. Set $b(r) = (b^1(r), ..., b^d(r))$, $f(r) = (f^{ij}(r))$, $g(x, r) = (g^1(x, r), ..., g^d(x, r))$ and write $|f(r)|^2 = \sum_{ij} |f^{ij}(r)|^2$.

**Theorem 2.11.** *For any $p \geq 2$, there exists a positive constant $C_p$ such that*

$$E\left[ \sup_{0 < s \leq t} |X(s)|^p \right] \leq$$

$$C_p \left\{ |x|^p + E\left[ \left( \int_0^t |b(r)|dr \right)^p \right] + E\left[ \left( \int_0^t |f(r)|^2 dr \right)^{p/2} \right] \quad (2.21)$$

$$+ E\left[ \left( \int_0^t \int_{\mathcal{Z}} |g(r, z)|^2 dr\nu(dz) \right)^{p/2} \right] + E\left[ \int_0^t \int_{\mathcal{Z}} |g(r, z)|^p dr\nu(dz) \right] \right\}$$

*holds for a semimartingale $X(t)$ represented by (2.20).*

**Remark.** If $X(t) = \int_0^t f(s)dW(s)$, the above inequality is known as a special case of Burkholder's inequality, which states that for any $p > 0$, there exist positive constants $c_p, C_p$ such that

$$c_p E\left[ \sup_{0 \leq s \leq t} |X(s)|^p \right] \leq E\left[ \langle X, X \rangle_t \right] \leq C_p E\left[ \sup_{0 \leq s \leq t} |X(s)|^p \right]$$

holds for any continuous martingale $X(t)$ with $X(0) = 0$. See Ikeda–Watanabe [10].

*Proof.* We give the proof in the case where $d = 1$ only. The inequality of the proposition is obvious if the right hand side is infinite. So we assume that the right hand side is finite. We shall obtain $L^p$ estimate for each term of the right hand side of (2.20). In the following arguments, constants $c_i$, $i = 1, 2, \ldots$ will be determined depending on $p$ only.

The inequality

$$E\left[ \sup_{0 < s \leq t} \left| \int_0^s b(r)dr \right|^p \right] \leq E\left[ \left( \int_0^t |b(r)|dr \right)^p \right]$$

is obvious. We shall consider the martingale $Y(t) = \sum_j \int_0^t f^j(r)dW^j(r)$. Set $F(x) = |x|^p$ where $p \geq 2$. It is a $C^2$ function with $F'(x) = p|x|^{p-2}x$ and $F''(x) = p(p-1)|x|^{p-2}$. Then by Itô's formula for continuous martingales, we have

$$|Y(t)|^p = p \int_0^t |Y(r)|^{p-2}Y(r)dY(r) + \frac{1}{2}p(p-1) \int_0^t |Y(r)|^{p-2}|f(r)|^2 ds.$$

Choose an increasing sequence of stopping times $\tau_n$ such that $P(\tau_n < T) \to 0$ as $n \to \infty$ and the stopped precess of $\int_0^t |Y(r)|^{p-2} Y(r) dY(r)$ are all bounded martingales with mean 0. Then taking the expectation of $|Y(t \wedge \tau_n)|^p$ and using Hölder's inequality, we obtain

$$
\begin{aligned}
E[|Y(t \wedge \tau_n)|^p] &= \frac{1}{2} p(p-1) E\left[\int_0^{t \wedge \tau_n} |Y(r)|^{p-2} |f(r)|^2 ds\right] \\
&\le \frac{1}{2} p(p-1) E\left[\sup_{0 < r \le t \wedge \tau_n} |Y(r)|^p\right]^{(p-2)/p} \\
&\quad \times E\left[\left(\int_0^{t \wedge \tau_n} |f(r)|^2 dr\right)^{p/2}\right]^{2/p}.
\end{aligned}
$$

Now, observe Doob's inequality $E[\sup_{0 < r \le t \wedge \tau_n} |Y(r)|^p] \le p^q E[|Y(t \wedge \tau_n)|^p]$. Then we have

$$
\begin{aligned}
E\left[\sup_{0 < r \le t \wedge \tau_n} |Y(r)|^p\right] &\le \frac{1}{2} p(p-1) p^q E\left[\sup_{0 < r \le t \wedge \tau_n} |Y(r)|^p\right]^{(p-2)/p} \\
&\quad \times E\left[\left(\int_0^{t \wedge \tau_n} |f(r)|^2 dr\right)^{p/2}\right]^{2/p}.
\end{aligned}
$$

Then we have

$$
E\left[\sup_{0 < r \le t \wedge \tau_n} |Y(r)|^p\right]^{2/p} \le \frac{1}{2} p(p-1) p^q E\left[\left(\int_0^{t \wedge \tau_n} |f(r)|^2 dr\right)^{p/2}\right]^{2/p}.
$$

Thus we obtain

$$
E\left[\sup_{0 < s \le t} |Y(s)|^p\right] \le (p(p-1) p^q / 2)^{p/2} E\left[\left(\int_0^t |f(r)|^2 dr\right)^{p/2}\right].
$$

We next consider $Y(t) = \int_0^t \int g(r, z) \tilde{N}(drdz)$. We apply Itô's formula II for $F(x) = |x|^p$. Since $Y(t)$ is purely discontinuous, we have

$$
|Y(t)|^p = Z(t) \tag{2.22}
$$
$$
+ \int_0^t \int_Z \{|Y(r-) + g|^p - |Y(r-)|^p - p|Y(r-)|^{p-2} Y(r-)g\} \hat{N}(drdz),
$$

where $Z(t)$ is a localmartingale. We may choose an increasing sequence of stopping times $\tau_n$ such that $P(\tau_n < T) \to 0$ as $n \to \infty$, $Z(t \wedge \tau_n)$ are martingales with mean 0 and the last term in (2.22) stopped at $\tau_n$ are integrable. To make the notation simple, we denote the stopped processes $Y(t \wedge \tau_n)$ etc by the same notations $Y(t)$, etc. Since,

$$|Y(r-) + g|^p - |Y(r-)|^p - p|Y(r-)|^{p-2}Y(r-)g$$
$$= \frac{1}{2}p(p-1)|Y(r-) + \theta g|^{p-2}g^2$$
$$\leq c_3|Y(r-)|^{p-2}g^2 + c_4|g|^p$$

holds for some $|\theta| < 1$, we have from (2.22)

$$E[|Y(t)|^p] \leq c_3 E\left[\int_0^t \int_Z |Y(r-)|^{p-2}g^2 dr\nu(dz)\right] \tag{2.23}$$
$$+ c_4 E\left[\int_0^t \int_Z |g|^p dr\nu(dz)\right].$$

We calculate the first term of the right hand side. It holds by Hölder's inequality

$$E\left[\int_0^t \int_Z |Y(r-)|^{p-2}g^2 dr\nu(dz)\right]$$
$$\leq E\left[\sup_{0<r\leq t} |Y(r-)|^p\right]^{1-2/p} E\left[\left(\int_0^t \int_Z g^2 dr\nu(dz)\right)^{p/2}\right]^{2/p}$$
$$\leq c_5 E\left[\sup_{0<r\leq t} |Y(r-)|^p\right] + c_6 E\left[\left(\int_0^t \int_Z g^2 dr\nu(dz)\right)^{p/2}\right].$$

In the last inequality, we used the inequality $ab \leq \frac{a^{p'}}{p'} + \frac{b^{q'}}{q'}$, where $a, b > 0$, $p', q' > 1$ and $\frac{1}{p'} + \frac{1}{q'} = 1$. Therefore, using Doob's inequality, we get from (2.23),

$$E\left[\sup_{0<s\leq t} |Y(s)|^p\right] \leq q^p c_3 c_5 E\left[\sup_{0<r\leq t} |Y(r-)|^p\right]$$
$$+ q^p c_3 c_6 E\left[\left(\int_0^t \int_Z |g|^2 dr\nu(dz)\right)^{p/2}\right]$$
$$+ q^p c_4 E\left[\int_0^t \int_Z |g|^p dr\nu(dz)\right].$$

Choose $c_5$ small enough such that $q^p c_3 c_5 < 1$ and move the term to the left. Note that $\sup_{0<s\leq t} |Y(s)| = \sup_{0<r\leq t} |Y(r-)|$ holds a.s. for any $t$. Then we get the inequality

$$E\left[\sup_{0<s\leq t} |Y(s)|^p\right] \leq c_7 E\left[\left(\int_0^t \int_Z |g|^2 dr\nu(dz)\right)^{p/2}\right]$$
$$+ c_8 E\left[\int_0^t \int_Z |g|^p dr\nu(dz)\right]. \tag{2.24}$$

So far we have chosen constants $c_7$, $c_8$ not depending on the form of martingales $Y(t)$ and stopping times $\tau_n$. Therefore the inequality holds up to stopping time $t \wedge \tau_n$ in (2.24) with the common constants $c_7$, $c_8$. Now let $n$ tend to infinity. Then we find that (2.24) holds for any $t$. The proof is complete.

**Corollary 2.12.** *For any $p \geq 2$, there exists a positive constant $C'_p$ such that*

$$E\left[\sup_{0<s\leq t}|X(s)|^p\right] \leq C'_p\left\{|x|^p + E\left[\int_0^t |b(r)|^p dr\right] + E\left[\int_0^t |f(r)|^p dr\right]\right. \quad (2.25)$$

$$+E\left[\int_0^t\left(\int_Z |g(r,z)|^2 v(dz)\right)^{p/2} dr\right] + E\left[\int_0^t\left(\int_Z |g(r,z)|^p v(dz)\right) dr\right]\right\}$$

*holds for any semimartingale represented by (2.20).*

We shall combine the above estimate and Itô's formula II. The following estimate will provide a useful tool for the study of stochastic flows in the next section.

**Corollary 2.13.** *For any $p \geq 2$, there exists a positive constant $C_p$ such that*

$$E\left[\sup_{0<s<t}|F(X(s))|^p\right] \quad (2.26)$$

$$\leq C_p\left\{|F(x)|^p + E\left[\int_0^t\left|\sum_i F'_{x_i}(X(r-))b^i(r)\right|^p dr\right]\right.$$

$$+E\left[\int_0^t\left(\sum_k|\sum_i F'_{x_i}(X(r-))f^{i,k}(r)|^2\right)^{p/2} dr\right]$$

$$+\frac{1}{2}E\left[\int_0^t\left|\sum_{i,j,k} F''_{x_i x_j}(X(r-))f^{ik}(r)f^{jk}(r)\right|^p dr\right]$$

$$+E\left[\int_0^t\left(\int_Z |F(X(r-)+g(r,z)) - F(X(r-))|^2 v(dz)\right)^{p/2} dr\right]$$

$$+E\left[\int_0^t\left(\int_Z |F(X(r-)+g(r,z)) - F(X(r-))|^p v(dz)\right) dr\right]$$

$$+E\left[\int_0^t\left(\int_Z |F(X(r-)+g(r,z)) - F(X(r-))\right.\right.$$

$$\left.\left.-\sum_i F'_{x_i}(X(r-))g^i(r,z)|v(dz)\right)^p dr\right]\right\},$$

*where $X(t)$ is a d-dimensional semimartingale represented by (2.20) and $F(x_1, ..., x_d)$ is a $C^2$-function.*

Finally we shall study the uniform convergence of a sequence of stochastic integrals with respect to a parameter. It is convenient to introduce a "$p$-Lipschitz norm" for a $p$-th integrable random field $\{X(x); x \in \mathbf{R}^d\}$

$$\|X\|_p = \sup_{x\in\mathbf{R}^d} E[|X(x)|^p]^{1/p} + \sup_{x\neq y, x, y\in\mathbf{R}^d} \frac{E[|X(x) - X(y)|^p]^{1/p}}{|x-y|}.$$

Let $b(x,r)$, $f(x,r)$, $g(x,r,z)$ and $b_n(x,r)$, $f_n(x,r)$, $g_n(x,r,s)$, $n = 1, 2, ...$ be a sequence of predictable processes. We consider a sequence of semimartingales with parameter $x$

$$X(x, t) = X(x) + \int_0^t b(x, r)dr + \int_0^t f(x, r)dW(r) + \int_0^t \int_Z g(x, r, z)\tilde{N}(drdz).$$

$$X_n(x, t) = X_n(x) + \int_0^t b_n(x, r)dr + \int_0^t f_n(x, r)dW(r)$$
$$+ \int_0^t \int_Z g_n(x, r, z)\tilde{N}(drdz).$$

**Theorem 2.14.** *Assume that the sequence* $\{b_n, f_n, g_n\}$ *satisfies*

$$\lim_{n\to\infty} \sup_{0<s\leq T} \|b_n(s) - b(s)\|_p = 0,$$

$$\lim_{n\to\infty} \sup_{0<s\leq T} \|f_n(s) - f(s)\|_p = 0,$$

$$\lim_{n\to\infty} \sup_{0<s\leq T} \left\| \left( \int_Z |g_n(\cdot, s, z) - g(\cdot, s, z)|^{p'} v(dz) \right)^{1/p'} \right\|_p = 0.$$

*for* $p' = 2$ *and* $p' = p$, *where* $p > d$. *Then if* $\|X_n - X\|_p \to 0$, *we have for any positive number* $N$,

$$\lim_{n\to\infty} E\left[ \sup_{|x|\leq N} \sup_{0<s\leq T} |X_n(x, s) - X(x, s)|^p \right] = 0. \tag{2.27}$$

*Proof.* We shall apply Theorem 4.4 in Appendix. Observe first that (2.25) implies

$$\lim_{n\to\infty} \sup_{0<s\leq T} \|X_n(s) - X(s)\|_p = 0.$$

Then we get (2.27) by Theorem 4.4.

## 3 Stochastic differential equation and stochastic flow

### 3.1 Semimartingale with spatial parameter and a SDE based on it

Let $(\Omega, \mathcal{F}, P)$ be a complete probability space equipped with a filtration $\{\mathcal{F}_t\}_{0\leq t\leq T}$. We assume that on the probability space an $m$-dimensional standard $\{\mathcal{F}_t\}$-Brownian motion $W(t) = (W^1(t), ..., W^m(t))$ and a $\{\mathcal{F}_t\}$-Poisson random measure $N((s, t] \times U)$ on a measurable space $(\mathcal{Z}, \mathcal{B})$ are defined.

Let $b(x, t) = (b^1(x, t), ..., b^d(x, t))$ and $f(x, t) = (f^{ij}(x, t))_{i=1,...,d, j=1,...,m}$ be predictable processes with spatial parameter $x \in \mathbf{R}^d$. Let $g(x, t, z) = (g^1(x, t, z), ..., g^d(x, t, z)), x \in \mathbf{R}^d, z \in \mathcal{Z}$ be a predictable process with parameters $x, z$. We will assume that these processes are of linear growth and are Lipschitz continuous, i.e., there exist positive constants $K, K(z)$ and $L, L(z)$ such that

$$\frac{|b(x,t)|}{1+|x|} \le K, \quad |b(x,t)-b(y,t)| \le L|x-y|, \tag{3.1}$$

$$\frac{|f(x,t)|}{1+|x|} \le K, \quad |f(x,t)-f(y,t)| \le L|x-y|,$$

$$\frac{|g(x,t,z)|}{1+|x|} \le K(z), \quad |g(x,t,z)-g(y,t,z)| \le L(z)|x-y|,$$

hold for all $x, y \in \mathbf{R}^d$ and $z \in Z$, a.e. $\Lambda \times P$, where $\Lambda$ is the Lebesgue measure on $[0, T]$. Further constants $K(z), L(z)$ satisfy

$$\int_Z (K(z)^p + L(z)^p)\nu(dz) < \infty \quad \forall p \ge 2. \tag{3.2}$$

We define an $\mathbf{R}^d$-valued process $X(x,t) = (X^1(x,t), ..., X^d(x,t))$ with parameter $x$ by

$$X^i(x,t) = \int_0^t b^i(x,r)dr \tag{3.3}$$

$$+ \sum_j \int_0^t f^{ij}(x,r)dW^j(r) + \int_0^t \int_Z g^i(x,r,z)\tilde{N}(drdz).$$

With vector–matrix notations, the above is written as

$$X(x,t) = \int_0^t b(x,r)dr + \int_0^t f(x,r)dW(r) + \int_0^t \int_Z g(x,r,z)\tilde{N}(drdz).$$

It is a semimartingale with spatial parameter $x$.

Let $\eta_t$ be an $\mathbf{R}^d$-valued adapted cadlag process. Then the stochastic processes $b(\eta_{t-}, t)$, $f(\eta_{t-}, t)$ and $g(\eta_{t-}, t, z)$ are predictable processes. We set

$$\int_{t_0}^t X(\eta_{r-}, dr) := \int_{t_0}^t b(\eta_{r-}, r)dr + \int_{t_0}^t f(\eta_{r-}, r)dW(r)$$

$$+ \int_{t_0}^t \int_Z g(\eta_{r-}, r, z)\tilde{N}(drdz).$$

It is well defined as an $\mathbf{R}^d$-valued semimartingale.

Now we shall consider a SDE based on $X(x,t)$. Suppose that we are given an $\mathcal{F}_{t_0}$-measurable $\mathbf{R}^d$-valued random variable $\xi_0$. Let $\xi_t = (\xi_t^1, ..., \xi_t^d)$ be an $\mathbf{R}^d$-valued $\{\mathcal{F}_t\}$-adapted cadlag process satisfying the equation

$$\xi_t = \xi_0 + \int_{t_0}^t X(\xi_{s-}, ds). \tag{3.4}$$

Then the process $\xi_t$ is called a *solution of the stochastic differential equation based on $X(x,t)$, starting from $\xi_0$ at time $t_0$*. The process $X(x,t)$ is called the *infinitesimal generator* of the solution $\xi_t$. Functionals $b(x,t)$, $f(x,t)$ and $g(x,t,z)$ are called *drift coefficients, diffusion coefficients* and *jump coefficients*, respectively.

**Theorem 3.1.** *Suppose that $b(x,t)$, $f(x,t)$, $g(x,t,z)$ satisfy (3.1) and (3.2). Then, if the initial data $\xi_0$ is p-th integrable, the equation has a unique solution in $L^p$.*

*Proof.* The existence of the solution can be verified by a standard method of successive approximation. We will find solutions in $L^p$ space. Given an $\mathcal{F}_{t_0}$-measurable random variable $\xi_0$, we define a sequence of $\mathbf{R}^d$-valued semimartingales by $\xi_t^0 = \xi_0$ and

$$\xi_t^n = \xi_0 + \int_{t_0}^t X(\xi_{s-}^{n-1}, ds), \quad n = 1, 2, \ldots \tag{3.5}$$

We shall compute the $L^p$-norm of the above. It holds

$$\xi_t^1 = \xi_0 + \int_{t_0}^t b(\xi_0, r)dr + \int_{t_0}^t f(\xi_0, r)dW(r) + \int_{t_0}^t \int_Z g(\xi_0, r, z)\tilde{N}(drdz).$$

By applying Theorem 2.11, we have

$$E\left[\sup_{t_0 \le r \le t} |\xi_r^1 - \xi_r^0|^p\right] \le C_p \left\{ E\left[\left(\int_{t_0}^t |b(\xi_0, r)|dr\right)^p\right]\right.$$
$$+ E\left[\left(\int_{t_0}^t |f(\xi_0, r)|^2 dr\right)^{p/2}\right]$$
$$+ E\left[\left(\int_{t_0}^t \int_Z |g(\xi_0, z)|^2 dr\nu(dz)\right)^{p/2}\right]$$
$$+ E\left[\int_{t_0}^t \int_Z |g(\xi_0, z)|^p dr\nu(dz)\right]\right\}.$$

Here, we set $|f|^2 = \sum_{i,j} |f^{ij}|^2$. By the linear growth properties of coefficients $b$, $f$, $g$, the above is dominated by $C_p' E[(1 + |\xi_0|)^p] < \infty$.

We have further

$$E\left[\sup_{t_0 \le r \le t} |\xi_r^{n+1} - \xi_r^n|^p\right] \le C_p \left\{ E\left[\left(\int_{t_0}^t |b(\xi_{r-}^n, r) - b(\xi_{r-}^{n-1}, r)|dr\right)^p\right]\right.$$
$$+ E\left[\left(\int_{t_0}^t |f(\xi_{r-}^n, r) - f(\xi_{r-}^{n-1}, r)|^2 dr\right)^{p/2}\right]$$
$$+ E\left[\left(\int_{t_0}^t \int_Z |g(\xi_{r-}^n, r, z) - g(\xi_{r-}^{n-1}, r, z)|^2 dr\nu(dz)\right)^{p/2}\right]$$
$$+ E\left[\int_{t_0}^t \int_Z |g(\xi_{r-}^n) - g(\xi_{r-}^{n-1})|^p dr\nu(dz))\right]\right\}.$$

Using Lipschitz conditions for $b$, $f$, $g$, we get

$$E\left[\sup_{t_0 \le r \le t} |\xi_r^{n+1} - \xi_r^n|^p\right] \le C \int_{t_0}^t E[|\xi_{r-}^n - \xi_{r-}^{n-1}|^p]dr$$
$$\le C \int_{t_0}^t E\left[\sup_{t_0 \le r \le s} |\xi_{r-}^n - \xi_{r-}^{n-1}|^p\right]ds.$$

Then we get the inequality

$$E\left[\sup_{t_0 \le r \le t} |\xi_{r-}^{n+1} - \xi_{r-}^n|^p\right] \le \frac{C^n}{n!} E\left[\sup_{t_0 \le r \le t} |\xi_{r-}^1 - \xi_{r-}^0|^p\right].$$

Then the sequence $\{\xi_r^n, t_0 \le r \le T\}$ converges in $L^p$ uniformly in $r$. Denote the limit as $\xi_r, r \le T$. It is a cadlag adapted process. Let $n$ tend to infinity in (3.5). Then $\xi_t$ satisfies (3.4). It is a desired solution of the SDE.

For the uniqueness of the solution, suppose that $\xi_t'$ is another solution belonging to $L^p$. Then we have $\xi_t - \xi_t' = \int_{t_0}^t X(\xi_{s-}, ds) - \int_{t_0}^t X(\xi_{s-}', ds)$. Then we get the estimate

$$E[|\xi_t - \xi_t'|^p] \le C \int_{t_0}^t E[|\xi_{s-} - \xi_{s-}'|^p]ds,$$

similarly as the above argument. This implies $E[|\xi_t - \xi_t'|^p] = 0$.

**Remark 1.** We can show that any solution of equation (3.4) belongs to $L^p$. Therefore the solution obtained by the successive approximation is the unique solution. Indeed, let $\xi_t$ be an arbitrary solution. Define a sequence of stopping times $\tau_n$ by $\tau_n = \inf\{t \in [t_0, T]; \; |\xi_t| \ge n\}$. Then $P(\tau_n < T) \to 0$ as $n \to \infty$. We can show that there exists a positive constant $C$ not depending on $n$ such that

$$E\left[\sup_{t_0 \le r \le t \wedge \tau_n} |\xi_r|^p\right] \le C\left(1 + E\left[\int_{t_0}^{t \wedge \tau_n} |\xi_{r-}|^p dr\right]\right).$$

(Cf. the proof of (3.6) in the next section.) Since $|\xi_r| \le n$ holds for $r < \tau_n$, the right hand side is finite a.s. Then we obtain $E[\sup_{t \le r \le \tau_n} |\xi_r|^p] \le Ce$, from the above functional inequality. Let $n$ tend do infinity. Since the constant $C$ does not depend on $n$, we find that $\xi_t$ is $p$-th integrable.

**Remark 2.** It is often considered a SDE with 'big jumps'. Let $U$ be a Borel subset of $\mathcal{Z}$ such that $v(U^c) < \infty$. Consider a semimartingale with spatial parameter;

$$\tilde{X}(x, t) = \int_0^t b(x, r)dr + \int_0^t f(x, r)dW(r) + \int_0^t \int_U g(x, r, z)\tilde{N}(drdz)$$
$$+ \int_0^t \int_{U^c} g(x, r, z)N(drdz).$$

where coefficients $b$, $f$, $g$ are Lipschitz continuous and of the linear growth in the above sense. However, $K(z)$, $L(z)$ do not necessarily satisfy (3.2) but they satisfy

$$\int_U (K(z)^p + L(z)^p)v(dz) < \infty.$$

The solution of the SDE based on the above $\tilde{X}(x, t)$ exists uniquely. However, it does not belong to $L^p$ in general. Such a SDE will be discussed in Section 3.5.

**Remark 3.** Let $v_1(x), ..., v_m(x)$ be Lipschitz continuous $\mathbf{R}^d$-valued functions and let $(Z^1(t), ..., Z^m(t))$ be an $m$-dimensional Lévy process. A SDE of the form

$$\xi_t = \xi_0 + \sum_{j=1}^{m} \int_{t_0}^{t} v_j(\xi_{s-})dZ^j(r)$$

where $Z^j$ are even general semimartingales, is studied in detail (see Protter [20]). Setting $X(x,t) = \sum_j v^j(x)Z^j(t)$, the above equation is written as (3.4). The equation is said to be of *separating type*.

Instead of the above SDE of the separating type, we will consider a "canonical SDE" written as

$$\xi_t = \xi_0 + \sum_{j=1}^{m} \int_{t_0}^{t} v_j(\xi_{s-}) \diamond dZ^j(r),$$

where the symbol $\diamond$ means the *canonical stochastic integral*. It will be discussed in Section 3.7.

### 3.2 Continuity

We denote by $\xi_t(x)$ the solution starting from $x$ at time $t_0$. We study the continuity of the solution $\xi_t(x)$ with respect to the initial condition $x$. Our idea is to apply Kolmogorov's criterion for the continuity of random field. See Theorem 4.1 in the Appendix. For this purpose, we claim the following.

**Theorem 3.2.** *Assume the same condition as in Theorem 3.1 for coefficients $b$, $f$, $g$. Then for any $p \geq 2$, there exists a positive constant $C_p$ such that*

$$E\left[\sup_{t_0 \leq s \leq t} (1 + |\xi_s(x)|)^p\right] \leq C_p(1 + |x|)^p, \quad \forall x \in \mathbf{R}^d \qquad (3.6)$$

$$E\left[\sup_{t_0 \leq s \leq t} |\xi_s(x) - \xi_s(y)|^p\right] \leq C_p|x - y|^p, \quad \forall x, y \in \mathbf{R}^d \qquad (3.7)$$

*hold for any $0 \leq t_0 < t \leq T$.*

*Proof.* We set $X(t) = \xi_t(x)$. It satisfies

$$X(t) = x + \int_{t_0}^{t} b(X(r-), r)dr$$

$$+ \int_{t_0}^{t} f(X(r-), r)dW(r) + \int_{t_0}^{t} \int_{Z} g(X(r-), r, z)\tilde{N}(drdz).$$

We apply Corollary 2.12. Then we have

$$E\left[\sup_{t_0 \leq s \leq t} |X(s)|^p\right]$$

$$\leq C_p'\left\{|x|^p + E\left[\int_{t_0}^{t} |b(X(r-), r)|^p dr\right] + E\left[\int_{t_0}^{t} |f(X(r-), r)|^p dr\right]\right.$$

$$+ E\left[\int_{t_0}^{t} \left(\int_{Z} |g(X(r-), r, z)|^2 v(dz)\right)^{p/2} dr\right]$$

$$\left. + E\left[\int_{t_0}^{t} \int_{Z} |g(X(r-), r, z)|^p v(dz)dr\right]\right\}.$$

Note the linear growth properties of $b$, $f$, $g$ in (3.1) and (3.2). Then we get

$$E\left[\sup_{t_0 \le s \le t} |X(s)|^p\right] \le C_p'\left\{|x|^p + (2K + K_2 + K_p)\int_{t_0}^t E[(1 + |X(r-)|)^p]dr\right\},$$

where $K_p = \int K(z)^p \nu(dz)$. Therefore,

$$E\left[\sup_{t_0 \le s \le t} (1 + |X(s)|)^p\right] \le C_p''\left\{(1 + |x|)^p\right.$$

$$\left. + (2K + K_2 + K_p)\int_{t_0}^t E\left[\sup_{t_0 \le s \le r}(1 + |X(s)|)^p\right]dr\right\}.$$

The above inequality implies (3.6).

We next set $Y(t) = \xi_t(x) - \xi_t(y)$. It satisfies

$$Y(t) = (x - y) + \int_{t_0}^t (b(\xi_{r-}(x), r) - b(\xi_{r-}(y), r))dr$$

$$+ \int_{t_0}^t (f(\xi_{r-}(x), r) - f(\xi_{r-}(y), r))dW(r)$$

$$+ \int_{t_0}^t \int_Z (g(\xi_{r-}(x), r, z) - g(\xi_{r-}(y), r, z))\tilde{N}(drdz).$$

Therefore we have by Corollary 2.12,

$$E[\sup_{t_0 \le s \le t} |Y(s)|^p] \le C_p'\left\{|x - y|^p + E\left[\int_{t_0}^t |b(\xi_{r-}(x), r) - b(\xi_{r-}(y), r)|^p dr\right]\right.$$

$$+ E\left[\int_{t_0}^t |f(\xi_{r-}(x), r) - f(\xi_{r-}(y), r)|^p dr\right]$$

$$+ E\left[\int_{t_0}^t \left(\int_Z |g(\xi_{r-}(x), r, z) - g(\xi_{r-}(y), r, z)|^2 \nu(dz)\right)^{p/2} dr\right]$$

$$\left. + E\left[\int_{t_0}^t \int_Z |g(\xi_{r-}(x), r, z) - g(\xi_{r-}(y), r, z)|^p \nu(dz)dr\right]\right\}.$$

Using the Lipschitz conditions for $b$, $f$, $g$, we get the inequality

$$E\left[\sup_{t_0 \le s \le t} |Y(s)|^p\right] \le C_p'\left\{|x - y|^p + (2L + L_2 + L_p)\int_{t_0}^t E\left[|Y(r-)|^p\right]dr\right\}$$

$$\le C_p''\left\{|x - y|^p + (2L + L_2 + L_p)\int_{t_0}^t E\left[\sup_{t_0 \le s \le r} |Y(s)|^p\right]dr\right\},$$

where $L_p = \int L(z)^p \nu(dz)$. Therefore we get the desired inequality (3.7).

Now we apply Kolmogorov–Totoki's theorem (Theorem 4.1 in Appendix) by setting $\gamma = \alpha = p > d$ and $X(x) = \{\xi_s(x), s \in [t_0, t]\}$, where the norm is the supremum norm with respect to $s \in [t_0, t]$. Then the random field $\xi_t(x)$ has a modification $\xi'_t(x)$ such that for any $x$ it is cadlag with respect to $t$ and for any $t$ it is continuous in $x$, a.s.

From now we denote the above modification by the same notation $\xi_t(x)$. Let $C = C(\mathbf{R}^d; \mathbf{R}^d)$ be the space of continuous maps from $\mathbf{R}^d$ into itself. Then $\xi_t$ may be considered as a $C$-valued cadlag proceess.

## 3.3 Differentiability

We will study the differentiability of $\xi_t(x)$ with respect to $x$. Let $e_i = (0, ..., 0, 1, 0, ..., 0)$ (1 is at the $i$-th component) be a unit vector in $\mathbf{R}^d$ and let $\lambda$ be a real number such that $\lambda \neq 0$. Set

$$N_t(x, \lambda) = \frac{1}{\lambda}\Big(\xi_t(x + \lambda e_i) - \xi_t(x)\Big).$$

It is continuous in $(x, \lambda) \in D$ for any $t$, where $D = \mathbf{R}^d \times (\mathbf{R} - \{0\})$. If we can show that the limit exists for all $t$, $x$ a.s. as $\lambda \to 0$, then $\xi_t(x)$ is differentiable with respect to $x$ at any $(t, x)$ a.s. We will prove this fact by applying Kolmogorov's criterion. Indeed, by taking $p\delta > d + 1$ in the following theorem (Theorem 3.3), we will find that the random field $N_t(x, \lambda)$, $(x, \lambda) \in D$ is uniformly continuous in $D$ and in fact it can be extended continuously to the closure of $D$. Then the extended random field $N_t(x, \lambda)$ is continuous in $\bar{D} = \mathbf{R}^{d+1}$. This means in particular that $N_t(x, 0) = \exists \lim_{\lambda \to 0} N_t(x, \lambda)$ a.s. and it is continuous in $x$, where the convergence takes place uniformly on bounded sets of $[0, T] \times \mathbf{R}^d$. This shows that $\xi_t(x)$ is continuously partial differentiable with respect to $x_i$ ($i$-th component of $x$) and $\nabla_{x_i}\xi_t(x) = N_t(x, 0)$ holds for any $(t, x)$ a.s. This argument holds for any component $x_i$ of $x = (x_1, ..., x_n)$. Thus $\xi_t(x)$ is continuously differentiable.

In this section we assume that coefficients $b, f, g$ of the SDE are differentiable and their derivatives $\nabla b, \nabla f, \nabla g$ are bounded and $\delta$-Hölder continuous: there exists positive constants $K'$, $K'(z)$ and $L'$, $L'(z)$ such that

$$|\nabla b(x, t)| \leq K', \quad |\nabla b(x, t) - \nabla b(y, t)| \leq L'|x - y|^\delta, \tag{3.8}$$
$$|\nabla f(x, t)| \leq K', \quad |\nabla f(x, t) - \nabla f(y, t)| \leq L'|x - y|^\delta,$$
$$|\nabla g(x, t, z)| \leq K'(z), \quad |\nabla g(x, t, z) - \nabla g(y, t, z)|^\delta \leq L'(z)|x - y|^\delta,$$

holds for all $x, y$ a.e. $\Lambda \times P$ and

$$\int_Z (K'(z)^p + L'(z)^p)\nu(dz) < \infty, \quad \forall p \geq 2. \tag{3.9}$$

**Theorem 3.3.** *Assume that coefficients of the SDE defining (3.3) are differentiable with respect to $x$ and satisfy (3.1), (3.2), (3.8) and (3.9). Then for any $p \geq 2$, there exists a positive constant $C_p$ such that [$E$ denoting expectation as usual]*

$$E\Big[\sup_{t_0 \leq s \leq t} |N_s(x, \lambda)|^p\Big] \leq C_p, \quad \forall(\lambda, x) \in D, \tag{3.10}$$

$$E\Big[\sup_{t_0 \leq s \leq t} |N_s(x, \lambda) - N_s(x', \lambda')|^p\Big] \leq C_p\{|x - x'|^{\delta p} + |\lambda - \lambda'|^{\delta p}\},$$

$$\forall((\lambda, x), (\lambda', x')) \in D^2 \tag{3.11}$$

*holds for any $t_0 \le t \le T$.*

*Proof.* Set $X'(t) = N_t(x, \lambda)$. It is written as

$$X'(t) = e_i + \int_{t_0}^t b'_\lambda(r)dr + \int_{t_0}^t f'_\lambda(r)dW(r)$$
$$+ \int_{t_0}^t \int_Z g'_\lambda(r, z)\tilde{N}(drdz), \tag{3.12}$$

where

$$b'_\lambda(r) = \frac{1}{\lambda}\Big(b(\xi_{r-}(x + \lambda e_i), r) - b(\xi_{r-}(x), r)\Big),$$
$$f'_\lambda(r) = \frac{1}{\lambda}\Big(f(\xi_{r-}(x + \lambda e_i), r) - f(\xi_{r-}(x), r)\Big),$$
$$g'_\lambda(r, z) = \frac{1}{\lambda}\Big(g(\xi_{r-}(x + \lambda e_i), r, z) - g(\xi_{r-}(x), r, z)\Big).$$

We shall apply Corollary 2.12. We first consider $b'_\lambda(r)$. Setting $\int_0^1 \nabla b(x + \theta y, r)d\theta = \bar{b}(x, y, r)$, we have

$$b'_\lambda(r) = \left(\int_0^1 \nabla b(\xi_{r-}(x) + \theta\lambda N_{r-}(x, \lambda), r)d\theta\right) N_{r-}(x, \lambda)$$
$$= \bar{b}(\xi_{r-}(x), \lambda N_{r-}(x, \lambda), r)N_{r-}(x, \lambda).$$

Since $|\bar{b}| \le c$ (bounded), we have

$$E\left[\int_{t_0}^t |b'_\lambda(r)|^p dr\right] \le cE\left[\int_{t_0}^t |X'(r-)|^p dr\right].$$

We have similarly

$$E\left[\int_{t_0}^t |f'_\lambda(r)|^p dr\right] \le cE\left[\int_{t_0}^t |X'(r-)|^p dr\right].$$

Next consider $g'$. Define $\bar{g}(x, y, r, z)$ from $g(x, r, z)$ as before. Then we have

$$g'_\lambda(r, z) = \bar{g}(\xi_{r-}(x), \lambda N_{r-}(x, \lambda), r, z)N_{r-}(x, \lambda).$$

Since $\int_Z |\bar{g}|^p \nu(dz) \le c_p$ (bounded) for $p \ge 2$, we can show similarly

$$E\left[\int_{t_0}^t \left(\int_Z |g'_\lambda|^2 \nu(dz)\right)^{p/2} dr\right] \le cE\left[\int_{t_0}^t |X'(r-)|^p dr\right],$$

and

$$E\left[\int_{t_0}^t \int_Z |g'_\lambda|^p \nu(dz)dr\right] \le cE\left[\int_{t_0}^t |X'(r-)|^p dr\right].$$

Consequently we have by Corollary 2.12,

$$E\left[\sup_{t_0 \le s \le t} |X'(s)|^p\right] \le c\left(1 + \int_{t_0}^t E\left[\sup_{t_0 \le s \le r} |X'(s)|^p\right]dr\right).$$

Therefore we get the $L^p$ estimate (3.10) of $N_t(x, \lambda)$.

We shall next study the $L^p$ estimate of $\tilde{X}(t) = N_t(x, \lambda) - N_t(x', \lambda')$. It is written as

$$\tilde{X}(t) = \int_{t_0}^t \left(b'_\lambda(r) - b'_{\lambda'}(r)\right)dr + \int_{t_0}^t \left(f'_\lambda(r) - f'_{\lambda'}(r)\right)dW(r)$$
$$+ \int_{t_0}^t \int_{\mathcal{Z}} \left(g'_\lambda(r, z) - g'_{\lambda'}(r, z)\right)\tilde{N}(drdz).$$

We shall consider the drift term of $\tilde{X}(t)$. It is written as

$$b'_\lambda(r) - b'_{\lambda'}(r)$$
$$= \bar{b}(\xi_{r-}(x), \lambda N_{r-}(x, \lambda), r)N_{r-}(x, \lambda)$$
$$- \bar{b}(\xi_{r-}(x'), \lambda' N_{r-}(x', \lambda'), r)N_{r-}(x', \lambda')$$
$$= \left\{\bar{b}(\xi_{r-}(x), \lambda N_{r-}(x, \lambda), r) - \bar{b}(\xi_{r-}(x'), \lambda' N_{r-}(x', \lambda), r)\right\}N_{r-}(x, \lambda)$$
$$+ \bar{b}(\xi_{r-}(x'), \lambda' N_{r-}(x', \lambda'), r)\left\{N_{r-}(x, \lambda) - N_{r-}(x', \lambda')\right\}.$$

Observe that $\bar{b}(x, y, r)$ is $\delta$-Hölder continuous with respect to $x$ and $y$. Then we have

$$\left|\bar{b}(\xi_{r-}(x), \lambda N_{r-}(x, \lambda), r) - \bar{b}(\xi_{r-}(x'), \lambda' N_{r-}(x', \lambda), r)\right|$$
$$\le L'\{|\xi_{r-}(x) - \xi_{r-}(x')|^\delta + |\xi_{r-}(x + \lambda e_i) - \xi_{r-}(x) -$$
$$(\xi_{r-}(x' + \lambda' e_i) - \xi_{r-}(x'))|^\delta\}$$
$$\le L'\{|\xi_{r-}(x) - \xi_{r-}(x')|^\delta + |\xi_{r-}(x + \lambda e_i) - \xi_{r-}(x' + \lambda' e_i)|^\delta\}.$$

Therefore, we have

$$|b'_\lambda(r) - b'_{\lambda'}(r)|$$
$$\le L'\{|\xi_{r-}(x) - \xi_{r-}(x')|^\delta + |\xi_{r-}(x + \lambda e_i) - \xi_{r-}(x' + \lambda' e_i)|^\delta\}|N_{r-}(x, \lambda)|$$
$$+ K'|\tilde{X}(r-)|,$$

where we used $|\bar{b}| \le K'$. Note that the $L^{2p}$ norm of $N_{r-}(x, \lambda)$ is uniformly bounded. Then we obtain the estimate

$$E\left[\int_{t_0}^t |b_\lambda'(r) - b_{\lambda'}'(r)|^P dr\right] \le c_1 \int_{t_0}^t E[|\xi_{r-}(x) - \xi_{r-}(x')|^{2\delta p}]^{1/2} dr$$

$$+ c_2 \int_{t_0}^t E[|\xi_{r-}(x + \lambda e_i) - \xi_{r-}(x' + \lambda' e_i)|^{2\delta p}]^{1/2} dr$$

$$+ c_3 \int_{t_0}^t E[|\tilde{X}(r-)|^P] dr$$

$$\le c_1'|x - x'|^{\delta p} + c_2'|\lambda - \lambda'|^{\delta p} + c_3 \int_{t_0}^t E[|\tilde{X}(r-)|^P] dr.$$

In the last inequality, we applied Theorem 3.2.

As to the diffusion coefficient, we have

$$E\left[\int_{t_0}^t \left|f'_\lambda(r) - f'_{\lambda'}(r)\right|^P dr\right] \le c_1'|x - x'|^{\delta p} + c_2'|\lambda - \lambda'|^{\delta p} + c_3' \int_{t_0}^t E[|\tilde{X}(r-)|^P] dr,$$

similarly for the case of drift.

Next we will estimate the coefficient of jumps. Again we have

$$\left|g_\lambda'(r, z) - g_{\lambda'}'(r, z)\right|$$

$$\le L'(z)\{|\xi_{r-}(x) - \xi_{r-}(x')|^\delta + |\xi_{r-}(x + \lambda e_i) - (\xi_{r-}(x' + \lambda' e_i)|^\delta\} N_{r-}(x, \lambda)$$

$$+ K'(z)|\tilde{X}(r-)|,$$

where $\int (L'(z)^P + K'(z)^P)\nu(dz) < \infty$ holds for any $p \ge 2$. Then we have

$$E\left[\int_{t_0}^t (\int_Z \left|g_\lambda'(r, z) - g_{\lambda'}'(r, z)\right|^2 \nu(dz))^{P/2} dr\right]$$

$$\le c_4\{|x - x'|^{\delta p} + |\lambda - \lambda'|^{\delta p}\} + c_5 E\left[\int_{t_0}^t |\tilde{X}(r-)|^P dr\right].$$

Similarly we obtain

$$E\left[\int_{t_0}^t \int_Z \left|g_\lambda'(r, z) - g_{\lambda'}'(r, z)\right|^P \nu(dz) dr\right]$$

$$\le c_6\{|x - x'|^{\delta p} + |\lambda - \lambda'|^{\delta p}\} + c_7 E\left[\int_{t_0}^t |\tilde{X}(r-)|^P dr\right].$$

Summing the three $L^P$ estimates, we obtain from Corollary 2.12,

$$E\left[\sup_{t_0 \le s \le t} |\tilde{X}(s)|^P\right] \le c_8\{|x - x'|^{\delta p} + |\lambda - \lambda'|^{\delta p}\} + c_9 E\left[\int_{t_0}^t |\tilde{X}(r-)|^P dr\right].$$

Therefore we get the inequality

$$E\left[\sup_{t_0 \le s \le t} |\tilde{X}(s)|^P\right] \le c_{10}\{|x - x'|^{\delta p} + |\lambda - \lambda'|^{\delta p}\}.$$

This proves the second assertion of the theorem.

**Theorem 3.4.** *Assume the same conditions as in Theorem 3.3 for the coefficients. Then the solution $\xi_t(x)$ is differentiable with respect to $x$ for any $t \geq t_0$ a.s. Further the derivative $\nabla \xi_t$ satisfies*

$$\nabla \xi_t(x) = I + \int_{t_0}^t \nabla X(\xi_{r-}(x), dr) \nabla \xi_{r-}(x), \tag{3.13}$$

*where $I$ is the identity matrix and*

$$\nabla X(x, t) = \int_0^t \nabla b(x, r) dr + \int_0^t \nabla f(x, r) dW(r) + \int_0^t \int_{\mathcal{Z}} \nabla g(x, r, z) \tilde{N}(drdz).$$

*Proof.* We have already shown the differentiability of $\xi_t(x)$.

For the proof of (3.13), observe the formula (3.12). Let $\lambda$ tend to 0 in the coefficients $b'_\lambda(r)$, $f'_\lambda(r)$ and $g'_\lambda(r, z)$. Since $\lambda N_{r-}(x, \lambda)$ converges to 0, we have

$$\lim_{\lambda \to 0} b'_\lambda(r) = \bar{b}(\xi_{r-}(x), 0, r) \nabla \xi_{r-}(x) = \nabla b(\xi_{r-}(x), r) \nabla \xi_{r-}(x).$$

Similarly we have

$$\lim_{\lambda \to 0} f'_\lambda(r) = \bar{f}(\xi_{r-}(x), 0, r) \nabla \xi_{r-}(x) = \nabla f(\xi_{r-}(x), r) \nabla \xi_{r-}(x),$$

$$\lim_{\lambda \to 0} g'_\lambda(z, r) = \nabla g(\xi_{r-}(x), r, z) \nabla \xi_{r-}(x).$$

Therefore we get the equality

$$\nabla \xi_t(x) = I + \int_s^t \nabla b(\xi_{r-}(x), r) \nabla \xi_{r-}(x) dr$$

$$+ \int_s^t \nabla f(\xi_{r-}(x), r) \nabla \xi_{r-}(x) dW(r)$$

$$+ \int_s^t \int_{\mathcal{Z}} \nabla g(\xi_{r-}(x), z, r) \nabla \xi_{r-}(x) \tilde{N}(drdz).$$

This proves (3.13).

### 3.4 Homeomorphic property

The solution of a jump SDE does not admit the homeomorphic property in general. Indeed, at the jump time $\tau$, the solution jumps from $\xi_{\tau-}(x)$ to $\xi_\tau(x) + g(\xi_\tau(x), \tau)$. In order that the homeomorphic property is preserved at the jump time, it is necessary that the map $\phi_{\tau, z} : y \to y + g(y, \tau, z)$ should be homeomorphic, i.e., the map $\phi_{\tau, z}$ should be one to one and onto. In the next section we will show conversely that if $\phi_{\tau, z}$ are homeomorphisms a.e. $\nu \otimes P$, then the solution $\xi_t$ admits the homeomorphic property (Theorem 3.10).

In this section we prove the homeomorphic property under a stronger condition. Let $\psi_{r, z}$ be the inverse map of $\phi_{r, z}$. We assume that the inverses $\{\psi_{r, z}\}$ are uniformly

Lipschitz continuous and of uniformly linear growth; i.e., there exist positive constants $\hat{L}$ and $\hat{K}$ such that

$$|\psi_{r,z}(x)| < \hat{K}(1+|x|), \quad |\psi_{r,z}(x) - \psi_{r,z}(y)| \le \hat{L}|x-y|, \quad \forall(r,z) \ a.e. \quad (3.14)$$

**Theorem 3.5.** *Assume the same condition for coefficients $b$, $f$, $g$ as in Theorem 3.3. Assume further that the maps $\phi_{r,z} : x \to x + g(x,r,z)$; $\mathbf{R}^d \to \mathbf{R}^d$ are homeomorphic and the inverse maps $\psi_{r,z}$ are uniformly Lipschitz continuous and of uniformly linear growth. Then the maps $\xi_t : C \to C$ are onto homeomorphisms for any $t$ a.s.*

**Remark.** If we can take Lipschitz constants $L(z)$ and linear growth constants $K(z)$ for jump coefficients $g(x,r,z)$ uniformly less than 1, say $L(z) < \epsilon$ and $K(z) < \epsilon$ for some $0 < \epsilon < 1$, then the inverse maps $\psi_{r,z}(x)$ are uniformly Lipschitz continuous and of uniformly linear growth. Indeed, since $|x - y - (\phi_{r,z}(x) - \phi_{r,z}(x))| \le L(z)|x-y|$ holds valid, we have

$$|\psi_{r,z}(x) - \psi_{r,z}(y) - (x-y)| \le L(z)|\psi_{r,z}(x) - \psi_{r,z}(y)|.$$

This implies

$$(1-L(z))|\psi_{r,z}(x) - \psi_{r,z}(y)| \le |x-y|.$$

Then $\psi_{r,z}$ is Lipschitz continuous with Lipschitz constant $(1-L(z))^{-1} \le (1-\epsilon)^{-1}$. Similarly if $K(Z) \le 1 - \epsilon$, then we have $|\psi_{r,z}(x)| \le (1-\epsilon)^{-1}(1+|x|)$.

For the proof, we need some lemmas. In the following Lemmas 3.6-3.9, we assume that the coefficients $b$, $f$, $g$ satisfy the same conditions as in Theorem 3.5.

**Lemma 3.6.** *Set $F(x) = (1+|x|^2)^{-1}$. Then we have the following.*
*1) $F(x + g(x,r,z)) \le \hat{K}^2 F(x)$ holds for any $x$ a.e. $\Lambda \times \nu \times P$.*
*2) For any $p \ge 2$, there exists a positive constant $c_p$ such that*

$$\int_Z |F(x + g(x,r,z)) - F(x)|^p \nu(dz) \le c_p F(x)^p, \quad \forall x \ a.e. \ \Lambda \times P. \quad (3.15)$$

*3) There exists a positive constant $c$ such that*

$$\int_Z |F(x + g(x,r,z)) - F(x) - \sum_i F'_{x_i}(x) g^i(x,r,z)|\nu(dz) \quad (3.16)$$

$$\le cF(x), \quad \forall x a.e. \Lambda \times P.$$

*Proof.* By condition (3.14), it holds $(1+|\psi_{r,z}(y)|^2) \le (1+\hat{K}^2)(1+|y|)^2$ for any $y$. Substitute $y = \phi_{r,z}(x)$. Since $\psi_{r,z}(\phi_{r,z}(x)) = x$, we have $1 + |x|^2 \le (1+\hat{K}^2)(1+ |\phi_{r,z}(x)|^2)$. Then we have

$$F(x + g(x,r,z)) = (1+|\phi_{r,z}(x)|^2)^{-1}$$
$$\le (1+\hat{K}^2)(1+|x|^2)^{-1} \le (1+\hat{K}^2)F(x).$$

We next prove (3.15). It holds $F(x+g) - F(x) = -(|g|^2 + 2(g,x))F(x+g)F(x)$. Since $|x| \leq F(x)^{-1/2}$, we have the inequality

$$|F(x+g) - F(x)| \leq |g|^2 F(x+g)F(x) + 2|g|F(x+g)F(x)^{1/2} \qquad (3.17)$$
$$\leq |g|^2(1+\hat{K}^2)F(x)^2 + 2|g|(1+\hat{K}^2)F(x)^{3/2}.$$

Now integrate $p$-th power of both sides with respect to the Lévy measure $\nu$. Then we have

$$\int_Z |F(x+g(x,r,z)) - F(x)|^p \nu(dz) \leq c_p F(x)^p.$$

This proves inequality (3.15).

We will consider the third inequality. Since $\sum_i F'_{x_i}(x)g^i = -2(x,g)F(x)^2$, we have

$$F(x+g) - F(x) - \sum_i F'_{x_i}(x)g^i$$
$$= -|g|^2 F(x)F(x+g) + 2(x,g)F(x)(F(x) - F(x+g)).$$

Therefore,

$$\left| F(x+g) - F(x) - \sum_i F'_{x_i}(x)g^i \right|$$

$$\leq |g|^2 F(x)F(x+g) + 2|x||g|F(x)|F(x+g) - F(x)|$$
$$\leq 5|g|^2(1+\hat{K}^2)F(x)^2 + 2|g|^3(1+\hat{K}^2)F(x)^{5/2}.$$

Now integrate both sides of the above by the Lévy measure $\nu$. Then we obtain (3.16).

**Lemma 3.7.** *For any $p \geq 2$ there exists a positive constant $C_p$ such that*

$$E\left[ \sup_{t_0 \leq s \leq t} (1 + |\xi_s(x)|^2)^{-p} \right] \leq C_p(1 + |x|^2)^{-p}, \quad \forall x \qquad (3.18)$$

*holds for any $t_0 \leq t \leq T$.*

*Proof.* Set $F(x) = (1 + |x|^2)^{-1}$ and $X(t) = \xi_t(x)$. Then,

$$X(t) = x + \int_{t_0}^t b(X(r-),r)dr + \int_{t_0}^t f(X(r-),r)dW(r)$$
$$+ \int_{t_0}^t \int_Z g(X(r-),r,z)\tilde{N}(drdz).$$

We shall apply Corollary 2.13. We want to estimate each term of the right hand side of (2.27). We first consider the drift term. Since $\sum_i F'_{x_i}(x)b^i(x,r) = -2(x,b(x,r))F(x)^2$, we have

$$\left| \sum_i F'_{x_i}(x)b^i(x,r) \right| \leq 2|x||b(x,r)|F(x)^2 \leq 2KF(x).$$

Here we used the property $|b(x, r)| \le K(1 + |x|^2)^{1/2}$ (The linear growth of $b(x, r)$). We have similarly,

$$\sum_k |\sum_l F'_{x_i}(x) f^{ik}(x)|^2 \le K' F(x).$$

Further, we have

$$\sum_{i,j,k} F''_{x_i x_j}(x) f^{ik}(x, r) f^{jk}(x, r) = 4F(x)^3 \sum_k (x, f^k(x, r))^2$$

$$-2F(x)^2 \sum_{ik} (f^{ik}(x, r))^2.$$

Using the linear growth property of $f^{jk}$ etc, we have

$$|\sum_{i,j,k} F''_{x_i x_j}(x) f^{ik}(x, r) f^{jk}(x, r)| \le cF(x).$$

Therefore

$$E\Big[\int_{t_0}^t |\sum_i F'_{x_i}(X(r-))b^i(X(r-), r)|^p dr\Big]$$

$$+E\Big[\int_{t_0}^t \Big(\sum_k |\sum_i F'_{x_i}(X(r-)) f^{ik}(X(r-), r)|^2\Big)^{p/2} dr\Big]$$

$$+\frac{1}{2}E\Big[\int_{t_0}^t |\sum_{i,j,k} F''_{x_i x_j}(X(r-)) f^{ik}(X(r-), r) f^{jk}(X(r-), r)|^p dr\Big]$$

$$\le cE\Big[\int_{t_0}^t (1 + |X(r-)|^2)^{-p} dr\Big].$$

We shall next consider the jump parts. We get from Lemma 3.6 (2),

$$E\Big[\int_{t_0}^t \Big(\int_Z |F(X(r-) + g(X(r-), r, z)) - F(X(r-))|^{p'} v(dz)\Big)^{p/p'} dr\Big]$$

$$\le CE\Big[\int_{t_0}^t (1 + |X(r-)|^2)^{-p} dr\Big],$$

for $p' = 2$ and $p' = p$. Next, we have from Lemma 3.6 (3),

$$E\Big[\int_{t_0}^t \Big(\int_Z |F(X(r-) + g(X(r-), r, z))$$

$$-F(X(r-)) - \sum_i F'_{x_i}(X(r-))g^i(X(r-), r, z)|v(dz)\Big)^p dr\Big]$$

$$\le C \int_{t_0}^t E[(1 + |X(r-)|^2)^{-p}] dr.$$

Summing up all these estimates, we get a functional inequality

$$E\left[\sup_{t_0 \le s \le t} (1 + |X(s)|^2)^{-p}\right] \le C\left\{(1 + |x|^2)^{-p} + \int_{t_0}^t E[(1 + |X(r-)|^2)^{-p}]dr\right\}.$$

This implies

$$E\left[\sup_{t_0 < s \le t} (1 + |\xi_s(x)|^2)^{-p}\right] \le C_p(1 + |x|^2)^{-p},$$

where the constant $C_p$ does not depend on $x$.

**Lemma 3.8.** Set $F(x) = (\delta + |x|^2)^{-1}$. Then we have the following.
1) $F(x - x' + g(x, r, z) - g(x', r, z)) \le (1 + \hat{L}^2)F(x - x')$ holds for all $x, x'$, a.e. $\Lambda \times \nu \times P$.
2) For any $p \ge 2$, there exists a positive constant $c'_p$ not depending on $\delta$ such that

$$\int_Z |F(x - x' + g(x, r, z) - g(x', r, z)) - F(x - x')|^p \nu(dz) \tag{3.19}$$

$$\le c'_p F(x - x')^p, \quad \forall x, x \ 'a.e. \ \Lambda \times P.$$

3) There exists a positive constant $c'$ not depending on $\delta$ such that

$$\int_Z |F(x - x' + g(x, r, z) - g(x', r, z)) - F(x - x') \tag{3.20}$$

$$- \sum_i F'_{x_i}(x - x')(g^i(x, r, z) - g^i(x', r, z))|\nu(dz) \le c'F(x - x'),$$

$$\forall x, x' \ a.e. \ \Lambda \times P.$$

*Proof.* We shall prove the first assertion. Since $\psi_{r,z}(y)$ is Lipschitz continuous, we have

$$|\psi_{r,z}(y) - \psi_{r,z}(y')| \le \hat{L}|y - y'|, \quad \forall y, y',$$

Now substitute $y = \phi_{r,z}(x)$ and $y' = \phi_{r,z}(x')$ in the above inequality and note the identities $x = \psi_{r,z}(\phi_{r,z}(x))$ and $x' = \psi_{r,z}(\phi_{r,z}(x'))$. Then we obtain $|x - x'| \le \hat{L}|\phi_{r,z}(x) - \phi_{r,z}(x')|$ for any $x, x'$. Then it holds $\delta + |x - x'|^2 \le (1 + \hat{L}^2)(\delta + |\phi_{r,z}(x) - \phi_{r,z}(x')|^2)$ for any $\delta > 0$, which implies

$$F(x - x' + g(x, z) - g(x', z)) = (\delta + |\phi_{r,z}(x) - \phi_{r,z}(x')|^2)^{-1}$$
$$\le (1 + \hat{L}^2)(\delta + |x - x'|^2)^{-1}$$
$$= (1 + \hat{L}^2)F(x - x'), \quad \forall x, x'.$$

Next we will prove (3.19). Set $w = x - x'$ and $k = g(x, r, z) - g(x', r, z)$. Then we have $F(w + k) \le (1 + \hat{L}(z)^2)F(w)$ by the first assertion. Also we have the inequality

$$|F(w + k) - F(w)| \le |k|^2(1 + \hat{L}^2)F(w)^2 + 2|k|(1 + \hat{L}^2)F(w)^{3/2}$$

as in Lemma 3.6. Now integrate the $p$-th power of both sides of the above with respect to the Lévy measure $\nu$. Then we have

$$\int_Z |F(w+k) - F(w)|^p \nu(dz)$$

$$\leq 2^p \int_Z |g(x,r,z) - g(x',r,z)|^{2p}(1+\hat{L}^2)^p \nu(dz) F(w)^{2p}$$

$$+4^p \int_Z |g(x,r,z) - g(x',r,z)|^p (1+\hat{L}^2)^p \nu(dz) F(w)^{3p/2}$$

$$\leq c_p' F(w)^p,$$

because

$$\int_Z |g(x,r,z) - g(x',r,z)|^p(1+\hat{L}^2)^{p'} \nu(dz) \leq c_p'' |x-x'|^p \leq c_p'' F(y)^{-p/2}.$$

The above inequality proves (3.19). Inequality (3.20) can be verified similarly as in the proof of Lemma 3.6.

**Lemma 3.9.** *For any $p \geq 2$ there exists a positive constant $C_p'$ such that*

$$E\left[ \sup_{t_0 \leq s \leq t} |\xi_s(x) - \xi_s(y)|^{-2p} \right] \leq C_p' |x-y|^{-2p}, \quad \forall x, y \tag{3.21}$$

*holds for any $t_0 \leq t \leq T$.*

*Proof.* Set $F(x) = (\delta + |x|^2)^{-1}$ and $Y(t) = \xi_t(x) - \xi_t(y)$. In the following arguments, all constants $c_i$ will be chosen independently of $\delta > 0$ and $x, y \in \mathbf{R}^d$. Using Lipschitz conditions for $b, f$ we can show

$$E\left[ \int_{t_0}^t |\sum_i F_{x_i}'(Y(r-))(b^i(\xi_{r-}(x), r) - b^i(\xi_{r-}(y), r))|^p dr \right]$$

$$+E\left[ \int_{t_0}^t |\sum_k |\sum_i F_{x_i}'(Y(r-))(f^{ik}(\xi_{r-}(x), r) - f^{ik}(\xi_{r-}(y), r))|^2|^{p/2} dr \right]$$

$$+\frac{1}{2}E\left[ \int_{t_0}^t |\sum_{i,j,k} F_{x_i x_j}''(Y(r-))(f^{ik}(\xi_{r-}(x), r) - f^{ik}(\xi_{r-}(y), r)) \right.$$

$$\left. \times (f^{jk}(\xi_{r-}(x), r) - f^{jk}(\xi_{r-}(y), r))|^p dr \right]$$

$$\leq c_1 E\left[ \int_{t_0}^t (\delta + |Y(r-)|^2)^{-p} dr \right].$$

We next consider jump parts. Using Lemma 3.8 (2), we can similarly show

$$E\Big[\int_{t_0}^t \Big(\int_Z |F\big(Y(r-) + g(\xi_{r-}(x), r, z) - g(\xi_{r-}(y), r, z)\big)$$

$$-F\big(Y(r-)\big)|^{p'} v(dz)\Big)^{p/p'} dr\Big]$$

$$\leq c_2 \int_{t_0}^t E[(\delta + |Y(r-)|^2)^{-p}]dr,$$

for $p' = p$ or $p' = 2$. Next, using Lemma 3.8 (3), we get

$$E\Big[\int_{t_0}^t \Big(\int_Z \Big|F\big(Y(r-) + g(\xi_{r-}(x), r, z) - g(\xi_{r-}(y), r, z)\big) - F\big(Y(r-)\big)$$

$$-\Big(\sum_i F'_{x_i}(Y(r-))(g^i(\xi_{r-}(x), r, z) - g^i(\xi_{r-}(y), r, z))\Big)\Big| v(dz)\Big)^p dr\Big]$$

$$\leq c_3 \int_{t_0}^t E[(\delta + |Y(r-)|^2)^{-p}]dr.$$

These computations yield

$$E\Big[\sup_{t_0 \leq s \leq t} (\delta + |Y(s-)|^2)^{-p}\Big] \leq c_4 \Big\{(\delta + |x - y|^2)^{-p}$$

$$+ \int_{t_0}^t E\Big[(\delta + |Y(r-)|^2))^{-p}\Big]dr\Big\}.$$

This implies

$$E\Big[\sup_{t_0 \leq s \leq t} (\delta + |Y(s)|^2)^{-p}\Big] \leq c_5(\delta + |x - y|^2)^{-p}.$$

Now let $\delta$ tend to 0 and observe $Y(s) = \xi_s(x) - \xi_s(y)$. Then we get the inequality

$$E\Big[\sup_{t_0 \leq s \leq t} |\xi_s(x) - \xi_s(y)|^{-2p}\Big] \leq c_5|x - y|^{-2p}.$$

Therefore we get the lemma.

*Proof of Theorem 3.5.* We first show that maps $\xi_t : \mathbf{R}^d \to \mathbf{R}^d$ are one to one for any $t$ a.s. Consider the random field

$$\eta_t(x, y) = \frac{1}{|\xi_t(x) - \xi_t(y)|}.$$

We can show that for any $p \geq 2$, there exists a positive constant $C = C(p)$ such that for any $\delta > 0$,

$$E\Big[\sup_{t_0 < s \leq t} |\eta_s(x, y) - \eta_s(x', y')|^{2p}\Big] \leq C\delta^{-4p}\{|x - x'|^{2p} + |y - y'|^{2p}\}$$

holds for any $x, x', y, y'$ such that $|x - y| \geq \delta$ and $|x' - y'| \geq \delta$. In fact, a simple computation yields

$$|\eta_t(x, y) - \eta_t(x', y)|^{2p}$$
$$\leq \eta_t(x, y)^{2p} \eta_t(x', y')^{2p} \times \{|\xi_t(x) - \xi_t(x')| + |\xi_t(y) - \xi_t(y')|\}^{2p}.$$

Take expectations for both sides and use Hölder's inequality. Then we obtain the inequality.

Now, by Kolmogorov's criterion, the random field $\eta_t(x, y)$ is continuous in the domain $D_\delta = \{(x, y); |x - y| \geq \delta\}$. Since this is valid for any $\delta$, we find that the random field $\eta_t(x, y)$ is continuous in the domain $\{(x, y) : x \neq y\}$. This proves that the maps $\xi_t; \mathbf{R}^d \to \mathbf{R}^d$ are one to one for any $t$, a.s.

We will next prove that the maps $\xi_t : \mathbf{R}^d \to \mathbf{R}^d$ are onto for any $t$ a.s. Set $\hat{x} = |x|^{-2}x$ if $x \neq 0$ and define

$$\eta_t(\hat{x}) = \frac{1}{1 + |\xi_t(x)|} \quad \text{if } \hat{x} \neq 0.$$

We set $\eta_t(0) = 0$. Then for each $p > 1$, there exists a positive constant $C = C(p)$ such that

$$E\left[ \sup_{t_0 \leq s \leq t} |\eta_s(x) - \eta_s(y)|^{2p} \right] \leq C|\hat{x} - \hat{y}|^{2p}.$$

In fact we have

$$|\eta_s(\hat{x}) - \eta_s(\hat{y})|^{2p} \leq \eta_s(\hat{x})^{2p} \eta_s(\hat{y})^{2p} |\xi_s(x) - \xi_s(y)|^{2p}.$$

Take the supremum with respect to $s$ and then take expectations for both sides. Then we have

$$E\left[ \sup_{t_0 \leq s \leq t} |\eta_s(\hat{x}) - \eta_s(\hat{y})|^{2p} \right] \leq C(1 + |x|)^{-2p}(1 + |y|)^{-2p}|x - y|^{2p} \leq C|\hat{x} - \hat{y}|^{2p}.$$

By Kolmogorov's criterion, $\eta_t(\hat{x})$ can be extended continuously to 0. This means that $\eta_t(\hat{x})$ converges to 0 for any $t$ a.s. as $\hat{x} \to 0$. But this implies $\lim_{|x| \to \infty} |\xi_t(x)| = \infty$ exists for any $t$ a.s. This establishes the onto property of the maps.

Let $t_0 < s < t$. Let $\xi'_t(x)$ be the solution starting from $x$ at time $s$. We define $\eta_r, t_0 < r < T$ by $\eta_r = \xi_r(x)$ if $r \leq s$ and $\eta_r = \xi'_r(\xi_s(x))$ if $r > s$. Substitute $y = \xi_s(x)$ in the equality $\xi'_t(y) = y + \int_s^t X(\xi'_{r-}(y), dr)$. Then $\eta_r$ satisfies

$$\eta_t = \xi_s(x) + \int_s^t X(\eta_{r-}, dr) = x + \int_{t_0}^t X(\eta_{r-}, dr).$$

Therefore $\eta_t = \xi_t$ holds. Consequently we have the cocycle property $\xi'_t \circ \xi_s = \xi_t$ for any $t_0 < s < t$ a.s.

Now for $s, t \in [t_0, T]$, we will define $\xi_{s,t}$ as follows. Let $\xi_t$ be the $C$-valued cadlag process with the initial condition $\xi_0(x) = x$. Let $\xi_t^{-1}$ be the inverse map. It is again a $C$-valued cadlag process. We set

$$\xi_{s,t}(x) = x \quad \text{if } t \leq s,$$
$$= \xi_t \circ \xi_s^{-1}(x), \quad \text{if } t \geq s.$$

Then it has the following properties:

1) $\xi_{s,t}$ is a cadlag process with respect to $t$ and also is a cadlag process with respect to $s$

2) $\xi_{s,t} : C \to C$ are onto homeomorphisms for all $s < t$ a.s.

3) $\xi_{s,t}(x)$ is the solution of the SDE based on $X(x, t)$ starting from $x$ at time $s$.

The third property follows from the fact that $\xi_{s,t} = \xi_t'$ holds a.s. The family of $C$-valued random variables $\{\xi_{s,t}; 0 \le t_0 \le s < t < T\}$ is called a *stochastic flow of homeomorphisms generated by* $X(x, t)$. Further if the map $\xi_{s,t} : \mathbf{R}^d \to \mathbf{R}^d$ are $C^k$-diffeomorphisms for all $s < t$, then it is called a *stochastic flow of $C^k$-diffeomorphisms*.

### 3.5 Stochastic flow of diffeomorphisms

So far we have studied the solution of a SDE through $L^p$ estimates. Conditions needed for this framework are that coefficients $b$, $f$, $g$ satisfy (3.1) and (3.2). In this section we will relax the latter condition (3.2). For $0 < \epsilon < 1$, set $U_\epsilon = \{z; K(z) < \epsilon, L(z) < \epsilon\}$. Instead of (3.2), we introduce a condition

$$\int_{U_\epsilon} (K(z)^2 + L(z)^2)\nu(dz) < \infty, \quad \text{and} \quad \nu(U_\epsilon^c) < \infty. \tag{3.22}$$

We define

$$X'(x, t) = \int_0^t b(x, r)dr + \int_0^t f(x, r)dW(r)$$
$$+ \int_0^t \int_{U_\epsilon} g(x, r, z)\tilde{N}(drdz), \tag{3.23}$$

$$X(x, t) = X'(x, t) + \int_0^t \int_{U_\epsilon^c} g(x, r, z)N(drdz), \tag{3.24}$$

and consider a SDE based on $X(x, t)$ and $X'(x, t)$. Observe that $X'(x, t)$ satisfies all the conditions required in Theorems 3.1 and 3.2. Further assuming that the maps $\phi_{r,z}$ are homeomorphisms, the solution of the SDE based on $X'(x, t)$ defines a stochastic flow $\xi_{s,t}'$ of homeomorphisms. (See Remark after Theorem 3.5.) We shall consider a SDE based on $X(x, t)$. Using the notation of the point process $p(t)$, the equation is written as

$$\xi_t = x + \int_{t_0}^t X(\xi_{r-}, dr) \tag{3.25}$$

$$= x + \int_{t_0}^t X'(\xi_{r-}, dr) + \sum_{t_0 < r \le t, r \in \mathbf{D}_p} g(\xi_{r-}, r, p(r))1_{U_\epsilon^c}(p(r)).$$

The solution of the above equation can be constructed by means of the stochastic flow of homeomorphisms $\xi_{s,t}'(x)$ generated by $X'(x, t)$ and jumps $g(\xi_{r-}(x), r, p(r))1_{U_\epsilon^c}(p(r))$. Let $0 = \sigma_0 < \sigma_1 < \cdots$ be a sequence of jump times of the Poisson process $N(t) = N((0, t] \times U_\epsilon^c), t \in [0, T]$. Then $P(\sigma_k < T) \to 0$ as $k \to \infty$. We let $\phi_{r,z}(x) = x + g(x, r, z)$ and define $\xi_{t_0,t}$ by

$$\xi_{t_0,t}(x) = \xi'_{\sigma_m,t} \circ \phi_{\sigma_m,p(\sigma_m)} \circ \cdots \circ \xi'_{\sigma_{n+1},\sigma_{n+2}} \circ \phi_{\sigma_{n+1},p(\sigma_{n+1})} \circ \xi'_{t_0,\sigma_{n+1}}(x), \qquad (3.26)$$

on the set $A_{n,m}(t_0, t) = \{\omega; \sigma_n \leq t_0 < \sigma_{n+1}, \sigma_m \leq t < \sigma_{m+1}\}$.

Then $\xi_{t_0,t}$ is a cadlag process with values in $C$ with respect to $t_0$ and $t$. Further, $\xi_{t_0,t} : \mathbf{R}^d \to \mathbf{R}^d$ are homeomorphisms for any $t_0$, $t$, since each of them are compositions of homeomorphisms $\xi'_{r,u}$ and $\phi_{r,z}$. Further the above $\xi_{t_0,t}(x)$ is a solution of (3.25). We will check it on the set $A_{n,n+1}(t_0, t)$. Since $dX = dX'$ holds on $(\sigma_n, \sigma_{n+1})$, we see that

$$x + \int_{t_0}^{\sigma_{n+1}-} X(\xi_{t_0,r-}(x), dr) = x + \int_{t_0}^{\sigma_{n+1}} X'(\xi'_{t_0,r-}(x), dr) = \xi'_{t_0,\sigma_{n+1}}(x).$$

Moreover, since

$$\int_{\{\sigma_{n+1}\}} X(\xi'_{t_0,r-}(x), dr) = \phi_{\sigma_{n+1},p(\sigma_{n+1})}(\xi'_{t_0,\sigma_{n+1}}(x)) - \xi'_{t_0,\sigma_{n+1}}(x),$$

we have

$$x + \int_{t_0}^{\sigma_{n+1}+} X(\xi_{t_0,r-}(x), dr) = \xi_{t_0,\sigma_{n+1}}(x).$$

Also, observe that $dX = dX'$ on $(\sigma_{n+1}, t]$. Then

$$\xi_{t_0,\sigma_{n+1}}(x) \quad + \int_{\sigma_{n+1}}^{t} X(\xi_{t_0,r-}(x), dr)$$

$$= \xi_{t_0,\sigma_{n+1}}(x) + \int_{\sigma_{n+1}}^{t} X(\xi_{\sigma_{n+1},r-}(\xi_{t_0,\sigma_{n+1}}(x)), dr)$$

$$= \{y + \int_{\sigma_{n+1}}^{t} X'(\xi'_{\sigma_{n+1},r-}(y), dr)\}|_{y=\xi_{t_0,\sigma_{n+1}}(x)}.$$

Furthermore,

$$y + \int_{\sigma_{n+1}}^{t} X'(\xi'_{\sigma_{n+1},r-}(y), dr) = \xi'_{\sigma_{n+1},t}(y).$$

Consequently we obtain

$$x + \int_{t_0}^{t} X(\xi_{t_0,r-}(x), dr) = \xi_{t_0,\sigma_{n+1}}(x) + \int_{\sigma_{n+1}}^{t} X(\xi_{t_0,r-}(x), dr)$$

$$= \xi'_{\sigma_{n+1},t}(\xi_{t_0,\sigma_{n+1}}(x)) = \xi_{t_0,t}(x).$$

We have thus proved the following theorem.

**Theorem 3.10.** *Assume that coefficients $b$, $f$, $g$ satisfy (3.1) and (3.22). Assume further that $\phi_{r,z}(x) = x + g(x, r, z)$ are homeomorphic for any $r$, $z$. Then the solution defined by (3.26) is a stochastic flow of homeomorphisms.*

Next we shall study the diffeomorphic property of the maps $\xi_{t_0,t}$. We introduce

$$\int_{U'_\epsilon} (K'(z)^2 + L'(z)^2)\nu(dz) < \infty, \quad \text{and} \quad \nu(U'^c_\epsilon) < \infty, \qquad (3.27)$$

where $U_\epsilon = \{z; K'(z) < \epsilon, L'(z) < \epsilon\}$. Then, under conditions (3.1), (3.22), (3.8) and (3.27), $\xi_{t_0,t}$ of (3.26) is a $C^1$-valued process, cadlag with respect to $s$ and $t$. Further, we can show that the Jacobian matrix $\nabla\xi_{t_0,t}$ satisfies (3.13).

**Theorem 3.11.** *Assume that coefficients of the equation satisfy (3.1), (3.22), (3.8) and (3.27). If $\phi_{r,z)}(x) := x + g(x, r, z)$ are homeomorphic and Jacobian matrix $I + \nabla g(x, r, z)$ is invertible for any $x$ a.e. $(r, z)$, then the solution defines a stochastic flow of $C^1$-diffeomorphisms.*

By Theorem 3.10, the solution defines a stochastic flow of homeomorphisms. Therefore, it is sufficient to show that the Jacobian matrix $\nabla\xi_{t_0,t}$ is invertible (locally diffeomorphic).

We define a matrix valued semimartingale $U(t)$ (with parameter $x$) by

$$U(t) = \int_{t_0}^t \nabla b(\xi_{t_0,r-}(x), r)dr + \int_{t_0}^t \nabla f(\xi_{t_0,r-}(x), r)dW(r)$$
$$+ \int_{t_0}^t \int_Z \nabla g(\xi_{t_0,r-}(x), r, z)\tilde{N}(drdz).$$

Then, in view of Theorem 3.4, $\Phi_t := \nabla\xi_{t_0,t}(x)$ satisfies the linear SDE

$$\Phi_t = I + \int_{t_0}^t dU(r)\Phi_{r-}.$$

Observe that $\Delta U(r) = \nabla g(\xi_{r-}(x), r, p(r))$. Now the Jacobian matrix of $\phi_{r,z}(x) := x + g(x, r, z)$ (with respect to $x$) is given by $I + \nabla g(x, r, z)$. It is invertible for any $x$ a.e. $(r, z)$ by the assumption of the theorem. Therefore the matrices $(I + \Delta U(r))$ are invertible a.s.

Associated with $U(t)$, we define a matrix valued process by

$$V(t) = -U(t) + [U, U]_t^c + \sum_{t_0 < r \le t} (I + \Delta U(r))^{-1}(\Delta U(r))^2.$$

The invertability of the matrix $\Phi_t \equiv \nabla\xi_{t_0,t}$ follows from the next lemma.

**Lemma 3.12 (Protter).** [20] *Let $\Psi_t$ be a matrix valued process satisfying*

$$\Psi_t = I + \int_{t_0}^t \Psi_{r-}dV(r).$$

*Then we have $\Psi_t\Phi_t = I$. In particular, $\Phi_t$ is invertible for any $t$.*

*Proof.* Apply Itô's formula I to the product of $\Psi_t$ and $\Phi_t$. Then we have

$$\Psi_t\Phi_t = I + \int_{t_0}^t \Psi_{r-}d\Phi_r + \int_{t_0}^t d\Psi_r\Phi_{r-} + [\Psi, \Phi]_t,$$

where

$$\int_{t_0}^{t} \Psi_{r-} d\Phi_r = \int_{t_0}^{t} \Psi_{r-} dU(r) \Phi_{r-},$$

$$\int_{t_0}^{t} d\Psi_r \Phi_{r-} = -\int_{t_0}^{t} \Psi_{r-} dU(r) \Phi_{r-}$$

$$+ \int_{t_0}^{t} \Psi_{r-} d[U, U]_r^c \Phi_{r-}$$

$$+ \sum_{t_0 < r \le t} \Psi_{r-} (I + \Delta U(r))^{-1} (\Delta U(r))^2 \Phi_{r-}.$$

Note that

$$[V, U]_t = -[U, U]_t + \sum_{t_0 < r \le t} (I + \Delta U(r))^{-1} (\Delta U(r))^3$$

$$= -[U, U]_t^c - \sum_{t_0 < r \le t} (\Delta U(r))^2 + \sum_{t_0 < r \le t} (I + \Delta U(r))^{-1} (\Delta U(r))^3$$

$$= -[U, U]_t^c - \sum_{t_0 < r \le t} (I + \Delta U(r))^{-1} (\Delta U(r))^2,$$

because $(\Delta U(r))^2 = \{(\Delta U(r))^2 + (\Delta U(r))^3\}(I + \Delta U(r))^{-1}$. Then,

$$[\Psi, \Phi]_t = \int_{t_0}^{t} \Psi_{r-} d[V, U]_r \Phi_{r-}$$

$$= -\int_{t_0}^{t} \Psi_{r-} d[U, U]_r^c \Phi_{r-}$$

$$- \sum_{t_0 < r \le t} \Psi_{r-} (I + \Delta U(r))^{-1} (\Delta U(r))^2 \Phi_{r-}.$$

Then all terms of $\int_{t_0}^{t} \Psi_{r-} d\Phi_{r-} + \int_{t_0}^{t} d\Psi_r \Phi_{r-} + [\Psi, \Phi]_t$ are cancelled to yield 0. Then we get $\Psi_t \Phi_t = I$.

### 3.6 Inverse flow and backward SDE

Let $\xi_{s,t}(x)$ be the stochastic flow of diffeomorphisms generated by the SDE

$$\xi_{s,t}(x) = x + \int_{s}^{t} X(\xi_{s,r-}(x), dr), \tag{3.28}$$

where

$$X(x, t) = \int_{t_0}^{t} b(x, r) dr + \int_{t_0}^{t} f(x, r) dW(r) + \int_{t_0}^{t} \int_{Z} g(x, r, z) \tilde{N}(drdz). \tag{3.29}$$

In this section we assume that the coefficients $b, f, g$ are deterministic functions. Then the flow $\xi_{s,t}$ has independent increments, i.e., $\xi_{t_i, t_{i+1}}, i = 0, ..., n - 1$, are independent whenever $t_0 < t_1 < \cdots < t_n$. We call $\xi_{s,t}$ a *Lévy flow*.

Let $\xi_{s,t}^{-1}$ be the inverse map of $\xi_{s,t}$. It has the backward flow property: $\xi_{s,t}^{-1} = \xi_{s,r}^{-1} \circ \xi_{r,t}^{-1}$. In this section we fix the time $t$ and consider $\xi_{s,t}^{-1}(x)$ as a stochastic process with time parameter $s$. It is a cadlag process, because it holds $\xi_{s,t}^{-1} = \xi_{t_0,s} \circ \xi_{t_0,t}^{-1}$ and $\xi_{t_0,s}$ is a cadlag process with respect to $s$. The object of this section is to find a backward SDE governing the inverse flow $\xi_{s,t}^{-1}$ or to find the backward infinitesimal generator $\hat{X}(x, t)$ of the inverse flow.

We need to define a backward integral. Let $Z(t)$ be the Lévy process and let $\mathcal{F}_{s,t}$ be the complete sub $\sigma$-field generated by $\{Z(u) - Z(v); s \le u \le v \le t\}$. Then it is a filtration to the backward direction, i.e, $\mathcal{F}_{s,t} \subset \mathcal{F}_{s',t'}$ if $(s, t] \subset (s', t']$ and is left continuous with respect to $s$.

In the following we will fix the time $t$. A stochastic process $f(r), 0 < r < t$ is called *backward adapted* if $f(r)$ is $\mathcal{F}_{r,t}$-measurable for any $r < t$. The backward predictable process can be defined similarly. If $f(r)$ is right continuous and is backward adapted, then it is backward predictable. A left continuous backward adapted process $Z(r)$ is called a *backward martingale* if it is integrable and satisfies

$$X(r) = E[X(s)|\mathcal{F}_{r,t}], \quad \forall s < r < t.$$

The *backward localmartingale* and *backward semimartingale* can be defined similarly.

We shall define a backward Itô integral. Let $W(t)$ be a standard Brownian motion of the previous section. Let $f(r)$ be a right continuous backward adapted process. Then the *backward Itô integral* is defined by

$$\int_s^t f(r)\hat{d}W(r) = \lim_{|\Pi| \to 0} \sum_k f(t_{k+1})(W(t_{k+1}) - W(t_k)),$$

where $\Pi = \{s = t_0 < t_1 < \cdots < t_n = t\}$ are partitions of the interval $[s, t]$. It is a continuous backward localmartingale. Next let $N(drdz)$ be a Poisson random measure. If $g(r, z)$ is a right continuous backward predictable process, we can define the backward Poisson integral by

$$\int_s^t \int_Z g(r, z)\tilde{N}(\hat{d}r, dz).$$

It is a left continuous backward localmartingale.

Now if $\xi_{s,t}$ is a Lévy flow, $\xi_{r,t}$ and its inverse $\xi_{r,t}^{-1}$ are backward adapted. Then we may define the backward integrals such as $\int_s^t f(\xi_{r,t}^{-1}(y), r)\hat{d}W(r)$ etc. We use the notation

$$\int_s^t X(\xi_{r,t}^{-1}(y), \hat{d}r) := \int_s^t b(\xi_{r,t}^{-1}(y), r)dr \tag{3.30}$$

$$+ \int_s^t f(\xi_{r,t}^{-1}(y), r)\hat{d}W(r) + \int_s^t \int_Z g(\xi_{r,t}^{-1}(y), r, z)\tilde{N}(\hat{d}r, dz).$$

It is left continuous with respect to $s$.

Now let $\psi_{r,z}$ be the inverse map of $\phi_{r,z} : x \to x + g(x, r, z)$, and set

$$h(x, r, z) = -\psi_{r,z}(x) + x. \tag{3.31}$$

Then we have

$$g(x, r, z) - g(\psi_{r,z}(x), r, z) = g(x, r, z) - \{\phi_{r,z}(\psi_{r,z}(x)) - \psi_{r,z}(x)\}$$
$$= g(x, r, z) - h(x, r, z).$$

**Theorem 3.13.** *Assume the same condition as in Theorem 3.5 for coefficients* $b$, $f$, $g$. *Assume further that the diffusion coefficient* $f(x, r)$ *is* $C^{2,1}$ *with respect to* $(x, t)$ *and the integral* $j(x, r) := \int_U |g(x, r, z) - h(x, r, z)| \nu(dz)$ *is bounded for some* $U \subset Z$ *such that* $\nu(U^c) < \infty$. *Then the inverse flow satisfies the following backward SDE*

$$\xi_{s,t}^{-1}(y) = y - \int_s^t \hat{X}(\xi_{r,t}^{-1}(y), \hat{d}r) \quad a.s., \tag{3.32}$$

*for any* $s < t$, *where*

$$\hat{X}(x, t) = X(x, t) - 2 \int_{t_0}^t c(x, r)dr - \int_{t_0}^t \int_Z \{g(x, r, z) - h(x, r, z)\} N(\hat{d}rdz), \tag{3.33}$$

*and*

$$c(x, r) = \frac{1}{2} \sum_{ij} \frac{\partial f^j(x, r)}{\partial x_i} f^{ij}(x, r). \tag{3.34}$$

**Remark.** 1) The stochastic integral in the right hand side of (3.32) is defined as a left continuous backward semimartingale with respect to $s$, while the process $\xi_{s,t}^{-1}(y)$ is right continuous with respect to $s$. Hence the equality holds a.s. for each fixed $s$. We may take a modification of the stochastic integral in the right hand side in such a way that it is left continuous with respect to $t$. Then the equality

$$\xi_{s-,t-}^{-1}(y) = y - \int_s^t \hat{X}(\xi_{r,t}^{-1}(y), \hat{d}r), \tag{3.35}$$

holds for any $s < t$ a.s.

2) Assume further that $h(x, r, z)$ satisfies $\int_Z |g(x, r, z) - h(x, r, z)| \nu(dz) < \infty$ and $\int_Z |h(x, r, z)|^2 \nu(dz) < \infty$ for any $(x, r)$. Then $\hat{X}(x, t)$ is rewritten as

$$\hat{X}(x, t) = \int_{t_0}^t b(x, r)dr + \int_{t_0}^t f(x, r)dW(r)$$
$$+ \int_{t_0}^t \int_Z h(x, r, z)\tilde{N}(drdz) \tag{3.36}$$
$$-2 \int_{t_0}^t c(x, r)dr - \int_{t_0}^t \int_Z (g(x, r, z) - h(x, r, z))\tilde{N}(drdz).$$

Therefore in the inverse flow, the diffusion coefficient $f$ is not changed, while the rule of jumps is changed from $g(x, r, z)$ to $h(x, r, z)$. Further the drift coefficient is changed from $b$ to $b - 2c - \int_Z (g - h)\nu(dz)$.

3) In the case where $\int_Z |g| \nu(dz) < \infty$, the integrals $\int_{t_0}^t \int_Z g \tilde{N}(drdz)$ and $\int_{t_0}^t \int_Z h \tilde{N}(drdz)$ are split into integrals with respect to $N(drdz)$ and $dr\nu(dz)$. Then the above $X$ and $\hat{X}$ are rewritten as

$$X(x,t) = \int_{t_0}^t b'(x,r)dr + \int_{t_0}^t f(x,r)dW(r) + \int_{t_0}^t \int_Z g(x,r,z)N(drdz),$$

$$\hat{X}(x,t) = \int_{t_0}^t b'(x,r)dr + \int_{t_0}^t f(x,r)dW(r)$$
$$+ \int_{t_0}^t \int_Z h(x,r,z)N(drdz) - 2\int_{t_0}^t c(x,r)dr,$$

where we set $b' = b - \int_Z g(z)\nu(dz)$. In this case the drift coefficients are changed from $b'$ to $b' - 2c$. We may observe that the correction term $2c$ appears because of Itô's SDE. Indeed, the flow $\xi_{s,t}$ satisfies the following Stratonovich SDE

$$\xi_t = \xi_0 + \int_{t_0}^t b''(\xi_r,r)dr + \int_{t_0}^t f(\xi_r,r) \circ dW(r) + \int_{t_0}^t \int_Z g(\xi_r,r,z)N(drdz),$$

where $b'' = b' - c$ and $\circ$ denotes the Stratonovich integral. Then the inverse flow satisfies

$$\xi_{s,t}^{-1}(y) = y - \int_s^t b''(\xi_{r,t}^{-1}(y),r)dr$$
$$- \int_s^t f(\xi_{r,t}^{-1}(y),r) \circ d\hat{W}(r) - \int_s^t \int_Z h(\xi_{r,t}^{-1}(y),r)N(\hat{d}rdz).$$

It will be shown in the next section that if we consider "canonical" SDE, the infinitesimal generator (in the canonical sense, to be stated in the next section ) of the inverse flow coincides with the infinitesimal generator of the original flow.

Let us come back to the proof of the theorem. Set $x = \xi_{s,t}^{-1}(y)$. Then equation (3.28) implies

$$y = \xi_{s,t}^{-1}(y) + \int_s^t X(\xi_{s,r}(x),dr) \Big|_{x=\xi_{s,t}^{-1}(y)}$$

Therefore the equation of the theorem is verified if we can show that

$$\int_s^t X(\xi_{s,r}(x),dr) \Big|_{x=\xi_{s,t}^{-1}(y)} = \int_s^t \hat{X}(\xi_{r,t}^{-1}(y),\hat{d}r), \quad a.s., \tag{3.37}$$

for any $s < t$. In the rest of the section, we give a proof of the above formula.

Let $\Pi_n = \{s = t_0 < t_1 < \cdots < t_{k_n} = t\}, n = 1, 2, ..$ be a sequence of partitions such that $|\Pi_n| \to 0$. For each $\Pi = \Pi_n$ we define a simple forward predictable process by

$$\xi_{s,t}^{\Pi}(x) = \sum_k \xi_{s,t_k}(x) 1_{(t_k,t_{k+1}]}(t).$$

By applying Theorem 2.14, we can show that the sequence $\{\int_s^t X(\xi_{s,r}^{\Pi_n}(x), dr)\}$ converges to $\int_s^t X(\xi_{s,r}(x), dr)$ in probability uniformly on compact sets with respect to $x$. Then we have

$$\int_s^t X(\xi_{s,r}(x), dr)\Big|_{x=\xi_{s,t}^{-1}(y)} = \lim_{n\to\infty} \int_s^t X(\xi_{s,r}^{\Pi_n}(x), dr)\Big|_{x=\xi_{s,t}^{-1}(y)}.$$

By the cocycle property, we have

$$\int_s^t X(\xi_{s,r}^{\Pi_n}(x), dr)\Big|_{x=\xi_{s,t}^{-1}(y)}$$

$$= \sum_{k=1}^{k_n} \{X(\xi_{s,t_k}(x), t_{k+1}) - X(\xi_{s,t_k}(x), t_k)\}\Big|_{x=\xi_{s,t}^{-1}(y)}$$

$$= \sum_{k=1}^{k_n} \{(X(\xi_{t_k,t}^{-1}(y), t_{k+1}) - X(\xi_{t_k,t}^{-1}(y), t_k)\}$$

$$= \sum_{k=1}^{k_n} \{(X(\xi_{t_{k+1},t}^{-1}(y), t_{k+1}) - X(\xi_{t_{k+1},t}^{-1}(y), t_k)\}$$

$$- \left[ \sum_{k=1}^{k_n} \{X(\xi_{t_{k+1},t}^{-1}(y), t_{k+1}) - X(\xi_{t_{k+1},t}^{-1}(y), t_k)\} \right.$$

$$\left. - \sum_{k=1}^{k_n} \{(X(\xi_{t_k,t}^{-1}(y), t_{k+1}) - X(\xi_{t_k,t}^{-1}(y), t_k)\} \right]$$

$$= I_1^{\Pi_n}(X) - I_2^{\Pi_n}(X).$$

The convergence

$$\exists \lim_{n\to\infty} I_1^{\Pi_n}(X) = \int_s^t X(\xi_{r,t}^{-1}(y), \hat{d}r)$$

is obvious from the definition of the backward Itô integral. We want to prove

**Lemma 3.14.**

$$\lim_{n\to\infty} I_2^{\Pi_n}(X) = 2 \int_s^t c(\xi_{r,t}^{-1}(y), r) dr$$

$$+ \int_s^t \int_Z \{g(\xi_{r,t}^{-1}(y), r, z) - h(\xi_{r,t}^{-1}(y), r, z)\} N(\hat{d}r dz).$$

*Proof.* Set

$$B = \int_0^t b(x,r)dr, \quad M_c = \int_0^t f(x,r)dW(r), \quad M_d = \int_0^t \int_Z g(x,r,z)\tilde{N}(drdz).$$

We shall consider the limits of $\{I_2^{\Pi_n}(B)\}$, $\{I_2^{\Pi_n}(M_c)\}$ and $\{I_2^{\Pi_n}(M_d)\}$ as $n \to \infty$. We have

$$|I_2^{\Pi_n}(B)| \to | \int_s^t b(\xi_{r-,t}^{-1}(y),r)dr - \int_s^t b(\xi_{r,t}^{-1}(y),r)dr|$$
$$= 0.$$

As to the second term $I^{\Pi_n}(M_c)$, we have

$$I_2^{\Pi_n}(M_c) = \sum_j \sum_k (f^j(\xi_{t_{k+1},t}^{-1}(y),t_{k+1}) - f^j(\xi_{t_k,t}^{-1}(y),t_k))$$

$$(W^j(t_{k+1}) - W^j(t_k))$$

$$= \sum_j \sum_k (f^j(\xi_{s,t_{k+1}}(x),t_{k+1}) - f^j(\xi_{s,t_k}(x),t_k))$$

$$(W^j(t_{k+1}) - W^j(t_k))\Big|_{x=\xi_{s,t}^{-1}(y)}.$$

where $f^j(x,t) = (f^{1j}(x,t),...,f^{dj}(x,t))$. Since $f^j$ are $C^{2,1}$ functions, $f^j(\xi_{s,t},t)$ is a (forward) semimartingale by Itô's formula. Therefore we have

$$\sum_k (f^j(\xi_{s,t_{k+1}}(x),t_{k+1}) - f^j(\xi_{s,t_k}(x),t_k))(W^j(t_{k+1}) - W^j(t_k))$$

$$\to [f^j(\xi_{s,t}(x),t),W^j(t)].$$

By Itô's formula II, $f^j(\xi_{s,t}(x),t)$ is written as

$$f^j(\xi_{s,t}(x),t) = f^j(x,s) + \sum_{i,k} \int_s^t \frac{\partial f^j}{\partial x_i}(\xi_{s,r}(x),r)f^{ik}(\xi_{s,r}(x),r)dW^k(r)$$
$$+ W_d(t) + A(t),$$

where $W_d(t)$ is a purely discontinuous localmartingale and $A(t)$ is a process of finite variation. We know $[W_d, W^j] = 0$ since $W^j$ and $W_d$ are orthogonal and $[A, W^j] = 0$ since $A(t)$ is a process of finite variation. Then we get

$$[f^j(\xi_{s,t}(x),t),W^j(t)] = \sum_{i,k}[\int_s^t \left(\frac{\partial f^j}{\partial x_i}f^{ik}\right)(\xi_{s,r}(x),r)dW^k(r),W^j(t)]$$

$$= \sum_i \int_s^t \left(\frac{\partial f^j}{\partial x_i}f^{ij}\right)(\xi_{s,r}(x),r)dr.$$

Here we used $[W^k,W^j]_t = \langle W^k, W^i \rangle_t = \delta_{ik}t$. Therefore we have

$$\lim_{n\to\infty} I_2^{\Pi_n}(M_c) = \int_s^t \Big(\sum_{i,j} \frac{\partial f^j}{\partial x_i} f^{ij}(\xi_{s,r}(x), r)\Big) dr \Big|_{x=\xi_{s,t}^{-1}(y)}$$

$$= 2\int_s^t c(\xi_{r,t}^{-1}(y), r) dr.$$

We next consider $I_2^{\Pi_n}(M_d)$. We split $M_d$ into the sum of $M_\epsilon$ and $N_\epsilon$, where

$$M_\epsilon(t) = \int_0^t \int_{V_\epsilon} g(x, r, z) \tilde{N}(drdz), \quad N_\epsilon = M_d - M_\epsilon.$$

where $V_\epsilon = \{z : K(z) < \epsilon\}$. Then $I_2^{\Pi_n}(M_\epsilon)$ is uniformly small if $\epsilon$ is sufficiently small. In fact we have for any $\delta > 0$,

$$\lim_{\epsilon\to 0} \sup_n P(|I_2^{\Pi_n}(M_\epsilon)| > \delta) = 0.$$

On the other hand, $N_\epsilon$ is a process of finite variation and the number of jumps are finite a.s. Then we have

$$\exists \lim_{n\to\infty} I_2^{\Pi_n}(N_\epsilon) = \sum_{r\in(s,t]} (\triangle X(\xi_{r,t}^{-1}(y), r) - \triangle X(\xi_{r-,t}^{-1}(y), r)) 1_{\{p(r)\in V_\epsilon^c\}} \qquad (3.38)$$

$$= \sum_{r\in(s,t]} (\triangle X(\xi_{r,t}^{-1}(y), r) - \triangle X \circ \xi_{r-,r}^{-1}(\xi_{r,t}^{-1}(y), r)) 1_{\{p(r)\in V_\epsilon\}}$$

$$= \int_s^t \int_{V_\epsilon^c} (g(\xi_{r,t}^{-1}(y), r, z) - g \circ \psi_{r,z}(\xi_{r,t}^{-1}(y), r, z)) N(\hat{d}r, dz).$$

Note that $g(r, z) - g(r.z) \circ \psi_{r,z} = g(r, z) - h(r, z)$. Then we obtain

$$\lim_{n\to\infty} I_2^{\Pi_n}(N_\epsilon) = \int_s^t \int_{V_\epsilon^c} \{g(\xi_{r,t}^{-1}(y), r, z) - h(\xi_{r,t}^{-1}(x), r, z)\} N(\hat{d}rdz).$$

It converges to

$$\int_s^t \int_Z \{g(\xi_{r,t}^{-1}(y), r, z) - h(\xi_{r,t.}^{-1}(y), r, z)\} N(\hat{d}rdz) < \infty,$$

as $\epsilon \to 0$. Then we obtain that $\lim_{n\to\infty} I_2^{\Pi_n}(M_d)$ exists and is equal to the above. Therefore we get the assertion of the lemma.

Now equality (3.37) holds valid by Lemma 3.14. Then we have the formula (3.32). The proof of Theorem 3.13 is completed.

## 3.7 Canonical SDE

The canonical SDE was introduced by S.I. Marcus [18]. It may be written as

$$d\xi_t = \sum_{j=1}^{m} v_j(\xi_r) \diamond dZ^j(r), \tag{3.39}$$

where $v_1, \ldots, v_m$ are vector fields on $\mathbf{R}^d$ and $Z(t)$ is an $m$-dimensional Lévy process. The precise definition is as follows.

$$d\xi_t = \sum_{j=1}^{m} v_j(\xi_t) \circ dZ_c^j(t) + \sum_{j=1}^{m} v_j(\xi_{t-})dZ_d^j(t) \tag{3.40}$$

$$+ \left\{ \mathrm{Exp}(\sum_j \Delta Z^j(t)v_j)(\xi_{t-}) - \xi_{t-} - \sum_j \Delta Z^j(t)v_j(\xi_{t-}) \right\},$$

where $Z_c(t)$ and $Z_d(t)$ are continuous and purely discontinuous parts of the Lévy processes, respectively and $\mathrm{Exp}(tv) = \varphi(t, x)$ is the solution flow of the ordinary differential equation generated by the vector field $v$

$$\frac{d\varphi(t)}{dt} = v(\varphi(t)), \quad \varphi(0) = x.$$

The above equation looks complicated. But the probabilistic meaning is clear. At the jump time of the driving process $Z(t)$, the solution flow flies from the state $\xi_{t-}(x)$ along with the integral curve $\mathrm{Exp}(tv), 0 \le t \le 1$ with infinite speed, where $v = \sum_j \Delta Z^j(t)v_j$ and lands at the position at $t = 1$, i.e.,

$$\xi_t(x) = \mathrm{Exp}(\sum_j \Delta Z_j(t)v_j)(\xi_{t-}(x)).$$

Therefore if the map $\xi_{t-} : \mathbf{R}^d \to \mathbf{R}^d$ is a homeomorphism, then the map $\xi_t$ should also be a homeomorphism, since the map $\mathrm{Exp}(v)$ is a homeomorphism.

If there is no jump part in the Lévy process $Z(t)$, i.e., if $Z(t)$ is a Brownian motion, then the canonical SDE coincides with the Stratonovich SDE. It is known that the Stratonovich SDE is coordinate free and we can define the Stratonovich SDE on any manifold. We may define the canonical SDE on any manifold $M$ provided that we are given vector fields $v_1, \ldots, v_m$ on $M$ and an $m$-dimensional Lévy process $Z(t)$. However we will not discuss the problem here. We refer to Fujiwara [6] and Applebaum–Kunita [1].

In this section we will study the property of the canonical SDE on Euclidean space, by applying results obtained in previous sections. For this purpose we shall transform the canonical equation to an Itô equation. Let us recall that the Lévy process $Z(t)$ admits the Lévy–Itô decomposition of Section 2.4 (Theorem 2.7). For simplicity, we assume that $W(t)$ in Lévy–Itô decomposition (2.8) is a standard Brownian motion. We assume that coefficients $v_1, \ldots, v_m$ are $C^2$ functions with bounded derivatives. Thus they are Lipschitz continuous. The term involving the Stratonovich integral is written as

$$\sum_j \int_{t_0}^t v_j(\xi_{r-}) \circ dZ_c^j(r) = \sum_j \int_{t_0}^t v_j(\xi_{r-}) dW^j(r) + \sum_j \int_0^t v_j(\xi_{r-}) b^j dr$$

$$+ \frac{1}{2} \sum_{l,k} \langle \int \frac{\partial v_j}{\partial x_k}(\xi_{r-}) v_l^k(\xi_{r-}) dW^l, W^i \rangle_t$$

$$= \sum_j \int_{t_0}^t v_j(\xi_{r-}) dW^j(r) + \sum_j \int_0^t v_j(\xi_{r-}) b^j dr$$

$$+ \frac{1}{2} \sum_{j,k} \int_{t_0}^t \frac{\partial v_j}{\partial x_k}(\xi_{r-}) v_j^k(\xi_{r-}) dr.$$

The jump part is written as

$$\sum_{j=1}^m \int_{t_0}^t v_j(\xi_{r-}) dZ_d^j(r) + \sum_{t_0 < r \le t} \left\{ \mathrm{Exp}(\sum_j \Delta Z^j(r) v_j)(\xi_{r-}) - \xi_{r-} \right.$$

$$\left. - \sum_j \Delta Z^j(r) v_j(\xi_{r-}) \right\}$$

$$= \int_{t_0}^t \int_{|z| \le 1} \left\{ \mathrm{Exp}(\sum_j z^j v_j)(\xi_{r-}) - \xi_{r-} \right\} \tilde{N}(drdz)$$

$$+ \int_{t_0}^t \int_{|z| \le 1} \left\{ \mathrm{Exp}(\sum_j z^j v_j)(\xi_{r-}) - \xi_{r-} - \sum_j z^j v_j(\xi_{r-}) \right\} \hat{N}(drdz)$$

$$+ \int_{t_0}^t \int_{|z| > 1} \left\{ \mathrm{Exp}(\sum_j z^j v_j)(\xi_{r-}) - \xi_{r-} \right\} N(drdz).$$

Therefore the canonical SDE is written as the Itô SDE:

$$\xi_t = x + \int_{t_0}^t X(\xi_{r-}, dr), \tag{3.41}$$

where

$$X(x,t) = X'(x,t) + \int_{t_0}^t \int_{|x|>1} \left\{ \mathrm{Exp}(\sum_j z^j v_j)(x) - x \right\} N(dsdz), \tag{3.42}$$

$$X'(x,t) = \sum_j v_j(x) W^j(t) + \int_{t_0}^t \int_{|x| \le 1} \left\{ \mathrm{Exp}(\sum_j z^j v_j)(x) - x \right\} \tilde{N}(dsdz)$$

$$+ t \left[ \sum_j b^j v_j(x) + \frac{1}{2} \sum_{j,k} \frac{\partial v_j}{\partial x_k}(x) v_j^k(x) \right.$$

$$\left. + \int_{|x| \le 1} \left\{ \mathrm{Exp}(\sum_j z^j v_j)(x) - x - \sum_j z^j v_j(x) \right\} v(dz) \right].$$

We first consider the case where $v$ is supported by a bounded set, say $\{|z| \leq c\}$. Then $X = X'$ holds. We show that the coefficients of the above equation satisfy Lipschitz conditions. The following is easily verified.

**Lemma 3.15.** *(1)* $|\mathrm{Exp}\, v(x) - x| \leq \|v\|_{lg} e^{\|v\|_{lg}} (1 + |x|)$.
*(2)* $|\mathrm{Exp}\, v(x) - x - (\mathrm{Exp}\, v(y) - y)| \leq \|v\|_{Lip} e^{\|v\|_{Lip}}$.
*(3)* $|\mathrm{Exp}\, v(x) - x - v(x) - (\mathrm{Exp}\, v(y) - y - v(y))| \leq \|\nabla v \cdot v\|_{Lip} e^{\|v\|_{Lip}} |x - y|$,
*where*

$$\|v\|_{lg} = \sup_x \frac{|v(x)|}{1 + |x|}, \qquad \|v\|_{Lip} = \sup_{x \neq y} \frac{|v(x) - v(y)|}{|x - y|}$$

*and*

$$(\nabla v \cdot w)^i = \sum_{j=1}^d \frac{\partial v^i}{\partial x^j}(x) w^j(x).$$

The diffusion coefficient $f(x) = (v_j^i(x))$ is clearly Lipschitz continuous, since $v_j$, $j = 1, 2, \ldots$ are Lipschitz continuous. We consider the jump coefficient. From the above lemma, we have

$$\left| \mathrm{Exp}(\sum_j z^j v_j)(x) - x - (\mathrm{Exp}(\sum_j z^j v_j)(y) - y) \right| \leq L(z)|x - y|,$$

where $L(z) = \sum_j |z^j| \|v_j\|_{Lip} e^{\sum_j |z^j| \|v_j\|_{Lip}}$. Therefore if the Lévy measure $v$ is supported by a bounded set, then $\int_{\mathbf{R}^m - \{0\}} L(z)^p v(dz) < \infty$. We have similarly $\int K(z)^p v(dz) < \infty$. We can show also that the drift coefficient

$$b = \sum_j b^j v_j + \frac{1}{2} \sum_{jk} \frac{\partial v_j}{\partial x_k} v_j^k + \int_{\mathbf{R}^m - \{0\}} \left\{ \mathrm{Exp}(\sum_j z^j v_j) - I - \sum_j z^j v_j \right\} v(dz) \quad (3.43)$$

is also Lipschitz continuous. Then the solution $\xi_t(x)$ of canonical SDE defines a flow of continuous maps.

Further, it holds $g(x, z) + x = \phi_z(x) = \mathrm{Exp}(\sum_j z^j v_j)(x)$ and $\phi_z$ are homeomorphisms. Its inverse map is given by $\psi_z(x) = \mathrm{Exp}(-\sum_j z^j v_j)(x)$. It is uniformly Lipschitz continuous and uniformly of linear growth, since $z$ is restricted to $\{z : |z| \leq c\}$. Consequently the solution $\xi_{s,t}$ of the canonical SDE defines a stochastic flow of homeomorphisms.

We next consider the case where the support of the Lévy measure is unbounded. Then the Lipschitz constants and linear growth constants $K(z)$ are unbounded. In this case we can construct the solution by the composition formula (3.26). It defines a stochastic flow of homeomorphisms.

Now assume that the first derivatives of $v_j$, $j = 1, \ldots, m$ and $\sum_{jk} \frac{\partial v_j}{\partial x_k} v_j^k$ are bounded and $\delta$-Hölder continuous. It is known that the Jacobian matrix of the map $\phi_z$ is invertible. Then the solution defines a stochastic flow of $C^1$-diffeomorphisms by Theorem 3.11.

Summing up the above discussions, we have

**Theorem 3.16.** *Assume that coefficients $v_j$, $j = 1, ..., m$ of the canonical SDE are of $C^2$-class and that the first derivatives of $v_1, ..., v_m$ and that of $\sum_{jk} \frac{\partial v_j}{\partial x_k} v_j^k$ are all bounded. Then the solution of the canonical SDE defines a stochastic flow of homeomorphisms.*

*Assume further that the first derivatives of the above are $\delta$-Hölder continuous, then the solution defines a stochastic flow of $C^1$-diffeomorphisms.*

**Remark.** Consider Ito's SDE of separating type, instead of the canonical SDE. We have $g(x, r, z) = \sum_j v_j(x)z^j$, so that $\phi_{r,z}(x) = x + \sum_j v_j(x)z^j$. It is homeomorphic if $|z|$ is sufficiently small. Therefore the solution defines a stochastic flow of homeomorphisms if the size of jumps of the Lévy process is sufficiently small. See Protter [20].

We shall consider the backward SDE for the inverse flow. We need a lemma.

**Lemma 3.17.** *The function $h(x, z) = -\text{Exp}(-\sum_j z^j v_j)(x) + x$ satisfies*

$$\int_{|z| \le c} |g - h| v(dz) < \infty, \quad \int_{|z \le c} |h(z)|^2 v(dz) < \infty$$

*for any $c > 0$.*

*Proof.* We have by Taylor's theorem,

$$g(x, z) - \sum_j z^j v_j(x) = \text{Exp}(\sum_j z^j v_j)(x) - x - \sum_j z^j v_j(x)$$

$$= \frac{1}{2} \sum_{j,k} z^j z^k \nabla v_j \cdot v_k \text{Exp}(\theta \sum_j z^j v_j)(x)$$

and

$$-h(x, z) + \sum_j z^j v_j(x) = \text{Exp}(-\sum_j z^j v_j)(x) - x + \sum_j z^j v_j(x)$$

$$= \frac{1}{2} \sum_{j,k} z^j z^k (\nabla v_j \cdot v_k) \text{Exp}(-\theta' \sum_j z^j v_j)(x),$$

where $0 < \theta, \theta' \le 1$. Therefore we have

$$|g(x, z) - h(x, z)| \le \frac{1}{2} \sum_{j,k} |z^j z^k| |R_{j,k}(x, z, \theta, \theta')|,$$

where $R_{j,k}(x, z, \theta, \theta')$ are bounded with respect to $z$. Then $g - h$ is integrable with respect to the Lévy measure on the set $\{|z| \le c\}$. The square integrability of $h$ will be obvious.

Now the inverse flow $\xi_{s,t}^{-1}$ satisfies the backward SDE based on $\hat{X}(x, t)$ of (3.36) in view of Remark 2) after Theorem 3.13. It is written as

$$\hat{X}(x, t) = \hat{b}(x)t + \sum_j v_j(x)W^j(t)$$

$$+ \int_{t_0}^t \int_{\mathbf{R}^m - \{0\}} \left\{ - \mathrm{Exp}(-\sum_j z^j v_j)(x) + x \right\} \tilde{N}(drdz)$$

where

$$\hat{b}(x) = b(x) - 2c(x) - \int_{\mathbf{R}^m - \{0\}} (g(x, z) - h(x, z))v(dz)$$

and $b(x)$ is given by (3.43). Then the drift term is written by

$$\hat{b}(x) = \sum_j b^j v_j(x) - \frac{1}{2} \sum_{j,k} \frac{\partial v_j}{\partial x_k}(x)v_j^k(x)$$

$$- \int_{\mathbf{R}^m - \{0\}} \left\{ \mathrm{Exp}(-\sum_j z^j v_j)(x) - x + \sum_j z_j v_j(x) \right\} v(dz).$$

Also we have the equality

$$- \int_s^t \hat{X}(\xi_{r,t}^{-1}(y), \hat{d}r)$$

$$= - \sum_j \int_s^t v_j(\xi_{r,t}^{-1}(y)) \circ \hat{d}Z_c(t) - \sum_j \int_s^t v_j(\xi_{r,t}^{-1}(y))\hat{d}Z_d(t)$$

$$- \sum_r \left\{ - \mathrm{Exp}(-\sum_j \Delta Z^j(r)v_j)(\xi_{r,t}^{-1}(y)) + \xi_{r,t}^{-1}(y) \right.$$

$$- \sum_j \Delta Z^j(r)v_j(\xi_{r,t}^{-1}(y)) \right\}$$

$$= \sum_j \int_s^t v_j(\xi_{r,t}^{-1}(y)) \circ \hat{d}(-Z_c)(t) + \sum_j \int_s^t v_j(\xi_{r,t}^{-1}(y))\hat{d}(-Z_d)(t)$$

$$+ \sum_r \left\{ \mathrm{Exp}(\sum_j \Delta(-Z^j)(r)v_j)(\xi_{r,t}^{-1}(y)) - \xi_{r,t}^{-1}(y) \right.$$

$$- \sum_j \Delta(-Z^j)(r)v_j(\xi_{r,t}^{-1}(y)) \right\}$$

$$= \sum_j \int_s^t v_j(\xi_{r,t}^{-1}(y)) \diamond \hat{d}(-Z^j)(r)$$

$$= - \sum_j \int_s^t v_j(\xi_{r,t}^{-1}(y)) \diamond \hat{d}Z^j(r).$$

Consequently we have the following.

**Theorem 3.18.** *Assume the same conditions as in Theorem 3.16. Then the inverse flow*
$\xi_{s,t}^{-1}(y)$ *satisfies*

$$\xi_{s,t}^{-1}(y) = x - \sum_j \int_s^t v_j(\xi_{r,t}^{-1}(y)) \diamond \hat{d}Z^j(r). \tag{3.44}$$

**Further generalizations.** Let $X(x, t)$, $x \in \mathbf{R}^d$, $t \in [0, T]$, be stochastic processes with values in $\mathbf{R}^d$ with parameter $x$. It is called a semimartingale with spatial parameter, if for each fixed $x$, it is an $\mathbf{R}^d$-valued semimartingale. The $\mathbf{R}^d$-valued processes $X(x, t)$ introduced in (3.3) etc. are examples of such semimartingales with spatial parameter.

Let $\eta_t$ be an adapted cadlag process with values in $\mathbf{R}^d$. Under some regularity conditions on $X(x, t)$, we can define Itô's stochastic integral $\int_0^t X(\eta_{r-}, dr)$ as a semimartingale. Then SDEs based on $X(x, t)$ can be defined. We may also define canonical stochastic integral $\int_0^t X(\eta_{r-}, \diamond dr)$ and canonical SDE. The regularity of the solution with respect to the initial data, the homeomorphic property and the diffeomorphic property have all been studied. We refer to Fujiwara–Kunita [7,8], Carmona–Nualart [3] and Applebaum–Tang [2].

# 4 Appendix. Kolmogorov's criterion for the continuity of random fields and the uniform convergence of random fields

We shall introduce Kolmogorov's criterion for a given random field to have a modification of a continuous random field.

**Theorem 4.1 (Kolmogorov–Totoki).** *Let $X(x)$, $x \in \mathbf{D}$ be a random field with values in a normed space $B$ where $\mathbf{D}$ is a domain in $\mathbf{R}^d$. Assume that there exist positive constants $\gamma$, $C$ and $\alpha > d$ satisfying*

$$E[\|X(x) - X(y)\|^\gamma] \le C|x - y|^\alpha, \quad \forall x, y \in \mathbf{D}. \tag{4.1}$$

*Then $X(x)$ has a continuous modification. Further, the continuous modification is uniformly continuous in $\mathbf{D}$ and it can be extended continuously to the closure $\bar{\mathbf{D}}$ of the domain $\mathbf{D}$, i.e., there exists a continuous random field $\tilde{X}(x)$, $x \in \bar{\mathbf{D}}$ such that $\tilde{X}(x) = X(x)$ holds a.s. for any $x \in \mathbf{D}$.*

In order to prove the theorem, we will introduce a modulus of continuity of a map from $\mathbf{D}$ to $B$. Let $\Pi_n$ be the set of all lattice points in $\mathbf{R}^d$ of the form $(i_1/2^n, ..., i_d/2^n)$, where $i_1, ..., i_d$ are integers. We set $\Pi = \cup_n \Pi_n$. It is a dense subset of $\mathbf{R}^d$. We let

$$\mathbf{D}_n = \{x \in \mathbf{D}; \exists x_n \in \Pi_n \cap \mathbf{D} \text{ such that } 0 \le x^i - x_n^i < 2^{-n}, i = 1, ..., d\}.$$

Then $\mathbf{D}_n$ is increasing with $n$ and $\cup_n \mathbf{D}_n = \mathbf{D}$ holds.

Given a map $f : \Pi \cap \mathbf{D} \to B$, we define for each $n$ a modulus of continuity and the modulus of $\beta$-Hölder continuity of $f$ by

$$\Delta_n(f) = \max_{x,y \in \Pi_n \cap \mathbf{D}, |x-y|=2^{-n}} \|f(x) - f(y)\|,$$

$$\Delta_n^\beta(f) = 2^{-\beta n} \Delta_n(f).$$

**Lemma 4.2.** *The inequality*

$$\|f(x) - f(y)\| \leq 2^{d+1} \left( \sum_{n=1}^{\infty} \Delta_n^{\beta}(f) \right) |x - y|^{\beta}, \quad \forall x, y \in \Pi \cap \mathbf{D} \qquad (4.2)$$

*holds for any map* $f : \Pi \cap \mathbf{D} \to B$.

*Proof.* Given a function $f$ on $\Pi \cap \mathbf{D}$, we define a sequence of simple functions $g_n$ : $\mathbf{D}_n \to B, n = 1, 2, \dots$ by $g_n(x) = f(x_n)$, where $x_n \in \Pi_n \cap \mathbf{D}$ is the point such that $0 \leq x^i - x_n^i < 2^{-n}$ holds for any $i = 1, \dots, d$. Then it holds

$$|g_{n+1}(x) - g_n(x)| \leq \Delta_{n+1}(f), \quad \forall x \in \mathbf{D}_n.$$

For any $x \in \mathbf{D}$, let $k$ be a positive integer such that $x \in \mathbf{D}_k$. Then

$$\sum_{n=k}^{\infty} \|g_{n+1}(x) - g_n(x)\| \leq \sum_{n=1}^{\infty} \Delta_{n+1}(f) \leq \sum_{n=1}^{\infty} \Delta_{n+1}^{\beta}(f) < \infty.$$

Then the sequence of simple functions $g_n(x)$ converges on $\mathbf{D}$. Let $g(x)$ be the limit function. Then $g(x) = f(x)$ holds valid for $x \in \Pi \cap \mathbf{D}$. In the sequel we prove the lemma for $g$ instead of $f$.

Now let $x, y$ be any points in $\mathbf{D}$. Then $x, y \in \mathbf{D}_n$ for a sufficiently large $n$. Take $k$ such that $2^{-(k+1)} \leq |x - y| < 2^{-k}$. Then for $n \geq k$,

$$\|g(x) - g_n(x)\| \leq \sum_{m=n+1}^{\infty} \Delta_m(f)$$

$$\leq 2^{-(n+1)} \sum_{m=n+1}^{\infty} \Delta_m^{\beta}(f)$$

$$\leq \left( \sum_{m=n+1}^{\infty} \Delta_m^{\beta}(f) \right) |x - y|^{\beta}.$$

Further since $2^{-(n+1)} \leq |x - y|$, we have

$$\|g_{n+1}(x) - g_{n+1}(y)\| \leq 2^d \Delta_{n+1}(f) \leq 2^d \Delta_{n+1}^{\beta}(f)|x - y|^{\beta}.$$

Therefore

$$\|g(x) - g(y)\| \leq \|g(x) - g_{n+1}(x)\| + \|g_{n+1}(x) - g_{n+1}(y)\|$$
$$+ \|g_{n+1}(y) - g(y)\|$$

$$\leq 2^{d+1} \left( \sum_{k=1}^{\infty} \Delta_k^{\beta}(f) \right) |x - y|^{\beta}.$$

The proof is complete.

By the above lemma, the map $f : \Pi \cap \mathbf{D} \to B$ satisfying $\sum_k \Delta_k^\beta(f) < \infty$ for some $\beta > 0$ is uniformly continuous on $\mathbf{D}$ and has a continuous extension $g : \bar{\mathbf{D}} \to B$ ,i.e., there exists a continuous map $g : \bar{\mathbf{D}} \to B$ such that $g(x) = f(x)$ holds for $x \in \Pi \cap \mathbf{D}$. The function $g$ is $\beta$-Hölder continuous.

We shall apply the above lemma to the random field $X(x)$. Observe that for each $\omega$, $X(\cdot, \omega)$ restricting $x$ to $\Pi \cap \mathbf{D}$ can be regarded as a map form $\Pi \cap \mathbf{D}$ to $B$. Then we have

$$\|X(x, \omega) - X(y, \omega)\| \le 2^{d+1} \left( \sum_k \Delta_k^\beta(X(\omega)) \right) |x - y|^\beta, \quad \forall x, y \in \Pi \cap \mathbf{D}. \quad (4.3)$$

**Lemma 4.3.** *Let $X(x), x \in \mathbf{D}$ be a random field satisfying the inequality (4.1). Let $\beta$ be a positive number satisfying $\beta\gamma < \alpha - d$. Then,*

$$E[(\sum_k \Delta_k^\beta(X))^\gamma]^{1/\gamma} \le \left( \sum_{k=1}^\infty 2^{-k\{(\alpha-d)/\gamma-\beta\}} \right) \cdot (2^d C)^{1/\gamma} < \infty. \quad (4.4)$$

*Proof.* We will consider the case $\gamma \ge 1$ only. Observe the inequality

$$(\Delta_k^\beta(X))^\gamma \le ( \sup_{x,y \in \Pi \cap \mathbf{D}, |x-y|=1/2^k} \|X(x) - X(y)\| 2^{k\beta})^\gamma$$
$$\le \sum (\|X(x') - X(y')\| 2^{k\beta})^\gamma,$$

where the summations are taken over all $x', y' \in \Pi \cap \mathbf{D}$ such that $|x' - y'| = 1/2^k$. Then the number of summations are at most $2^{(k+1)d}$. Therefore

$$E[\Delta_k^\beta(X)^\gamma] \le 2^{(k+1)d+k\gamma\beta} E[\|X(x') - X(y')\|^\gamma]$$
$$\le 2^{k(d+\gamma\beta-\alpha)} 2^d C.$$

In the last inequality, we applied the inequality (4.1). Therefore we get

$$E[(\sum_k \Delta_k^\beta(X))^\gamma]^{1/\gamma} \le \sum_{k=1}^\infty E[\Delta_k^\beta(X)^\gamma]^{1/\gamma}$$
$$\le \left( \sum_{k=1}^\infty 2^{-k\{(\alpha-d)/\gamma-\beta\}} \right) \cdot (2^d C)^{1/\gamma} < \infty. \quad (4.5)$$

The proof is complete.

*Proof of Theorem 4.1.* The random field $X(x)$ restricting $x$ on $\Pi \cap \mathbf{D}$ satisfies the inequality (4.3), where $\sum_k \Delta_k^\beta(X) < \infty$ holds a.s. Therefore $X(x), x \in \Pi \cap \mathbf{D}$ is uniformly $\beta$-Hölder continuous a.s. Then there exists a continuous random field $\tilde{X}(x)$ defined on $\bar{\mathbf{D}}$ such that $X(x) = \tilde{X}(x)$ holds a.s. for any $x \in \Pi \cap \mathbf{D}$. Since $X(x)$ is continuous in probability, the equality $X(x) = \tilde{X}(x)$ holds a.s for any $x \in \mathbf{D}$. Thus we have proved the theorem.

We shall consider the uniform convergence of a sequence of continuous random fields. Let $\gamma$ and $\alpha$ be positive numbers. Let $\{X_n(x)\}$ be a sequence of $\gamma$ integrable random fields. We define a $(\gamma, \alpha)$-Hölder norm for a random field $X(x)$ by

$$\|X\|_{\gamma,\alpha} = \sup_{x \in D} E[\|X(x)\|^\gamma]^{1/\gamma} + \sup_{x \neq y, x, y \in D} \frac{E[\|X(x) - X(y)\|^\gamma]^{1/\gamma}}{|x - y|^{\alpha/\gamma}}.$$

**Theorem 4.4.** *Let $\{X_n(x)\}$ be a sequence of continuous random fields such that*

$$\lim_{n \to \infty} \|X_n - X\|_{\gamma,\alpha} = 0$$

*holds for some $\alpha > d$ and $\gamma > 0$. Then we have for any $N$,*

$$\lim_{n \to \infty} E[\sup_{x \in D, |x| \leq N} \|X_n(x) - X(x)\|^\gamma] \to 0. \tag{4.6}$$

*Proof.* Take $\beta > 0$ such that $\gamma\beta < \alpha - d$. We want to prove

$$\lim_{n \to \infty} E[(\sum_k \Delta_k^\beta(X_n - X))^\gamma] = 0.$$

Set $C_n = \|X_n - X\|_{\gamma,\alpha}^\gamma$. Then it holds

$$E[\|X_n(x) - X(x)\|^\gamma] \leq C_n,$$
$$E[\|X_n(x) - X(x) - (X_n(y) - X(y))\|^\gamma] \leq C_n|x - y|^\alpha.$$

We have by Lemma 4.3,

$$E[(\sum_k \Delta^\beta(X_n - X))^\gamma]^{1/\gamma} \leq \left(\sum_{k=1}^\infty 2^{-k\{(\alpha-d)/\gamma-\beta\}}\right) \cdot (2^d C_n)^{1/\gamma} = C_n',$$

which converges to 0 as $n \to \infty$. Then we get from (4.3),

$$E[\sup_{|x| \leq N} \|X_n(x) - X(x) - X_n(x_0) + X(x_0)\|^\gamma]^{1/\gamma} \leq C_n' \sup_{|x| \leq N} |x - x_0| \to 0$$

for a fixed $x_0$. Now observe

$$E[\sup_{|x| \leq N} \|X_n(x) - X(x)\|^\gamma]^{1/\gamma}$$
$$\leq E[\|X_n(x_0) - X(x_0)\|^\gamma]^{1/\gamma}$$
$$+ E[\sup_{|x| \leq N} \|X_n(x) - X(x) - X_n(x_0) + X(x_0)\|^\gamma]^{1/\gamma}.$$

It converges to 0 again. Then we get the assertion of the theorem.

# References

[1]  D. Applebaum and H. Kunita, Lévy flows and Lévy processes on Lie groups, *J. Math. Kyoto Univ.* **33** (1993), 1103–1123.

[2]  D. Applebaum and F.Tang, The interlacing construction for stochastic flows of diffeomorphisms on Euclidean spaces, *Sankhya, Series A,* **63** (2001), 139–178.

[3]  R.A. Carmona and D. Nualart, Nonlinear Stochastic Integrators. Equations and Flows, Stochastic Monographs 6, Gordon and Breach Science Publishers, 1990.

[4]  C. Dellacherie and P.A. Meyer, *Probabilities and Potential B — Theory of Martingales,* North Holland, Amsterdam, 1982.

[5]  K.D. Elworthy, *Stochastic Differential Equations on Manifolds,* LMS Lecture Note Series, 70, Cambridge University Press, Cambridge, UK, 1982.

[6]  T. Fujiwara, Stochastic differential equations of jump type on manifolds and Lévy flows, *J. Math. Kyoto Univ.* **31**(1991), 99–119.

[7]  T. Fujiwara and H. Kunita, Stochastic differential equations of jump type and Lévy flows in diffeomorphisms group, *J. Math. Kyoto Univ.* **25** (1985), 71–106.

[8]  T. Fujiwara and H. Kunita, Canonical SDEs based on semimartingales with spatial parameters, Part I Stochastic flows of diffeomorphisms, *Kyushu J. Math.* **53** (1999), 265–300.

[9]  T. Fujiwara and H. Kunita, Canonical SDEs based on semimartingales with spatial parameters, Part II Inverse flows and backward SDEs, *Kyushu J. Math.* **53** (1999), 301–331.

[10]  N. Ikeda and S. Watanabe, *Stochastic Differential Equations and Diffusion Processes,* North Holland, Amsterdam. 1981.

[11]  K. Itô, Spectral type of the shift transformation of differential processes with stationary increments, *Trans. Amer. Math. Soc.* **81**(1956), 253–263.

[12]  J. Jacod and A.N. Shiryaev, *Limit Theorems for Stochastic Processes,* Springer, 1987.

[13]  H. Kunita, *Stochastic Flows and Stochastic Differential Equations,* Cambridge University Press, 1990.

[14]  H. Kunita, Stochastic flows with jumps and stochastic flows of diffeomorphisms. In: "Itô's Stochastic Calculus and Probability Theory." N. Ikeda et al., Eds., Springer, 1996, 197–211.

[15]  H. Kunita. Representation of martingales with jumps and applications to mathematical finance, Stochastic Analysis and Related Topics in Kyoto, Advanced Studies in Pure Mathematics, 41 (2004), 209–232.

[16]  H. Kunita, S. Watanabe, On square integrable martingales, *Nagoya Math. J.* **30** (1967), 209–245.

[17]  R. Léandre, Flot d'une équation differentielle stochastique avec semimartingale directrice discontinue. Séminaire Probab. XIX, Lecture Notes in Math. 1123(1985), 271–275.

[18]  S.I. Marcus, Modelling and approximations of stochastic differential equations driven by semimartingales, *Stochastics* 4(1981), 223–245.

[19]  P.A. Meyer, Un cours sur integrales stochastiques, Seminaire Proba. X, Lecture Notes in Math. 511 Springer, 1976, 246–400.

[20]  P. Protter, Stochastic Integration and Differential Equations. A New Approach, Applied Math. 21, Springer, 1992.

[21]  K. Sato, *Lévy Processes and Infinitely Divisible Distributions,* Cambridge University Press, Cambridge, UK, 1999.

[22]  D.W. Stroock, Markov Processes from K. Itô's Perspective, Annals of Mathematical Studies 155, Princeton and Oxford. (2003)

# Convolutions of Vector Fields-III: Amenability and Spectral Properties

M. M. Rao

Department of Mathematics, University of California, Riverside, Riverside, CA 92521 USA
rao@math.ucr.edu

## 1 Introduction

Although this is a continuation of the previous two parts, it may be studied independently of my earlier work (1980, 2001) and the necessary results will be briefly restated. An extended early section motivates the problems from a finite state space to the general case via a discussion of random walks, or equivalently convolution operators and their structural analysis. This naturally leads to a study of the latter operators on certain function spaces and function algebras. It also shows a need to consider the (algebraic) structure of the state space of random walks, namely an analysis of the underlying locally compact groups and the dependence on the spectral analysis of the associated convolution operators on function spaces built on them. In the nonabelian group case (of the state space of the walks) the analysis is intimately related to amenability of the group, which is the range or state space of the random walk.

The spaces that arise in this work include Beurling–Orlicz algebras which are usually subalgebras of $L^1(G)$. But convolution operators also act on certain spaces that are not necessarily algebras and the spectral properties of these operators are again related to amenability of the underlying groups. This interplay between the real and stochastic analysis thus becomes a special object of the following work. All the new results are presented essentially with full details and Orlicz spaces play a significant role in the analysis.

The next section contains a basic probabilistic motivation of the subject with a random walk problem, which indicates the need for spectral analysis in this study. Then Sections 3 and 4 treat the desired generalization with the state space of the walk as a locally compact group and its connection to amenability. The close relation of the latter property as it applies to the algebraic and analytic aspects of the walk with transience, periodicity and recurrence is exemplified here. Then Section 5 is devoted to a detailed study of spectral analysis of convolution operators which concentrates on amenability and several of its characterizations. The work here contains an essentially complete generalization of Day's (1964) paper for arbitrary locally compact groups in which he asked for a solution in the context of Orlicz spaces. The Beurling and Segal algebras are briefly treated in the last section that contains some further observations (relative to semigroups and algebras), raised by this study, but left for future analysis. Some special cases of the results of Sections 4 and 5 were announced in Rao (2001) without details because of space constraints there. Applications and more analysis of random walks in this general context will also be postponed.

## 2 Elementary Aspects of Random Walks

Let $\mathbb{Z}^k \subset \mathbb{R}^k$, $k \geq 1$, be the integer lattice representing the position of a particle making a random walk (r.w.). Thus if $X_i : \Omega \to \mathbb{Z}^k$ denotes the state of the particle starting at 0 at time $i$, then the probability

$$P[X_i = z] = p(0, z) = \begin{cases} \frac{1}{2k}, & |z|^2 = z_1^2 + \cdots + z_k^2 = 1 \\ 0, & \text{otherwise} \end{cases} \tag{1}$$

$i = 1, 2, \ldots$, of landing at the site $z \in \mathbb{Z}^k$, where $(\Omega, \Sigma, P)$ is the basic probability space. Then $S_n = \sum_{i=1}^n X_i$ is the position of the particle at time $n$ starting in state 0 at time 0. Let us set $S_0 = 0$, for convenience. The $X_n$ are independent random variables with the same distribution $p(i) = p(0, e_i)$ where $e_i = (0, \ldots, 0, 1, 0, \ldots, 0)$ and $P[X_n = e_i] = p(0, e_i) = p(i), n \geq 1$, combining the previous notations. The r.w. is *simple* if the $p(i)$ are as above, and *symmetric* if $p(i, j) = p(0, j - i)$ so that $p : \mathbb{Z}^k \times \mathbb{Z}^k \to [0, 1]$, $p(z, z') = p(0, z' - z)$ and $\sum_{z \in \mathbb{Z}^k} p(0, z) = 1$. Thus $p(z, z')$ is the one step transition probability from $z$ to $z'$, and $p(z, z') = p(0, z' - z)$ signifies that the transitions are *homogeneous* or *stationary*. A state $z \in \mathbb{Z}^k$ is *recurrent* if it is visited by the r.w. with probability 1, and it is *transient* if there is positive probability that it is *not* visited. The range of the r.w., namely $\mathbb{Z}^k$, is called the *state space*.

A classical result on the behavior of the r.w., going back to Chung, Fuchs, Ornstein and others states, if the walk has its values in $\mathbb{R}^k$, the following:

**Theorem 1.** *Let $S_n : \Omega \to \mathbb{R}^k$ be the r.w. $\left( S_n = \sum_{i=1}^n X_i \right)$ and $S_0 = 0$. Then 0 is a recurrent state of the r.w. iff for each $\varepsilon > 0$,*

$$\sum_{n=1}^{\infty} P[|S_n| < \varepsilon] = +\infty, \tag{2}$$

*and it is transient if the series converges. In case $k = 1$, the walk is recurrent if $\frac{S_n}{n} \to 0$ in probability, as $n \to \infty$, without regard to the existence of moments of the random variables $X_i$.*

The following illustration of this theorem is of interest. It was originally proved by G. Pólya in 1921 directly.

**Example 2.** A symmetric homogeneous r.w. is recurrent in $\mathbb{Z}^k$ if $k = 1$ or 2 and transient for $k \geq 3$. Indeed if $k = 1$, $S_n$ being the sum of Bernoulli random variables with values 1 and $-1$, has mean 0, so that by the weak (or strong) law of large numbers $\frac{S_n}{n} \to 0$ in probability (or a.e.), and so the result follows from the last part of the above theorem. For $k \geq 2$ let $p_n = P[|S_n| = 0]$ so that the particle visits '0' at step $n$ with probability $p_n$. Now for $k = 2$, the particle can visit '0' only if the positive and negative steps are equal so that $p_{2n} > 0$ and $p_{2n+1} = 0, n = 0, 1, 2, \ldots$ Using the multinomial probability theorem with equal probabilities $\frac{1}{4}$ for the four states, one has

$$p_{2n} = \sum_{j=0}^{n} \frac{(2n)!}{[j!(n-j)!]^2} \left(\frac{1}{4}\right)^{2n}$$

$$= 4^{-2n} \binom{2n}{n} \sum_{j=0}^{n} \binom{n}{j}^2 = 4^{-2n} \binom{2n}{n}^2. \tag{3}$$

Using Stirling's approximation $n! \sim \sqrt{2\pi n} \cdot n^n e^{-n}$, one finds $p_{2n} \sim \frac{1}{n}$ so that $\sum_{n=0}^{\infty} p_{2n} = +\infty$, and '0' is a recurrent state by the above theorem. Since the r.w. is homogeneous and symmetric, the same holds starting from any other state.

Next for $k = 3$, one can use a similar reasoning to get

$$p_{2n} = \sum_{0 \leq i, j, i+j \leq n} \frac{(2n)!}{[i!j!(n-i-j)!]^2} \left(\frac{1}{6}\right)^{2n}$$

$$= \sum_{\substack{0 \leq i+j \leq n \\ i, j \geq 0}} \left(\frac{n!}{i!j!(n-i-j)!} \frac{1}{6^n}\right)^2 \binom{2n}{n} \sim \frac{1}{n^{\frac{3}{2}}}, \tag{4}$$

again using Stirling's approximation, so that $\sum_{n=1}^{\infty} p_{2n} < \infty$, $p_{2n+1} = 0$. Thus '0' is a transient state, and as before the result holds for all states. The problem is briefly discussed in Spitzer ((1964), p.52), (cf., also Rao (1984), p.95).

One can define a r.w. with any finite subset of $\mathbb{Z}^k$ as the state space, which follows if an addition is suitably defined for it. For instance, if $k = 1$, and the subset is denoted $(0, 1, \ldots, a)$ so that $p(i, j) = p(0, j-i) = p$ and $p(j, i) = q(= 1-p) = 1 - p(i, j)$, then the r.w. stops when the particle reaches 0 or $a$, called the *absorbing barriers*, can reverse the directions termed the *reflecting barriers*, or partly absorbing and partly reflecting, the *elastic barriers*. One can define the "$n$-step" transition $p_n(x, y)(= p_n(0, y-x))$ when the particle starts at $x$ and lands at $y$ after $n$ transitions, so that it reaches $y$ from $x$ in $n$-steps with $x_1, \ldots, x_{n-1}$ as the intermediate stops before landing in $y$. Thus counting all the intermediate paths:

$$p_n(x, y) = \sum_{\{\text{all paths}\}} p(x, x_1) p(x_1, x_2) \ldots p(x_{n-1}, y). \tag{5}$$

This may be stated more simply as follows (cf. Spitzer (1964)). Let $p_0(x, y) = \delta_{xy}$ the Kronecker delta, and $p_1(x, y) = p(x, y)$ for $x, y \in \mathbb{Z}^k$. Define $p_n$ recursively as

$$p_n(x, y) = \sum_{\{\text{all paths}\}} p_1(x, x_i) p_{n-1}(x_i, y). \tag{6}$$

For integers $m \geq 0, n \geq 0$, this implies

$$p_{m+n}(x, y) = \sum_{\{\text{all paths}\}} p_m(x, x_i) p_n(x_i, y). \tag{7}$$

Note that $\sum_{y \in \mathbb{Z}^k} p_n(x, y) = 1$ and $p_n(x, y) = p_n(0, y - x)$. Letting $P = P_1 = (p_1(x, y))$, the stochastic matrix, if $P^0 = I$ (the identity), $P_n = (p_n(x, y))$, then $P_n = P^n$ and $P_{m+n} = P^m P^n$. The structure of $\mathbb{Z}^k$ or restriction to its subsets (subgroups) leads to a far reaching analysis. Since $P$ is generally an infinite matrix, let us motivate the later analysis by restricting the walk to a finite set.

If $P$ is an $n \times n$ transition probability matrix, then a column vector $x$ such that $Px = \lambda x$ for some $\lambda \in \mathbb{R}$ is a (right) eigenvector of $P$, associated with the scalar $\lambda$ called an eigenvalue of $P$, is a solution if $x$ has all components 1 and $\lambda = 1$. The elementary matrix theory implies that $r(P) = \max_i \{|\lambda_i|\}$, the spectral radius of $P$, with $\lambda_i$ as its eigenvalues, satisfies $r(P) = 1$ and the Frobenius theory of positive matrices implies that all of its eigenvalues are bounded by $r(P)$(here). The asymptotic analysis of a r.w. is of interest in this study and it is facilitated by the following classification.

A set $C \subset \mathbb{Z}^k$ is *closed* for the r.w. if $i \in C, j \notin C$ implies $p(i, j) = 0$. A state is *periodic* if it is recurrent and is visited at times $t, 2t, 3t, \ldots$ where $t > 1$ is the greatest integer with this property, and is *aperiodic* if $t = 1$. A walk is *irreducible* if all states are either transient or recurrent null (i.e., $\lim_{n \to \infty} p_n(i, i) = 0$ when $i$ is recurrent). With this classification one analyzes the walk for its limiting behavior.

First suppose the r.w. is symmetric and has a finite state space, say $N$ states, so that the transition $P$ is an $N \times N$ symmetric (or doubly stochastic) matrix. Let $P$ have distinct eigenvalues $\lambda_1, \ldots, \lambda_N$ with the corresponding eigenvectors $x_i \in \mathbb{R}^N, i = 1, \ldots, N$. The classical matrix results imply that the $x_i$ are orthogonal (for the inner product) and for convenience, normalize so that $(x_i, x_j) = \delta_{ij}$. It then follows that

$$(P^n)(i, j) = \sum_{k=1}^{N} \lambda_k^n x_k(i) x_k(j) \tag{8}$$

where $x_k(i)$ is the $i$th component of $x_k$. Taking $\lambda_1 = 1$ and $x_1(i) = 1, i = 1, \ldots, N$, as the corresponding eigenvector so $Px_1 = x_1$, it is verified that $r(P) = 1$, and $|\lambda_k| < 1, \ k = 2, \ldots, N$, and that

$$\lim_{n \to \infty} (P^n)(i, j) = \delta_{ij}. \tag{9}$$

[See Feller (1968), p.431, for details of this well-known case.]

Now if $N = +\infty$, then $P$ is an infinite matrix (or operator) on $\ell^2$, the Hilbert space of sequences, corresponding to $\mathbb{R}^N$ in the special case. However $x = (1, 1, \ldots) \notin \ell^2$ and $Px$ is not defined. A refined analysis is therefore called for. Using the group structure of $\mathbb{Z}^k$, and $P: \ell^2 \to \ell^2$ as a bounded linear operator, satisfying suitably modified (8) and (9), the problem was generalized by Kesten (1959a), utilizing the fact that $\mathbb{Z}^k$ is a locally compact group in its discrete topology. For a deep analysis of r.w. on $\mathbb{Z}^k$, one may consult Spitzer's (1964) monograph.

Thus abstracting the above case, let $G$ be a countable (not necessarily abelian) group with $A = (a_1, a_2, \ldots)$ as a set of its generators. (In $\mathbb{Z}^k$ above one has $u_1, \ldots, u_k$ the unit vectors as its generators.) Using multiplication as the group operation, a r.w. with state space as $G$ moves one step either to the right or to the left (denoted as inverse multiplication). Thus the probability $p$ on $G$ is such that from a state '$a$' to '$a\sigma$' in one step is determined by ('$e$' being the identity element of the group $G$):

$$p(\sigma) = p(a, a\sigma)(= p(e, a^{-1}a\sigma)) \tag{10}$$

and the walk is *symmetric* if also $p(\sigma) = p(\sigma^{-1})$ so that $2\sum_{i=1}^{\infty} p(a_i) = 1$. If $p_i = p(a_i)$, then the probability from $a_i$ to $a_j$ in one step is given by $m_{ij} = p(a_i a_j^{-1})(= p(e, a_j - a_i)$ in the classical abelian case considered before). For the symmetric r.w. $m_{ij} = p(a_i^{-1}a_j) = p((a_i^{-1}a_j)^{-1}) = p(a_j^{-1}a_i) = m_{ji}$ so that $M = (m_{ij})$ is symmetric. Let $\ell_G^2 = \{x = (x_1, x_2, \ldots): G \to \ell^{\infty}, \sum_{i \geq 1} |x_i|^2 < \infty, x_i = x(a_i)\}$. Then $M: \ell_G^2 \to \ell_G^2$, and is defined by:

$$\begin{aligned}
(Mf)(i) &= \sum_{j \geq 1} m_{ij} f(j), \quad f \in \ell_G^2 \\
&= \sum_{a_j \in A} p(a_i^{-1}a_j) f(a_j) \; (= \sum_{a_j \in A} f(a_i^{-1}a_j) p(a_j)).
\end{aligned} \tag{11}$$

However, $f$ and $p$ can be regarded as functions on $G$, and (11) can then be expressed alternatively as:

$$\begin{aligned}
(Mf)(i) &= \sum_{a_k \in G} p(a_i^{-1}a_k) f(a_k) \\
&= \sum_{a_j a_k = a_i} p(j) f(k) \\
&= (p * f)(i)
\end{aligned} \tag{12}$$

where $p * f$ is the convolution of $f \in \ell_G^2$ and the probability measure $p$ on $G$. This implies that $M = (p*)$ is a convolution operator on $\ell_G^2$ which is not a group algebra (unlike $l_G^1$) but $p * f$ is well-defined since (12) is finite for all $a_i \in G$. If $\lambda_0 = \sup\{|\lambda|: \lambda \in \sigma(M)\}$ where $\sigma(M)$ is the spectrum of $M$, then set $\lambda_0 = r(M)$, the spectral radius. Since $M$ is symmetric on $\ell_G^2$ one has:

$$\begin{aligned}
r(M) \leq \|M\| &= \sup\{\|Mf\|_2: \|f\|_2 \leq 1\} \\
&= \sup\{|(Mf, f)|: \|f\|_2 \leq 1\}, \text{ by classical analysis,} \\
&\quad (\cdot, \cdot) \text{ is the inner product of } \ell_G^2,
\end{aligned}$$

$$\leq \sup_i \left\{ \sum_{j \geq 1} p_{ij} \right\} = 1, \tag{13}$$

since $\sum_{j \geq 1} p_{ij} = 1$ being a transition probability. Thus the spectral radius is bounded by one. This property is a connecting link between the probabilistic aspect of the problem and the spectral analysis of the convolution operator by a probability measure on $G$, leading to a profound study of such mappings with the structure of its underlying group. *Thus the r.w. on a locally compact group is defined through convolution*, as a key part of the study, and it is made more explicit in what follows.

## 3 Role of the Spectrum of Convolution Operators

A new element in the problem from $\mathbb{Z}^k$ to $G$ is its noncommutativity for the analysis of which somewhat more sophisticated tools are required. To see the type of techniques required, let us recall a pair of results obtained by Kesten (1959 a,b) for a countable group $G$:

(i) The transition operator $M: \ell_G^2 \to \ell_G^2$ of symmetric r.w. discussed above, has the property that the spectral radius $r(M) = 1$ if $G$ is either finite or abelian, generalizing the known case of $G \subset \mathbb{Z}^k$.

(ii) For any countable group $G$, $r(M) = 1$ iff it satisfies a certain condition of Følmer which is equivalent to the existence of an *invariant mean* in the sense that there is a positive linear functional $m: B(G) \to \mathbb{R}$, where $B(G)$ is the Banach space of scalar bounded functions on $G$, such that $m(1) = 1$ and $m(L_s f) = m(f), s \in G$, where $(L_s f)(x) = f(s^{-1}x)$, $f \in B(G)$, $s, x \in G$, $L_s$ denoting the left translation operator.

The cases of interest here are those groups $G$ that admit left invariant means and they are called (left) *amenable*. It is known that, for locally compact groups, they admit left invariant means iff they have right invariant means. So hereafter only the 'left' class is analyzed, and that qualification will be omitted.

The second result above connects amenability with spectral radius of transition operators of r.w. and the recurrence properties. Since by (12) of the preceding section the convolution operator $M = p*$, by a probability measure $p$ on $G$, is the key object of investigation, some related properties between $M$ and $G$ will be studied with respect to amenability. This property of a group plays an important role also in certain other areas, especially in noncommutative harmonic analysis. Let us restate here the concepts of recurrence and transience for a general locally compact group $G$ of a r.w. on a probability space $(\Omega, \Sigma, P)$ with state space $G$ and $\mu$ as its distribution or image measure. Thus a *symmetric random walk* is a sequence of random variables $X_n: \Omega \to G, n \geq 1$, such that (a) $P(X_n \in A) = \mu(A), A \in \mathcal{B}(G)$, the Borel $\sigma$-algebra, $n \geq 1$, and (b) $Y_n = X_n X_{n-1} \ldots X_1 (= X_n Y_{n-1}): \Omega \to G$, where $Y_{n-1}$ and $X_n$ are independent. Note that if $G$ is abelian, then this is written as $Y_n = \sum_{k=1}^n X_k = X_n + Y_{n-1}$, which corresponds to the partial sum of a sequence of independent identically distributed symmetric random variables, with group operation as addition. The existence of such a walk is of course a consequence of the classical Kolmogorov theorem where $(\Omega, \Sigma, P) = \otimes_i (G, \mathcal{B}, \mu)_i$, and $X_n: \Omega \to G, n \geq 1$ as coordinate variables. [See, e.g. Rao (1995), Chapter 1, Sections 2-3, for several generalized forms of this assertion.] The existence result holds even if $\mu$ is not symmetric.

With this setting, a r.w. measure $\mu$ is called *transient* if starting from the identity $x_0 = e \in G$ (i.e., $\mu(X_0 = e) = 1$), it visits a compact set $A \subset G$ infinitely often (i.o.) with $\mu$-measure 0, and is *recurrent* if it visits each neighborhood of 'e' infinitely often with $\mu$-measure 1. In symbols this may be written, $P[X_n \in U, i.o.] = 0$ for each relatively compact neighborhood $U$ of $e$ in $G$, and $P[X_n \in U, i.o.] = 1$ respectively. Here if $A_n = [X_n \in U]$, then $A_n, i.o.$ means $\limsup_n A_n (= \cap_{k \geq 1} \cup_{n \geq k} A_n)$. Thus a walk $Y_n, n \geq 1$, is transient if $Y_n \to \infty$, as $n \to \infty$ with probability one, or $P[Y_n \in U, i.o.] = 0$. Observe that such a walk with state space $(G, \mathcal{B})$ corresponds to convolution of $\mu$ and the point measure $\delta_g, g \in G$ or $\delta_g * \mu: A \mapsto (\delta_g * \mu)(A)$, and thus it is described by a sequence $(X_n, Y_n)$ of pairs satisfying the above conditions. The following result, essentially due to Furstenberg (1973) is illustrative.

**Proposition 1.** *Let $\mu$ be a symmetric probability measure on a group $(G, \mathcal{B})$ as above with full support. Then one has:*

*(a) $G$ is not amenable implies that the r.w. $\{(X_n, Y_n), n \geq 1\}$, or $\mu$, is transient.*
*(b) The r.w. or $\mu$ is recurrent iff $f * \mu (= \int_G f(y^{-1}x)d\mu(x), y \in G) = f$ for any uniformly right continuous $f: G \to \mathbb{R}$ signifies that $f = constant$, or equivalently, for each relatively compact neighborhood $U$ of $e \in G$, one has*

$$\sum_{n \geq 0} \mu^n(U) = \infty \tag{1}$$

*where $\mu^n$ denotes the n-fold convolution of $\mu$ with itself (and when (1) holds $G$ is necessarily amenable).*

A r.w. with initial measure $\mu$ (i.e., of $X_1$) on the state space $G$ is called *adapted* if it is supported by $G$ (and is *irreducible* if the smallest closed subgroup containing the support of $\mu$ is $G$). The walk is *aperiodic* if $\mu$ is adapted and its support is not contained in a coset of $G$ relative to a closed normal subgroup. (Cf., Revuz (1984), p. 98.) One has the following simple statement. The monographs (Revuz (1984), p. 196 and Spitzer (1964), Sec.II.7) may be consulted on these ideas:

**Proposition 2.** *If $G$ is a discrete group and $\mu$ is the measure of r.w. $\{X_n, n \geq 1\}$ which is adapted, irreducible and periodic with period $d > 1$, then the periodic class containing $e \in G$, is a subgroup $G_0$ and the $d - 1$ periodic classes are the (right) cosets of $G_0$ which form a cyclic decomposition of $G$.*

**Proof.** Let $\mu^k$ be the $k$th order convolution of $\mu$ with itself. Then one has $\mu^k(ab^{-1})$ to be the probability of the r.w. moving from $a$ to $b$ in $k$ steps ($p_k(a, b)$ in the notation of the last section). Let $G_0, G_1, \ldots, G_{d-1}$ be the cyclic classes with $e \in G_0$ and $x \in G_i$ implies $\mu^{kd+r}(x^{-1}) > 0$ for some $k \geq 0$ and $0 \leq r < d$. By definition $\mu^{kd}(ab^{-1}) > 0$ iff $a$ and $b$ belong to the same class (by periodicity), and $G_0, G_1, \ldots, G_{d-1}$ forms a partition since $X$ (or $\mu$) is irreducible. Thus to establish the proposition, it suffices to verify

(i) $G_0$ is a group, and
(ii) each $G_k$ is a (right) coset of $G_0$.

For (i) let $a, b \in G_0$. Since $x \in G_r$ iff $\mu^{kd+r}(x^{-1}) > 0$ for some integers $0 \leq r < d$ and $k \geq 1$, one must have $\mu^{md}(a^{-1}) > 0$ and $\mu^{nd}(b^{-1}) > 0$ for some integers $m, n \geq 1$. Then

$$\mu^{(m+n)d}((ab)^{-1}) = \sum_{t \in G} \mu^{md}(t^{-1}b^{-1}a^{-1})\mu^{nd}(t)$$

$$\geq \mu^{md}(a^{-1})\mu^{nd}(b^{-1}) > 0,$$

if $t = b^{-1}$. Thus $((ab)^{-1})^{-1} = ab \in G_0$. Taking $a = b$ here shows $\mu^{kd}(aa) > 0$ for some integer $k \geq 1$ (since $ab \in G$), and

$$\mu^{(k+m)d}(a) = \sum_{t \in G} \mu^{kd}(t^{-1}a)\mu^{md}(t)$$

$$\geq \mu^{kd}(aa)\mu^{md}(a^{-1}) > 0,$$

if $t = a^{-1}$. Thus $a^{-1} \in G_0$ and $G_0$ is a (sub)group.

For (ii), to show that $G_k$ is a (right) coset of $G_0$, consider $G_0t = \{at : a \in G_0\}$. If $a_1 = at, b_1 = bt$ with $ab \in G_0$, one has $a_1, b_1 \in G_0t$ and $ba^{-1} \in G_0$. So for some $m \geq 1$, $\mu^{md}(ab^{-1}) > 0$. Then

$$\mu^{md}(a_1b_1^{-1}) = \mu^{md}(ab^{-1}) > 0,$$

and $a_1, b_1$ belong to some class $G_{r(t)}$ for some $0 \leq r(t) < d$, which implies $G_0t \subset G_{r(t)}$. If $s \neq t$ but $r(s) = r(t)$, then $(G_0t) \cup (G_0s) \subset G_{r(t)}$ implies $\mu^{kd}(rs^{-1}) > 0$ for some $k \geq 1$. Hence $ts^{-1} \in G_0$ and $ts^{-1} = a \in G_0$, whence $s = t$ and $G_0t = G_0s = G_0e$ since $G_0$ is a group. Thus $G_{r(t)} = Gt$ is a (right) coset of $G_0$, and then $G = \cup_{i=0}^{d-1} G_i$ is the desired partition.     □

The above properties indicate a relation between the structure of r.w. with a locally compact group $G$ as the state space that is (not) amenable. Also a classification of walks depends on the type of $G$ and there is a richer theory in the amenable case. Thus one needs to characterize the latter property to apply it for a finer analysis of r.w.'s. This will be the major theme of the ensuing work starting with the next section. It is useful to remark that amenability of $G$ does not necessarily imply recurrence of r.w. on it as Example 2 of Section 1 (Pólya's theorem) shows since all abelian locally compact and compact general groups are known to be amenable. So it is a nontrivial task to analyze and obtain properties of locally compact nonabelian groups from the point of view of spectral analysis of convolution operators on certain function spaces and study amenability of these from it. In fact a large amount of work is available on this subject, and for a wide survey of amenability analysis on different types of groups, one may refer to Paterson (1988) where an extensive bibliography, as of its publication date, is included. But first it is of interest to characterize amenability, using certain algebraic methods introduced in the study by Johnson (1972), which proves to be fruitful. See also Pier (1984) on amenability in general.

## 4 Amenable Function Algebras and Groups

Let $\mu$ be a (left) Haar measure of a locally compact group $G$. Then the Lebesgue space $L^1(G, \mu)[= L^1(G)]$ is well-known to be a group algebra under convolution as multiplication. If moreover $G$ is an amenable group, then $L^1(G)$ has several additional properties, and a problem of importance here is its converse, namely, to find those properties of $L^1(G)$ that characterize the amenability of $G$. An interesting contribution to this problem appears in Johnson (1972), and it will be extended here to show that certain classes of subalgebras of $L^1(G)$, with stronger topology, can also characterize amenability of $G$. Other methods through spectral analysis of convolution operators on $L^p(G)$ and certain more inclusive Banach function spaces will be treated in the next section. The algebraic part is considered here. It is necessary to recall some general concepts before presenting the actual result. (See Bonsall and Duncan (1973) for the standard background material.)

Let $A$ be a Banach algebra. Then a Banach space $X$ is a left $A$-module if for each $a \in A$ and $x \in X$, one has $ax \in X$ and the mapping $(a, x) \rightarrow ax$ is continuous in the sense that there is a constant $K > 0$ such that $\|ax\| \leq K\|a\|\|x\|$. A right $A$-module $(a, x) \rightarrow xa$ is similarly defined, and $X$ is an $A$-bimodule if it is both a left and a right module. For instance if $A = L^1(G)$ and $X = L^\varphi(G)$, an Orlicz space, then $X$ is a left $A$-module and is an $A$-bimodule if $G$ is unimodular. Now for an $A$-bimodule $X$, a bounded linear mapping $D: A \rightarrow X$ is called a *derivation* if $D(ab) = a(Db) + (Da)b$. Let $Z(A, X)$ be the subspace of all derivations of $B(A, X)$, the Banach space of all bounded linear operators from $A$ into $X$ under uniform norm. Also $A$ can be given other multiplications such as the Lie and Jordan products. The Lie multiplication $\tilde{D}_x: A \rightarrow X$ is $\tilde{D}_x a = ax - xa$, for each $x \in X$ (a bimodule). Then $\tilde{D}_x \in Z(A, X)$ since

$$\begin{aligned} \tilde{D}_x(ab) &= x(ab) - (ab)x \\ &= axb - (ab)x + x(ab) - axb \\ &= a[xb - bx] + [xa - ax]b = a\tilde{D}_x b + (\tilde{D}_x a)b, \quad x \in X. \end{aligned} \quad (1)$$

Let $Z_1(A, X)$ be the subspace of all derivations of the type (1), called *inner derivations*. This is a closed subspace. An immediate question is to find conditions on $A$ in order that both the sets coincide, i.e., the quotient space $H^1(A, X) = Z(A, X)/Z_1(A, X) = \{0\}$. This is an additive group. It is a surprising fact, discovered by B. E. Johnson (1972), that this happens for $A = L^1(G)$, the group algebra, iff $G$ is amenable. Now the (additive) group $H^1(A, X)$ was already studied by G. Hochschild in the middle 1940s for all associative algebras $A$, which is now called the (first) *Hochschild cohomology group*. The interesting point here is that the apparently unrelated concepts of amenability of $G$ and the (algebraic) cohomology property of a group algebra are found to be linked, enabling a characterization of the former. This will be elaborated here for a large class of subalgebras of the group algebra $L^1(G)$ that already characterize amenability of $G$.

For completeness, let us also remark that the concept of an $n$th order Hochschild cohomology group $H^n(A, X)$, $n \geq 1$, exists, generalizing the above definition with $A$ replaced by its tensor product $A^n$, and extending the concept of $A$-bimodule to $A^n$-bimodule of $X$. This will be useful for the direct sum decomposition theory of algebras, but is not needed here and hence no further discussion is included. (See Runde (2002) for details.)

Returning to analyze some classes of subalgebras of $\mathcal{A} = L^1(G)$, the group algebra, it is appropriate to recall that Orlicz spaces are to be used here. Consider a continuous Young function $\varphi \colon \mathbb{R} \to \mathbb{R}$ such that $\varphi(2x) \leq K\varphi(x)$, $x > 0$, $\varphi(-x) = \varphi(x)$, (denoted $\varphi \in \Delta_2$) and $\varphi'(0) > 0$ where $\varphi'$ is the right (or left) derivative of $\varphi$. This class includes $\varphi(x) = |x|$ as well as $\varphi(x) = (e_0 + |x|)\log(e_0 + |x|) - e_0$, where $e_0$ is the base of the natural 'log'. Then the Orlicz space $L^\varphi(G)$ with $\mu$ as (left) Haar measure on $G$, is closed under convolution as multiplication and is a Banach algebra, under the gauge norm $\| \cdot \|_{(\varphi)} \colon f \mapsto \inf\{k > 0 \colon \int_G \varphi\left(\frac{f(x)}{k}\right) d\mu \leq 1\}$. If $\varphi(x) = |x|^p$, $p \geq 1$, $L^\varphi(G)$ becomes the classical Lebesgue space $L^p(G)$. Moreover, $L^\varphi(G) \subset L^1(G)$, the embedding being topological. This includes the second example of $\varphi$. The above general statement follows from the fact that for such a $\varphi$,

$$|x|\varphi'(0) = \int_0^{|x|} \varphi'(0)dt \leq \int_0^{|x|} \varphi'(t)dt = \varphi(x).$$

since $\varphi'$ is increasing, and hence

$$\int_G |f| d\mu \leq \frac{1}{\varphi'(0)} \int_G \varphi(f) d\mu < \infty, \quad f \in L^\varphi(G). \tag{2}$$

The following simple assertion on $L^\varphi(G)$ will be needed for much of the analysis. Let $\mathcal{M}^\varphi(G)$ be the closed subspace of $L^\varphi(G)$, determined by simple functions:

**Proposition 1.** *Let $\varphi$ be a Young function such that $\varphi \in \Delta_2$ and $\varphi'(0) > 0$. Then for a locally compact group $G$ with a (left) Haar measure $\mu$, the space $(L^\varphi(G), \| \cdot \|_{(\varphi)})$ is closed under convolution, and has a left approximate identity (i.e., there is a net $\{u_V, V \subset G\} \subset L^\varphi(G)$, $V \subset G$ is a relatively compact open neighborhood of $e \in G$, directed by inclusion, $\|u_V\|_{(\varphi)} \leq c < \infty$, and for each $\varepsilon > 0$, there is $V_\varepsilon \subset G$ as above, such that $\|f * u_{V_\varepsilon} - f\|_{(\varphi)} < \varepsilon$, and $\|u_{V_\varepsilon} * f - f\|_{(\varphi)} < \varepsilon$, for each $f \in L^\varphi(G)$, '*' denoting convolution).*

**Proof.** This is essentially known, and a brief argument is included for completeness and the result is often referred to below. Since by (2) $L^\varphi(G)$ is closed under convolution, for $f, g \in L^\varphi(G)$, one has

$$\|f * g\|_{(\varphi)} \leq K\|f\|_{(\varphi)}\|g\|_{(\varphi)}, \tag{3}$$

for some $K > 0$. Also $\varphi$ being continuous, the set $C_c(G)$ of continuous compactly supported functions on $G$ is dense in $L^\varphi(G)$. Hence for any symmetric relatively compact neighborhood $V_0$ of $e \in G$, there is $0 \leq u_{V_0} \in C_c(G)$ such that $\|u_{V_0}\|_{(\varphi)} = 1$ by the structure of $L^\varphi(G)$. So $u_{V_0} * f \in L^\varphi(G)$ for any $f \in L^\varphi(G)$. Taking an element $h \in L^\psi(G)$, where $\psi$ is a complementary Young function to $\varphi$, and noting that $\varphi \in \Delta_2 \Rightarrow (L^\varphi(G))^* = L^\psi(G)$, one has with the duality pairing $\langle \cdot, \cdot \rangle$ the following;

$$|\langle u_{V_0} * f - f, h\rangle| = \left|\int_G \left[\int_G u_{V_0}(t) f(t^{-1}s) d\mu(t) - f(s)\right] h(s) d\mu(s)\right|$$

$$\leq \int_G u_{V_0}(t) \left|\int_G (f(t^{-1}s) - f(s)) h(s) d\mu(s)\right| d\mu(t)$$

$$\leq 2 \int_G u_{V_0}(t) \|f(t^{-1}\cdot) - f(\cdot)\|_{(\varphi)} \|h\|_{(\psi)} d\mu(t), \qquad (4)$$

by the Hölder inequality for the $(L^\varphi(G), L^\psi(G))$ pair (cf. Rao and Ren (1991), p. 58). Since $\varphi \in \Delta_2$, its norm is absolutely continuous, and hence for $\varepsilon > 0$, there is a neighborhood $V_1$ of $e \in G$ such that

$$\|f(t^{-1}\cdot) - f\|_{(\varphi)} < \frac{\varepsilon}{2}, \quad t \in V_1. \qquad (5)$$

Taking $V = V_0 \cap V_1$, one gets from (4) and (5) that

$$|\langle u_V * f - f, h\rangle| \leq \varepsilon \|h\|_{(\psi)} \int_G u_V(t) d\mu(t). \qquad (6)$$

Computing the supremum on $\|h\|_{(\psi)} \leq 1$ in (6) and noting that it determines the norm $\|\cdot\|_{(\varphi)}$, one has

$$\|u_V * f - f\|_{(\varphi)} \leq \varepsilon \|u_V\|_1 \leq c_1 \varepsilon \|u_V\|_{(\varphi)} \leq c_1 \varepsilon, \quad \text{by}(2). \qquad (7)$$

It follows that $\|u_V\|_{(\varphi)} = c_2 < \infty$ for all such $u_V$, and $\{u_V, V \subset G\}$ is a bounded approximate identity in $L^\varphi(G)$.

A similar argument works for right approximate identity also in $L^\varphi(G)$ since by definition the right translation $(R_t f)(s) = f(st^{-1})\Delta(t^{-1})$ where $\Delta(\cdot)$ is the modular function and $\|R_t f - f\|_{(\varphi)} < \varepsilon$ for a suitable neighborhood $\tilde{V}$ of $e \in G$. Thus $L^\varphi(G)$ has a bounded (two sided) approximate identity. $\qquad \square$

**Remark.** If $\mathcal{M}^\varphi(G) \subset L^\varphi(G)$ is the closed linear span of $C_c(G)$ when $\varphi$ is any continuous Young function with $\varphi'(0) > 0$, then the norm in $\mathcal{M}^\varphi(G)$ is absolutely continuous, and the above proposition holds if $L^\varphi(G)$ is replaced by $\mathcal{M}^\varphi(G)$, by dropping the condition that $\varphi \in \Delta_2$ in addition. It is also known that $(\mathcal{M}^\varphi(G))^* = L^\psi(G)$. Thus $\varphi(t) = e^{|t|} - 1$ is included in this extension. The norm $\|\cdot\|_{(\varphi)}$ in $\mathcal{M}^\varphi(G)$ (but not in $L^\varphi(G)$) works, (cf. Rao and Ren (1991)), and it is a group algebra.

Let $\mathcal{A} = L^\varphi(G)$ and $\mathcal{B} = L^1(G)$ with $\varphi'(0) > 0$ and $\varphi \in \Delta_2$. Then $\mathcal{A}$ and $\mathcal{B}$ both have bounded (two sided) approximate identities by the above proposition (for $\mathcal{B}, \varphi(x) = |x|$), and $\mathcal{A} \subset \mathcal{B}$, the inclusion being continuous. To analyze the group structure of $\mathcal{H}^1(\mathcal{A}, \mathcal{X})$ in the present extended context, the following concept will be employed.

**Definition 2.** Let $\mathcal{A}$ be a Banach algebra and $\mathcal{X}$ an $\mathcal{A}$-bimodule. Then $\mathcal{A}$ is termed *amenable* if the set of all derivations from $\mathcal{A}$ to the dual $\mathcal{X}^*$ of $\mathcal{X}$ are inner where $\mathcal{X}^*$ becomes an $\mathcal{A}$-bimodule under the actions: for all $x \in \mathcal{X}, y \in \mathcal{X}^*$ and $a \in \mathcal{A}$ define $ay$ and $ya$ by:

$$\langle x, ya \rangle = \langle ax, y \rangle, \quad \text{and} \quad \langle x, ay \rangle = \langle xa, y \rangle,$$

so that the 'left' and 'right' in $\mathcal{X}$ become respectively the 'right' and 'left' for $\mathcal{X}^*$.

Using this notion some auxiliary facts from V. Runde (2002) will now be stated in a form to obtain the desired result. These results are already in a compressed form in Johnson (1972) but are detailed in Runde (2002).

Let $\varphi$ be a continuous Young function with $\varphi'(0) > 0$ so that $L^\varphi(G)$ is closed under convolution, and its subspace $\mathcal{M}^\varphi(G)$ determined by simple (or bounded) functions has a bounded approximate identity by Proposition 1, and the Remark following it. But by Proposition 2.1.5 of Runde ((2002), p.39), for each Banach $\mathcal{M}^\varphi(G)$-bimodule $E$, $\mathcal{H}^1(\mathcal{M}^\varphi(G), E^*) = \{0\}$ iff the same holds for a subclass of the $E$ which satisfy $E = \{axb : a, b \in \mathcal{M}^\varphi(G), x \in E\}$, called *pseudo-unital* so that one can restrict to pseudo-unital $\mathcal{A}$-bimodules where $\mathcal{A} = \mathcal{M}^\varphi(G)$. Consequently the direct part of Johnson's theorem, namely if $G$ is amenable then $L^1(G)$ is an amenable algebra in the sense of Definition 2, which does not use any special structure of $L^1(G)$ other than that it is a Banach algebra with a bounded (two sided) approximate identity applies. There the argument uses only the functional analytic facts that the weak*-closed convex hull of $\{(\tilde{D}\delta_g)\delta_{g^{-1}} : g \in G\}$ is compact where $\tilde{D}$ is an extension of $D \in \mathcal{Z}^1(L^1(G), E^*)$ with $\tilde{D} \in \mathcal{Z}^1(M(G), E^*)$, $\delta_g$ being the point (or Dirac) measure at $g \in G$ and $M(G)$ the set of regular signed measures on $G$. That $\tilde{D}$ is inner is proved through a fixed point theorem (due to M. M. Day). But all these results are also valid if $L^1(G)$ is replaced by $\mathcal{M}^\varphi(G)$ which has bounded (two sided) approximate identities. Hence the algebra $\mathcal{M}^\varphi(G)$ is an amenable subalgebra of $L^1(G)$.

In the opposite direction, suppose $\mathcal{A} = \mathcal{M}^\varphi(G)$ is amenable and $\mathcal{B} = L^1(G)(\supset \mathcal{A})$. If $I$ is the identity operator, it defines a continuous homomorphism since $\mathcal{A}$ is continuously embeddable in $\mathcal{B}$, and that $\mathcal{A}$ is dense in the latter. So by Proposition 2.3.1 of Runde ((2002), p.46) $\mathcal{B}$ is also amenable. But then the converse part of Johnson's theorem implies that $G$ is amenable.

Summarizing the above discussion, the following extended version of Johnson's theorem can be stated:

**Theorem 3.** *Let $G$ be a locally compact group, $\varphi$ a continuous Young function with $\varphi'(0) > 0$ and $\mathcal{M}^\varphi(G)(\subset L^\varphi(G))$ be the closed subalgebra that is densely contained in $L^1(G)$. Then $G$ is amenable iff the convolution algebra $\mathcal{M}^\varphi(G)$ (of $L^1(G)$) is amenable.*

This is the original statement when $\varphi(x) = |x|$, but applies also if

$$\tilde{\varphi}(u) = (e_0 + |u|) \log(e_0 + |u|) - e_0, \quad |u| > 2,$$

and $\varphi(u) = \chi_{[0,2]}(|u|) + (\tilde{\varphi}\chi_{[|u|>2]})(|u|)$, then $\mathcal{M}^\varphi(G)(= L^\varphi(G))$ is contained in $L^1(G)$ with a stronger (norm) topology. Thus the extension includes more 'group algebras' that characterize amenability of $G$. In particular, for $\varphi(x) = e^{|x|} - 1$, one has $\mathcal{M}^\varphi(G) \underset{\neq}{\subset} L^\varphi(G) \underset{\neq}{\subset} L^1(G)$ and bounded approximate identities exist in $\mathcal{M}^\varphi(G)$ and $L^1(G)$, the former being dense in the latter. Thus the extension has applicational potential including more spaces.

In this context one should also compare $\mathcal{M}^\varphi(G)$ with a class of subalgebras of $L^1(G)$, called *Segal algebras*. A Segal algebra is a dense subalgebra $S(G)$ of $L^1(G)$

such that (a) $f \in S(G)$ implies $L_a f \in S(G)$, and $\|L_a f\|_S = \|f\|_S$ for a norm $\|\cdot\|_S$ under which $(S(G), \|\cdot\|_S)$ is a Banach algebra, (b) the embedding of $S(G)$ in $L^1(G)$ is continuous, and (c) for each $\varepsilon > 0$, $f \in S(G)$, there is a neighborhood $V$ of $e$ such that $\|L_a f - f\|_S < \varepsilon, a \in V$. An important example is $\mathcal{M}^\varphi(G), \varphi'(0) > 0, \varphi$ continuous. If $\varphi(x) = e^{|x|} - 1$, then $L^\varphi(G) \subset L^1(G)$ for which (a) and (b) but not (c) hold so that it is not a Segal algebra, although $\mathcal{M}^\varphi(G)$ will be. The basic theory of Segal algebras for abelian $G$ is given in Reiter ((1968), Sec.6.2). The problem of characterizing amenability of a general locally compact group $G$ by the amenability of a general Segal algebra is unsolved, and its solution will supplement the above work.

Next a spectral analysis of the amenability question outlined in Section 2 above will be treated in detail.

## 5 Spectra of Convolution Operators and Amenability

The preceding work deals with the algebraic aspect of characterizing amenability of a locally compact group. However, some key properties of a r.w. on such groups are also determined by the spectral properties of the associated convolution operators on various classes of Banach (function) spaces as noted in Section 3 above. A comprehensive characterization of amenability of these groups based on the spectral radius of a convolution operator will be obtained here when the spaces are of Orlicz type based on a Haar measure. The interplay of spectral analysis and the structural properties of groups will be central for this work, and ultimately it contains the desired generalization raised in Day's (1964) paper for groups.

Let us first recall a few results on convolution operators on a locally compact group $G$ to use them in the work that follows. If $f, g \in C_c(G)$, with uniform norm, then $C_c(G) \subset L^1(G)$, and $f * g \in C_c(G)$, the convolution operation being uniformly continuous relative to the (left) uniform structure of $G$. If $f \in C_c(G)$ and $g \in L^1(G)$, then $f * g \in L^1(G)$ and is (uniformly) left continuous. An interesting condition on $G$ introduced by Dieudonné (1960), using the $L^p(G)$ spaces, can be stated for Orlicz spaces as follows. (It is a technical condition but is closely related to amenability.)

**Definition 1.** Let $\varphi$ be a Young function, $\varphi \in \Delta_2$ and $G$ a locally compact group. Then $G$ is said to have *property $P_\varphi$* if for each $\varepsilon > 0$, and compact set $K \subset G$, there is an element $0 \le f_\varepsilon \in C_c(G)$ supported by $K$, $\|f_\varepsilon\|_{(\varphi)} = 1$, and satisfying $\|L_x f_\varepsilon - f_\varepsilon\|_{(\varphi)} < \varepsilon, x \in K$, where $L_x$ is the usual left translation operator of $x$ on $L^\varphi(G)$.

If $\varphi(x) = |x|^p, p \ge 1$, this property denoted $P_p$, was originally introduced by Dieudonné (1960). Also recall that $\varphi$ is an $N$-function if it is a Young function and $\frac{\varphi(x)}{x} \uparrow \infty$ as $x \uparrow \infty$ and $\frac{\varphi(x)}{x} \downarrow 0$ as $x \downarrow 0$. The following observation provides a useful relation between $P_\varphi$ and $P_1$ for later applications.

**Proposition 2.** *If $\varphi$ is an $N$-function, $\varphi \in \Delta_2$, then $P_\varphi$ and $P_1$ are equivalent properties for $G$. In particular $P_p, 1 < p < \infty$, is included.*

**Proof.** The argument is a modification of the classical one (cf. Reiter (1968), p.168). The convexity of $\varphi$ implies for $0 < s < t$,

$$
\begin{aligned}
\varphi(t - s) &= \int_0^{t-s} \varphi'(u)du = \int_0^t \varphi'(u) - \int_{t-s}^t \varphi'(u)du \\
&= \varphi(t) - \int_0^s \varphi'(v + t - s)dv \\
&\leq \varphi(t) - \int_0^s \varphi'(v)dv, \quad \text{since } \varphi' \uparrow, \\
&= \varphi(t) - \varphi(s).
\end{aligned}
$$

This implies, by interchanging $s$, $t$, and noting $\varphi \colon \mathbb{R} \to \mathbb{R}^+ (\varphi(-x) = \varphi(x))$:

$$
\varphi(s - t) \leq |\varphi(s) - \varphi(t)|. \tag{1}
$$

Hence if $0 \leq f \in L^1(G)$ and $\varepsilon > 0$, $K \subset G$ compact such that $\|L_x f - f\|_1 < \varepsilon, x \in K$, then $P_1$ holds. Let $g = \varphi^{-1}(f)$. Thus $\varphi(g) \in L^1(G)$ and $\varphi \in \Delta_2 \Rightarrow g \in L^\varphi(G)$. Taking $s = g(x)$ and $t = (L_y g)(x)$ in (1) and integrating, one has

$$
\int_G \varphi(L_y g - g)(x)d\mu(x) \leq \int_G |(L_y f - f)(x)|d\mu(x) < \varepsilon, \quad y \in K. \tag{2}
$$

Thus $P_\varphi$ is valid for $G$.

For the opposite direction, suppose $P_\varphi$ is valid. By convexity of $\varphi$ again,

$$
\varphi(t) = \varphi(s) + \int_0^t \varphi'(u)du \leq \varphi(s) + (t - s)\varphi'(t), \quad 0 < s < t,
$$

from which one deduces for all $s, t \in \mathbb{R}$, ($\vee = \max$)

$$
|\varphi(s) - \varphi(t)| \leq |t - s|\varphi'(t) \leq |t - s|\varphi'(s \vee t), \quad \text{since } \varphi' \uparrow .
$$

Now let $g \in L^\varphi(G)$ be such that for compact $K \subset G$ and $\varepsilon > 0$, one has $\|L_x g - g\|_{(\varphi)} < \varepsilon, x \in K$ as before (cf. Definition 1). If $\varphi(g) = f$, then $f \in L^1(G)$ and

$$
\begin{aligned}
\|L_x f - f\|_1 &= \int_G |L_x \varphi(g) - \varphi(g)| (t)d\mu(t) \\
&= \int_G |\varphi(L_x g) - \varphi(g)| (t)d\mu(t), \quad \text{since } L_x \varphi(g) = \varphi(L_x g), \\
&\leq \int_G |L_x g - g| \varphi'(L_x g \vee g)(t)d\mu(t) \\
&\leq 2\|L_x g - g\|_{(\varphi)} \|\varphi'(L_x g \vee g)\|_{(\psi)}, \tag{3}
\end{aligned}
$$

by the Hölder inequality for Orlicz spaces and the key fact that for $\varphi \in \Delta_2$, $\varphi$ an $N$-function imply $\int_G \psi(\varphi'(g))d\mu < \infty$ (cf. Rao and Ren (1991), Proposition 8 on p.79). Further the norms are invariant so that $\|L_x g\|_{(\varphi)} = \|g\|_{(\varphi)}$ and the same is true for the complementary norm $\| \cdot \|_{(\psi)}$. Hence from (3) and the $P_\varphi$-condition, one has

$$\|L_x f - f\|_1 \leq 2\|L_x g - g\|_{(\varphi)} \|\varphi'(L_x g \vee g)\|_{(\psi)} \to 0, \tag{4}$$

as $L_x g \to g$ in norm and $\varphi'(L_x g \vee g)$ is in a ball of $L^\psi(G)$. This shows at once that $\Gamma_\varphi$ implies $P_1$. Thus they both are equivalent. $\qquad\square$

**Remarks 1.** If $G$ is amenable, then Reiter has shown that $G$ has property $P_1$ and equivalently $P_p$. The above proposition shows that it is also equivalent to $P_\varphi$ which is weaker than $P_p$ since there are $N$-functions $\varphi \in \Delta_2$ that are not $p$th powers for any $p > 1$.

    2.  In the definitions of $P_1$ and $P_\varphi$, the element $f \in L^1(G)$ [or $L^\varphi(G)$] can be restricted to a (left) uniformly continuous one on $G$ having a nonmeager support. This follows from the fact that $f$ may be replaced by $\tilde{f} = f * g$ where $g \in C_c(G)$. For such $\tilde{f}$ with $\|\tilde{f}\|_{(\varphi)} > 0$ one can find an open set contained in $\operatorname{supp}(\tilde{f})$. (See, e.g., Hewitt and Ross (1963), p.275.) This will be utilized below without comment.

    3.  In the above work the fact that $L_x : L^\varphi(G) \to L^\varphi(G)$ is an isometry was used several times. This follows at once from the definition of its norm given by $\|f\|_{(\varphi)} = \inf\{k > 0: \int_G \varphi\left(\frac{|f|}{k}\right) d\mu \leq 1\}$ where $\mu$ is as usual a left Haar measure.

    The following result is also invoked in the ensuing computations and it extends (13) of Section 2 above, obtained for countable groups.

**Proposition 3.** *Let $\varphi$ be an $N$-function, $\varphi \in \Delta_2$, and $f \in L^\varphi(G)$. If $T_f = f*: L^\varphi(G) \to L^\varphi(G)$ is the convolution operator, then it is continuous and if $r_\varphi(T_p) = \sup\{|\lambda|: \lambda \in spec(T_f)\},[\ spec(T_f) = \{\alpha: (\alpha I - T_f)^{-1} \text{ does not exist}\}],$ is the spectral radius of $T_f$ (and $spec\ (T_f)$ denotes the spectrum of $T_f$), then one has $r_\varphi(T_f) \leq \|T_f\|_{(\varphi)} \leq 1, \text{ if } \|f\|_{(\varphi)} \leq 1$.*

**Proof.** By the equivalence of $P_1$ and $P_\varphi$, if $f \in L^\varphi(G)$ and $\tilde{f} = \varphi(f)$ as in Proposition 2 above, then $\tilde{f} \in L^1(G)$ (because $\varphi \in \Delta_2$) and one has $T_f g$ defined for all $g \in L^\varphi(G)$. Moreover,

$$\|T_f g\|_{(\varphi)} \leq \|\tilde{f}\|_1 \|g\|_{(\varphi)}, \quad \text{by Proposition 11 in Rao (1980).} \tag{5}$$

Hence $\|T_f\| = \sup\{\|T_f g\|_{(\varphi)}: \|g\|_{(\varphi)} \leq 1\} \leq \|\tilde{f}\|_1 = 1$, if $\|f\|_{(\varphi)} \leq 1$. Now $T_f$ is a linear operator determined by an element in the unit ball of the convolution algebra $L^1(G)$, which is complete. The inequality between the spectral radius and the operator norm is however a standard property of Banach algebras. $\qquad\square$

    Another special result that eventually helps characterize amenability through spectral analysis of convolution operators will be established. Let $P(G) \subset L^1(G)$ be the positive part of the unit sphere so that $f \in P(G)$ iff $f \geq 0$ and $\int_G f d\mu = 1$. The first set of conditions is used for a preliminary simplification of the main result. The following is a technical observation.

**Proposition 4.** *Let $\varphi \in \Delta_2$ be a Young function, $f_n \in L^\varphi(G), n \geq 1$. Consider the set $Z_\varphi \subset G$ defined for the $f_n, n \geq 1$, sequence as:*

$$Z_\varphi = Z_\varphi (\{f_n, n \geq 1\}) = \left\{x \in G: \lim_n \|L_x f_n - f_n\|_{(\varphi)} = 0\right\}. \tag{6}$$

*Then $Z_\varphi$ is a subgroup and for any Young function $\tilde{\varphi} \in \Delta_2$ if $F_n = \tilde{\varphi}^{-1}(\varphi(f_n))$ one has $Z_\varphi(\{f_n, n \geq 1\}) = Z_{\tilde{\varphi}}(\{F_n, n \geq 1\})$.*

**Proof.** By the $\Delta_2$-condition $\varphi^{-1}$ and $\tilde{\varphi}^{-1}$ are continuous functions and since $\tilde{\varphi}(F_n) = \varphi(f_n)$, one has $\varphi(L_x f_n) = \tilde{\varphi}(L_x F_x) \in L^1(G)$. Further

$$\|\varphi(f_n) - L_x\varphi(f_n)\|_1 = \|\varphi(f_n) - \varphi(L_x f_n)\|_1$$
$$\leq 2\|L_x f_n - f_n\|_{(\varphi)}\|\varphi'(L_x f_n \vee f_n)\|_{(\psi)} \to 0, \qquad (7)$$

as $n \to \infty$ for all $x \in Z_\varphi(\{f_n, n \geq 1\})$. This implies $\|\tilde{\varphi}(L_x F_n) - \tilde{\varphi}(F_n)\|_1 \to 0$ because $\varphi(L_x f_n) = \tilde{\varphi}(L_x F_n)$. Thus $x \in Z_{\tilde{\varphi}}(\{F_n, n \geq 1\})$, so that $Z_\varphi(\{f_n, n \geq 1\}) \subset Z_{\tilde{\varphi}}(\{F_n, n \geq 1\})$. The argument being symmetric, the opposite inclusion and hence equality between these sets holds for $\varphi, \tilde{\varphi} \in \Delta_2$. It is seen that $f_n = \varphi^{-1}(\tilde{\varphi}(F_n))$ for the $F_n$ sequence above.

To verify the group property, let $x, y \in Z_\varphi$. Since clearly $L_{xy}(f) = L_x(L_y f)$. We must have $xy \in Z_\varphi$ as well. In fact,

$$\|L_{xy} f_n - f_n\|_{(\varphi)} \leq \|L_x f_n - f_n\|_{(\varphi)} + \|L_{xy} f_n - L_x f_n\|_{(\varphi)}$$
$$= \|L_x f_n - f_n\|_{(\varphi)} + \|L_x(L_y f_n - f_n)\|_{(\varphi)}$$
$$= \|L_x f_n - f_n\|_{(\varphi)} + \|L_y f_n - f_n\|_{(\varphi)},$$
$$\text{by the invariance of } \|\cdot\|_{(\varphi)},$$
$$\to 0, \qquad \text{as } n \to \infty \text{ since } x, y \in Z_\varphi. \qquad (8)$$

Thus $xy \in Z_\varphi$. Since $x = e \in Z_\varphi$ trivially, one has

$$\|L_{x^{-1}} f_n - f_n\|_{(\varphi)} = \|L_{x^{-1}x} f_n - L_{x^{-1}} f_n\|_{(\varphi)}$$
$$= \|L_{x^{-1}}(L_x f_n - f_n)\|_{(\varphi)}$$
$$= \|L_x f_n - f_n\|_{(\varphi)} \to 0 \text{ as } n \to \infty. \qquad (9)$$

Thus $x^{-1} \in Z_\varphi$ and it is a subgroup of $G$. $\qquad\qquad\qquad\qquad\qquad \square$

Some key equivalence relations leading to amenability will now be given:

**Proposition 5.** *Consider the following conditions on a locally compact group $G$, where $M_1^+(G) = \{0 \leq \nu \in (L^1(G))^* : \|\nu\| = \nu(G) = 1\}$, the set of probability measures, and let $T_\nu = \nu * : L^\varphi(G) \to L^\varphi(G), \varphi \in \Delta_2$ an N-function so that one has*

$$(T_\nu f)(t) = \int_G f(t^{-1}x)\nu(dx), f \in L^\varphi(G),$$

*thus $T_{\delta_x} f = L_x f, f \in L^\varphi(G)$, and $\delta_x \in M_1^+(G)$):*

1. *The spectrum of $T_\nu$ contains $1$ for each $\nu \in M_1^+(G)$.*
2. *The spectral radius of $T_\nu$ satisfies $r_\varphi(T_\nu) = 1, \nu \in M_1^+(G)$.*
3. *$\|T_\nu\|_{\varphi \to \varphi} = \sup\{\|T_\nu f\|_{(\varphi)} : \|f\|_{(\varphi)} \leq 1\} = 1, \nu \in M_1^+(G), \varphi(x) = |x|^p, 1 < p < \infty$.*
4. *$G$ is (left) amenable.*

5. If $K_n \subset K_{n+1} \subset G, n \geq 1$, are compact sets, then there is a sequence $\{f_n, n \geq 1, 0 \leq f_n, \|f_n\|_{(\varphi)} = 1\}$ such that

$$\|L_x f_n - f_n\|_{(\varphi)} \to 0 \text{ as } n \to \infty \text{ uniformly in } x \in K_m, m \geq 1.$$

6. Same as 5. with 'uniformity' omitted.

   Then 1. $\Rightarrow$ 2. $\Rightarrow$ 3. and 4. $\Rightarrow$ 5. $\Rightarrow$ 6. $\Rightarrow$ 1. Moreover, all are mutually equivalent if $\varphi(x) = |x|^p, 1 < p < \infty$ throughout.

The proof depends on the following special case of a result due to Berg and Christensen (1974), which is obtained as a consequence of M. Riesz's convexity theorem. It is stated in the form directly applicable to the case at hand. (See also Leptin (1968).) A substitute of this result applicable for the $L^\varphi(G)$ is given in Lemma 7 below.

**Lemma 6.** Let $T_\nu: L^p(G) \to L^p(G), 1 < p < \infty, \nu \in M_1^+(G)$, be a convolution operator $(T_\nu = \nu *$ as defined above) with supp $\nu = S_\nu \subset G$, such that $e \in S_\nu$ and $S_\nu$ generates a dense subgroup of $G$. Then $\|T_\nu\|_{p \to p} = 1$ implies that $G$ is amenable, where $\| \cdot \|_{p \to p}$ is the operator norm on $B(L^p(G))$.

*Proof of Proposition 5.* From Proposition 3 above, it is immediate that 1. $\Rightarrow$ 2. $\Rightarrow$ 3., the last is actually for $N$-functions $\varphi \in \Delta_2$. Now suppose that 3. holds for some $\varphi(x) = |x|^p, 1 < p < \infty$. It is asserted that 3. $\Rightarrow$ 4. This will be established using Lemma 6, and the fact that if $G = \bigcup_\alpha G_\alpha$ where $\{G_\alpha, \alpha \in A\}$ is a directed system of closed amenable subgroups of $G$, then $G$ itself is amenable, so that 4. follows. The latter conclusion is a consequence of a known result [see, Greenleaf (1969), Theorem 2.3.4].

Now let $\nu \in M_1^+(G), supp \ \nu = S_\nu$, and $G_\nu$ be the subgroup of $G$ generated by $S_\nu$ so that $e \in G_\nu, G \supset \cup\{G_\nu, \nu \in M_1^+(G)\}$. It is asserted that $\{G_\nu, \nu \in M_1^+(G)\}$ is a directed set (by inclusion) and that the union is actually $G$ itself. First observe that $\nu$ is decomposable (relative to Haar measure) into $\nu = \nu_c + \nu_s + \nu_d$, the continuous, singular and discrete parts. In fact $M_1^+(G) = (M_c^+ \oplus M_s^+ \oplus M_d^+)(G)$, if $G$ is nondiscrete (cf. Hewitt and Ross (1963), p. 273). To verify our assertions, consider $\nu \in M_1^+(G)$, and $x \in S_\nu$. Let $G_\nu$ be the subgroup generated by $S_\nu$ so that $x, e \in G_\nu$. If $V_e, V_x$ are relatively compact open sets containing $e$ and $x$, define $\tilde{\nu}$ by the equation:

$$\tilde{\nu}: A \mapsto \int_A \left( \chi_{V_e} + \chi_{V_x} \right) (\mu(V_e) + \mu(V_x))^{-1} d\mu, \tag{10}$$

for each Borel set $A \subset G$. Then $\tilde{\nu} \in M_c^+(G)$ and $e, x \in S_{\tilde{\nu}} \subset G_\nu$. If $\nu$ is discrete, then let $\tilde{\nu} = \frac{1}{2}(\delta_e + \delta_x) \in M_d^+(G)$ where $\delta_{(\cdot)}$ is the Dirac measure, and $e, x \in S_{\tilde{\nu}}$ trivially. Finally if $\nu$ is singular (relative to $\mu$), then from the fact that for any relatively compact open $U_x \subset G$ containing $x$, one can find a pair of singular measures $\bar{\nu}_1, \bar{\nu}_2$ such that $\tilde{\nu}: A \mapsto \frac{(\bar{\nu}_1 + \bar{\nu}_2)(A)}{(\bar{\nu}_1 + \bar{\nu}_2)(G)}$, is a singular probability measure such that $e, x \in S_{\tilde{\nu}}$. Thus in all cases there exists a $\nu \in M_1^+(G)$ such that $e, x \in S_\nu \subset G_\nu$. Then by 3., $\|T_\nu\|_{p \to p} = 1$ and the subgroup $G_\nu$ satisfies the hypothesis of Lemma 6 with $\varphi(u) = |u|^p, 1 < p < \infty$, so that $G_\nu$ is (left) amenable. Moreover, $G = \cup\{G_\nu: \nu \in M_1^+(G)\}$ since $x \in G$ implies $x \in G_\nu$ for some $\nu \in M_1^+(G)$ as constructed above.

It is asserted that the collection $\{G_\nu\}_\nu$ forms a directed system. Indeed for a pair $\nu_1, \nu_2 \in M_1^+(G)$, let $\nu(\cdot) = \frac{\nu_1+\nu_2}{(\nu_1+\nu_2)(G)}(\cdot) \in M_1^+(G)$. Then $e \in G_\nu$ where $G_\nu$ is the subgroup generated by $S_\nu$, and $S_{\nu_1} \cup S_{\nu_2} \subset G_{\nu_1} \cup G_{\nu_2} \subset G_\nu$. Thus the $G_\nu$ system is directed for the inclusion ordering. Consequently by the earlier comment each $G_\nu$ and hence $G$ itself is amenable, establishing 4.

4. $\Rightarrow$ 5. From the structure theory of $L^\varphi(G)$-spaces, when $\varphi \in \Delta_2$ and $G$ is amenable, then $G$ has property $P_\varphi$, by Propositions 1 and 2 above, and for each compact $K \subset G$ and $\varepsilon > 0$, there exists an $f_{K,\varepsilon} \geq 0$, $\|f_{K,\varepsilon}\|_{(\varphi)} = 1$, such that

$$\|L_x f_{K,\varepsilon} - f_{K,\varepsilon}\|_{(\varphi)} < \varepsilon, \quad x \in K. \tag{11}$$

Thus for the given compact sequence $K_n \subset K_{n+1} \subset G$ and $\varepsilon = \frac{1}{n}$, there are $0 \leq f_n$, $\|f_n\|_{(\varphi)} = 1$ and $f_n$ (left) continuous such that (11) becomes $\|L_x f_n - f_n\|_{(\varphi)} < \frac{1}{n}, x \in K_n, n \geq 1$. This implies 5.

5. $\Rightarrow$ 6. In the construction of $f_n$ above, one can take $f_n$ simply Borel measurable, keeping the rest of the argument. In fact since $C_c(G)$ is norm dense in $L^\varphi(G)$ when $\varphi \in \Delta_2$, and the result of 5. immediately gives 6.

6. $\Rightarrow$ 1. For this implication it suffices to verify that for $\nu \in M_1^+(G)$, $\|T_\nu f_n - f_n\|_{(\varphi)} \to 0$ as $n \to \infty$ for each compact sequence $K_n \subset K_{n+1} \subset G$. This is because a Haar measure $\mu$ on $G$ is localizable, (Borel) regular, and is uniquely determined by its action on compact sets (cf. Hewitt and Ross (1963), p.133, or Rao (1987), Theorem 6 on p. 453).

Thus for a fixed $\nu \in M_1^+(G)$ of $\sigma$-compact support, and $(K_n, f_n)$ of 6., with $K_n \uparrow$ supp$(\nu)$, let $\nu_n = \nu|K_n$ so that $\nu_n \uparrow \nu$. Then with $T_{\nu_n}$ as the resulting convolution operator on $L^\varphi(G)$, one has

$$\begin{aligned}
\|T_\nu f_n - f_n\|_{(\varphi)} &\leq \|(T_\nu - T_{\nu_n})f_n\|_{(\varphi)} + \|T_{\nu_n} f_n - f_n\|_{(\varphi)} \\
&\leq \|T_\nu - T_{\nu_n}\| \|f_n\|_{(\varphi)} + \|T_{\nu_n} f_n - f_n\|_{(\varphi)} \\
&= (\nu - \nu_n)(G) + \|T_{\nu_n} f_n - f_n\|_{(\varphi)}.
\end{aligned} \tag{12}$$

It thus suffices to show that the last term tends to zero as $n \to \infty$, since $\nu_n \uparrow \nu$ on $\bigcup_{n\geq 1} K_n$. Now let $\tilde{G} = \bigcup_{n\geq 1} K_n$, and $\tilde{\nu} = \nu|\tilde{G}$ so that $\nu_n \uparrow \tilde{\nu}$ and it is asserted that for the $K_n$-sequence the following holds for all large enough $n(\geq n_0)$

$$\|T_{\nu_n} f_n - f_n\|_{(\varphi)} < \frac{1}{n} + (\tilde{\nu}(G) - \nu_n(G))$$

$$\leq \frac{1}{n_0}(1 - \nu_n(G)). \tag{13}$$

To establish this, consider the norm on the left side of (13). Here one uses the inverse Hölder inequality for the $L^\varphi(G)$-spaces. Thus let $\psi$ be the complementary Young function of $\varphi$, and $h \in L^\psi(G)$ with $\|h\|_{(\psi)} \leq 1$. Using the duality pairing $(L^\varphi(G), L^\psi(G))$ and the fact that $L^\varphi(G)$-norm is determined by the unit ball of $L^\psi(G)$, one has, on assuming $\nu(\tilde{G}) = 1$ for convenience, the following computation:

$$|\langle T_{\nu_n} f_n - f_n, h\rangle| = \left| \int_G (f_n * \nu_n - f_n)(x) h(x) d\mu(x) \right|$$

$$= \left| \int_G \int_{K_n} f_n(x^{-1}y) h(x) d\nu_n(\mu) d\mu(x) \right.$$

$$\left. - \int_{K_n} \int_G f_n(x) h(x) d\mu(x) \frac{d\nu_n(y)}{\nu_n(K_n)} \right|, \text{ since supp } \nu_n \subset K_n,$$

$$= \left| \int_{K_n} \langle L_x f_n - \frac{f_n}{\nu_n(K_n)}, h \rangle d\nu_n(x) \right|$$

$$\leq 2 \int_{K_n} \left\| L_x f_n - \frac{f_n}{\nu_n(K_n)} \right\|_{(\varphi)} \|h\|_{(\psi)} d\nu_n(x), \text{ by the}$$

Hölder inequality for $L^\varphi(G)$-spaces,

$$\leq \frac{2\|h\|_{(\psi)}}{\nu_n(K_n)} \int_{K_n} \|(\nu_n(K_n) L_x f_n - L_x f_n +$$

$$L_x f_n - f_n\|_{(\varphi)} \chi_{K_n} d\nu_n(x)$$

$$\leq \frac{2\|h\|_{(\varphi)}}{\nu_n(K_n)} \int_{K_n} \left[ (1 - \nu_n(K_n)) \|L_x f_n\|_{(\varphi)} \right.$$

$$\left. + \|L_x f_n - f_n\|_{(\varphi)} \right] d\nu_n(x)$$

$$= \frac{2\|h\|_{(\varphi)}}{\nu_n(K_n)} \left[ (1 - \nu_n(K_n)) \nu_n(K_n) + \|L_x f_n\|_\varphi \nu_n(K_n) \right],$$

since $\| \cdot \|_{(\varphi)}$ is translation invariant,

$$= 2\|h\|_{(\varphi)} \left[ (1 - \nu_n(K_n)) + \frac{1}{n} \right]$$

$$\to 0 \text{ as } n \to \infty \text{ since } \nu_n(K_n) \uparrow \nu(\tilde{G}) = 1.$$

Thus (13) holds and so 1. is established.

It is now obvious that if $\varphi(u) = \frac{|u|^p}{p}, 1 < p < \infty$, then all the statements are equivalent as asserted.    $\square$

The crucial part of this equivalence is the implication of 3. $\Rightarrow$ 4. Here $\varphi(u) = \frac{|u|^p}{p}, 1 < p < \infty$ is critically used for Lemma 6 in invoking the Riesz convexity theorem asserting that a contraction in $L^1(G)$, $L^\varphi(G)$ and $L^p(G)$ with unit norms has the same property for all $1 \leq p \leq \infty$. The same conclusion does not follow from the extended Riesz convexity theorem for Orlicz spaces which is also available. This is because between a pair of Lebesgue spaces $L^p$ and $L^{p'}$ there exist Orlicz spaces $L^\varphi$, without even $\varphi \in \Delta_2$. Consequently an alternative approach seems desirable. Such a possible procedure is employed in Day's (1964) work using the geometric property of the $L^p(G)$, $1 < p < \infty$, namely its uniform convexity. He even remarked that it may be extended to Orlicz spaces. The essential point is that it replaces Lemma 6 above, and is given here employing Day's idea.

**Lemma 7.** *Let $\varphi$ be a Young function satisfying the Milnes–Akimovič condition (MA): for each $\varepsilon > 0$, there exist a $K_\varepsilon > 1$ and an $x_1(\varepsilon) \geq 0$ such that*

$$\varphi'((1+\varepsilon)x) \geq K_\varepsilon \varphi'(x), \quad x \geq x_1(\varepsilon). \tag{14}$$

*If the convolution operator $T_\nu$, $\nu \in M_1^+(G)$, on $L^\varphi(G) \to L^\varphi(G)$, satisfies $\|T_\nu\|_{\varphi \to \varphi} = 1$, then there is a sequence $0 \leq f_n \uparrow$, $supp(f_n) = K_n$, compact, and $\|f_n\|_{(\varphi)} = 1$ such that as $n \to \infty$*

$$\|L_x f_n - f_n\|_{(\varphi)} \to 0, \quad \text{uniformly in } x \in K_m, m \geq 1.$$

*Consequently the spectral radius satisfies $r_\varphi(T_\nu) = 1$, $\nu \in M_1^+(G)$.*

**Proof.** By hypothesis for each $\nu \in M_1^+(G)$, one has the operator norm as

$$\begin{aligned}
1 = \|T_\nu\|_{\varphi \to \varphi} &= \sup\{\|T_\nu f\|_{(\varphi)} : \|f\|_{(\varphi)} = 1\} \\
&= \lim_n \|T_\nu f_n\|_{(\varphi)} \tag{15}
\end{aligned}$$

for some sequence $\{f_n, n \geq 1\}$ in the unit sphere. In this selection one can choose $f_n$ such that (by the lattice property of $L^\varphi(G)$), $0 \leq f_n \uparrow$ and since $C_c(G)$ is dense in $L^\varphi(G)$, it may be arranged so that $supp(f_n) = K_n \subset K_{n+1}$ and the supports compact. Now it is asserted that for such a sequence $\{f_n, n \geq 1\}$ one has $\|L_x f_n - f_n\|_{(\varphi)} \to 0$ as $n \to \infty$ for each $x \in K_m, m \geq 1$. In fact, if this is false, then there exist $m_0 \geq 1$, $\varepsilon > 0$ such that $\|L_x f_n - L_y f_n\|_{(\varphi)} \geq \varepsilon$ for $x \in K_{m_0}$ and $y \in K_n, n \geq 1$. For otherwise, $\|L_x f_n - f_n\|_{(\varphi)} \to 0$ and $\|L_y f_n - f_n\|_{(\varphi)} \to 0$ for all $x \in K_m$ and $y \in K_m$, which is supposed not to be true. Thus $\|L_x f_n - L_y f_n\|_{(\varphi)} \nrightarrow 0$. Now (15) holds for each $\nu \in M_1^+(G)$. But molecular measures, and hence discrete measures, are abundant here (even linearly dense in this space for the variation norm). Consequently (15) also holds for a discrete measure, say $\nu_0$. Now $\|T_{\nu_0} f_n\|_{(\varphi)}$ may be calculated using the additional hypothesis of the conditions (MA) which implies that $L^\varphi(G)$ is uniformly convex (cf. e.g., Rao and Ren (1991), p. 288), and this will yield the desired contradiction.

For a uniformly convex Banach space $\mathcal{X}$, if $x_n$, $y_n$ are in its unit ball and $\|x_n - y_n\| \geq \varepsilon > 0$, then for any $0 < \alpha = 1 - \beta < 1$, $\limsup_n \|\alpha x_n + \beta y_n\| < 1 - \eta_\varepsilon$ for some $0 < \eta_\varepsilon < 1$. This is usually stated for $\alpha = \beta = \frac{1}{2}$, but the above is seen to be an equivalent statement as follows.

Indeed, if $\liminf_n \|x_n - y_n\| = \varepsilon > 0$ and $\|\alpha x_n + \beta y_n\| \to 1$ for some $0 < \alpha = 1 - \beta < 1$, then $\|\frac{x_n + y_n}{2}\| \to 1$ as well. For otherwise, choose $n_0(\varepsilon) > 0$ such that $n \geq n_0 \Rightarrow \|x_n - y_n\| \geq \varepsilon$ and there exists an $\eta_\varepsilon > 0$ such that $\|\frac{x_n + y_n}{2}\| < 1 - \eta_\varepsilon, n \geq n_0$. It may be assumed that $0 < \alpha < \frac{1}{2}$. Then by the Hahn–Banach theorem there is a continuous linear functional $f_0$, $\|f_0\| = 1$ and $f_0(\alpha x_n + \beta y_n) = \|\alpha x_n + \beta y_n\|$, and the following computation holds.

$$\begin{aligned}
\left\|\frac{x_n + y_n}{2}\right\| - \|\alpha x_n + \beta y_n\| &\geq f_0\left(\frac{x_n + y_n}{2}\right) - f_0(\alpha x_n + \beta y_n) \\
&= f_0(x_n)\left(\frac{1}{2} - \alpha\right) + f_0(y_n)\left(\frac{1}{2} - \beta\right) \\
&= \left(\frac{1}{2} - \alpha\right)(f_0(x_n) - f_0(y_n)), \ [\because \alpha + \beta = 1], \\
&= \left(\frac{1}{2} - \alpha\right) f_0(x_n - y_n).
\end{aligned}$$

Since $f_0$ in the unit ball of the adjoint space is arbitrary one has

$$\left\|\frac{x_n + y_n}{2}\right\| - \|\alpha x_n + \beta y_n\| \geq (\frac{1}{2} - \alpha)\|x_n - y_n\| \geq (\frac{1}{2} - \alpha)\varepsilon.$$

Consequently,

$$(1 - \eta_\varepsilon) - (\frac{1}{2} - \alpha)\varepsilon \geq \|\alpha x_n + \beta y_n\| \to 1,$$

the last limit by hypothesis. Then this gives

$$0 \geq \eta_\varepsilon + (\frac{1}{2} - \alpha)\varepsilon > 0$$

which is impossible. Thus $\|\frac{x_n + y_n}{2}\| \to 1$ as well.

Now let $x_n = L_x f_n$, $y_n = L_y f_n$ and $\mathcal{X} = L^\varphi(G)$ with $\alpha$, $\beta$ of the above paragraph as $\alpha = \frac{v_0(x)}{v_0(x) + v_0(y)}$, $\beta = \frac{v_0(y)}{v_0(x) + v_0(y)}$. (Recall that $v_0$ is discrete here.) Then one has on evaluating the convolution

$$\|T_{v_0} f_n\|_{(\varphi)} = \left\|v_0(x)(L_x f_n) + v_0(y)(L_y f_n) + \sum_{\{z_i \neq x, y\}} v_0(z_i) L_{z_i} f_n\right\|_{(\varphi)}$$

$$\leq \|v_0(x)(L_x f_n) + v_0(y)(L_y f_n)\|_{(\varphi)} + \sum_{\{z_i \neq x, y\}} v_0(z_i)\|L_{z_i} f_n\|_{(\varphi)}$$

$$\leq (1 - \eta_\varepsilon)(v_0(x) + v_0(y)) + \sum_{\{z_i \neq x, y\}} v_0(z_i) \cdot 1$$

$$\leq (1 - \eta_\varepsilon)(v_0(x) + v_0(y)) + 1 - (v_0(x) + v_0(y))$$

$$< 1 - \eta_\varepsilon(v_0(x) + v_0(y)) < 1,$$

where $\eta_\varepsilon$ is chosen as in the above paragraph. Hence $\lim_n \|T_{v_0} f_n\|_{(\varphi)} \leq 1 - \eta_\varepsilon < 1$. But this is in contradiction to the original choice of the $f_n$-sequence. Therefore $\|L_x f_n - f_n\|_{(\varphi)} \to 0$ for $x \in K_k, k \geq 1$ as $n \to \infty$ must hold. This implies that $\|T_v f_n - f_n\|_{(\varphi)} \to 0$ for each $v \in M_1^+(G)$ so that $(I - T_v)^{-1}$ does not exist in $L^\varphi(G)$ and $1 \in \mathrm{spec}(T_v)$ for each $v$. It follows that $r_\varphi(T_v) = 1$.    □

**Remark.** In the case that $\|T_v\|_{p \to p} = 1$, $1 < p < \infty$, one has by the Riesz convexity theorem $\|T_v\|_{p \to p} = 1$ for all $1 \leq p \leq \infty$, and in particular for $p = 2$ this implies (and is equivalent to) another property, called $P'$, introduced by R. Godement motivated by the work on representation theory. (It may be recalled that a locally compact group $G$ has *property $P'$* if the constant function 1 can be uniformly approximated on compact sets by positive definite functions from $C_c(G)$.) Reiter ((1968), p.173) has presented a proof of the equivalence of properties $P'$ and $P_1$ hence all $P_p$ by Proposition 2, and even $P_\varphi$ for $\varphi \in \Delta_2$.

With all the preceding work of this section, the following comprehensive result can be presented on a characterization of amenability of locally compact groups in terms of spectral properties of convolution operators.

**Theorem 8.** *For a locally compact group G the following conditions are mutually equivalent:*

(i) *G is (left) amenable, i.e., there exists a left invariant mean $l: L^\infty(G) \to \mathbb{R}$ so that $l(L_x f) = m(f), x \in G$, where $(L_x f)(y) = f(x^{-1}y)$, and $l(1) = 1$.*

(ii) *G is strongly (left) amenable, in the sense that there is a net $f_\alpha \in P(G) \subset L^1(G)[0 \le f_\alpha, \|f_\alpha\|_1 = 1]$ such that $(T_\nu f_\alpha - f_\alpha) \to 0$ in the weak\*, or $\sigma(L^\infty, L^1)$-topology for each $\nu \in M_1^+(G)$.*

(iii) *For a $\varphi \in (MA)$ and each $\nu \in M_1^+(G)$, the convolution operator $T_\nu = \nu*$, $T_\nu: L^\varphi(G) \to L^\varphi(G)$, has one in its spectrum.*

(iv) *Condition (iii) holds for a $\varphi(u) = |u|^p, 1 < p < \infty$.*

(v) *For $T_\nu$ of (iii), its spectral radius satisfies $r_\varphi(T_\nu) = 1$.*

(vi) *Same as (v) with $\varphi(u) = |u|^p, 1 < p < \infty$.*

(vii) *For each N-function $\varphi$ satisfying the condition (MA) and for each $\nu \in M_1^+(G)$, the convolution operator $T_\nu: L^\varphi(G) \to L^\varphi(G)$ satisfies $\|T_\nu\|_{\varphi \to \varphi} = 1$.*

(viii) *Same as (vii) with $\varphi(u) = |u|^p, 1 < p < \infty$.*

(ix) *For each compact increasing sequence of sets $K_n \subset G$, and an N-function $\varphi$ of $\Delta_2$-class, there is a sequence $\{0 \le f_n, supp(f_n) \subset K_n, \|f_n\|_{(\varphi)} = 1, f_n$ uniformly left continuous$\}$, $\|L_x f_n - f_n\|_{(\varphi)} \to 0$ as $n \to \infty$ uniformly in $x \in K_m, m \ge 1$.*

(x) *Same as (ix) with $\varphi(u) = |u|^p, 1 < p < \infty$.*

(xi) *Each $\sigma$-compact subgroup of G is contained in a (strongly) left amenable $\sigma$-compact open and closed subgroup of G.*

**Proof.** Most of the assertions or equivalences are consequences of Proposition 5 and Lemma 7. Those that are not directly covered will be considered here. That (i) $\Longleftrightarrow$ (ii) is a known result (cf., e.g., Greenleaf (1969), p. 33). It was shown above that (i) $\Longleftrightarrow$ (iii)$\Rightarrow$(iv), as well as (i) $\Longleftrightarrow$ (v)$\Rightarrow$(vi). Proposition 3 and Reiter's theorem give (i) $\Longleftrightarrow$ (viii) and clearly (vii)$\Rightarrow$(viii). Also Proposition 5 shows that (i) $\Longleftrightarrow$ (ix)$\Rightarrow$(x). It remains to show that (x)$\Rightarrow$(xi)$\Rightarrow$(i), and this will be established now. The necessary construction is an extension of that of Day's (1964) on discrete (semi-) groups, to the general group case which however is somewhat involved.

By hypothesis of (x), for a compact sequence $K_n \subset K_{n+1} \subset G$, there is $0 \le f_n \in L^p(G), \|f_n\|_p = 1, f_n$, left continuous, $supp(f_n) \subset K_n$ and $\|L_x f_n - f_n\|_p \to 0$ as $n \to \infty$, for each $x \in K_m, m \ge 1$. Consequently, if $g_n = f_n^{\frac{1}{p}}$, then $\cup_n K_n \subset Z_1(\{g_n, n \ge 1\})$ for $0 \le g_n \in L^1(G)$, $supp(g_n) \subset K_n$, where $Z_1(\{g_n, n \ge 1\}) = \{x \in G : \lim_n \|L_x g_n - g_n\|_1 = 0\}$. In fact from definition of $g_n, \|g_n\|_1 = 1, supp(g_n) \subset K_n$ and by Proposition 2, $\|L_x g_n - g_n\|_1 \to 0$, as $n \to \infty$, $x \in K_m, m \ge 1$ implies $x \in Z_1(\{g_n, n \ge 1\})$. Also observe that for an open subgroup $H$ of $G$, if $\mu$ is a left Haar measure on $G$, then the restriction $\mu|H$ is also a Haar measure on $H$ and for $f \in L^1(H)$, since $supp(f) \subset H$, one has $\|f\|_{L^1(H)} = \|f\|_{L^1(G)}$.

Starting with the actual agreement, let $H \subset G$ be a subgroup generated by a $\sigma$-compact subset as in (xi). It is asserted that $H$ is contained in a $\sigma$-compact open (hence also closed) strongly amenable subgroup $H_1 \subset G$. Then by hypothesis if $A_1$ is a $\sigma$-compact set, there exist $K_{1n} \subset K_{1,n+1} \subset G$, (compact) such that $A_1 = \cup_{n \ge 1} K_{1n}$ and as seen above, left continuous $f_{1n} \geq 0$, $\|f_{1n}\|_1 = 1$ and $\|L_x f_{1n} - f_{1n}\|_1 \to 0$, $x \in K_{1m}$, $m \geq 1$ as $n \to \infty$, and hence $A_1 \subset Z_1(\{f_{1n}, n \geq 1\})$. Also one can express $\cup_{n \geq 1}$ supp$(f_{1n}) = \cup_{n \geq 1} K_{1n}$, $\tilde{K}_{1n} \uparrow\subset G$ compact since supp$(f_{1n})$ is $\sigma$-compact ($f_{1n} \in L^1(G)$). Let $A_2 = A_1 \cup \cup_{n \geq 1}$ supp$(f_{1n})$, which is $\sigma$-compact. So it can be expressed as $A_2 = \cup_{n \geq 1} K_{2n}$, $K_{2n} \uparrow\subset G$ compact. Since supp$(f_{1n})$ has nonempty interior (being left continuous), there exist $0 \leq f_{2n} \in L^1(G)$, left continuous, $\|f_{2n}\|_1 = 1$ and $\|L_x f_{2n} - f_{2n}\|_1 \to 0$ uniformly in $x \in K_{2m}$, $m \geq 1$, as $n \to \infty$. By construction $K_{2n}(= K_{1n} \cup \tilde{K}_{1n}) \supset K_{1n}$, $n \geq 1$. Suppose one has constructed $A_1 \subset A_2 \subset \cdots \subset A_m$ with $A_m = \cup_{n \geq 1} K_{mn}$, $K_{mn} \uparrow$ each compact and $0 \leq f_{mn} \in L^1(G)$, left continuous, $\|f_{mn}\|_1 = 1$ such that $\|L_x f_{mn} - f_{mn}\|_1 \to 0$ as $n \to \infty$, $x \in K_{mn}$ with supp$(f_{mn})$ $\sigma$-compact. Then consider $A_{m+1} = A_m \cup \cup_n$ supp$(f_{mn}) = \cup_n K_{m+1,n}$ (say), $K_{m+1,n} \uparrow$ and compact. Choose $0 \leq f_{m+1,n} \in L^1(G)$, left continuous with all the other properties as above. Then $K_{mn} \subset K_{m,n+1}$ and $K_{mn} \subset K_{n+1,n}$ all compact, and the corresponding $0 \leq f_{mn} \in L^1(G)$, $\|L_x f_{mn} - f_{mn}\|_1 \to 0$ as $n \to \infty$, $x \in K_{nm'}$, $m' \geq 1$. If one now considers the diagonal sequence, then $K_{nn} \subset K_{n+1,n+1}$ compact, $0 \leq f_{nn} \in L^1(G)$ of unit norms and $\|L_x f_{nn} - f_{nn}\|_1 \to 0$ for each $x \in K_{mm}$, $m \geq 1$. Hence by Proposition 4, $\cup_n K_{nn} \subset Z_1(\{f_{nn}, n \geq 1\})$, and $\cup_n A_n = \cup_n K_{nn}$.

Let $H_1$ be the subgroup generated by $\cup_n A_n$. It is $\sigma$-compact and since $A_1$ generates $H$, one has $H \subset H_1$ and also $A_n$ contains an open set for each $n \geq 1$ so that $H_1$ contains an open set, whence it is an open (and so closed) subgroup of $G$. Thus it is a $\sigma$-compact open subgroup. It is asserted that $H_1$ is amenable. For this (cf. Greenleaf (1969), Theorem 2.4.3 on p. 33) it suffices to find a net $0 \leq g_n \in L^1(H_1)$, $n \geq 1$, $\|g_n\|_1 = 1$ and $\|L_x g_n - g_n\|_1 \to 0$, $x \in H_1$.

Set $g_n = f_{nn}$, $n \geq 1$, the diagonal sequence constructed above. Then $0 \leq g_n \in L^1(H_1) \subset L^1(G)$, (supp$(g_n) \subset A_n \subset H_1$) and $\|g_n\|_{L^1(H_1)} = \|g_n\|_{L^1(G)}$. Further $(L_x g_n)(y) = g_n(x^{-1}y)$ so that $x^{-1}y \in$ supp$(g_n)$ implies $y \in x$ (supp $(g_n)) \subset H_1$ and supp$(f_n - L_x f_n) \subset$ supp$(g_n) \cup$ supp$(L_x g_n) \subset H$, as well as $\|L_x g_n - g_n\|_1 \to 0$ as $n \to \infty$, $x \in K_{mm}$, $m \geq 1$. Therefore, $\cup_n K_{mm} \subset Z_1(\{g_n, n \geq 1\})$, and since the latter is a group by Proposition 4, it follows that the group generated by $\cup K_{nn} \subset Z_1(\{g_n, n \geq 1\})$. Hence $H_1 \subset Z_1(\{g_n, n \geq 1\})$ and $\|L_x g_n - g_n\|_1 \to 0$ for all $x \in H_1$ implying that $H_1$ is strongly amenable. Thus (xi) holds. Note that by the preceding work (cf. Lemma 7) the same holds if ($\varphi(u) = |u|^p$, $1 < p < \infty$ is replaced by $\varphi$ satisfying the condition (MA). So (x) $\Rightarrow$ (xi) as well.

(xi)$\Rightarrow$(i). Since the collection of all such $\sigma$-compact open subgroups form a directed set and fills up $G$, as seen in the proof of Proposition 5 (4.) the group $G$ is (left) amenable. Thus (i) holds, and the proof is finished.    $\square$

**Remark.** In the above work, while dealing with $L^\varphi(G)$, it was assumed that $\varphi \in$ (MA) to invoke uniform convexity of the space. However, it is also known that a reflexive Orlicz space $L^\varphi(G)$ is isomorphic to $L^{\tilde{\varphi}}(G)$ which is uniformly convex (cf. Rao and Ren (1991), p. 297, Theorem 2 there), and $L^\varphi(G)$ is reflexive iff $\varphi \in \Delta_2 \cap \nabla_2$. Since for the amenability characterization of $G$ the uniform convexity of $L^{\tilde{\varphi}}(G)$ is essential (cf. Lemma 7), and because of the above noted (topological) isomorphism can be used for all convergence statements, it appears that the work can be stated for this generalized class. In the case of the $L^p(G)$, however the two spaces $L^\varphi(G)$ and $L^{\tilde\varphi}(G)$ coincide because $\varphi(u) = |u|^p, 1 < p < \infty$ implies that $L^\varphi(G)$ is simultaneously reflexive, uniformly convex, and smooth.

The preceding theorem generalizes Day's (1964) work for locally compact groups (and Orlicz spaces at the same time). For discrete semi-groups it was considered by Truitt (1967) who has extended Day's theorems to the uniformly convex Orlicz *sequence spaces*. The general case is naturally more involved, as already expressed by Day in his (1964) paper.

In view of Proposition 2 above, the following structure theorem on amenable groups due to Reiter ((1969), p. 185) will be of interest. Recall that a topological group $G$ is *solvable* if there exists a finite chain of closed normal subgroups $\{G_k\}_{k\geq 1}$ such that $G = G_0 \supset G_1 \supset \cdots \supset G_{n+1} = \{e\}$ and the quotient group $G_k/G_{k+1}$ is abelian for $0 \leq k \leq n$. One can then state the desired result as:

**Theorem 9.** *Let $\tilde G$ be the connected component of the identity element of $G$, a locally compact group, so that $\tilde G$ is a closed normal subgroup. If the quotient group $G/\tilde G$ is either compact or abelian (or more generally solvable) then $G$ is amenable iff $\tilde G/R_0$ is compact where $R_0$ is the radical of $\tilde G$ so that it is the largest connected solvable normal subgroup of $\tilde G$, and is closed.*

As a consequence of this theorem, all connected, semi-simple noncompact Lie groups with finite center (e.g., the usual $GL_n(\mathbb{R})$) are not amenable. (A proof of the above theorem and consequences are fully discussed in Reiter's book ((1969), Sec.8.7).) This is of interest in the study of r.w.'s on such groups since they can only be transient, as noted in Proposition 3.1.

# 6 Beurling and Segal Algebras for Amenability

Both Beurling and Segal algebras are two classes of Banach algebras closely related to the group algebra $L^1(G)$, each with a rich structure. A few of their properties in the context of Orlicz spaces will be indicated as they raise interesting problems for a future investigation related to the preceding work.

At the end of Section 4 above it was noted that certain classes of Segal algebras on $G$ can characterize amenability of the latter. Recall that a Segal algebra $S(G)$ is a densely contained Banach algebra of the group algebra $L^1(G)$, with a stronger translation invariant norm $\|\cdot\|_S$ relative to which the embedding is continuous and the translation satisfies $\|L_a f - f\|_S \to 0$ as $a \to e, f \in S(G)$. An important example is $M^\varphi(G)$ where $\varphi$ is a continuous Young function with its (right) derivative $\varphi'(0) > 0$. Also note that a Banach algebra $\mathcal{A}$ is amenable if for each $\mathcal{A}$-bimodule $\mathcal{X}$, each continuous derivation $D: \mathcal{A} \to \mathcal{X}$ is inner. It was seen in Section 4 that the $M^\varphi(G)$ are of the above type and are amenable (in particular $L^1(G)$ which correspond to $\varphi(x) = |x|$) iff $G$ is an amenable group. A natural question now is to investigate whether the amenability of a general Segal algebra $S(G) \subset L^1(G)$ which is not necessarily related to Orlicz spaces (in fact certain subalgebras of continuous function spaces qualify to be in this class) characterize the corresponding property of $G$. This is an interesting problem to analyze and it complements some aspects of Johnson's (1972) work. Another algebra, also motivated by the latter, is the following.

This is a versatile and somewhat mysterious convolution algebra, namely a Beurling algebra, which depends on a (submultiplicative or subadditive) weight function. More precisely, let $w: G \to [1, \infty)$ be an upper semi-continuous function such that $w(xy) \leq w(x)w(y)$, and if $\tilde{\mu}: A \mapsto \int_A w(x)d\mu(x)$ $(A \subset G$ Borel) $\mu$ being a (left) Haar measure on $G$, then $L^1(G, \tilde{\mu})$ is a Banach algebra under convolution for the weight $w$, called a *Beurling algebra* (cf. Reiter (1968), p. 83) and the norm satisfies

$$\|L_a f\|_{1,w} = \int_G f(a^{-1}x)d\tilde{\mu}(x) \leq w(a)\|f\|_{1,w}, (\|R_a f\|_{1,w} \leq w(a)\|f\|_{1,w}), \quad (1)$$

$L_a (R_a)$ being the left (right) translation. Similarly if $\varphi$ is a continuous Young function with $\varphi'(0) > 0$, then it can be verified that $L^\varphi(G, \tilde{\mu})$ is a Beurling algebra (cf. e.g., Rao and Ren (2002), p. 34 and p. 382) and is contained continuously and densely in $L^1(G)$ if $w$ is bounded, (cf. Rao (2001), p. 3604). More general convolution algebras were introduced by Beurling (1964). However, the choice of $w(\cdot)$ in these algebras gives considerable flexibility and aids the applicational potential to large classes of problems in harmonic analysis and others. It will be of interest to study amenability of these algebras with a view to discovering new characterizations of amenability of the group $G$. However, unlike the classical $L^1(G)$ case, it may be observed that amenability of a Beurling algebra does not necessarily imply a similar property of the underlying group $G$ for *unbounded* weight functions as seen below.

Recall that (cf. Definition 4.2) a Banach algebra $\mathcal{A}$ is *amenable* if for each $\mathcal{A}$-bimodule $\mathcal{X}$ one has the cohomology group $\mathcal{H}^1(\mathcal{A}, \mathcal{X}^*) = \{0\}$. It is called *weakly amenable* if each continuous derivation $D: \mathcal{A} \to \mathcal{X}$, an $\mathcal{A}$-bimodule, vanishes. If $\mathcal{A}$ is commutative and $\mathcal{X}$ is a commutative bimodule (so $ax = xa$, for all $a \in \mathcal{A}, x \in \mathcal{X}$), then $\mathcal{H}^1(\mathcal{A}, \mathcal{X}) = \{0\}$ iff each derivation of $\mathcal{A}$ into $\mathcal{X}$ vanishes. Thus in the commutative case amenability of an algebra implies weak amenability, but the two concepts are generally different. The following result due to Bade, Curtis and Dales (1987) shows that Beurling algebras relative to a weight function of different growth rates distinguish the above concepts leading to interesting nontrivial new studies on amenability.

Consider a Beurling algebra $L^1_w(G)$, denoted $L^1(G, \tilde{\mu})$ above, to be called $\ell^1(w)$ when $G = \mathbb{Z}$, the integers, and $\mu$ is the counting measure, $w: G \to [1, \infty)$ being a weight. For instance $w_\alpha(n) = (1 + |n|^\alpha), \alpha \geq 0$ is such a function, on $\mathbb{Z}$, an amenable group. This $w_\alpha$ with $\alpha > 0$ qualifies to be an example of the following general (negative) result due to these authors.

**Proposition 1.** *Consider the Beurling algebra $\ell^1(w)$ where $w(n) \to \infty$ as $|n| \to \infty$. Then $\ell^1(w)$ is not amenable although $\mathbb{Z}$ is an amenable group. If on the other hand $\frac{w(n)w(-n)}{n} \to 0$ as $n \to \infty$, then it is weakly amenable. In case the weight function satisfies the following boundedness condition*

$$\frac{w(m+n)(1+|n|)}{w(m)w(n)(1+|m+n|)} \leq K < \infty, \quad m, n \in \mathbb{Z}, \quad (2)$$

*then $\ell^1(w)$ is not weakly amenable either.*

For the particular weight function $w_\alpha(\cdot)$ defined above, these authors show that $\ell^1(w_\alpha)$ is weakly amenable if $0 < \alpha < \frac{1}{2}$ but not amenable, and for $\alpha \geq \frac{1}{2}$ it is not even weakly amenable. Thus all possibilities are present for this class of convolution algebras.

The problem of interest then is to find a suitable growth condition on the weight function $w$ (other than boundedness) in order that the corresponding Beurling algebra $L_w^1(G)$ *characterizes amenability of a locally compact group $G$*. On the other hand if $G$ is amenable what conditions on $w$ are needed in order that $L_w^1(G)$ is (weakly or strongly) amenable. More generally the same problem can be raised for the Beurling–Orlicz algebra $\mathcal{M}^\varphi(G, \tilde{\mu}), \varphi'(0) > 0$, defined analogously. There are several other problems on these algebras related to some variations of amenability. Many of these, such as 'strong amenability' [Johnson (1972)] 'Connes-amenable' [Runde (2002)] can be studied particularly for operator algebras. (The latter author defines 'weak amenability' of a Banach algebra $\mathcal{A}$, somewhat differently from the one given above, which follows Bade et al, by taking for $\mathcal{X} = \mathcal{A}$ itself so that this *restricted weak amenability* of $\mathcal{A}$ is studied if $\mathcal{H}^1(\mathcal{A}, \mathcal{A}^*) = \{0\}$ without the qualification "restricted".) For the general Beurling algebras, the following types of growth conditions on the weight function are found useful in studies of harmonic analysis:

(i) The Beurling–Domar type growth:

$$\sum_{n \geq 1} \frac{\log w(x^n)}{n^2} < \infty, \quad x \in G.$$

(ii) The Shilov type condition:

$$w(x^n) = O(|x|^\alpha) \text{ as } |n| \to \infty \text{ for some } \alpha(= \alpha_x) > 0, x \in G,$$

and

$$\liminf_{|n| \to \infty} \frac{w(x^n)}{|n|} = 0, \quad x \in G.$$

[See Reiter (1968), p. 132 for further discussion.] These conditions have not yet been found adequate for amenability characterizations of $G$.

An important problem for the present applications is to obtain new and more detailed results on random walks in terms of the work of Section 3, using amenability. This can also be considered for Markov chains, as detailed, for instance, in Revuz (1984). Other questions on the ergodic and limiting behavior of convolution powers related to such processes will be of interest for future investigations. Extensions of the above work for general topological semi-groups (as in Day (1964), and its extension by Truitt (1967) for the countable case) and Skantharajah (1992) for (double coset or) hypergroup extensions will also be of interest in the context of Orlicz spaces generalizing the above work.

# References

[1] Bade, W. G., P. C. Curtis, Jr., and H.G. Dales, Amenability and weak amenability for Beurling and Lipschitz algebras, *Proc. London Math. Soc.(3)* **55** (1987), 359–377.

[2] Berg, C., and J. P. R. Christensen, On the relation between amenability of locally compact groups and norms of convolution operators, *Math. Ann.* **208** (1974), 149–153.

[3] Beurling, A., Construction and analysis of some convolution operators, *Ann. Inst. Fourier Grenoble (2)* **14** (1964), 1–32.

[4] Bonsall, F. F., and J. Duncan, *Complete Normed Algebras*, Springer, New York, 1973.

[5] Day, M. M., Convolutions, means, and spectra, *Illinois J. Math.* **8** (1964), 100–111.

[6] Dieudonné, J., Sur le produit de composition-II, *J. Math. Pures Appl.* **39** (1960), 275–292.

[7] Feller, W., *Introduction to Probability Theory and its Applications-I*, 3rd Ed., Wiley and Sons, New York, 1968.

[8] Furstenberg, H., Boundary theory and stochastic processes on homogeneous spaces, in: *Harmonic Analysis on Homogeneous Spaces*, Proc. Symp. Pure Math. **26** (1973), Amer. Math. Soc., 193–229.

[9] Greenleaf, F. P., *Invariant Means on Topological Groups*, Van Nostrand-Reinhold Co., New York, 1969.

[10] Hewitt, E., and K. A. Ross, *Abstract Harmonic Analysis-I*, Springer, New York, 1963.

[11] Johnson, B. E., Cohomology in Banach algebras, *Memoirs of Amer. Math. Soc.* **127** (1972), 1–96.

[12] Kesten, H., Symmetric random walks on groups, *Trans. Amer. Math. Soc.* **92** (1959a), 336–354.

[13] Kesten, H., Full Banach mean values on countable groups, *Math. Scand.* **7** (1959b), 146–156.

[14] Leptin, H., On locally compact groups with invariant means, *Proc. Amer. Math. Soc.* **19** (1968), 489–494.

[15] Paterson, A. L. T., *Amenability*, Amer. Math. Soc. Surveys, Providence, RI, 1988.

[16] Pier, J.- P., *Amenable Locally Compact Groups*, Wiley-Interscience, New York, 1984.

[17] Rao, M. M., Convolutions of vector fields-I, *Math. Zeits.* **174** (1980) 63–79, *and II* (2001), *Nonlinear Anal.* **47**, 3599–3615.

[18] Rao, M. M., *Probability Theory with Applications*, Academic Press, New York, 1984.

[19] Rao, M. M., *Measure Theory and Integration*, Wiley-Interscience, New York, 1987.

[20] Rao, M. M., *Stochastic Processes: General Theory*, Kluwer Academic, Boston, 1995.

[21] Rao, M. M., and Z. D. Ren, *Theory of Orlicz Spaces*, Marcel Dekker Inc., New York, 1991.

[22] Rao, M. M., and Z. D. Ren, *Applications of Orlicz Spaces*, Marcel Dekker, Inc., New York, 2002.

[23] Revuz, D., *Markov Chains*, (Revised Edition), North-Holland, Amsterdam, The Netherlands, 1984.

[24] Reiter, H., *Classical Harmonic Analysis and Locally Compact Groups*, Oxford University Press, Oxford, UK, 1968.

[25] Runde, V., *Lectures on Amenability*, Lect. Notes in Math. **1774**, Springer, New York, 2002.

[26] Skantharajah, M., Amenable hypergroups, *Illinois J. Math.* **36** (1992), 15–46.

[27] Spitzer, F., *Principles of Random Walk*, D. Van Nostrand Co., New York, 1964.

[28] Truitt, C. C. B., *An extension to Orlicz spaces of theorems of M.M. Day on "Convolutions, means, and spectra,"* Ph.D. thesis, University of Illinois, Urbana, IL, 1967.

# Index